International Conference on Advances in Materials Processing & Manufacturing Applications (iCADMA 2020)

Held online due to COVID-19

IOP Conference Series: Materials Science and Engineering
Volume 1017

Jaipur, India
5 - 6 November 2020

ISBN: 978-1-7138-3321-5
ISSN: 1757-8981

Printed from e-media with permission by:

Curran Associates, Inc.
57 Morehouse Lane
Red Hook, NY 12571

Some format issues inherent in the e-media version may also appear in this print version.

Institute of Physics
Dirac House, Temple Back
Bristol BS1 6BE UK

Phone: 44 1 17 929 7481
Fax: 44 1 17 920 0979

techtracking@iop.org

Additional copies of this publication are available from:

Curran Associates, Inc.
57 Morehouse Lane
Red Hook, NY 12571 USA
Phone: 845-758-0400
Fax: 845-758-2633
Email: curran@proceedings.com
Web: www.proceedings.com

TABLE OF CONTENTS

IOP Conf. Series: Materials Science and Engineering 1017 (2021) 011001 doi:10.1088/1757-899X/1017/1/011001

Preface

It is with immense pleasure to introduce the proceedings of the **International Conference on Advances in Materials Processing and Manufacturing Applications (iCADMA 2020)** which was held online at Malaviya National Institute of Technology Jaipur, India during 5-6, November 2020.

The conference consisted of eight keynote lectures and 23 technical sessions given by renowned researchers and scholars working in the areas of Materials, Manufacturing, Optimization, Industrial Engineering, and Tribological studies from all over the world. To enrich the quality of program, some selected papers were invited to be presented. Distinguished researchers from India, Austria, New Guinea, and Hungary were invited to deliver eight keynote lectures sharing the latest research discoveries in their respective areas of expertise. In the technical sessions, authors made oral presentations and each presentation was confined to 10 minutes, followed by question and answer. One excellent presenter from each technical session was chosen for the best paper award depending on the technical merit, presentation and application.

Few glimpse of the inaugural ceremony of the conference at MNIT, Jaipur, Rajasthan India.

The conference was organised through video conferencing using Zoom. Authors shared the screen and attendees could discuss their work via online platform. Overall, the conference was held successfully organised through the platform. The conference committee were dedicated to support for the success of the conference.

All accepted papers were subjected to peer-review by conference committee and invited international reviewers. The papers were selected based on their quality and their relevance to the conference.

I would like to express special gratitude to members of the conference committee and organizers of the conference. I would also like to thank the reviewers for their valuable time and advice which helped in improving the quality of the papers selected for presentation at the conference and for publication in the proceedings. Finally, I want to thank the authors, the chairpersons, sponsors, and all the participants for their support for iCADMA 2020.

Dr. Amar Patnaik

Convener & Organizing Secretary
iCADMA 2020

Chief Patron& Principal General Chair:

Prof. Udaykumar R. Yaragatti

Director, MNIT Jaipur

Prof. S L Soni

Director, NIT Uttarakhand

Patron & Conference Chairman : Prof. A P S Rathore

Dean International Affairs, MNIT Jaipur

Prof. G S Dangayach

Dean Planning and Development, MNIT Jaipur

Co- Patron & Organizing Chairman : Prof. M. L. Mittal

Professor & Head

Mechanical Engineering Department, MNIT Jaipur

Prof. Dilip Sharma

Professor

Mechanical Engineering Department, MNIT Jaipur

Prof. Jyotirmay Mathur

Professor, Mechanical Engineering Department &

Head MIIC, MNIT Jaipur

Prof. M K Banerjee

Former Professor, MNIT Jaipur

Dr. Vinod Singh Yadav

Coordinator, TEQIP-III, NIT Uttarakhand

General Chair:

Prof. Ernst Kozeschnik

Institute of Materials Science and Technology, TU Wien, Austria.

Dr. Erwin Povoden-Karadeniz

Institute of Materials Science and Technology, TU Wien, Austria.

Convener & Organizing Secretary:

Dr. Amar Patnaik

Associate Professor

Mechanical Engineering Department, MNIT Jaipur

Dr. M L Meena

Associate Professor

Mechanical Engineering Department, MNIT Jaipur

Dr. Vikas Kukshal

Assistant Professor

Mechanical Engineering Department, NIT Uttarakhand

Dr. Yogesh Kumar Prajapati

Assistant Professor,

Mechanical Engineering Department, NIT Uttarakhand

IOP Conf. Series: Materials Science and Engineering 1017 (2021) 011001 doi:10.1088/1757-899X/1017/1/011001

Joint Secretary:

Dr. Dinesh Kumar, Associate Professor, MNIT Jaipur

Dr. Amit K Singh, Assistant Professor, MNIT Jaipur

Dr. Gunjan Soni, Assistant Professor, MNIT Jaipur

Dr. Monica Sharma, Associate Professor, MNIT Jaipur

Dr. Jinesh K Jain, Associate Professor, MNIT Jaipur

Dr. Mukesh Kumar, Assistant Professor, MNIT Jaipur

Dr. Anoj Meena, Assistant Professor, MNIT Jaipur

Dr. Pankaj K Gupta, Assistant Professor, MNIT Jaipur

Technical Program Committee:

Prof. Rakesh Jain,Professor, MNIT Jaipur

Prof. G D Agarwal, Professor, MNIT Jaipur

Prof. Himanshu Chaudhary, Professor, MNIT Jaipur

Mr. Amit Pancharya, Associate Professor, MNIT Jaipur

Dr. Harlal Singh Mali, Associate Professor, MNIT Jaipur

Dr. Nirupam Rohatgi, Associate Professor, MNIT Jaipur

Dr. Rajeev Agrawal, Associate Professor, MNIT Jaipur

Dr. T C Gupta, Associate Professor, MNIT Jaipur

Dr. Anup Malik, Assistant Professor, MNIT Jaipur

Dr. Amit Arora, Assistant Professor, MNIT Jaipur

Dr. Manish Kumar, Assistant Professor, MNIT Jaipur

Dr. Ram Dayal, Assistant Professor, MNIT Jaipur

Dr. Tapas Bajpai, Assistant Professor, MNIT Jaipur

International Advisory Board

Dr. Nasir Abdul Jalil, Sunway University, Malaysia

Dr. Abdul Khaliq, King Fahd University of Petroleum and Minerals Dhahran-31261, Saudi Arabia

Prof. M. Sarwar, Northumbria University, United Kingdom

Dr. Robert C. Creese, Industrial and Management Systems Engineering West Virginia University

Prof Manoj Gupta, National University of Singapore

Prof. John Lippold, The Ohio State University, Columbus,

Prof. Sujeet K. Sinha, National University of Singapore, Singapore

Prof. Tahir Shah, University of Bolton, U.K.

Prof. Kunigal.N.Nagabhushan, Centers and Institutes at North Corolina A & T State University, USA

Prof. Kunigal Shivakumar, North Carolina A&T State University, Greensboro, USA

Prof. Dhananjay Kumar, North Carolina A&T State University, Greensboro, USA

Prof. Nikhil Gupta, NYU-Tandon School of Engineering, USA

Prof. Sandra Maria Fernandes Carvalho, University of Minho, Portugal

Prof. Seeram Ramakrishna, National University of Singapore, Singapore

Prof. Rajamani Lakshminarayana, National University of Singapore, Singapore

Prof. Gyou-Jin Cho, Sungkyunkwan University (SKKU), South Korea

Prof. Shi-Hoon Choi, Sunchon National University, Suncheon, South Korea

Dr. Ramesh C Poonia, Norwegian University of Science and Technology (NTNU), Norway

National Advisory Board

Prof. Bhabani K Satapathy, IIT Delhi

Prof. Kamraj M, IIT Madras.

Prof. I K Bhat, Vice Chancellor, Manav Rachna University,Faridabad

Prof. B B Biswal, Director, NIT Meghalaya

Prof. Rakesh Sehgal, Director, NIT Srinagar

Prof. Lalit Kumar Awasthi, Director, Dr B R Ambedkar, NIT Jalandhar

Prof. Karunesh Kumar Shukla, Director, NIT Jamshedpur

Prof. Vinod Yadava, Director, NIT Hamirpur

Prof. Mahesh Chandra Govil, Director, NIT Sikkim

Prof. S S Mahapatra, NIT Rourkela

Prof. S K Biswas, Ex Chief Scientist Central Glass & Ceramic Research Institute, Kolkata

Dr. Lalit Guglani, Director & Head,CIPET: Centre for Skilling & Technical Support (CSTS), Aurangabad,

Prof. Devendra Kumar, IIT BHU

Prof. Alok Satpathy, NIT Rourkela

Prof. Inderdeep Singh , IIT Roorkee

Prof. Apurbba Kumar Sharma, IIT Roorkee

Prof. Naveet Arora, IIT Roorkee

Prof. Shishir Sinha, IIT Roorkee

Prof. Vikas Rastogi, Delhi Technological University, Delhi

Prof. A A Shaikh, SV NIT, Surat

Prof. Saroj Kumar Pradhan, CET Bhubaneswar

IOP Conf. Series: Materials Science and Engineering 1017 (2021) 011001 doi:10.1088/1757-899X/1017/1/011001

Shri SachinTyagi, Assistant Director, Ordnance Factory Medak, Telangana

Dr. Sisir Mantry, Institute of Minerals and Materials Technology, Bhubaneswar

Dr. Rahul Mulik, IIT Roorkee

Dr. Jaiveer Singh, IIT, Jodhpur

Dr. Vinod Kumar, IIT Indore

Dr. Akshay Dvivedi, IIT Roorkee

Dr. Prashant K. Jain, IIITDM Jabalpur

Dr. Rohit Bhakar, MNIT Jaipur

Dr. Saroj Kumar Meher, Indian Statistical Institute, Bangalore Centre, Bangalore

Dr. Mrityunjay Doddamani, NIT Surathkal

Dr. Rajesh Kumar Sharma, NIT Hamirpur

Dr. Maheshwar Dwivedi, BML Munjal University, Gurugram

Dr. K Venkateswarlu, Materials Science Division at NAL, Bangalore

Dr. Kishore Debnath, NIT Meghalaya

Dr. Pramanedra Kumar Bajpai, NSIT New Delhi

Dr. Rosalin Sahoo, IIT BHU, Varanasi

Mr. Shyam Kishore, Scientist, DRDO Bangalore

IOP Conf. Series: Materials Science and Engineering 1017 (2021) 011002 doi:10.1088/1757-899X/1017/1/011002

Peer review declaration

All papers published in this volume of *IOP Conference Series: Materials Science and Engineering* have been peer reviewed through processes administered by the Editors. Reviews were conducted by expert referees to the professional and scientific standards expected of a proceedings journal published by IOP Publishing.

- **Type of peer review:** Single-anonymous
- **Describe criteria used by Reviewers when accepting/declining papers. Was there the opportunity to resubmit articles after revisions?** YES
- **Conference submission management system:** Easychair
- **Number of submissions received:** 100
- **Number of submissions sent for review:** 100
- **Number of submissions accepted:** 40
- **Acceptance Rate (Number of Submissions Accepted / Number of Submissions Received X 100):** 40 %
- **Average number of reviews per paper:** 02
- **Total number of reviewers involved:** 50
- **Any additional info on review process (ie plagiarism check system):** Plagiarism checked through turnitin software
- **Contact person for queries:** Dr. Amar Patnaik
Associate Professor
Mechanical Engineering Department
MNIT, Jaipur-302017, Rajasthan, India
E-mail: apatnaik.mech@mnit.ac.in

IOP Conf. Series: Materials Science and Engineering 1017 (2021) 012001 doi:10.1088/1757-899X/1017/1/012001

Modelling of WEDM Parameters Using Taguchi & Fuzzy Logic

Siddharth Sharma[1*], Ashish Goyal[1], Love Kishore Sharma[2]

[1]Department of Mechanical Engineering, Manipal University Jaipur, Dehmi Kalan, Rajasthan, 303007, India
[2]Department of Mechanical Engineering, Arya College Of Engineering & Information Technology,Kukas,302028 Jaipur, India

Email: siddharthsharma987@gmail.com, ashish.goyal@jaipur.manipal.edu, sharma.ajmer31@gmail.com

Abstract. This work highlights the Fuzzy Logic approach for predicting best operating parameters of WEDM. This process is attracting the industries/researchers due to precision in machining. The only limitations that exist are high tool costing and lower value of finished surface. Four parameters are opted for conducting experiments viz Ton, Toff, Wire Feed & Current and their impacts are calculated for SRR and Ra. The DOE is done using Taguchi (L9) OA. ANOVA test for identification of parameters with high significance is considered.

1. Introduction

WEDM (Wire Electric Discharge Machining) process uses thin wire electrode along with dielectric fluid to machine difficult to cut contour in materials by repeated sparks of short duration. The wire tension is important factor to avoid occurrence of taper on machined surface. Temperature reaches nearly 10,000C in machining zone. The movement of wire is continuous to cut the material with spark (electrical discharge) in the vicinity between electrode and material. The wire material also gets damaged after certain duration due to high/ temperature generated during machining. Also it suffers forces of electromagnetic type, electric field type, pressure energy of bubbles, shock due to pressurised dielectric and wire driving force. Thus wire tension becomes unstable after certain duration of operation. Thus short circuit occurs and abnormal discharges leading to loss of-surface integrity, also cutting is also affected severely [1]. The machining gap is classified into 3 categories viz Open status, discharge status and short circuit. Open Status- the gap that exists without breaking down. Both the wire & work piece are well insulated. Due to higher resistance in gap the discharge current value is zero, hence low machining rate. Discharge status- is best condition for machining. With establishment of discharge-channel in machining zone the discharge operating voltage & current appears. Short-Circuit- resistance generated is of low value with weak high frequency component [2]. Some drawbacks include energy consumption, null material ablation and poor surface quality. The process finds application in industries viz aerospace, biomedical etc. **D-3 steel**, 1.2080 (Werkstoff), inherits air hardening, %carbon high, high-chromium. It possesses excellent resistance (wear) and shows dimensional stability along with higher compressive strength. D3 Steel finds application for forming dies, blanking, Press tools Punches. The wire electrode for performing cutting operation on workpiece travels a predefined path as described by end user through program (CNC). Fig. 1 shows the wire path travel for cutting operation or WEDM working process.

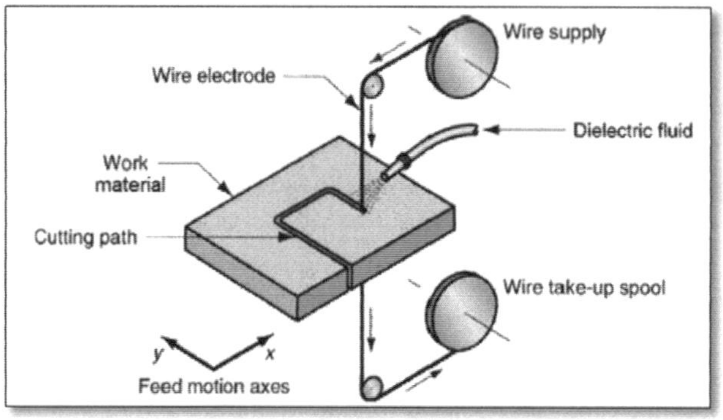

Figure 1 Wire Travel Path [3]

2. Literature

Kumar et. al. [4] used MONEL-400 alloy to investigate the effect of responses on MRR, Dimensional Deviation, spark gap. The results show that Ip (Current) is significant factor followed by other factors. Lodhi & Agarwal [5] worked on D3 for studying wire offset behaviour and concluded that high values of Ip and Ton (Pulse-on-Time) leads to higher wire offset values. Singh et al [6] prepared AA7075/SiC (SiC 5%wt.) to analyse Fuzzy model and shows that MRR is affected by Toff (Pulseoff-Time) while Ra by Ton. Goswami & Kumar [7] selected Nimonic 80A and studied responses MRR &WWR(WireWearRatio). It was seen that interaction terms (Ton x Toff &Ton x Ip) are significant for TWR while for MRR, Ton & Toff are of significance. Nain et .al [8] proposed ANN (Artificial Neural Network) and Fuzzy for studying behaviour of Ra &Waviness on Udimet L605 and concluded that the results for former are superior than later. Vellingiri et al [9] prepared alloy of LM13 SiC composite and LM-13 Alloy, and found that MRR for LM13Alloy is 36.654 mm^3/min, Ra 2.311 µm. MRR for LM13 SiC composite is 28.251 mm^3/min, Ra 4.72 µm. Goyal et al [10] worked on $Ni_{49}T_{51}$ using brass electrode and concluded that surface under investigation contains cracks and uneven crater formation also Ton and Ip are significant parameters for MRR.

3. Research Methodology

3.1. Taguchi Method

Genichi Taguchi proposed a concept which studies the weight of inputs on the way they deviates means and variance from actual and the performance of process is indicated. Classical approach for DOE(design of experiment) is complex and difficult to apply. The experimental sets increase when parameters and their corresponding level increases. To avoid this OA(orthogonal array) are introduced, which defines design consisting of small digit of experiment sets for examination [11]. These OA are basically FFD(full factorial design) which eliminates interaction focus on end goal (response) estimation For any process their exists input factors (controlled by observer), which have directly effects magnitude of responses and noise factors (cannot be controlled e.g. environmental factors) which have negative influence on quality characteristics of any process. Finally design is obtained, which is least sensitive to variations offered in uncontrolled factors. S/N (signal/noise) ratio is dependent on type quality characteristics. In our research the ratios considered are:

Larger-Better: for maximizing SRR

$$S/N(\eta) = -10\log_{10}\left[1/n\sum_{i=1}^{n} 1/y_i^2\right] \quad (1)$$

Smaller-Better: for minimizing Ra

IOP Conf. Series: Materials Science and Engineering 1017 (2021) 012001 doi:10.1088/1757-899X/1017/1/012001

$$S/N(\eta) = -10\log_{10}\left[1/n\sum_{i=1}^{n} y_i^2\right] \qquad (2)$$

3.2. Fuzzy Logic

FLS (Fuzzy Logic System) reasoning is centred on the model of fuzzy sets. Fuzzy logic consists of a set of "if-then" statements based on combining fuzzy sets. FLS performs a nonlinear correlation of multiple inputs dataset to output data. It contains 4 main parts viz, fuzzifier, rules (if/then), inference-engine, defuzzifier. Membership functions (MF`s) are used for steps (fuzzification/defuzzification) to correlate the input (non-fuzzy) parameters to fuzzy language terms. The various MF`s, defined as triangular, trapezoidal etc.[12]. The selection of each MF is context dependent and based on user experience. In present case the model for knowledge base is created using IF/THEN and joined by AND connective. The Fuzzy value is attained by interface engine performing a fuzzy reasoning.

Table 1 Fuzzy-Rules (If-Then)

S.no	**Ton**	**Toff**	**(W.F.)**	**Ip**	**SRR**	**Ra**
1	L	L	L	L	M	VVH
2	L	M	M	M	SM	S
3	L	H	H	H	VS	VVS
4	M	L	M	H	MH	MH
5	M	M	H	L	VVS	H
6	M	H	L	M	VH	VS
7	H	L	H	M	S	VH
8	H	M	L	H	VVH	M
9	H	H	M	L	H	MS

(L=Low, V=Very, S=Small, M=Medium, H=High)

4. Experimentation

The performance /inputs parameters for our research work are opted after thorough analysis of literature, which provides brief description of overall WEDM operational parameters and enables the researchers for selecting the data. The workpiece dimensions are 65mmx65mmx0.75mm. The wire(brass) [13] used for conducting experiment is 0.25mm diameter. The variation in hardness number of specimen (D3 steel) is done by heat treatment (hardening & tempering) and oil quenching prior to experimentation, so that the material hardness is considerably altered (56 HRC).Machine equipment, Electronica Tool Master 6S at CIPET.

The efficiency of any process is calculated in terms SRR/MRR (Stock/Material Removal Rate) is calculated using difference of weight method and unit is gm/min. the mathematical expression is;

$$SRR = \frac{(Wi - Wf)}{MachineOperatingtime} \qquad (3)$$

Wi = initial specimen weight
Wf = final specimen weight
The surface roughness can be described as Ra, unit (μm), where Ra is mean roughness of machining surface.

$$Ra = \int_0^L |Y(X)dx| \quad (4)$$

L= length for evaluation, Y=profile curve ordinate

Ra measurement is done using roughness tester (Mitutoyo SG-410)

Table 2 Orthogonal Array (L9)

S.no	**Ton** (μsec)	**Toff**(μsec)	**W.F.** (mm/sec)	**Current** (A)
1	2	7	6	6.5
2	2	8	6.5	7
3	2	9	7	7.5
4	3	7	6.5	7.5
5	3	8	7	6.5
6	3	9	6	7
7	4	7	7	7
8	4	8	6	7.5
9	4	9	6.5	6.5

For generating fuzzy output in fuzzy interface engine various rules are defined to predict values as per membership functions. The “If/Then” rules justifies all the necessary logics of input variables in and accordingly generates outputs.

5. Results

5.1. Taguchi & Fuzzy Logic

Table 3 Results

S.no	**Experimental**		**Fuzzy Logic**		**Error % (Actual & Fuzzy Values)**	
	SRR	**Ra**	**SRR**	**Ra**	**SRR**	**Ra**
1.	0.6386	3.259	0.6551	3.2166	2.5186994	1.3010126
2.	0.5912	2.5685	0.6183	2.4283	4.3829856	5.4584388
3.	0.5169	2.1512	0.5447	2.1939	5.1037268	1.9849386
4.	0.6716	3.0516	0.6919	2.8437	2.93395	6.8128195
5.	0.4987	3.1865	0.5192	2.9823	3.9483821	6.408285
6.	0.7421	2.4677	0.7655	2.2897	3.0568256	7.2131945
7.	0.566	3.2589	0.5815	3.1207	2.6655202	4.2406947
8.	0.7768	2.9056	0.79098	2.7053	1.7927128	6.8935848
9.	0.6986	2.784	0.7287	2.5677	4.1306436	7.7693966

$$\text{Error\%} = \left(\frac{\text{ExperimentalValue} - \text{FuzzyValue}}{\text{ExperimentalValue}} \right) \times 100 \quad (5)$$

IOP Conf. Series: Materials Science and Engineering 1017 (2021) 012001 doi:10.1088/1757-899X/1017/1/012001

Maximum error % is calculated using eq. 5 and is found to be 7.7% for SRR while for Ra it is 5.1% only. The plot between target responses and fuzzy values for SRR and Ra is shown in fig.2 & fig3.

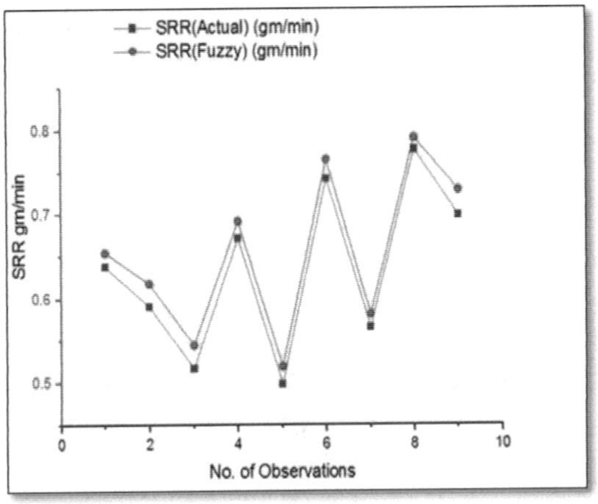

Figure 2 Actual vs Fuzzy output (SRR)

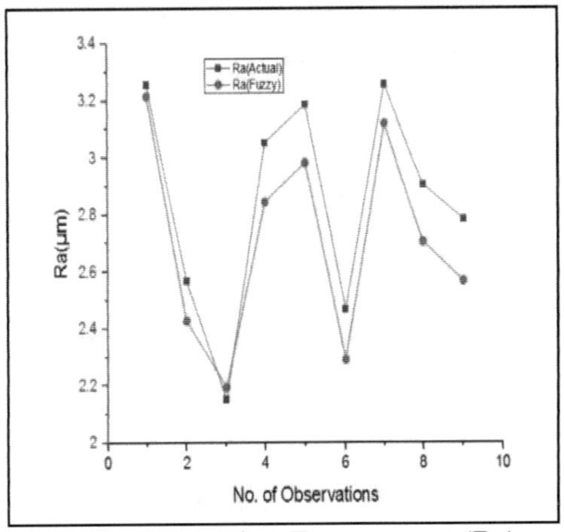

Figure 3 Actual vs Fuzzy output (Ra)

Table 4 Larger-Better for SRR

Level	Ton (A)	Toff(B)	Feed(C)	Ip (D)
1	-4.731	-4.099	-2.893	-4.351
2	-4.031	-4.267	-3.713	-4.033
3	-3.418	-3.813	-5.573	-3.794
Delta	1.313	0.455	2.68	0.557
Rank	2	4	1	3

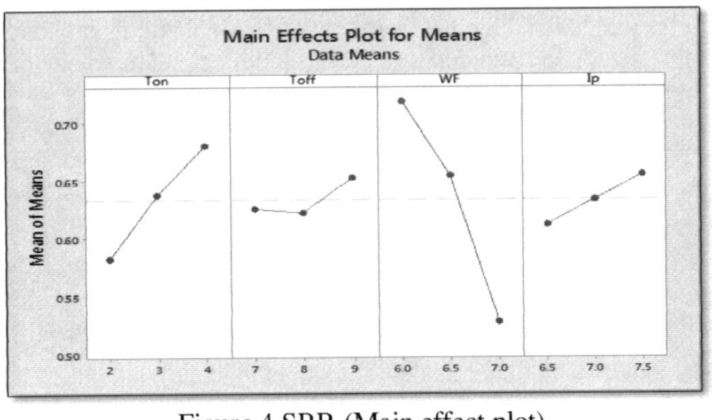

Figure 4 SRR (Main effect plot)

Table 4 shows s/n ratio for SRR. Also the rank of each parameter is specifically described.C1A2D3B4 is best sets/combination of parameters for achieving target value effectively. From fig. 7 it is stated that SRR increases with higher Ton values, as more time is offered for heat energy to melt and evaporate material. Higher Toff values also leads to higher SRR as more time duration is available' for proper flushing of debris attached to exposed layer of specimen. Lower value of SRR is obtained at high wire feed as short circuit phenomenon occurs and irregular spark discharges also takes place.

Higher Ip values gives high SRR as again high heat is generated which supports machining of work piece.

Table 5 Smaller-Better for Ra

Level	Ton(A)	Toff(B)	WF(C)	Ip(D)
1	-8.37	-10.071	-9.124	-9.74
2	-9.201	-9.175	-8.926	-8.767
3	-9.473	-7.798	-8.994	-8.536
Delta	1.104	2.274	0.198	1.204
Rank	3	1	4	2

Table 6 shows s/n ratio for Ra. Also the rank of parameters is specifically described as .B1D2A3C4.

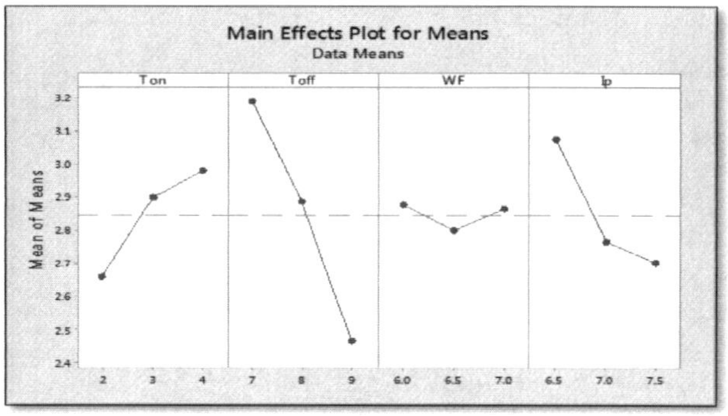

Figure. 5 Ra (Main effect plot)

From fig 8 it can be verified that Ra increases with higher Ton, since it becomes higher with high removal of material. Higher Toff leads to eradication of debris on workpiece surface by pressurized fluid (dielectric) thereby decreasing Ra. Increase in feed of wire leads to irregular cutting of material due to abrupt spark discharges, which causes formation of thermal stress on work piece. Increasing Ip also leads to lower Ra as heat energy (for melting & vaporizing of material) is directly proportional to current.

5.2. ANOVA

95% confidence level is considered for statistical evaluation of SRR and Ra. The significance of any process parameter is justified by analysis of P-value, which is considered best when P< α, where (α=.05), as in Table 6&7. Similarly overall model significance is justified by analyzing R^2 and adj. R^2, for SRR (R^2 96.7% and adj. R^2 (93.41%) while for Ra (R^2 94.98% and adj. R^2 89.96%). Thus overall models are statistically good. The % contribution of every individual parameter on either response is depicted using pie charts as in figures

Table 6 ANOVA (SRR)

Source	Contribution	Adj. SS	Seq. MS	F-Value	P-Value
Regression	96.70%	0.073647	0.018412	29.34	0.003
Ton	19.01%	0.014475	0.014475	23.07	0.009
Toff	1.45%	0.001104	0.001104	1.76	0.255
WF	72.58%	0.055277	0.055277	88.09	0.001
Ip	3.66%	0.002791	0.002791	4.45	0.103

Error	3.30%	0.00251	0.000627
Total	100.00%		

Table 7 ANOVA (Ra)

Source	Contribution	Adj SS	Seq MS	F-Value	P-Value
Regression	94.98%	1.1488	0.2872	18.93	0.007
Ton	12.96%	0.15675	0.156752	10.33	0.032
Toff	64.69%	0.78236	0.782359	51.57	0.002
WF	0.02%	0.00021	0.000212	0.01	0.912
Ip	17.32%	0.20948	0.209478	13.81	0.021
Error	5.02%	0.06069	0.015172		
Total	100.00%				

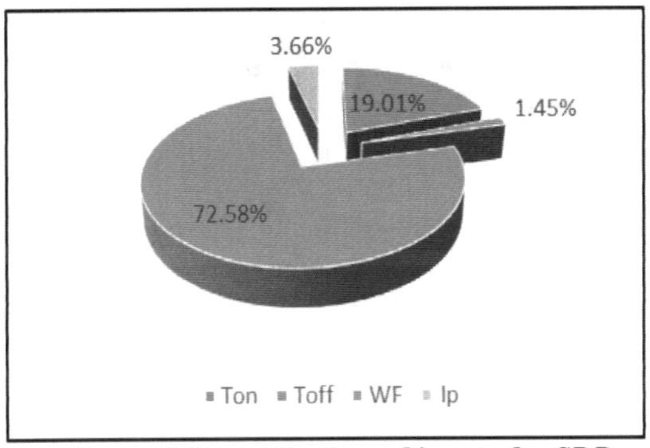

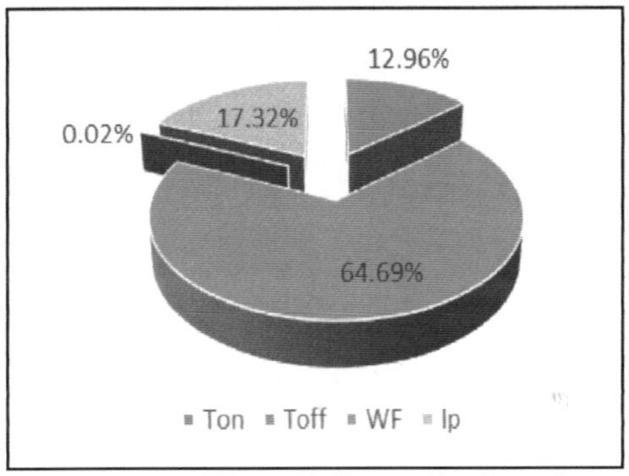

Figure 6 %contribution of inputs for SRR Figure 7 %contribution of inputs for Ra

6. Conclusion & Future Scope of Work

The following statements can be successfully drawn after implementing Taguchi L9 and Fuzzy system on D3(heat treated) specimen:

1) The suitable combination of inputs as per Taguchi is found to be for C1A2D3B4 SRR and B1A2D3C4 for Ra.

2) For SRR (R^2 96.7% and adj. R^2(93.41%) while for Ra (R^2 94.98% and adj. R^2 89.96%)

3) Fuzzy Logic is applied and the maximum % error between target value & fuzzy result sets is found to be 7.7 % for SRR and 5.1% for Ra.

4) Other output parameters viz, taper angle, tool wear analysis, FEM analysis, wire offset etc can be done using other optimizing techniques.

References

[1] Lingling L., Zhidong L., Xiefeng L., and Mingming L., 2014, "Non-even wire tension in high-speed wire electrical discharge machining," *The International Journal of Advanced Manufacturing Technology*, 78, pp.503-510.

[2] C. Li, Bai J., Ding J., and Fan Y., 2014, "Gap current voltage characteristics of energy-saving pulse power generator for wire EDM," *The International Journal of Advanced Manufacturing Technology*, 77, pp.1525-1531.

[3] Bisaria H. and Shandilya P., 2015, "Machining Of Metal Matrix Composites By Edm And Its Variants : A Review," Ch-23, pp267-282.

[4] Arunkumar N. E., Ganesh M., and Vivekanandan N.,2020 "Optimization of machining parameters in WEDM of Monel 400 using Taguchi technique," *Material Today Proc.*,22, pp. 2199–2206.

[5] Kumar B. and Agarwal S., 2020, "Optimization of wire off set in WEDM process by Taguchi Method," *Material Today Proc.*, 22, pp. 2366–2371,.

[6] Kumar A., Roy K., Das S., and Das S., 2020, "WEDM investigation and fuzzy logic modelling of AA7075 / SiC metal matrix composites," *Material Today Proc.*, 26, pp. 1988-1994,.

[7] Goswami A. and Kumar J., 2014, " Investigation of surface integrity , material removal rate and wire wear ratio for WEDM of Nimonic 80A alloy using GRA and Taguchi method," *Eng. Sci. Technol. an Int. J.*, 17, no. 4, pp. 173–184,.

[8] Nain S. S., Sihag P., and Luthra S., 2018, "Performance Evaluation of Fuzzy-Logic and BP-ANN Methods for WEDM of Aeronautics Super Alloy SC," *MethodsX*,5,pp.890-908.

[9] Vellingiri S.,Soundararajan R., Mohankumar N., Nithyananthakumar K., and Muthuselvam K., 2020 "Exploration on WEDM process parameters effect on LM13 alloy and LM13 / SiC composites using Taguchi method," *Materials Today Proc.*.

[10] Goyal A., Rahman H. UR, and Ghani S. A. C.,2020 "Experimental investigation & optimisation of wire electrical discharge machining process parameters for Ni49Ti51 shape memory alloy," *J. King Saud Univ. - Eng. Sci.*, pp. 1–7,.

[11] Goyal A., Pandey A., and Sharma P., 2017, "Machinability of inconel 625 aerospace material using cryogenically treated WEDM," *Solid State Phenom.*, 266 SSP, pp. 38–42.

[12] Choudhuri B., Sen R., Ghosh S. K., and Saha S. C.,2018, "Modelling of Surface Roughness and Tool Consumption of WEDM and Optimization of Process Parameters Based on Fuzzy-PSO," *Materials Today Proc.*, 5, no. 2, pp. 7505–7514.

[13] Goyal A., 2017,"Investigation of material removal rate and surface roughness during wire electrical discharge machining (WEDM) of Inconel 625 super alloy by cryogenic treated tool electrode," *J. King Saud Univ. - Sci.*, 29, no. 4, pp. 528–535.

Uncontaminating Concrete Mix Using Sugarcane Bagasse Ash: A Review on its Hardened Properties

R B Philip[1], G N Das[1], S K Kuriakose[1], D Jacob[1], A S Prabha[1]

[1]Department of Civil Engineering, Providence College of Engineering, Chengannur, Kerala.

E-mail: annareshma50@gmail.com

Abstract. Concrete is one among the most used material for construction other than water. Cement is the important constituent of concrete. During the production of cement high amount of carbon dioxide is emitted and is responsible for global warming. High amount of cement production is causing environmental problem. Use of agro-waste materials like bagasse ash, increases the strength characteristics of concrete and also causes proper disposal of the waste materials. In this paper, use of sugarcane bagasse ash, an agro-waste product, as a partial replacement of the cement in concrete mix and its effect on the strength parameters of concrete are also identified on the basis of compressive strength, spilt tensile strength and flexural strength. This waste usage is very economical and also have positive impacts on minimizing environmental pollutions.

1. Introduction

We are aware that a lot of damage is done to environment in the manufacture of cement. It emits carbon dioxide associated with other chemicals. The researchers have shown that every one ton of cement manufacture releases its half ton of carbon dioxide, so there is an important need to control the use of cement. On the other hand, material wastes such as sugarcane bagasse is difficult to dispose which causes environmental hazards. The bagasse ash gives high early strength to concrete and also decreases the permeability of concrete [1]. The silica present in bagasse ash reacts with cement components during hydration and improves properties like chloride resistance, corrosion resistance etc. [2]. Hence the use of bagasse ash in concrete not only reduces the environmental pollution, but also improves the properties of concrete thereby reducing the cost. It also improves the durability of the concrete [3].

Because of expeditious developments in agricultural and industrial sectors, large quantities of waste materials are being produced in these sectors. Clearance of these wastes is becoming serious environmental issue these days. Most of the discarded waste results in filling the landfills, which reduces the availability of useful land. These may inturn result in polluting the environment. Most of the industrial byproducts which include, silica fume, fly ash, coal, GGBS etc. could be used for social and economical benefits, by successfully transforming it into cementitious materials which find applications especially in construction fields [4]. Recently, the forestry and agricultural remains are used as biomass fuels. The resulting ashes from the fumes are the final waste products and proper management and utilization of these waste materials are now emerging as a wide area of research. These biomass ashes like palm oil fuel ash, sugarcane bagasse ash, bamboo stem ash, risk husk ash, elephant grass ash etc. are now being utilized as Supplementary Cementitious Materials (SCM) [5]. From the researches and studies done till now, it could be inferred that incorporating the waste ashes in cementitious materials could improve or maintain the mechanical characteristics and properties of concrete [5]. In addition to these the incorporation of these biomass ash in cement production can considerably reduce the greenhouse gas generation involved with the process of cement production and also considerably making these processes economical. This results in reducing the waste disposal pressure and environmental pollutions leading to a sustainable development.

2. Material

2.1. Sugarcane bagasse ash

Sugarcane bagasse ash (SCBA) is the main byproduct from sugar factories which is obtained after the burning of sugarcane bagasse. Sugarcane bagasse is the waste product resulting after extracting the useful or economical sugar from sugarcanes. The dumping of these materials are causing a large environmental problem for the sugar manufacturing factories.

3. Physical properties

The composition and physical properties of SCBA will vary with factors, in particular, the sugar cane variety, combustion duration and combustion temperature, purity of sugarcane bagasse, locations and methods of ash collection, type of cooling, equipments used, ash fineness etc. The colour of SCBA is dark to light black. A better carbon content is specified by a dark black colour of SCBA. Due to the crystallization along with its decomposition at higher temperatures, the bagasse ash seems to be gray over 800 °C and seems to be white over 900 °C [6].

4. Chemical composition

The chemical composition among different SCBA samples is remarkably different. However, most of the samples possesses components of silica, alumina, and other metallic elements [3,9]. As per the specifications of ASTM C618-08a, natural pozzolans having SiO_2, Al_2O_3, Fe_2O_3 together tallying the whole mass over 50% could be designated as a class C pozzolan and class F pozzolans which have the ratio over 70% [5]. The combined effect of relative stiffness and the pozzolanic activity created by SCBA is due to the presence of silica [7,9].

5. Application

- Sugarcane bagasse, a voluminous by product in the sugar mills following the extraction of juice from the sugarcane, was generally being used as a fuel to the fireside furnaces which was within the corresponding sugar mill. This yield about 8% to 10% ash which contain high amounts of unburnt matter, alumina, silica, iron oxides and calcium oxides [5,6].
- Earlier SCBA was utilized as a fertilizer or it was dumped into landfills. This led to a serious environmental concern. But in the recent years, researches on SCBA focused mainly on its usage as construction materials, because of its pozzolanic features [8].
- Studies indicates that SCBA, is a form of propitious construction material and its usage as a pozzolanic material, include its usage in alkali - activated binders, fillers and aggregates in construction medium [2,8].
- The impression of SCBA in fresh concrete and hardened concrete highlights its properties, which include the physical characteristics, microstructure, mechanical characteristics and sturdiness [5].
- It is now used as a biofuel for warmth, energy and electricity generation and also in the manufacturing of pulp and building materials.
- SCBA is used as a soil stabilization material especially in the case of laterite soil.

6. Effect of sugarcane bagasse ash on workability

Reddy et al. (2017) revealed from the workability slump cone test results that cement could be very beneficialy replaced with SCBA upto a maximum value of 10%. In their work, proportions of concrete mix designs for M20 grade was carried out according to the specifications of IS:10262-2009. Different mixes with 5%, 10%, 15% and 20% SBCA replacements were tested for workability and concluded that, the partial replacement of cement in concrete with SCBA, enhances the workability, whereby the use of superplasticizer is not much essential [11].

Mangi et al. (2017) conducted slump test with and without SCBA, to verify the consistency of normal concrete mix. From the results, after executing the slump test on M15 and M20 concrete grades, a continuous slump rise was observed when replaced with SCBA. Here the values observed fall in between low degree and medium degree of workability. Hence the usage of superplasticizer is not very crucial [1].

Priya et al. (2016) conducted both slump test and compaction factor test for workability with 0% to 25% replacement of SCBA in cement. In both the tests the workability values increased with the increase in percentages of SCBA, depicting that the SCBA content can reduce the water requirement of concrete [8].
The comparison of slump values are shown in Table 1

Table 1. Comparison of slump values of concrete mixes

Si. No	Authors	Mix	Replacement with SCBA (%)	Average slump value (mm)
1	Mangi et al. (2017) [1]	M15	0%	31.94
			5%	37.48
			10%	44.64
2	Mangi et al. (2017) [1]	M20	0%	27.89
			5%	42.50
			10%	51.12
3	Priya et al. (2016) [8]	M25	0%	70
			5%	85
			10%	90
			15%	100
			20%	125
			25%	160

7. Effect of bagasse ash on its hardened properties

7.1. Compressive strength

Priya et al. (2016) replaced cement by 5%, 10%, 15% and 20% respectively and tested for compression strength at 28 days. The loading rate of compression testing machine was 0.5 tonnes per second. It was detected that compressive strength of M0 concrete mix had a higher value than M5 concrete mix and concluded that the amount of amorphous silica present was higher than the CaOH produced during the hydration reaction and this would reduce the strength of specimen. The result showed that, the concrete mix with 10% of SCBA replacement and with 28 days curing, indicated the maximum strength in comparison with the concrete mixes with other percentage replacements of SCBA [8].

From the studies of Amin (2015) the compressive strength of 3 days, 7 days and 28 days with different SCBA replacement levels has increased and has attained comparable values to those of the control concrete specimens. Concrete grade used was M25. The 3 days compressive strength of 20% SBCA was same as that of 7 days compressive strength of control specimen however the 28 strength being slightly higher. The optimum percentage replacement of SBCA was found to be 20% [10].

The comparison of compressive strength of various specimens in the above mensioned works are shown in Table 2.

Table 2. Comparison of average compressive strength of specimens (M25)

		Average compressive strength (N/mm^2)	
Si. No	SCBA replacement (%)	Priya et al. (2016) [8]	Amin (2011) [10]
1	0%	33.281	36
2	5%	37.351	42
3	10%	38.077	43
4	15%	36.769	42
5	20%	35.752	40
6	25%	30.956	35

Kvjd (2015) had done an experimental research on partially replaced concrete mixes with SCBA in HCL solutions. He casted M35 grade blocks with SCBA mix and tried curing it in normal water and HCL solutions. Finally, it was found that there was a decrease in strength for concrete specimens in HCL solutions as compared to concrete specimen cured in normal water. Compressive strength increased for 7 days, 28 days and 60 days when cured in normal water but it after curing 28 and 60 days in HCL solutions [2]. Table 3 describes the variation in results both in the case of normal water and in HCL.

Table 3. Compressive strength results (in normal water and HCL), Kvjd (2015) [2]

Sample Designation	Replacement of SCBA in cement	Compressive strength					
		7 days		28 days		60 days	
		Normal water	HCL	Normal water	HCL	Normal water	HCL
C0	0%	35	32.5	45	42.5	51.38	45
C1	5%	39	37.33	48.8	43.5	49.6	46.5
C3	10%	39.5	39	52.3	51.33	52.5	48.16
C4	15%	35.5	33.16	48.5	40.5	50.66	44
C5	20%	31.66	31.66	44.83	40	45.5	39.5
C6	25%	32.83	20.16	43.66	30.66	40.3	26

7.2. Split tensile strength

Priya et al. (2015) studied that the split tensile strength of concrete was more than M0 concrete mix upto 20% replacement SCBA in cement. Further, increase in replacement upto 25% led to the reduction in tensile strength due to dilution effect [8].

Bisanal et al (2020) experimented the split tensile strength of cylindrical concrete specimens of grade M30 grade with 5%, 10%, 20% & 30% of bagasse ash substitutions with cement. The 7 days and 28 days split tensile strengths of these specimens increased considerably upto 20% replacements of SCBA [12]

Batool et al (2020) conducted split tensile strength tests with 5%, 10%, 15%, 20%, 25%, and 30% replacement of SCBA in concrete and it was depicted that maximum tensile strength was given by specimen with 10% SCBA replacement. The results implied that there was a significant reduction in tensile strength for samples with higher SCBA substitution [13].

Modani et al. (2013) examined the aspects of SCBA, on the tensile strength of concrete. They casted 45 cubes and 15 cylindrical specimens. SCBA was replaced by 0%, 10%, 20%, 30% & 40% by volume of sand based on the mix volume. The water-cement ratio was 0.40 and the superplasticizer dosage was 0.8%. The 28 days split tensile strength was noted. The test was implemented on cylindrical specimens with 150mm diameter & 300mm height (as per Indian standards). Form the test they detected that the tensile strength development of these mixes decreased with the increase in percentage replacements of SCBA [4].

7.3. Flexural strength

The results obtained from the studies of Priya et al. (2016) specify the difference in flexural strength of specimens with different cement substitutions. The value of flexural strength of M2 mix of 10%

SCBA is higher than that of M5 mix with 25% replacement. Since the flexural strength values of SCBA replaced concrete was more, it could be employed in slabs, beams etc. [8].

Amin et al (2020) investigated the contribution of finely Ground Bagasse Ash (GBA) in the production of Engineered Cementitious Composites (ECCs) along with Polyvinyl Alcohol (PVA). 10%, 20% & 30% replacements were made to cement with bagasse ash. Nine rectangular beam specimens of size 320 mm length, 40 mm height and 12 mm thickness were casted for testing flexural strength using four point loading and tested at end of 14, 28 & 91 days. The results depicted that, the influence of increasing percentage of bagasse ash on the flexural strength reduction was decreasing with aging from 14 days to 28 days and then from 28 days to 91 days [14].

Bisanal et al (2020) experimented the flexural strength of M30 concrete beams with 5%, 10%, 20% & 30% of bagasse ash substitutions. The 28 day flexural strength of these beams increased considerably upto 20% replacements of SCBA [12].

8. Conclusion

From the papers reviewed, it could be inferred that in its unadulterated form, the bagasse ash proved to be a possible ingredient of concrete, since it can be a beneficial replacement for fine aggregate and cement.

- In economic point of view, SCBA when replaces cement saves capital especially in the construction industry. Since sugar bagasse ash is a by-product material, it could be used as a replacement material for cement inorder to reduces the level of carbon dioxide emissions by the cement manufacturing industry.
- Cement with partial replacements of SCBA enhances the workability property of fresh concrete and use of super plasticizers could be considerably reduced.
- In concrete, replacement of cement with upto 20% SCBA performs well compared to ordinary concrete in terms of strength parameters like compressive strength, split tensile strength and flexural strength.
- Permeability of concrete also reduces with the addition of SCBA resulting in increased durability of structures.
- In addition, it resolves the scrapping problems which are associated with the sugar manufacturing industries hence keeping the environment pollution free and sustainable.

References

[1] Mangi S A *et al.*,2017, "Utilization of sugarcane bagasse ash in concrete as partial replacement of cement," *IOP Conf. Ser. Mater. Sci. Eng.*, **271(1)**, 012001.

[2] Kvgd B, 2015, "Partial replacement of cement in concrete with sugarcane Bagasse Ash,".

[3] Yogitha B *et al.*,2020 "Progress of sugarcane bagasse ash applications in production of Eco-Friendly concrete - Review," *Mater. Today Proc.*

[4] Modani P O and Vyawahare M R, 2013, "Utilization of bagasse ash as a partial replacement of fine aggregate in concrete," *Procedia Eng.*, **51**, 25–29.

[5] Balagopal V and Viswanathan T S, 2020, "A Comprehensive Exploration on Sustainable Supplementary Cementitious Materials Derived from Agro-Wastes," *Int. J. Eng. Adv. Technol.*, **9(3),** 2482–2489.

[6] Xu Q *et al.*, 2018, "Characteristics and applications of sugar cane bagasse ash waste in cementitious materials," *Materials (Basel).*, **12(1)**, 1–19.

[7] Ganesan K *et al.*, 2007, "Evaluation of bagasse ash as supplementary cementitious material," *Cem. Concr. Compos.*, **29(6)**, 515–524.

[8] Lakshmi Priya K and Ragupathy R, 2016, "Effect of Sugarcane Bagasse Ash on Strength Properties of Concrete," *Int. J. Res. Eng. Technol.*, **5(4)**, 159–164.
[9] Rukzon S and Chindaprasirt P, 2012, "Utilization of bagasse ash in high-strength concrete," *Mater. Des.*, **34**, 45–50.
[10] Amin N U,2011, "Use of bagasse ash in concrete and its impact on the strength and chloride resistivity," *J. Mater. Civ. Eng.*, **23(5)**.
[11] Reddy M *et al.*, 2017, "Utilization Of Sugarcane Bagasse Ash (SCBA) In Concrete By Partial Replacement Of Cement," *IOP Conf. Ser. Mater. Sci. Eng.*, **271(1)**, 12–16.
[12] Bisanal M G *et al.*, 2020, "Study On Sugar Cane Bagasse Ash In Concrete By Partial Replacement Of Cement," **4(1**1), 160–162.
[13] Batool F *et al.*, 2020, "Characterization of Sugarcane Bagasse Ash as Pozzolan and Influence on Concrete Properties," *Arab. J. Sci. Eng.*, **45(5)**, 3891–3900.
[14] Amin M N *et al.*, 2020, "Role of Sugarcane Bagasse Ash in Developing Sustainable Engineered Cementitious Composites," **7**, 1–12.

A6061/B4C MMCs fabrication, experimental investigation and prediction of properties

Shyn C S[1], Rajesh R[1], Dev Anand M[2]

[1]Department of Mechanical Engineering, Noorul Islam Centre for Higher Education, Kumaracoil, Tamil Nadu-629180, India
[2]Director Research, Noorul Islam Centre for Higher Education, Kumaracoil, Thuckalay, Tamil Nadu-629180, India

E-mail address: shyncs2015@gmail.com

Abstract. The growing demand for light composite materials is the new trend in engineering solicitations. The objective is to fabricate light weight composite material through less expensive method. In this work, nine different composites have been fabricated with A6061 as base matrix and reinforced with 3%/6%/9%wt%B4C having particle sizes of 30μm, 40μm and 60μm respectively. The investigation of the actual density and hardness of the fabricated composite is the major task. The densities of the fabricated composites were investigated with Two-way Analysis of Variance(ANOVA). Also the densities of the composites were predicted with Taguchi, regression and Poisson method. From all of these fabricated composites 60μm size, 9% wt B4C particle reinforced A6061/B4C composite shows lighter density and this value were nearest to Poisson predicted value.

1. Introduction

Metal matrix composites(MMCs)[1] shows an essential part in the latest engineering applications. In aluminium metal matrix composites(AMMCs)[2] the base metal is aluminium alloy and the addition of reinforcement makes composites. The commonly used reinforcement particles during the fabrication of AMMCs include boron carbide(B4C), silicon carbide(SiC), titanium carbide(TiC)[2]–[4]. Stir casting method is the liquid state fabrication process in AMMCs[5]. The density[6] of the fabricated composites can be measured using Archimedes[7] test. Also rule of mixture[8] is used for the analysis of theoretical density. Once the actual density and theoretical density obtained , the porosity percentage of the composite can be identified[9]. Micro hardness [10] has been obtained through Brinell Hardness test. Also the hardness is obtained using Rockwell test[11]. The X-ray diffraction (XRD) method is used for the phase analysis[12]. Scanning electron microscopy (SEM) is used for the characterization study[13]. The core aspects of this research paper are divided into the following sections: section 2 focuses on experimentation of A6061MMCs reinforced with B4C particles. Section 3 explains the result and discussion of mechanical properties of A6061+(wt%B4C)MMCs. Section 4 concludes the study with future scope of A6061/B4C MMCs.

2. Experimentation

2.1. Matrix material and reinforcement particles

The aluminium alloy 6061(A6061) has been used in the present investigation as base alloy with T6 grade. It is in the form of rod with the size of 20mm. The boron carbide reinforcement particles with 3%wt, 6%wt, 9%wt and different particle size having mesh size 230(60µm), 325(40µm) and 450(30µm) were used. Aluminium 6061 alloy .
and boron carbide were purchased from Coimbatore Metal suppliers, Coimbatore, Tamil Nadu, India. The table 1 shows the composition of A6061 alloy [14]. The physical properties of A6061 and B4C are shown in table 2 [15] and table 3 respectively [16].

Table 1. Composition of A6061 alloy[14]

Component	Magnesium	Silicon	Iron	Copper	Zinc	Titanium	Manganese	Chromium	Others	Aluminium
Weight %(wt%)	0.8 -1.2	0.4 - 0.8	Max. 0.7	0.15 - 0.40	Max. 0.25	Max. 0.15	Max. 0.15	0.04 - 0.35	0.05	Balance

Table 2. Physical properties of A6061[8]

Property	Unit	Value
Density	g/cm^3	2.7
Melting point	^{0}C	580
Modulus of elasticity	GPa	70 - 80
Poisson's ratio		0.33
UTS	MPa	310
Hardness	HRB	60

Table 3. Physical properties of B4C[17]

Property	Unit	Value
Density	g/cm^3	2.52
Melting point	^{0}C	2445
Modulus of elasticity	GPa	450

IOP Conf. Series: Materials Science and Engineering 1017 (2021) 012003 doi:10.1088/1757-899X/1017/1/012003

2.2. Fabrication

Stir casting technique had used for the manufacture of composite, the reinforcement materials added as per the design of experiments. The boron carbide particles with 3%wt, 6%wt, 9%wt and different particle size having mesh size 230(60μm), 325(40μm) and 450(30μm) have been used as reinforcement particles. The present study comprises of 2 control factors with 3 levels. The different possible designs are L9, L18, L27, L36, L54 under single level and mixed 3 level. The base DOF for this experiment is calculated as,

DOF =((number of control factors)× (number of level - 1))

DOF_T for the Taguchi's design based on DOE = (number of orthogonal array in Taguchi design- 1)

This value of DOF_T must be greater than or equal to base DOF. Also for the reduced number experiments the Taguchi's L9 orthogonal array is selected. Nine experiments were executed separately for the stir casting setup. In the current work all the analysis were carried out using Minitab statistical software. The various parameters for this experiment were shown in table 4. The persistent parameters used in this fabrication process have been presented in table 5.

Table 4. Parameters for experiment

	Levels		
Parameters	1	2	3
wt% of B4C	3	6	9
Particle Size(*μm*)	30	40	60

Table 5. Stir casting parameters

Sl No	Parameter	Unit	Value
1	Preheat Temperature of Reinforcement Particles	^0C	300
2	Melting Temperature	^0C	750
3	Reinforcement Particles adding temperature	^0C	650
4	Stiring RPM	rpm	300
5	Stiring time	Minutes	5

The procedure of stir casting has been presented in figure 1. The reinforcement particles were preheated to 300^0C. After that it is mixed to liquid metal in the furnace and temperature being raised to 850^0C. Then the molten metal is poured in to preheated (250^0C) cast iron mould and permitted to set at 26^0C. The fabricated casting size is 100×100×10mm of square prism. The figure 2 shows stir casting arrangement[18]. The details of fabricated cast samples were displayed in table 6.

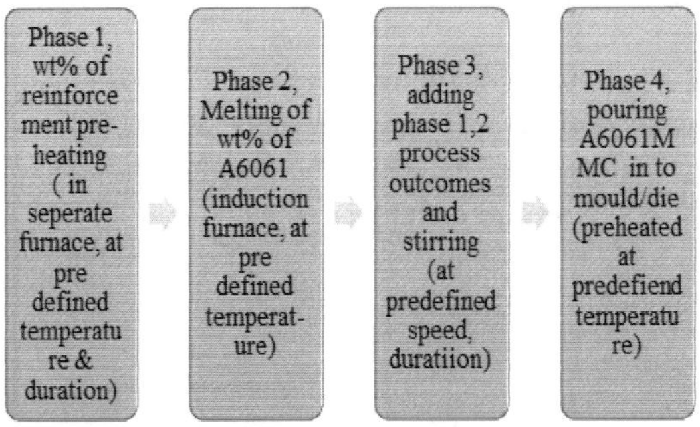

Fig. 1. Stir casting procedure

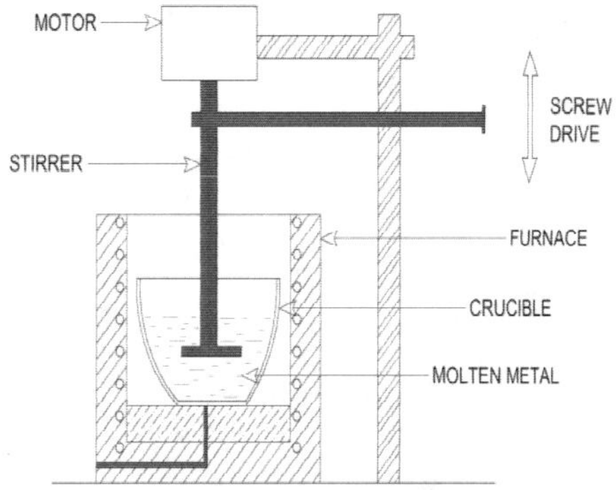

Fig. 2. Stir casting arrangement

Table 6. Fabricated samples

Exp. No.	Base Matrix Alloy	Reinforcement Particle	wt% of Reinforcement Particle	Size of particle(μm)	Metal Matrix Composite fabricated
1	A6061	B4C	3	30	A6061+3% B4C
2	A6061	B4C	6	40	A6061+6% B4C
3	A6061	B4C	9	60	A6061+9% B4C
4	A6061	B4C	3	30	A6061+3% B4C
5	A6061	B4C	6	40	A6061+6% B4C
6	A6061	B4C	9	60	A6061+9% B4C
7	A6061	B4C	3	30	A6061+3% B4C
8	A6061	B4C	6	40	A6061+6% B4C
9	A6061	B4C	9	60	A6061+9% B4C

IOP Conf. Series: Materials Science and Engineering 1017 (2021) 012003 doi:10.1088/1757-899X/1017/1/012003

2.3. Archimedes test

The Archimedes test gives the density and the porosity levels of the fabricated samples. The main factors controlling the properties are percentage porosity as well as its size. The air entering during pouring and stirring may cause porosity. For achieving good mechanical properties, the porosity should be very negligible value. The equation 1 gives the fabricated composites density [19].

$$\rho = \left(\frac{w}{w-w_1}\right) \times \rho_w \quad (1)$$

where ρ= density of fabricated part
ρ_w= density of pure water
w= weight of composite in air
w_l= weight of composite in water

The density theoretical (ρ_{th}) was observed by the rule of mixture

$$\rho_{th}=\rho_m V_m+\rho_r V_r \quad (2)$$

where ρ_m theoretical density of fabricated sample, V_m sample volume fraction, ρ_r B4C theoretical density and V_r is the B4C volume fraction[20].

2.4. Rockwell hardness test

The hardness has been obtained from Rockwell hardness experiment. The B scale hardness values are observed under the load of 100kg, with three different locations in the surface of the sample. The wire EDM machine has been used to prepare the specimens of size 15mm*15mm*5mm for the test. A 1/16" (2.11mm) diameter steel ball has been used under a load of 100Kgf. Three set of readings at different locations have been taken on the composite surface to reduce errors.

2.5. Microstructure studies

Microphotographs disclose uniform dispersal of B4C particles throughout fabricated composites. The porosity has found to be less and better bonding has been observed between B4C particles and A6061 alloy.

2.6. XRD Analysis

The test composites prepared with the dimensions 15mm×15mm×2mm have been analysed with x-ray diffraction technique.

3. Result & discussion

3.1. Density

Experimental outcomes are investigated by following the standard way and the results are shown in table 7. The actual density is observed from the Archimedes test and theoretical density from rule of mixture. The figure 3 shows disparity in the actual and theoretical density along with wt% of the B4C in the composite fabricated. It is observed that the relative density reduced a little with the wt% of B4C particles from 3 to 9. The relative density is obtained from the equation 3.

$$\rho_{rel}(\%) = \left(\frac{\rho}{\rho_{th}}\right) \times 100 \qquad (3)$$

The percentage porosity is obtained from the equation (4).

$$Porosity(\%) = \left(1 - \frac{\rho}{\rho_{th}}\right) \times 100 \qquad (4)$$

The variations of relative density and percentage porosity with wt% boron carbide were shown in figure 4 and figure 5 respectively.

Table 7. Experimental results

Exp. No	wt% of particles	Particle size	Actual density	SN Ratio actual density	Predicted values of actual density			% Error of actual density			Rockwell Hardness
	%	µm	g/cc		Taguchi Method	Regression Method	Poisson Method	Taguchi Method	Regression Method	Poisson Method	HRB
1	3	30	2.6461	-8.4742	2.6527	2.6521	2.6503	0.2511	0.2252	0.1587	69
2	3	40	2.6324	-8.4089	2.6329	2.6386	2.6456	0.0194	0.234	0.5014	73
3	3	60	2.618	-8.3355	2.6108	2.6116	2.6353	0.2733	0.246	0.6608	81
4	6	30	2.6497	-8.4588	2.6481	2.6446	2.6366	0.06	0.1917	0.4944	75
5	6	40	2.6271	-8.3934	2.6283	2.6311	2.632	0.0448	0.153	0.1865	79
6	6	60	2.6058	-8.3201	2.6062	2.6041	2.6217	0.0158	0.0645	0.6102	86
7	9	30	2.6429	-8.4247	2.6378	2.6372	2.6095	0.1913	0.2164	1.2638	70
8	9	40	2.6197	-8.3593	2.618	2.6237	2.605	0.0645	0.1519	0.5611	72
9	9	60	2.5892	-8.286	2.5959	2.5967	2.5948	0.2605	0.2889	0.2163	77

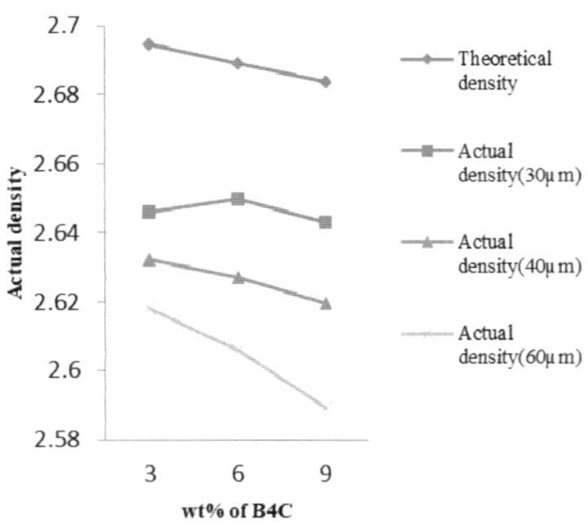

Fig. 3. Actual density Vs wt% of B4C

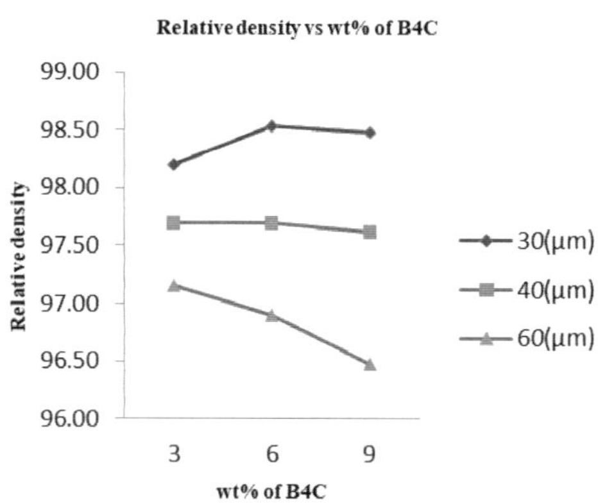

Fig. 4. Relative density Vs wt% of B4C

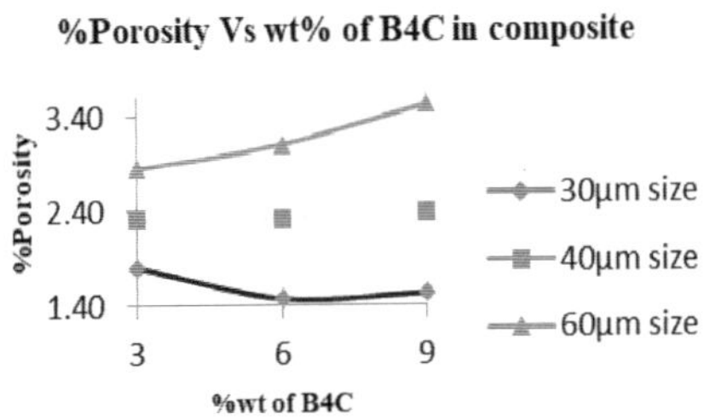

Fig. 5. Percentage porosity Vs wt% B4C

Table 8. S/N ratios and parameters

Two-way ANOVA: density of composite versus wt% of particle, particle size

Source	DF	SS	MS	F	P
wt% of particles	2	0.0003489	0.0001744	4.02	0.11
Particle size	2	0.0026359	0.001318	30.37	0.004
Error	4	0.0001736	0.0000434		
Total	8	0.0031584			

Two – way ANOVA

The optimum level arrangements of input parameters like wt% of particles and particle size of each process parameters have been determined using ANOVA technique. The two way ANOVA have been used to discover the parameters, which are considerably affect the density. The analysis has been conducted with MINITAB software. The performance characteristics used are lower than better criteria. The S/N ratios of different parameters are given in table 8. The P value expresses the importance of the most dependent parameter and is less than 0.05 which indicates that experiments performed are within standard acceptable limits of 5%.

Regression analysis

Regression analysis of density versus wt% of particles, Particle size has been conducted.
The regression equation is
Density = 2.70 – [(0.00248) × (wt% of particles)] – [(0.00135) × (Particle size)]

The residual plots for density is shown in figure 6.

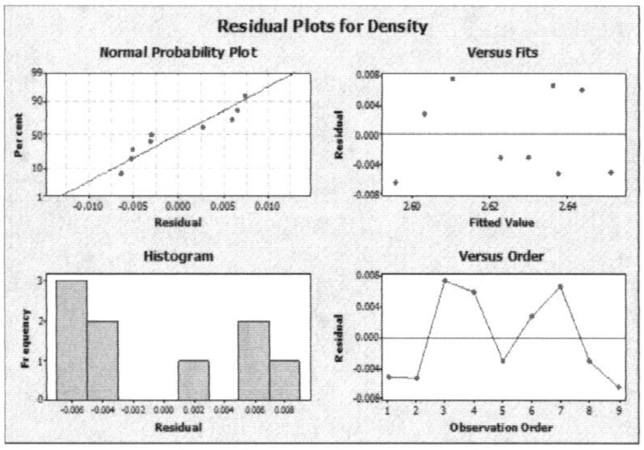

Fig. 6. Residual plots for density

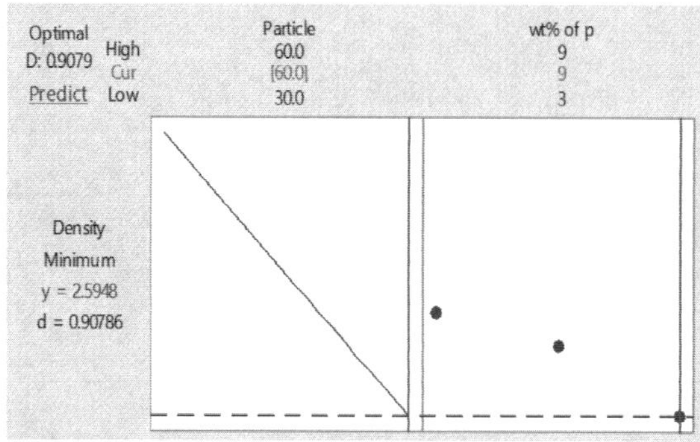

Fig. 7. Predicted density of composite

Prediction of density

The densities of the fabricated composites have been predicted using taguchi method, regression method and Poisson prediction method. The table 8 shows the experimental results of density and predicted densities. The figure 7 shows the optimal value of density predicted using Poisson method. Figure 8 and figure 9 show the various predicted density versus the wt% B4C and particle size of B4C respectively.

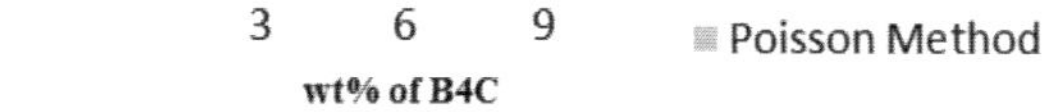

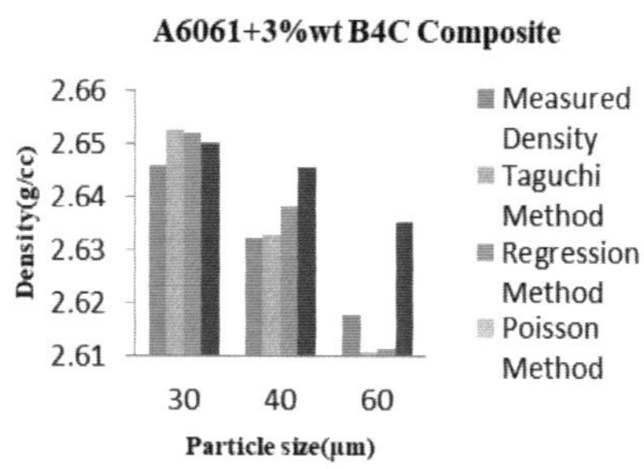

Fig. 8. Variation of predicted densities with wt% of B4C Fig. 9. Variation of predicted densities Vs particle size

3.2. Hardness

The rockwell harness values of the composites versus particle size and wt% of B4C in the A6061/B4C MMCs fabricated have been shown in figure 10. From this graph, the HRB values rise with the particle size and wt%B4C in the fabricated composite.

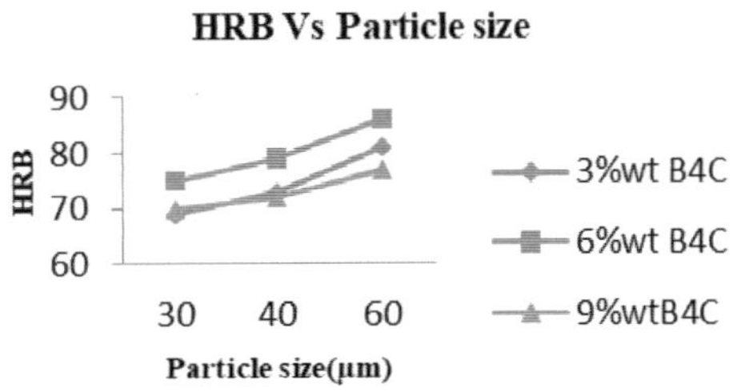

Fig. 10. Variation of HRB Vs particle size

3.3. Microstructure

The uniform distributions of particles have been observed with optical microscope. The microstructure of A6061 alloy and A6061/9% wt B4C MMCs with reinforcement particle size 60µm have been shown in Figure (11-12). Also the greater hardness is related with minor porosity of the A6061/B4C MMCs.

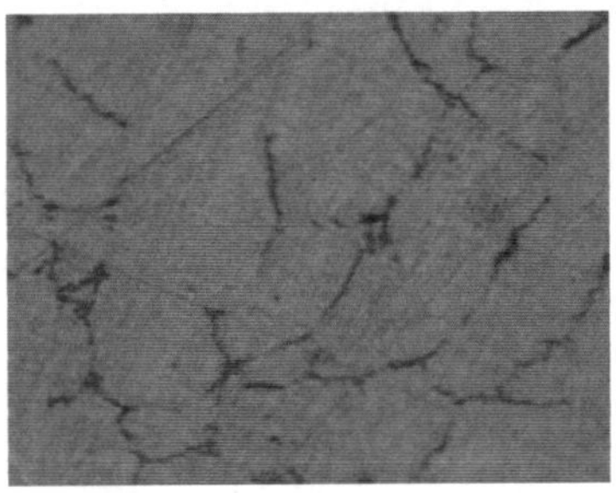

Fig. 11. Micro structure of A6061 alloy

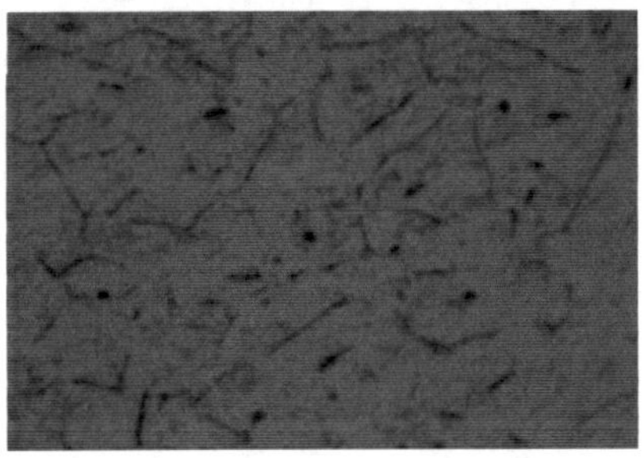

Fig. 12. Microstructure of A6061/9%wt B4C MMC

3.4. XRD analysis

Figure 13 shows the XRD patterns for optimal density A6061/9%wt B4C composites prepared through stir casting method. The XRD analysis gives, a clear picture of the unwanted material existing in the fabricated material. In figure 13, the diffraction angle θ is between 0^0 and 80^0, here four peeks are obtained for A6061 corresponding to 39.085^0, 45.309^0, 65.608^0, 78.79^0 angles. The peaks for B4C have been observed for twelve peeks, such as 17.270^0, 20.598^0, 22.971^0, 32.38^0, 35.264^0, 36.293^0, 38.013^0, 40.788^0, 42.382^0, 43.184^0, 57.191^0 and 75.043^0. From the XRD analysis, it is clear that the B4C particles are uniformly dispersed in base alloy. On comparing the above XRD results with A6061 alloy results as shown in figure 14, the peaks observed for A6061 as 38.905^0, 44.83^0, 65.608^0 and 78.79^0. It is observed that a small shift in peaks occurred for the fabricated composite at 39.085^0 and 45.309^0.

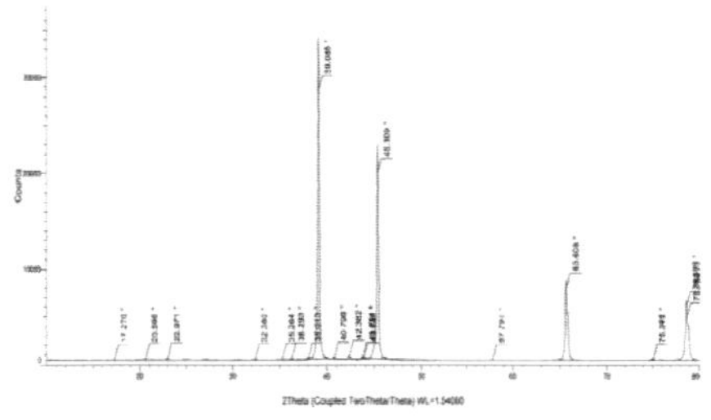

Fig.13. XRD pattern of A6061/9%wt B4C composite

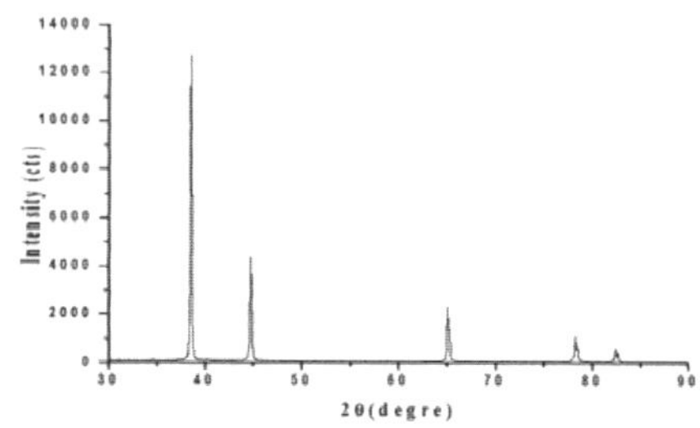

Fig. 14. XRD pattern of A6061 alloy

4. Conclusion

The major findings of the current research are:-

- A6061/B4C composites can be easily made by stir casting technique.
- According to Archimedes principle, the fabricated composite shows higher density than base alloy.
- The actual density of fabricated composites reduces with rise in particle size, wt% B4C and A6061/9%wtB4C with 60μm particle size composite shows the lowest density of 2.5892g/cc.
- The 60μm size 9%wt B4C composite possesses low relative density with respect to base alloy.
- The percentage porosity turns out to be bigger due to wt% of B4C and 6%wt B4C with 30μm size composite shows 1.41% porosity.
- The predicted density of 60μm, 9%wt B4C reinforced A6061/B4C composite with Poisson method gives minimum density of 2.5948g/cc and is nearer to the actual density.
- Rockwell Hardness of composite increases as the particle size increases. The A6061/B4C MMC with 6% wt B4C with 60μm shows a highest hardness of 86HRB. Also 3%wt B4C with 30μm size shows lowest of 69HRB among the nine fabricated composites.
- Microstructure and XRD analysis of the fabricated material indicate that B4C particles are properly dispersed.

Acknowledgements

The authors are thankful to Dr. M Dev Anand, Noorul Islam Centre for Higher Education, for their valuable technical support.

Funding

This research received no specific grant from any funding agency in the public, commercial, or not-for-profit sectors.

Conflict of interest

None.

Ethical approval

The authors state that the research was conducted according to ethical standards.

References

[1] I. O. P. C. Series and M. Science, "Investigation on Mechanical properties of Al-6061 Alloy based MMC Investigation on Mechanical properties of Al-6061 Alloy based MMC," 2018.

[2] R. Soundararajan, A. Ramesh, S. Sivasankaran, and M. Vignesh, "ScienceDirect Modeling and Analysis of Mechanical Properties of Aluminium Alloy (A413) Reinforced with Boron Carbide (B 4 C) Processed Through Squeeze Casting Process Using Artificial Neural Network Model and Statistical Technique," *Mater. Today Proc.*, vol. 4, no. 2, pp. 2008–2030, 2017.

[3] A. Ali, K. Suban, M. Perumal, and A. Ayyanar, "Microstructural analysis of B 4 C and SiC reinforced Al alloy metal matrix composite joints," 2017.

[4] G. V. R. Kumar, "Effect of Wear on Surface Hardness of AA2024 Alloy / TiC Composites," no. December, pp. 1–6, 2011.

[5] "Review of A6061 Metal Matrix Composites fabricated through stir casting route and applications."

[6] M. Gao, H. Kang, Z. Chen, E. Guo, P. Peng, and T. Wang, "Author ' s Accepted Manuscript," *Mater. Sci. Eng. A*, 2018.

[7] S. Gudipudi, S. Nagamuthu, K. Subbu, S. Prakasa, and R. Chilakalapalli, "Engineering Science and Technology , an International Journal Enhanced mechanical properties of AA6061-B 4 C composites developed by a novel ultra-sonic assisted stir casting," *Eng. Sci. Technol. an Int. J.*, no. xxxx, 2020.

[8] V. Bharath, M. Nagaral, V. Auradi, and S. A. Kori, "Preparation of 6061Al-Al 2 O 3 MMC's by Stir Casting and Evaluation of Mechanical and Wear Properties," *Procedia Mater. Sci.*, vol. 6, no. Icmpc, pp. 1658–1667, 2014.

[9] H. Reza, S. Abolkarim, M. Haddad, and Y. Huang, "Investigation of microstructure and mechanical properties of Al6061-nanocomposite fabricated by stir casting," *J. Mater.*, vol. 55, pp. 921–928, 2014.

[10] V. Umasankar, "Experimental evaluation of the influence of processing parameters on the mechanical properties of SiC particle reinforced AA6061 aluminium alloy matrix composite by powder processing," *J. Alloys Compd.*, vol. 582, pp. 380–386, 2014.

[11] A. Nirala, S. Soren, N. Kumar, V. K. Dwivedi, and D. R. Kaushal, "Materials Today : Proceedings A comprehensive review on stir cast Al-SiC composite," *Mater. Today Proc.*, no. xxxx, 2019.

[12] M. Sambathkumar, P. Navaneethakrishnan, K. Ponappa, and K. S. K. Sasikumar, "Mechanical and corrosion behavior of Al7075 (Hybrid) metal matrix composites by two step stir casting process," *Lat. Am. J. Solids Struct.*, vol. 14, no. 2, pp. 243–255, 2017.

[13] S. Grigoriev, P. Peretyagin, A. Smirnov, W. Solís, L. A. Díaz, A. Fernández, and R. Torrecillas, "Effect of graphene addition on the mechanical and electrical properties of Al2O3-SiCw ceramics," *J. Eur. Ceram. Soc.*, vol. 37, no. 6, pp. 2473–2479, 2017.

[14] G. Requena, D. Telfser, C. Hörist, and H. P. Degischer, "Creep behaviour of AA 6061 alloy and AA 6061 metal matrix composite," *Mater. Sci. Technol.*, vol. 18, no. 5, pp. 515–521, 2002.

[15] S. S. Mohamed and S. A. Abdallah, "An Investigation on the Mechanical and physical Properties of AL6061 / SiCp / Gr Metal Matrix Composites," no. 3, pp. 1–5, 2018.

[16] E. Ghasali, A. Pakseresht, A. Rahbari, H. Eslami-Shahed, M. Alizadeh, and T. Ebadzadeh, "Mechanical properties and microstructure characterization of spark plasma and conventional sintering of Al-SiC-TiC composites," *J. Alloys Compd.*, vol. 666, pp. 366–371, 2016.

[17] K. Rajkumar, S. Santosh, S. Javed Syed Ibrahim, and A. Gnanavelbabu, "Effect of Electrical discharge machining parameters on microwave heat treated Aluminium-Boron carbide-Graphite composites," *Procedia Eng.*, vol. 97, pp. 1543–1550, 2014.
[18] R. Sharma, S. J. P, K. Kakkar, K. Kamboj, and P. Sharma, "A Review of the Aluminium Metal Matrix Composite and its Properties," *Int. Res. J. Eng. Technol.*, vol. 4, no. 2, pp. 832–842, 2017.
[19] C. H. Gireesh, K. G. D. Prasad, K. Ramji, and P. V Vinay, "ScienceDirect Mechanical Characterization of Aluminium Metal Matrix Composite Reinforced with Aloe vera powder," *Mater. Today Proc.*, vol. 5, no. 2, pp. 3289–3297, 2018.
[20] B. P. Kumar and A. K. Birru, "Microstructure and mechanical properties of aluminium metal matrix composites with addition of bamboo leaf ash by stir casting method," *Trans. Nonferrous Met. Soc. China (English Ed.*, vol. 27, no. 12, pp. 2555–2572, 2017.

IOP Conf. Series: Materials Science and Engineering 1017 (2021) 012004 doi:10.1088/1757-899X/1017/1/012004

Network in Sequential Form: Combine Tree Structure Components into Recurrent Neural Network

Swaleha Zubair[1] and Anjani Kumar Singha[1]

[1] Aligarh Muslim University, India
[2] Aligarh Muslim University, India

Email id: anjani348@gmail.com

Abstract. Smaller units nestled in larger units-a natural language that is hierarchically ordered. Smaller sections are replaced because of the completion of the bigger constituencies. The basic LSTM architecture does not have a clear choice for modeling the hierarchy of components as separate neurons do not require knowledge to be monitored on various time scales. The idea in this paper is to introduce inductive bias by grouping the neurons. The modifications we create to the master input vector are changed in all neurons following the order of the specified neuron and forget windows. Four activities on which we can achieve strong care unattended sorting, logical inference, and language modeling and guided syntax evaluation.

1. Introduction

In recent years (Bowman's et al . , 2006; Yoga tama year et al . , 2017; Sheen et al. ,2016; Jacob et .al. , 2019; Choir[1,2] et al.,2017;)[1,2,3] the deep NN techniques have a strong inference that tree structures too can make better representations of the natural language. A supervised algorithm [3, 4] is a straightforward way to guess the respective tree structure. The results obtained showed the way to turn words into semi-semantic sentences (Soccer et al., 2013; Bowman's et al., 2015), and, despite previous words (Wu et al., 2017), the forth coming ones are guessed. Yet there are some drawbacks to this method too. I some of the languages have comprehensively explicit data for supervised analysis parser training for a few of our parts we have, certain rules that need broken (e.g. in social site) on more use of languages tend to change gradually, so that new syntax rules may evolve. Recent attempts include the implementation of a trial solution, which faces more difficulty throughout the learning cycle induced by inducing. When going forward, few techniques are comparatively more difficult to implement and understand, such as the Parsing-Reading-Predict Networks (PRPN) [3, 4] shown in et.al (2017) . In recent years, the study of neural networks, mainly use the techniques that use parse structures of tree to more information of sentences [5]. Syntactic parsers may provide a direct way to predict a sentence. The parse trees that output from it will words of semantics into sentences form and grammar form, when it the defined of learning a specific method of syntax through broad natural and structured [4, 5] collection of non-access texts to transform structured data into pro leaves a strong dynamic behind. Many such sentence to induce a trivial or even left-branching parse tree [5, 6] (Williams et al., 2019), [8,5, 6,7]) or observations of learning dynamics caused by branched RL learning guidelines (Yoga tama et al., 2016),. Moreover, in a specific model like PRPN [6, 7] which was told in Sheen et al,. (2017), there are some hard forms of training.

2. Related Work

Prior research involves exploiting the materials of structures for the proper language tasks. Soccer et al . , (2011), use unsupervised learning or supervised learning for the estimation of trees on [9, 8] adept-unlabeled or labeled parse tree et al. ,(2015); Soccer et al. , and Tai et al. , (2016), specifically modeled parse tree construction by using a reasonable outer description to define the data. Afterwards Bowman et al., (2017), [9,7] used some guidance from the monitored definition to guide a web of stack-augmented nerve organs. In theoretical terms, Recurrent Neural Network may modulate recognition generated by CFG, in addition to CSG, in addition to LSTM s. The structural knowledge generated in LSTM is, however, beneficial. Suncor et.al. , (2018), [9] has shown that RNNG which has a strong overt bias against one of the semantic structures, overcomes Long Short-Term Memory (Linen et al . , (2018)), on the subject-verb agreement task [9, 10]. We build and function in this own function with a modern, largest-scale space associated with the grammatical test recently proposed by Linen and Marvin et al. , (2017), [7, 10] the effect that the demonstration caused by tree-structured models is more apparent. Interesting is the work generated by recursive plans, Shi et al., (2018), [11,10] says that although approved grammar woods are not as good as pyramidal structures, relevant works would therefore work. Some way that the basic problems of efficiently inferring some kind of construction by the resulting data remain starting and ending good query. The grammar associated with learning the level construction of the sentence by text are called semantic induction (Chen et al.,1991; Cohen et al., 2017), [10, 11]. Until research incorporates syntactic construction within the semantic model system [11,12]. There have been undertaking to trigger many mechanisms for greater possibilities using Neural Network models, including in recent studies (Grefenstette et al. 2019, Mikonos and Julian et al. (2014, 2015). Generally, these works maintain the RNN core model with a stack and focus on finding solutions to yoga Tama et el .al, (2019), operating on the semantic model. In addition to the function of semantic evaluation (Linen et al. , 2017), [9, 10] does not view the context in which the modeling structure is organized Sheen et al . , (2015),[11,12,13] implements the replication of Parsing-Reading-Predict Systems, which aims to perform tree simply by the process of grammar recreation.

3. Proposed Method

In this part, we present an innovative recurrent neural network unit, ON- Long Short-Term Memory. The innovative model uses a structure similar to the regular Long Short-Term Memory, indicated as:

$$f_t = \sigma\,(w_{f.}x_t + u_{f.}v_{t-1} + b_f) \quad (1)$$

$$j_t = \sigma\,(w_J x_t + u_{J.}v_{t-1} + b_J) \quad (2)$$

$$k_t = \sigma\,(w_k x_t + u_{k.}v_{t-1} + b_k) \quad (3)$$

$$\hat{C}_t = \tanh\,(w_c x_t + u_{c.}v_{t-1} + b_c) \quad (4)$$

$$V_t = k_t \circ \tanh(C_t) \text{ ,where } \circ \text{ is any operation} \quad (5)$$

With the LSTM, we change the update function (c_t) using a new function (f_t) which is explained in following sections. Where, the forget gates function (f_t,) input gates function (j_t) used to control the wipe out, output gates operation (k_t) on cell states (c_t) as before and bias function b. Since, the gates of each neurons are independent of LSTM.

3.1. Function of Activation

Qmax () in order to an implement to particular recurrence neural network, we present a functionality:

$$\hat{G} = Qmax(..) = Qsum(QSoftMax\ (..)) \quad (6)$$

Where, Qsum denotes the cumulative sum [14, 15]. For assumption of the Boolean gates G= (1; 1; 0; 0) we produced Ĝ.The binary gates divide unit state into two different segments like the 1-segment and the 0-segment. Therefore, applying different semantic can different updating rules on two segments to differentiate (LSTM), information. It is denoted by d1 in random variables by representing index first in G

$$P(d) = \text{SoftMax}(\ldots) \quad (7)$$

The represents variable d1, split into two segment and compute probability of the i-th values in G being 1 from find the values of probability of any values before the i-th divide point, which is d≤ i=(d=0)∨(d=1) ∨--------∨(d=i).Computing distribution function is evaluating by this formula

$$P(G_i = 1) = P(d \leq i) = \Sigma_J \leq 1\, p(d=J) \quad (8)$$

Where, G is has to discrete variable.

3.2. Mechanism of Structured

Following Qmax activation (), we bring out a new Master -Forget- Gate (MFG) $\hat{h}_t$, and input Master-Gate(MG) $\tilde{I}_t$.

$$\hat{h}_t = \text{Qmax}(w_{\hat{f}} + x_t + u_{\hat{f}}.v_{t-1} + b_{\hat{f}}) \quad (9)$$

$$\tilde{I}_t = 1 - \text{Cumax}(w_f x_t + u_f.v_{t-1} + b_f) \quad (10)$$

Using the properties from activation function Qmax (), the resulting results in the MSG are monotonically non decreasing from 0 to 1 and input MG are non increasing from 1 to 0.These MG control by update of cell state to new update master gate rule.

$$w_t = \hat{h}_t \circ \tilde{I}_t \quad (11)$$

$$f_t = f_t \circ w_t + (\hat{h}_t - w_t) = \hat{h}_t (f_t \circ j_t + 1 - \tilde{I}_t) \quad (12)$$

$$\hat{h}_t = f_t \circ w_t + (\tilde{I}_t - w_t) = \tilde{I}_t \circ (\tilde{I}_t \circ \hat{h}_t + 1 - \hat{h}_t) \quad (13)$$

$$C_t = \hat{h}_t \circ C_{t-1} + \tilde{I}_t \circ \hat{C}_t \quad (14)$$

In order to update new master rule, we suppose that new master gate are in Boolean form
Applying the MSG entering $\hat{h}_t$, the individual eliminates of semantic the model. postulate $\hat{h}_t$ = (0, 0, 1, 1) and the divide into cell d_t^fCurrently the equation (12) and (14), the stored information d_t^f neuron and unit cell c_{t-1} fully deleted .
The MG input $\tilde{I}_t$ to control the script semantic of the model. Suppose that split point d_t^f of equation (13) and (14) the large d_t^f means that start input X^t contain long term.
The particular product from the two MG wt represents the certain overlap of f_t and $\tilde{I}_t$

Figure1. Single complexity model test and **validate** for the Penn Treebank vocabulary model task.

Model	Parameters	Validation	Test
Zaremba et al. (2014) - LSTM (large)	66M	82.2	78.4
Gal & Ghahramani (2016) - Variational LSTM (large, MC)	66M	–	73.4
Kim et al. (2016) - CharCNN	19M	–	78.9
Merity et al. (2016) - Pointer Sentinel-LSTM	21M	72.4	70.9
Grave et al. (2016) - LSTM	–	–	82.3
Grave et al. (2016) - LSTM + continuous cache pointer	–	–	72.1
Inan et al. (2016) - Variational LSTM (tied) + augmented loss	51M	71.1	68.5
Zilly et al. (2016) - Variational RHN (tied)	23M	67.9	65.4
Zoph & Le (2016) - NAS Cell (tied)	54M	–	62.4
Shen et al. (2017) - PRPN-LM	–	–	62.0
Melis et al. (2017) - 4-layer skip connection LSTM (tied)	24M	60.9	58.3
Merity et al. (2017) - AWD-LSTM - 3-layer LSTM (tied)	24M	60.0	57.3
ON-LSTM - 3-layer (tied)	25M	**58.29 ± 0.10**	**56.17 ± 0.12**
Yang et al. (2017) - AWD-LSTM-MoS*	22M	56.5	54.4

4. Experiment

We evaluate the four model: Unsupervised constituent parsing, targeted semantic, language model analysis (Linen and Marvin et al . , 2019), [2, 17] and logical interface (Bowman et el . , (2018)).

4.1. Language of Model

Model of world range is a macroscopic language evaluation from model's stamina the work out with many old age depictions. Our model evaluates simply by counting complexity on certain penn-tree-bank (PTB), for more accurate we precisely stick to the major parameters, regulation and easy way optimization in AWD-LSTM. This work used Tri-layered ON-LSTM models along the 1200 unit in a prefaced layer and embedment association of size 500 units. For grasp gates the loud element is M=20. The particular tool moves from 26 to 27 million along with matrices intended for computing predominant gates. (see Figure.1) performs greater standards of LSTM.

4.2. Unsupervised Parsing

The unsupervised constituency parsing relates valuable tree typed model with these marginalized human authorities. Looking at the well experienced configurations listed at Hut el al (2018) [16, 17, 18]. We obtain a perfect structure to the language modeling task in addition to its WSJ10 dataset and WSJ analyzed set. It has 7633 lines; we filter WSJ dataset with the limit of only 15 words or less and right punctuation marks and excluding null elements. While the WSJ analyze contains 3000 paragraphs with varying lengths. That is WSJ10 test set consists of sentences from the education, test and validation arranged regarding PTB dataset, here WSJ test utilizes the pairs of similar group phrases as the PTB analyze set. This is a tree structured paragraph from previous models, initialization is done by hiding state with zero vectors, and insertion of paragraphs into model is made, as we require paragraph modeling activity .In each time we complete an estimation

$$\hat{\eta}_t^f = E[d_t^f] = \sum_{i=1}^{D_m} i P_f(d_t = i) = \sum_{i=1}^{D_m} \sum_{j=1}^{i} P_f(d_t = i) = D_m - \sum_{i=1}^{D_m} \hat{h} \quad (15)$$

Where P_f is the probability normal distribution and D_m is the size of hidden layer. Given $\hat{\eta}_t^f$, use the top-down parsing algorithm within Sheen et al. ,(2017),for unsupervised parsing tree. We first sort the any d_t^f in non-increasing order .For d_t^f the particular sorted sequence order, we are divided the sentence into language ((X <j); (XI; (X>j))). After that, we recursive repeat this particular operation for constituent (X <j) and (X> j), until each constituent contains paragraphs. The performance of (see Figure.2). Typically the second layer of ON- Long Short-Term Memory got state-of-the-art unsupervised parsing tree output on WSJ test set, while the first layer and four layer usually do not used.

4.3. Evaluation of Syntactic Language

Model of semantic proposed asks to defined in Linen & Marvin (2017,2018) [3, 18, 19].It actually a group of task that will evaluate language models together, the different structured-CSG phenomena: reflexive anaphora and Verb agreement items, negative polarity, identified a huge count small phrases in English language, each containing the new in ungrammatical and grammatical terms. The grammatical phase using the similar configuration proposed in Marvin as well as linzen. Our language models consists of ON LSTM model [18,19] designed on the baseline of LSTM language model containing new 100 million words of Google. Vocabulary models semantic has more than two levels 700 cells and a mass scale 129 units, a new rate of dropout 4, rate of leering 30, and trained for epoch 50. The embedding of input has 300 and even embedding of output has 700 demission.

5. Results within RS

Spinn and St-Gumbel are assessed with WSJ. We run the design on a particular F1 norms. The relativity and better portion of truth matters of the variety that corresponds with the matters among model parses. The model along with best F1 report to ADJP, natural language processing, PP. The baselines of WSJ10 are from Linzen and manning. The result of italics is more serious than the random one. We can realize that some are particular to NPI test.

Figure .2. WSJ test. & full WSJ10 results F1 for unlabeled parsing

Model	Training Data	Training Object	Vocab Size	Parsing F1 WSJ10		Parsing F1 WSJ		Depth WSJ	Accuracy on WSJ by Tag ADJP	NP	PP	INTJ
				μ (σ)	max	μ (σ)	max					
PRPN-UP	AllNLI Train	LM	76k	66.3 (0.8)	68.5	38.3 (0.5)	39.8	5.8	28.7	65.5	32.7	*0.0*
PRPN-LM	AllNLI Train	LM	76k	52.4 (4.9)	58.1	35.0 (5.4)	42.8	6.1	37.8	59.7	**61.5**	**100.0**
PRPN-UP	WSJ Train	LM	15.8k	62.2 (3.9)	70.3	26.0 (2.3)	32.8	5.8	24.8	54.4	17.8	*0.0*
PRPN-LM	WSJ Train	LM	10k	70.5 (0.4)	71.3	37.4 (0.3)	38.1	5.9	26.2	**63.9**	24.4	*0.0*
ON-LSTM 1st-layer	WSJ Train	LM	10k	35.2 (4.1)	42.8	20.0 (2.8)	24.0	5.6	38.1	23.8	18.3	**100.0**
ON-LSTM 2nd-layer	WSJ Train	LM	10k	65.1 (1.7)	66.8	47.7 (1.5)	**49.4**	5.6	**46.2**	61.4	55.4	*0.0*
ON-LSTM 3rd-layer	WSJ Train	LM	10k	54.0 (3.9)	57.6	36.6 (3.3)	40.4	5.3	44.8	57.5	47.2	*0.0*
300D ST-Gumbel	AllNLI Train	NLI	–	–	–	19.0 (1.0)	20.1	–	*15.6*	*18.8*	*9.9*	59.4
w/o Leaf GRU	AllNLI Train	NLI	–	–	–	22.8 (1.6)	25.0	–	18.9	24.1	*14.2*	51.8
300D RL-SPINN	AllNLI Train	NLI	–	–	–	*13.2 (0.0)*	*13.2*	–	*1.7*	*10.8*	*4.6*	50.6
w/o Leaf GRU	AllNLI Train	NLI	–	–	–	*13.1 (0.1)*	*13.2*	–	*1.6*	*10.9*	*4.6*	50.0
CCM	WSJ10 Full	–	–	–	71.9	–	–	–	–	–	–	–
DMV+CCM	WSJ10 Full	–	–	–	77.6	–	–	–	–	–	–	–
UML-DOP	WSJ10 Full	–	–	–	**82.9**	–	–	–	–	–	–	–
Random Trees	–	–	–	31.7 (0.3)	32.2	18.4 (0.1)	18.6	5.3	17.4	22.3	16.0	40.4
Balanced Trees	–	–	–	43.4 (0.0)	43.4	24.5 (0.0)	24.5	4.6	22.1	*20.2*	*9.3*	55.9
Left Branching	–	–	–	*19.6 (0.0)*	*19.6*	*9.0 (0.0)*	*9.0*	12.4	–	–	–	–
Right Branching	–	–	–	56.6 (0.0)	56.6	39.8 (0.0)	39.8	12.4	–	–	–	–

6. Inference

This work proposes unordered neurons, RNN is turned for novel learning purpose. on relevant particular idea ,we purpose a novel of RNN cells, the particular on- Long Short-Term Memory which new activation function Rectified Linear Unit(ReLU)&include mechanism of mathematical function i.e Qmax(.).This composition operation applied certain tree structure closer to Recurrent Neural Network by allocating hidden layer neurons with short term and long information. The performance of unsupervised learning model constituency parsing tree appearance which on - Long Short-Term Memory to generated inheriting structure of Natural Language Processing an approached with human expert annotation. The inductive bias sanctions ON-LSTM helped to achieve better results on language modeling and logical inference tasks

7. References

[1] Zoph, B. and Le, Q.V., 2016. Neural architecture search with reinforcement learning. *arXiv preprint arXiv:1611.01578.*

[2] Grave, E., Joulin, A. and Usunier, N., 2016. Improving neural language models with a continuous cache. *arXiv preprint arXiv:1612.04426.*

[3] Bod, R., 2006, July. An all-subtrees approach to unsupervised parsing. In *Proceedings of the 21st International Conference on Computational Linguistics and 44th Annual Meeting of the Association for Computational Linguistics* (pp. 865-872).

[4] Bowman, S.R., Potts, C. and Manning, C.D., 2014. Recursive neural networks can learn logical semantics. *arXiv preprint arXiv:1406.1827.*

[5] Bowman, S.R., Manning, C.D. and Potts, C., 2015. Tree-structured composition in neural networks without tree-structured architectures. *arXiv preprint arXiv:1506.04834.*

[6] Charniak, E., 2001, July. Immediate-head parsing for language models. In *Proceedings of the 39th annual meeting of the Association for Computational Linguistics* (pp. 124-131).

[7] Chelba, C. and Jelinek, F., 2000. Structured language modeling. *Computer Speech & Language*, *14*(4), pp.283-332.

[8] Shen, Y., Lin, Z., Huang, C.W. and Courville, A., 2017. Neural language modeling by jointly learning syntax and lexicon. *arXiv preprint arXiv:1711.02013*.

[9] Choi, J., Yoo, K.M. and Lee, S.G., 2017. Learning to compose task-specific tree structures. *arXiv preprint arXiv:1707.02786*.

[10] Chomsky, Noam, Ch ., 1956, "Three models for the description of language." *IRE Transactions on information theory* 2, no. 3: 113-124..

[11] Noam Chomsky,No., 1956, Aspects of the Theory of Syntax. The MIT Press, Cambridge. URL http://www.amazon.com/Aspects-Theory-Syntax-Noam-Chomsky/dp 0262530074.

[12] Junyoung Chun, SungjinAhn, and YoshuaBengio,Ju., 2016, Hierarchical multiscale recurrent neural networks arXiv preprint arXiv:1609.01704.

[13] Shay B Cohen, Dipanjan Das, and Noah A Smith,Sh,2011, Unsupervised structure prediction with nonparallelmultilingual guidance. In Proceedings of the Conference on Empirical Methods in NaturalLanguage Processing, pp. 50–61.

[14] StanislasDehaene, FlorentMeyniel, Catherine Wacongne, Liping Wang, and Christophe Pallier.St.,2015,The neural representation of sequences: from transition probabilities to algebraic patterns andlinguistictrees. Neuron, 88(1):2–19.

[15] Chris Dyer, AdhigunaKuncoro, Miguel Ballesteros, and Noah A Smith, Recurrent neural networkgrammars,Cd.,2016, In Proceedings of the Conference of the North American Chapter of the Associationfor Computational Linguistics: Human Language Technologies. pp. 199–209.

[16] Salah El Hihi and YoshuaBengio. Hierarchical,Sa.,1996, recurrent neural networks for long-term dependencies.In Advances in neural information processing systems. pp. 493–499.

[17] Yarin Gal and ZoubinGhahramani,Ya.,2016, A theoretically grounded application of dropout in recurrentneural networks. In Advances in neural information processing systems. pp. 1019–1027.

[18] BarretZoph and Quoc V,Ba.,2016, Le. Neural architecture search with reinforcement learning. ArXivpreprintarXiv 1611.01578.

Determination of Standard Oxygen Transfer Rate in Venturi Aeration System

Anamika Yadav[1], Avinash Kumar[1], Sudipto Sarkar[1]

[1]Department of Agricultural Engineering, Triguna Sen School of Technology, Assam University, Silchar, Assam, India

E-mail: anamika.iit26@gmail.com

Abstract. The oxygen transfer is the process of atmospheric oxygen diffusion from a gas stage to a liquid stage in the form of dissolved oxygen (DO). The previous research works have successfully proven that the DO concentration level effortlessly rises by adopting a venturi as an aeration method in aquatic bodies. Present experimental study was conducted on laboratory scale aeration tank with two different sizes of venturi (40 and 80 mm throat length). The objective of the current study was to determine the oxygen transfer coefficient at different discharge rates through venturi aeration system. The system was operated with three different discharges of 0.50, 0.58 and 0.67 liter/second (l/s) to circulate 200 liter volume of water, was aimed for circulation in a closed loop system. During the operations, water moves through the inlet section into the contract section of the venturi, which increases the rate of water flow and reduces the pressure as per the Bernoulli's theorem. In order to study the oxygen transfer coefficient at varying discharge, the system worked by a single hole and accordingly applied to the larger number of holes for both the throat lengths of venturi. It was observed that the oxygen transfer coefficient proportionally linked to the DO diffusion with increasing rate of discharge. The oxygen transfer coefficient found to be increased by 31% and 55% at 40 and 80 mm throat length respectively, with increasing the discharge. The rate of DO diffusion found higher causes more surface area of contact between air and water through maximum air entrainment. Moreover aeration efficiency also increases with the rate of increasing discharge for the maximum size of throat length. The maximum SOTR values was obtained 0.0035, 0.0041, and 0.0050 kgO_2/h at varying discharge of 0.50, 0.58 and 0.67 l/s respectively for 40 mm throat length. For the 80 mm throat length, the SOTR values varies from 0.0042, 0.0062 and 0.0091 kgO_2/h at increasing discharge of 0.50, 0.58 and 0.67 l/s respectively with increasing number of holes.

1. Introduction

The dissolved oxygen concentration (DO) is very crucial for the production of aquatic species in aquacultural systems. By incorporating any aerator in water course, sufficient amount of oxygen can be infused which promotes growth and productivity [1]. Such methods are incredibly helpful in maintaining sufficient DO concentration which positively impact the survival of aquatic species from around 29.0 % to 81.8 % [2] and also for enhancing water quality [3]. The aeration method naturally or artificially blends air and water together and raises the interface area between them, to allow more oxygen distribution at that moment and also to prevent stratification to water bodies [4]. Venturi aeration is very apt technology that stimulates high yielding and healthy global environment by maximizing the DO concentration in intrinsic water bodies [5, 6]. It has perfect combination of productivity, low cost, reliable production, high volume of distribution, low power consuming, more efficient, required less than 20 % differential to initiate vacuum and also the venturi system have a long operating life [7]. Venturi has three different sections: converging, throat, and diverging section, as shown in figure 1. Each section of venturi has a significant aspect in the aeration process. The venturi has been developed to mix air and water before inserting into the mass liquid at its maximum restricted point [6]. The study of [8] has asserted that, in accordance with the Bernoulli principle, when the flowing liquid has been accelerated to the high velocity stage, the static pressure drastically false at the constricted point of venturi. The throat section (constricted portion) of venturi produces vacuum, which draws air into the liquid. A negative pressure drags atmospheric air through the holes presents in the throat or in the way of outlet section to the bulk liquid [9].

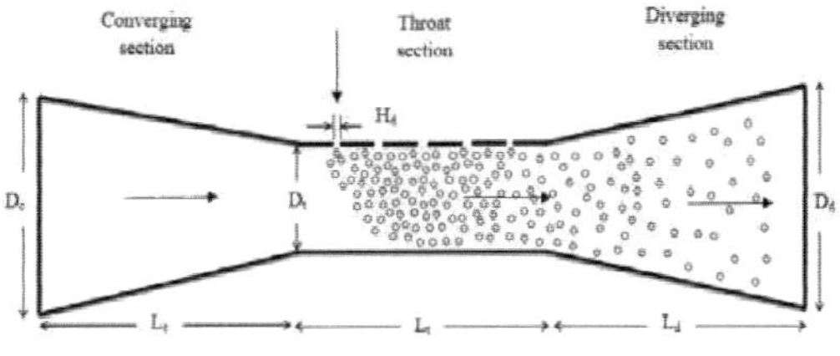

Figure 1. Schematic view of venturi section

A bulk of experimental studies has been conducted on the numerous venturi designs under different circumstances. Air entrainment and oxygen transfer relies on numerous geometric parameters of the venturi designs adopted by various researchers, such as, length of the throat, diameter of air hole, flow velocity, inlet and outlet angles [5, 10, 11 and 12]. Many of the researchers have performed their work by selecting the appropriate parameters intended for performance evaluation of the aeration systems; in the study of [13] and [14] they worked by using the overall oxygen transfer coefficient as a typical parameter of the aeration system whereas the study by [15] have elected the individual standard oxygen transfer efficiency. The effect of geometrical parameters on the performance of a venturi type bubble generator has been conducted by [16]. Three different significant parameters were analyzed including the diameter of air injection hole, number of air holes and the varying angle. The effect of a converging angle on the air entry rate can be described by the sum of radial flow generated in the suction chamber and the pressure force flowing through the nozzle and throat portion [17]. The experimental research conducted by [5] on the air entry rate and oxygen transfer efficiency with air holes along the length of converging-diverging waterway in the venturi nozzle, mainly the impact of variable numbers, positions, and open/close status of the air holes. Additional study on the plunging water jet by [18] have been conducted on the air entrainment features of a plunging water jet system utilizing rectangular nozzles of rounded ends. Furthermore, a combine research by utilizing a venturi nozzle with plunging water jet aeration system was examined by [9] they have analyzed the impact on the intensity of air entry rate by changing the diverging cone angle and diverging length of venturi tube. In the study of [19] examined the influences of contract nozzle angle of a vertical plunging water jet on air entrainment. Emiroglu and Baylar asserted that air entrainment occurs when there is a change from laminar flow to turbulent flow in jet, characteristically described in terms of Reynolds numbers. However the entrainment rate not individually depends on the Reynolds number but similarly on the nozzle model [20]. The experimental study was conducted by [21] on the performance of venturi with varying numerous geometric parameters to determine the standard aeration efficiency value. It can be seen from the result that the highest value of standard aeration efficiency was obtained 1.166 kg O_2/kWh for venturi aeration with a nozzle diameter of 14 mm, aeration depth of 60 cm and the venturi tube angle of 45°. To acquire the advantages in order to improve the aeration efficiency, two separate aerator modules structures (parallel and series connected) were tested by [22]. The aeration efficiency calculated on the basis of oxygen transfer coefficient, and the values attained by [22] were 0.07, 0.06, and 0.06 kg O_2/kWh for the module where three venturies are connected in series, and 0.14 and 0.10 kg O_2/kWh for the module where two venturies were connected parallel. It was observed by [23] in his study that, the aeration efficiencies varies with the position of venturi nozzles in the water column at various depths.

In the present study, the adaptation of venturi of two different sizes has been made for aeration to observe the experimental investigation of oxygen transfer coefficient under laboratory conditions at varying discharge rates.

2. Oxygen Transfer Mechanism

Two-film theory [24, 25 and 26] has been a significant theory used for evaluating the oxygen transfer rate. The process of oxygen diffusion includes the transition of atmospheric oxygen mass transfer from gas stage to liquid stage, which is a three step process as shown in figure 2. The following assumptions were made for the development of two-film theory: laminar flow along both sides of the gas-liquid interface, steady state conditions and instantaneous establishment of equilibrium conditions between gas-liquid phases at the interface [25]. The mass transfer through diffusion is typically given by the equation (1) as:

$$\frac{dM}{dt} = -D_L.A.\frac{dC}{dy} \tag{1}$$

where, $\frac{dM}{dt}$ is the rate of mass transfer by diffusion (kg/s); A is the cross-sectional area through which diffusion occurs (m^2); D_L is the oxygen molecular diffusion coefficient in water (m^2/s) and $\frac{dC}{dy}$ represents the concentration gradient of oxygen in the direction perpendicular to the cross-sectional area ($kg/m^3/s$). The differential form of oxygen mass transfer coefficient at T °C can be expressed as following according to the two-film theory:

$$\frac{dC}{dt} = K_L a_T(C^*_\infty - C) \tag{2}$$

where, K_La_T is the total volumetric oxygen transfer coefficient at some T °C temperature; t is the time; C is the liquid-phase oxygen concentration; and C^*_∞ is the equilibrium liquid-phase oxygen concentration.

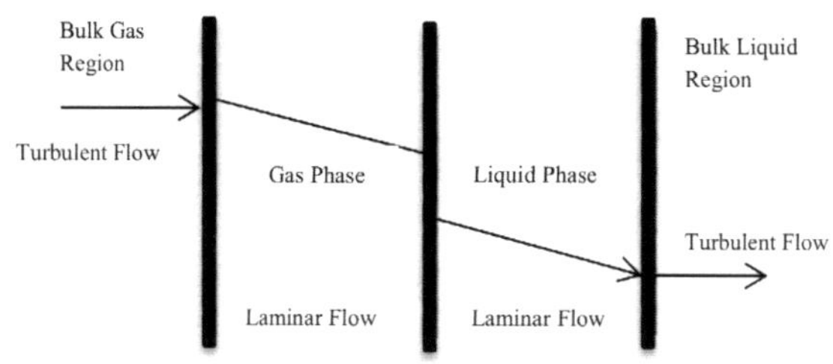

Figure 2. Two film theory of gas mass transfer through diffusion

The combined form of equation (2) can be stated in exponential form and expressed as follows [27]:

$$K_L a_T = \frac{\ln(C_s - C_0) - \ln(C_s - C_t)}{t} \tag{3}$$

where, ln signifies a natural logarithm and C_s, C_0 and C_t are the dissolved oxygen concentrations in parts per million (ppm). C_s is the saturation oxygen concentration, C_0 is the initial oxygen concentration and C_t is the oxygen concentration at time t. K_La is also standardized to a standard temperature value of 20 °C to deliver a clear source for comparing different systems. Equation (4) defines the performance of venturi aeration in oxygen transfer efficiency with the effect of temperature [28].

$$K_L a_{20} = K_L a_T \times \theta^{(20-T)} \tag{4}$$

The aerator capability to transfer off oxygen from atmosphere to water is evaluated through standard oxygen transfer rate (*SOTR*) by [28]. *SOTR* is the amount of oxygen transfer into the water body per unit time at the standard conditions of 20 °C water temperature, 0 mg/l initial *DO* concentration, one atmospheric pressure, and clear tap water [29]. *SOTR* can be determined as follows:

$$SOTR = K_L a_{20} \times (C_s - C_0) \times V \times 10^{-3} \quad (5)$$

where, C_s is the saturation concentration of *DO* at 20 °C, *V* is the volume of water (m^3), *SOTR* is the standard oxygen transfer rate (kgO_2/h).

3. Materials and Methods

The experimental set-up and the method employed to fulfill the requirement of the present study was organized in the Agricultural Engineering Department of Assam University Silchar, India for the assessment of the venturi performance on *SOTR*. The detailed information of the experimental setup, specification of venturi, the experimental procedure and the experimental design is given below.

3.1 Experimental Setup

A 200 liter capacity of water tank having dimensions of 90 cm × 55 cm × 49 cm (length × breadth × depth) was used for determining the venturi aeration performance on the *SOTR*. The experimental setup involves a water tank, axial flow pump, venturi unit, valves, and water meter linked with the pipe fittings for circulation of water through a closed loop as shown in figure 3. The venturi unit is supported by another supporting structure and is being operated by a 1.02 HP axial flow pump.

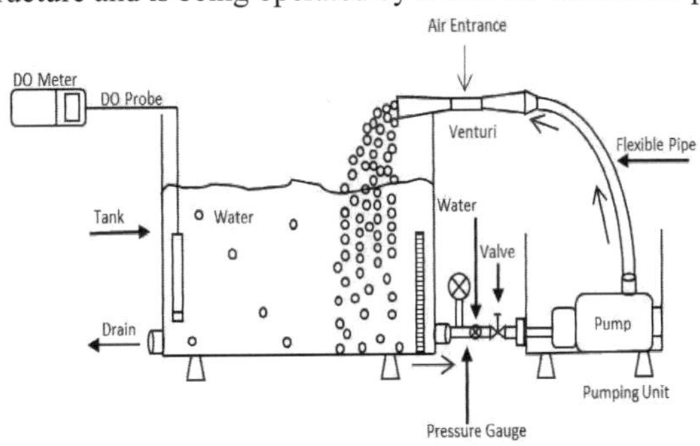

Figure 3. Line diagram of experimental setup

3.2 Specifications of Venturi

The work began with the fabrication of different venturi parts. The fabricated venturi was made of 5 mm thick aluminum, which could be dismantled from each other as shown in figure 4 with three substantial parts (converging, throat and diverging section). The length and diameter for both the sections of converging and diverging were kept equal as 113 mm and 60 mm respectively. The converging and diverging angles were kept at 10°. Two different sizes of throat length i.e. 40 mm and 80 mm were selected with 5 and 13 number of holes with 2 mm diameter respectively; the holes are kept at an equal interval of 5 mm, which signify the difference between two adjacent air holes as presented in figure 5. The diameter of throat section was kept 20 mm constant.

IOP Conf. Series: Materials Science and Engineering 1017 (2021) 012005 doi:10.1088/1757-899X/1017/1/012005

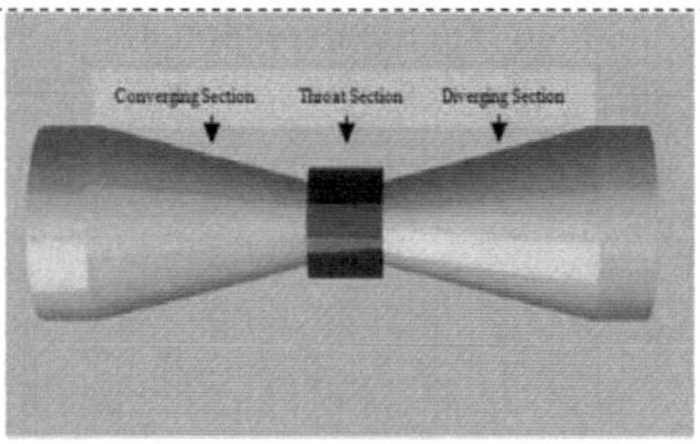

Figure 4. Fabricated venturi section

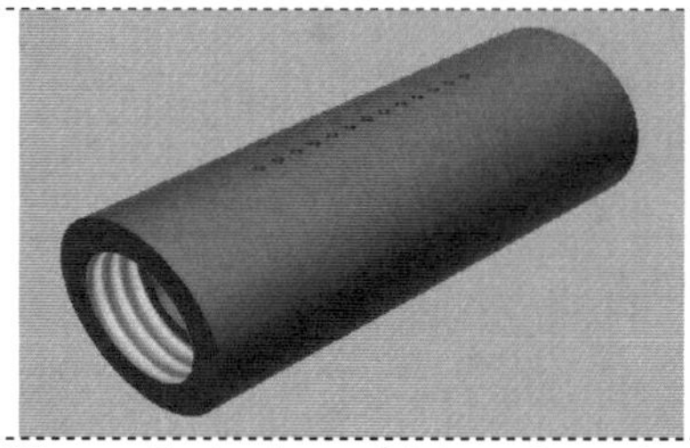
Figure 5. A view of injected number of hole on the throat section

3.3 Experimental Procedure

The venturi operated by an axial flow pump, which is used to circulate the clean tap water throughout the experiments of venturi under laboratory conditions. Water flowing starts from converging to the throat section, which allows the water flow rate to increase and pressure decreases. Due to the pressure variation the transition of oxygen from air to water continues through the air holes further moves towards the outlet section [30]. For each experiment, the non-steady re-aeration study was carried out to deoxygenate the water [31]. In every experiment, for reducing each mg/l of *DO*, the clean tap water used to deoxygenate through a solution of 10 mg sodium sulphite (Na_2SO_3) and 0.1 mg cobalt chloride ($CoCl_2$) as a catalytic agent per liter water tank volume [3, 25]. Subsequently for the *DO* concentration measurements during the experiments, the EXTECH dissolve oxygen meter 407510 is mounted at the base of the tank by the installation of luminescent dissolved oxygen (LDO) sensor. During the experiments, *DO* readings and temperature were taken regularly at equal intervals up to *DO* concentration reaches 80% saturation. For each noted value of the *DO* concentration, the *DO* deficit values were also determined. The oxygen transfer coefficient was determined from the most appropriate gradient line produced by graphical compression between natural logarithms for *DO* deficits on the Y axis, and the aeration period on the X-axis, which has been adjusted by temperature correction factor of 20 °C at the examining water temperature [32]. Moreover, the values of K_La_T, K_La_{20} and *SOTR* were calculated by using equation (3), (4) and (5) respectively. Since this study was carried out in the lower part of the Assam (Silchar) and a large number of impurities i.e. iron and fluoride are found in groundwater beyond its permissible limit. Even after the purification of groundwater, some impurities remained in the tap water. During the experiments, some difficulties were faced to get the clean tap water and the purification of tap water again conducted.

3.4 Experimental Schedule

In the present study, the determination of oxygen transfer coefficient has been done with two sets of venturi (40 mm and 80 mm throat length) at three different pump discharge rates i.e. 0.50, 0.58 and 0.67 l/s, respectively. The discharge rate of pump was varying with valve fixed in all the sets of experiments. Mainly the experiments were performed with a single hole and consequently expanded to a larger number of holes for both the throat length of venturi. A total 54 number of experiments were conducted with varying the discharge rate. The schedule of experiments for venturi aerator is given in Table 1.

Table 1. Schedule of experiments of venturi aerator.

Set of experiments	Pump Discharge (l/s)	Throat Length (mm)	Variation of number of air holes
Set-I	0.50	40	1 to 5
		80	1 to 13
Set-II	0.58	40	1 to 5
		80	1 to 13
Set- III	0.67	40	1 to 5
		80	1 to 13

4. Result and Discussion

The experimental results representing the deviations of ln (Do_s-DO) over time to determine the oxygen transfer coefficient at different discharge rate is shown in figure 6. It can be seen in the figure 6 that, the *DO* absorption rises with declining the *DO* deficit (Do_s-DO) until it reaches to the saturation level at which the oxygen dissolution rate is equivalent to the intake rate of dissolved oxygen. It was also found that at maximum discharge, the distribution of *DO* concentration is reached more rapidly than the lower discharge rate.

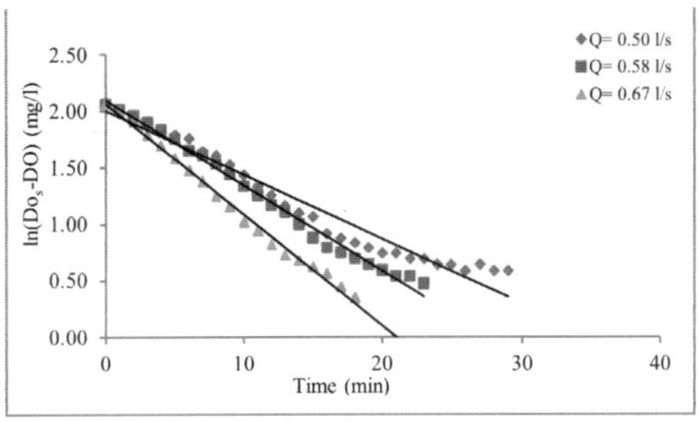

Figure 6. Determination of oxygen transfer coefficient

Moreover, the effect of discharge on the oxygen mass transfer coefficient for 40 mm and 80 mm throat length is presented in figure 7 (a & b). The magnitude of overall oxygen mass transfer coefficient rises with increasing the air holes for each throat length of venturi. Also, a similar trend of increment was found for the oxygen mass transfer coefficient at increasing discharge rate. In general, the overall oxygen transfer coefficient is the combined form of area interface between two phases (air and water phase) and the oxygen transfer coefficient. In this specific venturi aeration, the air entrainment rate fluctuates with number of holes, which leads to increase more surface area interaction between air and water. This could be due to the large amount of holes that maximize the performance of the aeration device and creates higher ambient oxygen dispersion. Hence, the collective result has contributed to a consistent increase in overall oxygen transfer coefficient for venturi aeration with increasing discharge rate, which delivers a proportionate relation with *DO* diffusion. The overall oxygen transfer coefficient raises 31% and 55% from 0.50 to 0.67 l/s discharge range for 40 and 80 mm throat length respectively. Table 2 provides the comprehensive findings for K_La_T, K_La_{20} and *SOTR* values according to the experimental sets.

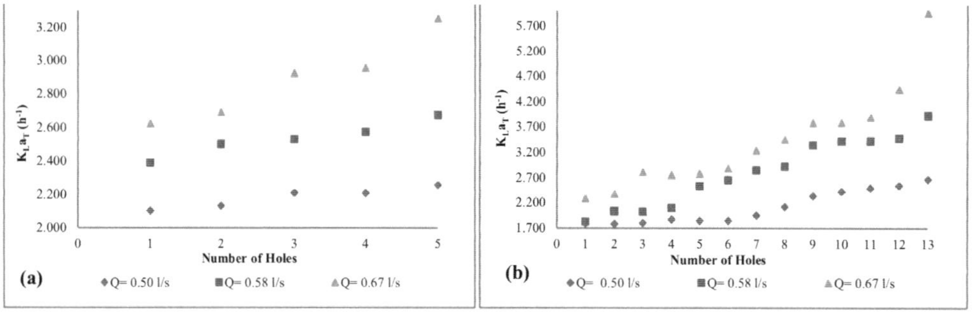

Figure 7. Effect of discharge on the oxygen mass transfer coefficient (a) 40 mm throat length and (b) 80 mm throat length

In order to see the impact of discharge on *DO* diffusion, the each set of experiments were carried out by varying the discharge with two different throat lengths as mentioned in Table 1. Figure 8 and 9 shows the experimental result on the *DO* diffusion over aeration time at three different operating conditions for 40 and 80 mm throat length. It can be found from the figure (8 and 9) that *DO* concentration proportionally rises with time. The maximum *DO* diffusion occurs with the maximum discharge (Q= 0.67 l/s) for both the throat lengths of venturi with in minimum time. The experiments were conducted until the *DO* reaches 80 % of saturation level at the specific temperature of experiment. The high concentration of *DO* was found at 6.30 mg/l in 19 min for discharge 0.67 l/s with maximum 13 numbers of air holes and 80 mm throat length. Similarly, at each operating discharge, it was observed that the *DO* diffusion follows almost similar trend for each hole of 40 and 80 mm throat length.

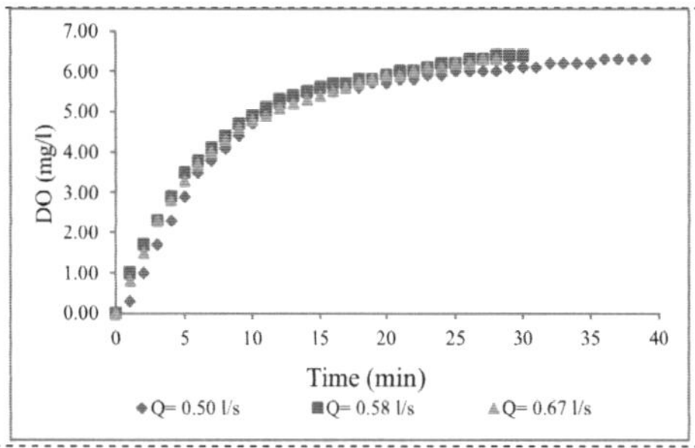

Figure 8. DO diffusion over time at varying discharge rate for 40 mm throat length

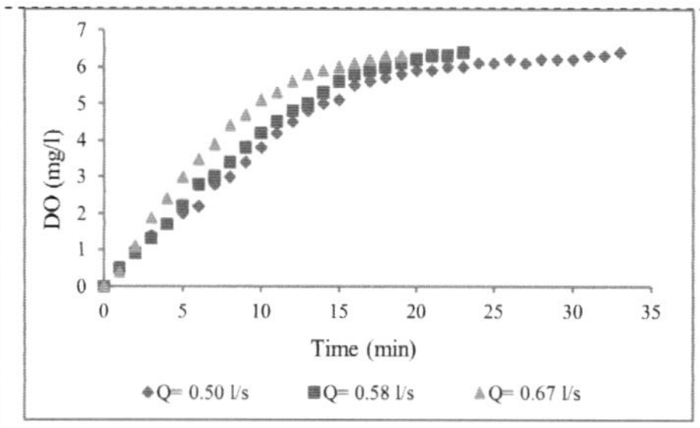

Figure 9. DO diffusion over time at varying discharge rate for 80 mm throat length

It can be observed from Table 2 that the *SOTR* values are equally growing for each experimental set with the increasing number of holes. The variation of *SOTR* with different water discharge (0.50, 0.58 and 0.67 l/s) for two different size of venturi is illustrated in figure 10 and 11. The result indicates that the minimum *SOTR* was obtained with the minimum discharge of 0.50 l/s for both the sizes of venturi. It can be seen that the *SOTR* keeps on increasing with increasing discharge throughout the experiments. Also a similar trend was found in relation to the number of holes of 40 and 80 mm throat length. For the particular 40 mm throat length, the *SOTR* range varies between 0.0033-0.0035 kgO_2/h, 0.0037-0.0041 kgO_2/h, 0.0041-0.0050 kgO_2/h at increasing discharge of 0.50, 0.58 and 0.67 l/s respectively with increasing number of holes.

The maximum *SOTR* values were calculated 0.0035, 0.0041, and 0.0050 kg O_2/h at different discharge of 0.50, 0.58 and 0.67 l/s respectively for 5 numbers of hole and 40 mm throat length. Also for the 80 mm throat length, the *SOTR* values varies from 0.0028-0.0042 kgO_2/h, 0.0029-0.0062

kgO_2/h, 0.0036-0.0091 kgO_2/h at increasing discharge 0.50, 0.58 and 0.67 l/s respectively with increasing number of holes. It can be said that the maximum number of holes are able to improve the *SOTR* for all discharge rate. And the maximum *SOTR* values found as 0.0042, 0.0062 and 0.0091 kgO_2/h at discharge 0.50, 0.58 and 0.67 l/s respectively with 13 number of hole. With increasing the discharge rate from 0.50 to 0.67 l/s leads to an increase in the *SOTR* by 30.14 % and 54.12 % for 40 and 80 mm throat length.

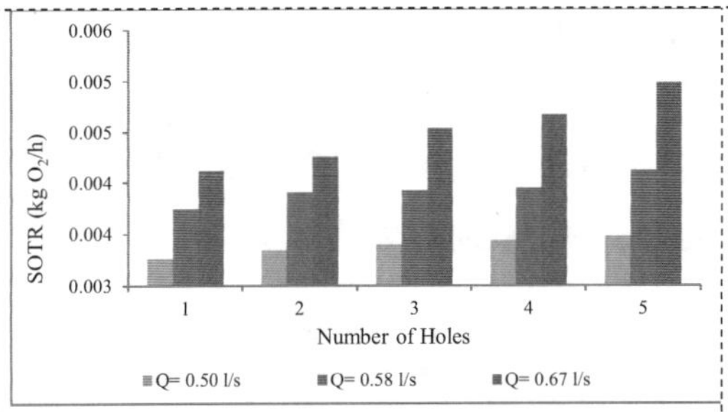

Figure 10. Variation of SOTR with different discharge rate at particular throat length 40 mm

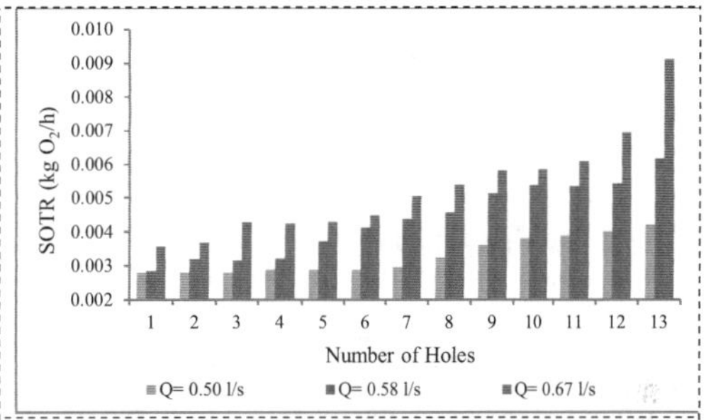

Figure 11. Variation of SOTR with different discharge rate at particular throat length 80 mm

Table 2. Experimental result of venturi aeration

Throat Length	Number of holes	K_La_T (h^{-1})			K_La_{20} (h^{-1})			SOTR (kgO_2/h)		
		0.50 l/s	0.58 l/s	0.67 l/s	0.50 l/s	0.58 l/s	0.67 l/s	0.50 l/s	0.58 l/s	0.67 l/s
40	Hole 1	2.100	2.389	2.623	1.800	2.062	2.269	0.0033	0.0037	0.0041
	Hole 2	2.129	2.501	2.692	1.843	2.154	2.348	0.0033	0.0039	0.0043
	Hole 3	2.209	2.528	2.924	1.875	2.162	2.501	0.0034	0.0039	0.0045
	Hole 4	2.206	2.572	2.954	1.895	2.174	2.575	0.0034	0.0039	0.0047
	Hole 5	2.255	2.674	3.251	1.919	2.273	2.748	0.0035	0.0041	0.0050
80	Hole 1	1.797	1.824	2.283	1.529	1.573	1.966	0.0028	0.0029	0.0036
	Hole 2	1.786	2.033	2.375	1.534	1.755	2.034	0.0028	0.0032	0.0037
	Hole 3	1.801	2.022	2.810	1.543	1.739	2.352	0.0028	0.0032	0.0043
	Hole 4	1.871	2.101	2.753	1.581	1.763	2.338	0.0029	0.0032	0.0042
	Hole 5	1.845	2.534	2.782	1.585	2.047	2.368	0.0029	0.0037	0.0043
	Hole 6	1.841	2.653	2.889	1.590	2.263	2.467	0.0029	0.0041	0.0045
	Hole 7	1.954	2.849	3.244	1.636	2.402	2.774	0.0030	0.0044	0.0050
	Hole 8	2.121	2.925	3.455	1.797	2.519	2.965	0.0033	0.0046	0.0054
	Hole 9	2.335	3.347	3.778	1.983	2.815	3.207	0.0036	0.0051	0.0058
	Hole 10	2.422	3.420	3.785	2.086	2.949	3.221	0.0038	0.0053	0.0058
	Hole 11	2.493	3.430	3.893	2.142	2.944	3.341	0.0039	0.0053	0.0061
	Hole 12	2.545	3.482	4.450	2.208	2.985	3.828	0.0040	0.0054	0.0069
	Hole 13	2.671	3.928	5.954	2.305	3.395	5.025	0.0042	0.0062	0.0091

5. Conclusion and Future Work

The aeration performance in laboratory was estimated at varying discharge rate with two different size of venturi. It can be concluded that the performance of the venturi varies with the increasing discharge rate. The overall oxygen transfer coefficient was found maximum at higher discharge rate with the large size of throat length. Also the respective values of *SOTR* follows a similar trend for both the sized of venturi with increasing discharge rate. Accordingly the aeration can be improved with increasing the size of venturi for the maximum number of holes. Therefore it can be concluded that number of holes on the throat section of venturi plays a major role in the aeration. This work can be further extended by altering the size and angle of the converging and diverging section to determine the influences on the aeration efficiency of the venturi aeration system.

References

[1] Bagatur T, Baylar A 2006 Experimental Studies on Air Entrainment and Oxygen Content Downstream of Sharp-Crested Weirs *Turkey Water and Environment Journal* **20**(4) 210-216
[2] Pawar N A, Jena J K, Das P C, Bhatnagar D D 2009 Influence of duration of Aeration on Growth and Survival of Carp Fingerlings during High Density Seed Rearing *Aquaculture* **290** 263-268
[3] Boyd C E 1998 Pond Water Aeration Systems *Aquacultural Engineering* **18**(1) 9-40
[4] Boyd C E, Martinson D J 1984 Evaluation of Propeller-Aspirator Pump Aerators *Aquaculture* **36**(3) 283-292
[5] Baylar A, Emiroglu M E 2003 Air Entrainment and Oxygen Transfer in a Venturi *Proceedings of the Institution of Civil Engineers-Water & Maritime Engineering* **156**(3), 249-255
[6] Baylar A, Ozkan F 2006 Applications of Venturi Principle to Water Aeration Systems *Environmental Fluid Mechanics* **6** 341-357
[7] Ozkan F, Baylar A, Tugal M 2006a The Performance of Two Phase Flow Systems in Pond Aeration International *Journal of Science and Technology* **1**, 65-74
[8] Rathinakumar V, Dhinakaran G, Suribabu C R 2014 Assessment of Aeration Capacity of Stepped Cascade System for Selected Geometry *International Journal of Chemical Technology and Research* **6**(1) 254-262
[9] Baylar A, Ozkan F, Ozturk M 2005 Influence of Venturi Cone Angles on Jet Aeration Systems Process *Institute of Civil Engineering-Water Management* **158** 9-16
[10] Baylar A, Ozkan F, Tugal M 2006 The Performance of Two Phase Flow Systems in Pond Aeration International *Journal of Science & Technology* **1**(1) 65-74
[11] Baylar A, Ozham F, Unsal M 2007 Determination of the Optimal Location of the Air Hole in Venturi Aerators *Clean-soil Air Water* **35**(3) 246-249
[12] Gourich B, El Azher N, Vial C, Belhaj M 2007 Influence of Operating Conditions and Design Parameters on Hydrodynamics and Mass Transfer in Emulsion Loop-Venturi Reactor *Chemical Engineering and Processing* **46**(2) 139-149
[13] Kulkarni A, Shah Y T, Kelkar B G 1987 Gas Holdupi Bubble Column with Surface-Active Agents: A Theoretical Model *AICHE* **33** 690-693
[14] Dudley J 1995 Mass Transfer in Bubble Columns: A Comparison of Correlations *Water Resources* **29** 1129-1138
[15] Capela S, Gillot S, Heduit A 1999 Oxygen Transfer Under Process Conditions: Comparison of Measurement Methods *Proceedings of the WEF 72nd Annual Conference and Exposition; New Orleans, LA*
[16] Li J J, Song Y C, Yin J L, Wang D Z 2017 Investigation on the Effect of Geometrical Parameters on the Performance of a Venturi Type Bubble Generator *Nuclear Engineerign and Design* **325** 90-96
[17] Yadav R L, Patwardhan A W 2018 Design aspects of ejectors: Effects of Suction Chamber Geometry *Chemical Engineering Science* **63** 3886-3897
[18] Bagatur T, Sekerdag N 2003 Air-Entrainment Characteristics in a Plunging Water Jet System Using Rectangular Nozzles with Rounded Ends *Water SA* **29**(1) 35-38
[19] Yamagiwa K, Ito A, Tajima K, Yoshida M 2000 Effect of Nozzle Contraction Angle on Air Entrainment Rate of a Vertical Plunging Liquid Jet *Journal of Chemical Engineering of Japan* **33**(5) 805-807

[20] Emiroglu M E, Baylar A 2003a Role of Nozzles with Air Holes in Air Entrainment by a Water Jet *Water Quality Research Journal Canada* **38**(4) 785-795
[21] Ghomi M R, Sohrabnejad M, Ovissipour M R 2009 An Experimental Study of Nozzle Diameters, Aeration Depths and Angles on Standard Aeration Efficiency (SAE) in a Venturi Aerator *Water Practice and Technology* **4**(3) 1-8
[22] Zhu J, Miller C F, Dong C, Wu X, Wang L, Mukhtar S 2007 Aerator Module Development using Venturi Air Injectors to Improve Aeration Efficiency *Applied Engineering in Agriculture* **23**(5) 661-667
[23] Dong C, Zhu J, WU X, Miller C F 2012 Aeration Efficiency Influenced by Venturi Aerator Arrangement, Liquid Flow Rate and Depth of Diffusing Pipes *Environmental Technology* **33**(11) 1289-1298
[24] Lewis W K, Whitman W G 1924 Principles of Gas Absorption *Industrial and Engineering Chemistry Research* **16**(12) 1215-1220
[25] Metcalf & Eddy, Inc. 2003 *Wastewater Engineering: Treatment and Reuse* Boston:McGraw-Hill
[26] Eckenfelder W W, O'Conner D J 1961 *Biological Waste Treatment* Pergamon Press, Oxford
[27] Anonymous 1997 *Standard Guidelines for In- process Oxygen Transfer Testing* ASCE New York **64**
[28] Eckenfelder W W, Ford D L 1970 *Water Pollution Control* Pemberton Press, Jenkins Publication, Austin Taxes
[29] Anonymous 1993 *Measurement of oxygen transfer in clean Water, 2nd Edition* ASCE New York **45**
[30] Yadav A, Kumar A, Sarkar S 2020 An Experimental Study to Evaluate the Efficacy of Air Entrainment Holes on the Throat of Venturi Aeration System *Aquaculture International* **28** 1057-1068
[31] ASCE 2007 *Measurement of Oxygen Transfer in Clean Water ASCE Standard No. ASCE/EWRI 2-06* Virginia: American Insitute of Civil Engineers
[32] Anonymous 1980 *American Water Works Association, and Pollution Control Federal (16th ed.)* APHA Washington DC

Adopting zero accident vision in India: a conceptual framework

Jai Narain[1], Makkhan Lal Meena[1], Govind Sharan Dangayach[1], Amar Patnaik[1]

[1]Department of Mechanical Engineering, Malaviya National Institute of Technology Jaipur, Rajasthan-302017, India

Email: jainarain@mnit.ac.in

Abstract. The research concentrates on the five success factors of Zero Accident Vision (ZAV) methodologies, which are the pillar of ZAV. As safety is the priority of any industry, so, the research aims to contribute a new idea for accident prevention in high-risk industries. The research is based on the survey format in which the rating of workers are taken on important factors regarding accident prevention and awareness, and from the basis of that result, we find the necessary critical factors. From the result, it can be concluded that what are the factors which are necessary for the implementation of ZAV. As earlier there is limited work that has been done in the field of ZAV, so here a framework is proposed for ZAV to prevent the accident that will lead the industries towards zero accident vision.

1. Introduction:

Accidents are unforeseen events, which not only causes harm to human, environment or property, unintentionally and unexpectedly but also reduced the productivity of the nation. Initially, it was deemed that the accidents are inevitable during the works. However, now various theories established that all accidents can be avoided by establishing proper safety management systems. As per International Labour Organization statistics, of the 2.34 million work-related deaths per year, a massive 2.02 million work-related deaths are attributed by various type of work-related diseases and remaining 321,000 due to accidents. Additionally, the inadequate avoidance of work-related diseases has intense negative effects on workers, their families and society. As per International Social Security Association statistics, 270 million workers suffer non-fatal workplace accidents along with 160 million new cases of occupational illness every year, attributing a huge financial burden of compensation, health care, and rehabilitation. This financial burden cause approximately 4% of world GDP for work-related injuries only, the cost may be as high as 10% of GDP for some developing countries.

The costs of occupational accidents may not be measured only in the financial term but can be better judged by grouping into three categories: direct costs, indirect costs, and human costs.... Direct costs, which include medical costs are relatively easy to evaluate without any extraordinary evaluation methods. Whereas, indirect costs include salary, administrative costs, loss of productivity which are related to the lost opportunities for the injured employee, the employer, the co-workers, and the community. As compared to direct costs, indirect costs are quite challenging to evaluate and are hardly protected. Indirect costs also include human costs which are related to the cost of change in the quality of lives of workers and their surrounding people.

1.1 Vision zero accidents

Initially, it was the mindset that accidents happen due to human errors, and a strong bureaucratic system can prevent these. Almost all the nations have the government mechanism to prosecute the wrongdoers, but unfortunately, there is not any decrement in the rate of accidents. One more aspect which is very critical in accident prevention is 'NON REPORTING' which is the outcome of this punishment-based government mechanism. The accidents and the reasons for accidents are being hidden and the actual cause of accidents, neither known to the stack holder nor the researchers. Due to this reason, the accidents are repeated for the same cause. Improvement in the safety of complex industrial systems is one of the most difficult tasks for organizations. There are so many methodologies developed in the field to help to face this challenge, but the Zero Accident Vision (ZAV) is a promising approach developed in industry, but unfortunately, it is not better addressed by the safety science research community.

ZAV is based on the assumption that all the serious accidents are preventable. There are some conflicts also based on ZAV, as some researcher feels that taking the goal as zero accident is impossible to achieve [1, 12], they argue that to achieve such perfection will increase bureaucratic methods in industrial management. But I feel that ZAV is a concept which if adopted by a company, motivates it to reduce its accidents from what its level is by following best practices of accident prevention. It never can be set as a goal, which if not achieved shall become a punishable offence, but it can be set as a target for perfection. In ZAV the participation from workers to management is on the self-esteemed and motivational basis, for the betterment of society. During the accident investigation focus should be on the reason for the accident and not on who is responsible and what should be the punishment. Every reason of accident should be investigated and the efforts should be made to search the mechanism or method to reduce the accident for same reason in the future. All the accident report should be made public for awareness.

1.2 Present scenario of Indian manufacturing companies

As India is a developing country therefore, the manufacturing sector plays a very important role in the lead it towards a developed nation. But for any manufacturing company safety is the prime requirement. Here, the data is taken from the labor bureau from the year 2000 to 2014 which shows the fatal and non-fatal frequency rates of accidents.

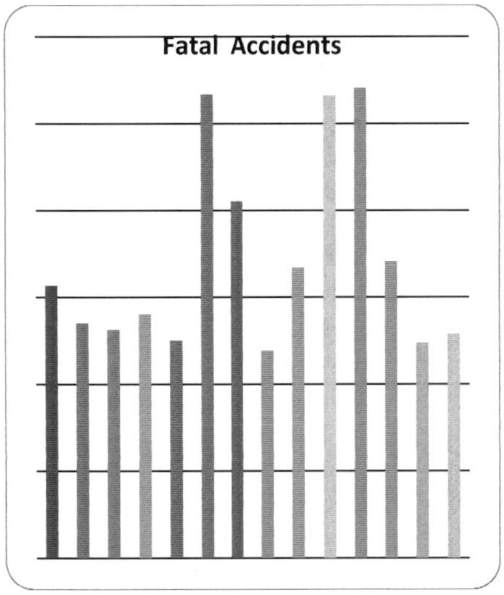

Fig. 1. Fatal frequency rate

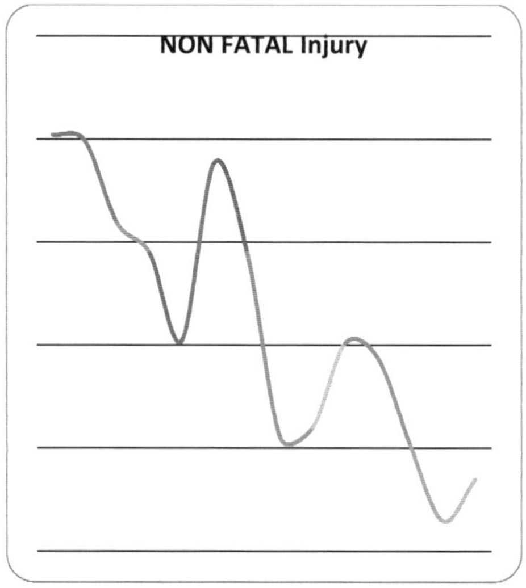

Fig. 2. Non-fatal injury graph

(As per Std Reference Note of DGFASLI/ Ministry of Labour Govt. of India on their website https://dgfasli.gov.in/others/standard-reference-note)) Show the number of injuries from the year 2000 to 2014 is fluctuating, .

2. Literature survey

This section is divided into four major subsections as below:

2.1. Causes of accidents

The following works describe the main causes, which are responsible for the occurrence of accidents, whether they are minor or major. The category of hazards which commonly occur in the industries is described [5]. They divided it into four categories (obvious hazard, trivial hazard, emerging hazard, hidden hazard) based on the participatory action research methodology. An obvious hazard refers to unguarded machinery, electrical wires lying in the water, uneven ground, etc. Trivial hazard refers to screwdriver on the floor, cable across access way, and broken light switches, etc. Examples of emerging hazards are loose rung on the ladder, unlabelled chemical bottles repetitive works while hidden hazard refers to underground electrical cables, unexpected ground movements, etc.

Rozenfeld et al. [34] described some factors which influence the accident happening chances are: frequent work rotation, change in weather, unskilled workers. The causes of the risk associated with the construction site, and the associated factors are uncertainty, the complexity of threats, manual operations, congestion, time pressure, Construction complexity, fragmentation of workforce, and working height. Verma et al. [36] described the risk involved in the mining sector are depends on the working condition of workers and the environment, condition of machines which are being used, and near-miss management; these are the primary causes.

2.2. Identification of future accident and proneness areas

The risk levels have been categorized proneness areas on three different levels. Level 1- Traditional risk areas (TR) – in an automated assembly line, noise is the main risk. Level 2- New and increasing risk (NIR)- nowadays, robots and mobile components are the main cause for the NIR. Level 3- New risk (NR) with the invention of new material, some drawbacks have also arisen like toxicity and hazardous effect on the environment on their grave stage. Albert et al. [2] have suggested the type of hazard associated in the particular fields in industries is basically due to gravity, motion, mechanical, electrical, sound, pressure, temperature, chemical, radiation, and biological. Here they proposed two methods to prevent the accidents i.e. predictive technique and retrospective technique. They suggested that hazard which remains not under consideration or unmanaged can potentially result in destructive accidents. Hazard recognition is critical to developing situational awareness and identifying and responding to danger. In this study, data is taken from six construction projects with 18 construction crew revealed that the hazardous energy source with the highest recognition levels, including gravity 67%.

Kontogiannis et al. [24] have described the proactive method to identify the future accident with the help of the model. He suggests that with the help of the safety management team, we first anticipate the common operational picture of the process, then if it feels safe, we go further, and while implementation, we monitor the threats, and then we send the actual feedback to input. For the enhancement of safety, we propose a strategic safety management approach in which strategic and cultural integration to enhance learning and continuous improvement. He proposed the stages for delivering the safety training, the first stage is to risk screening for critical task second stage was task skill and knowledge for training and the third stage is VR interaction feature, mode, simulation, learning experience training method and then training output.

2.3. Frameworks available for methods of accident analysis

The following researchers describe the accident analysis methods- Goerlandt et al. [18] describe the qualitative method of risk analysis. In the planning phase, he has done accident modelling and followed by estimation of accident consequences. In the hazard identification phase, he had estimated possibilities of accident and risk. Albery et al. [3] described the method which is categorized in two parts first is risk matrix (categorization of hazard), and the second is the functional Resource analysis method. A risk matrix is the current approach utilised for the classification of hazards. They suggested two methods for safety analysis. The first approach was used for the evaluation of risk influenced by various hazards identified and management of those hazards and organizational learning. Conversely, a risk matrix is an approach utilised for the evaluation of risks based on the likelihood and consequence of adverse outcomes. There must be a collaboration with stakeholders having expertise in each category to practically evaluate the likelihood and consequence. Additionally, it should be also understood why things go right knowing the variability in the system which may cause errors.

De Silva et al. [11] described the societal risk factors, a hazard that indirectly affects the public with the help of risk priority numbers. He has suggested a process for risk criteria. First was the measure of major accident hazards. Second is the selection of a method to derive risk criteria. The third was the explanation of the concept of risk; fourth was the selection of a measure of risk. Next is the selection of an appropriate demonstration of risk and framework to derive the risk criteria. Nascimento et al. [30] have developed an offshore helicopter transportation safety model with the help of task analysis techniques. They followed the probabilistic risk assessment (PRA) and human reliability analysis (HRA) processes to develop a novel task analysis-based process to describe the safety-critical components of an industry in the aviation sector and implement the process to generate the offshore helicopter transportation system Model.

2.4. Research related to safety management

Li et al. [28] have described the challenges of managers in safety and safety status and hazard level in the project. He has proposed the accident caution model and tracking techniques. He has suggested the method of how the near misses become hazard; the first step is to define and classify the hazard second is to formulate the state transition and nest is the process the hazard state. He has proposed two models. The first is the accident caution model, and the second is the location tracking model and procedure for recognizing states. Moreover, next is the modelling system of the state.

Tchiehe et al. [35] have described the risk identifying method and to asses them based on severity, occurrence, and tolerance level. For generating the risk acceptability matrix, he categorizes them based on equity, utility, and technology. He has proposed a risk assessment process first is acceptability and tolerability, the second is the formation of risk acceptability matrix, the third approach of risk acceptability, and next is an influential factor risk acceptability. Granerud et al. [20] focused on OHSAS certification is based on the establishment of a reporting and feedback system. This was a systematic means for employers to handle challenges and reduce haphazard attitudes to risk and problems in the work environment. An OHSMS certified company should specify its health and safety goals and plan the way how to achieve it. To examine how OHSMS influence the management of health and safety, they utilize data from Danish company and five cases selected to illustrate the ci variation, and it is observed certified companies also get low ci score due to the routine problem.

Legg et al. [26] gave the key sources on how to manage safety in small and medium enterprises (SME). Fera & Macchiaroli [16] has described how to improve safety condition in industry, by assessing the effectiveness and reliability of new model FERA and with the help of AHP model, we judge the choices relative to each other, and then we make the risk assessment model. He followed the AHP method to express the relative judgment from work team selection, main event identification, frequency calculation, consequences calculation, criticality calculation, improvement action prioritization, and verification of improvement.

Garretson et al. [17] have described the idea of sustainable and safe practice in the industry. He suggested the key factors which are necessary for sustainable production. Zhao & Liu [40] evaluate the severity and vulnerability they design and implement a tool for the regional risk assessment process. They introduced the geographic information system is a tool to analyze, manage, and display data. Gonzalez et al. [19] research is based on the survey and uses a quantitative approach for data collection. He suggests that workshops and training are effective tools. Xu et al. [39] focused on signs and symptoms associated with coal workers using hand-held vibrating tools. The study tells us that by characterization of the working process and characterization of testing of an analysis of particular emerging risk, we will be able to find an accurate and effective result. Zwetsloot et al. [43] have suggested the strategies for the zero-accident vision implementation. He has suggested six strategies, which are commitment, safety integration, innovation, prevention culture, ethics, networking, and co-creation. Zwetsloot et al. [43] have conducted a ZAV survey on 27 companies, and the result is that most of the company's scores relatively lower. Only 22% of companies score 3 or more out of 5. Langeland [25] focused on the prevention of accidents and fatalities on the road site.

2.5. Key factors for the success of ZAV

Following are the key factors that are necessary for the success of ZAV.

2.5.1. Commitment

The achievement or success of ZAV as a goal is crucially dependent on Commitment to it. The safety commitment connotes that to what extent the organization earnestly and conscientiously intends to prevent accidents and improve safety performance. This is an intellectual plan from organizational management for organizational safety culture [43, 9, 15]. This is characterized by the sense of supporting the attitude of organization and dynamic engagement in ZAV operation. It is initiated firstly by believing in the organizational goal and then adopting it at the individual level to fulfill the obligation. This is credibly reached by adhering to the principle of the ethical duty of care and concern towards the workforce rather than the policy framework relying on legislative provisions.

2.5.2. Communication

To realize the full potential of ZAV it is of utmost importance that the organizational management should pass on their strategy for ZAV to personnel in the workforce to be implemented through their concerted efforts guided by the management. This act of establishing interactions with the organizational workforce is termed as a communication climate [22, 41]. The ZAV can turn into reality only through the harmonious exchange of ZAV commitment and plans with the workforce. The communication channel should serve to improve the safety culture within the organization while sharing the organizational faith in ZAV with personnel that all accidents can be prevented through their active participation [8, 42]. The communication climate deals with enrich reliability in ZAV via a dynamic involvement of all organizational levels rather than entrusting the vision wherein subordinates are well informed and devoted to percept [4].

2.5.3. Culture

The ZAV can be comprehended through a well-defined safety culture, which refers to the attitude that addresses the issues at the workplace while involving both employers and personnel. This calls upon a certain sense of belief/value and perceptions in the organization about their safety benefits. Such culture demands long-lasting safety performance wherein ZAV should be thought of as impeccable endeavours in the workplace to prevent accidents with responsibility and accountability. The safety culture is derived from a positive persuasion, belief and pertaining behavior towards safety [33]. The practice and promotion of safety culture across the workforce leading to proactive safety culture, render ZAV possible rather than a reactive approach [23]. Such a defined safety culture ensures safety performance while adequately dealing with unforeseen risks [37]. Besides, the commitment to safety culture has also been translated to a safety climate that is believed to bring real benefits of safety and influenced by the policies and procedures of the organization. The safety climate assessment commits to improving workplace safety by various means in a more localized fashion [23].

2.5.4. Learning

The ZAV can be significantly empowered through comprehensive learning about workplace safety at the organizational level. The organizational workforce at each level should be adequately trained to render the workplace safe for all. The complete organization (employers, managers, and workers) should be substantially educated and aware regarding the concerned workplace risk and hazard and the related safety procedures or management as no workplace is completely safe. The learning process should endow the workforce with effective safety management skills to cope with situations on their own at the time of need with competence. Such learning can then effectively contribute to the successful implementation of ZAV. Learning from incidents/accidents and participation in post investigations at the individual level should provide opportunities for the workforce to adopt safety protocols to realize a hazard-free workplace [14]. The organizational and individual propensity to learn is strongly recommended to achieve ZAV [14].

3. Method of study

The study is based on the data collected in the questionnaire format with the help of Google form through the mail. The questionnaire consists of 47 factors that are essential to know and must implement in any manufacturing company regarding safety. These 47 factors are combined in 8 common factors according to the similarity in hazard and area. The questionnaire is sent to several companies over 300 employees and about 250 peoples have responded. Table 1 shows the summary of questionnaire.

Table 1. Summary of the questionnaire

S. No	Code	Hypothesis	Major Parameters	Sub Parameters	Reference
1	MSA	H1	Regarding health & safety knowledge and skills	> National & Company safety policy >Equipment and safe operating procedure >Emergency response plan >Occupational health and safety plan > Advancement in health, safety and emergency response plans at the sites >Attend health and safety meeting regularly • Attended various safety training such as chemical safety, fire safety, first aid, handling of hazardous materials, operation of numerous equipment at the site and reviewed MSDS	[21]
2	MSB	H2	Guide lines for personal safety	>Practice of relevant personal protective equipment such as safety goggles, steel toed boot, hard hat, hearing and breathing protection etc >You wear appropriate clothing for environmental protection >Aware of hazardous in your surroundings > Communication equipment available at all time	[31]

3	MSC	H3	Hazards linked to job	> >Risk of slippery floors due to spills of water and lubricants > Injuries from moving machinery parts >Injuries caused due to fall into tanks, clarifiers etc. > Falls from ladders, stairs and elevated platforms >Chemical burns caused by corrosive liquids, splashes, and other hazardous substances. > Burns by steam and hot vapors, fire explosion etc. > Electric shocks caused by contact of faulty electrical equipment >>Risk of cut or injuries by sharp objects >Risk of eye injuries by flying minute particles developed from grinders and other similar rotating equipment >Fire and explosion owed by formation of undesirable gases > Acute poisoning caused by chemical present in the waste > Damage of eyes due to splashes of corrosive liquid	[38]
4	MSD	H4	Physical hazards	> Exposure of excessive noise levels from mechanical Equipments > Exposure to vibration due to power tools and sitting surface > Exposure to UV radiations > Potentially health detrimental factors > Exposure to dust	[27]
5	MSE	H5	Chemical /biological hazardous	> Chronic poisoning by inhalation or indigestion of chemical used >Danger to skin via waste water, acid and other chemical solutions > Latex allergy caused by latex gloves > Diseased caused • By infectious agent present in waste water and other waste • By contact of toxin released by infectious agents • By insects proliferating in the sludge drying bed	[29]
6	MSF	H6	Ergonomics, and organizational factors	>Risk of injuries to muscle system due to over exertion or uncomfortable working posture > Visual discomfort and eye problems caused by inadequate illumination >Visual discomfort due to inadequate lighting > Uneasiness due to prolong wear of PPE > Discomfort due to smell of waste	[10]

7	MSG	H7	Safety guidelines for confined space entry	>Acquainted with the various constituents of an entry permit >Familiarity regarding completeness of pre-entry activities on the enrty permit such as air quality test, hazards control etc. >Check information required by the entry permit, e.g. PPE and emergency equipment needed, air monitoring provisions, list of authorized entrants, operating procedures, MSDS of materials to be used >Evaluation of various information such as requirement of PPE kit and emergency equipment, operating manuals, and material safety and data sheet >Well-informed about work activities and list change in conditions and procedures if any on the permit	[6]
8	MSH	H8	Safety measures about hazardous materials	> >Well informed with MSDS > Hazardous material symbols > >List recurrent origins of fire >List various types of fire >Knowledge to douse various types of fire >Experienced to handle fire extinguisher > Fire safety guidelines	[32]

4. Results and discussion

Based on the responses to 47 factors, we have found the opinion of the employees and the factors that are strongly agreed or disagreed by the workers. These factors are divided into 10 groups which are:

1. MSA- M/S regarding health and safety knowledge & skills.
2. MSB- M/S guidelines for personal safety.
3. MSC- M/S hazards related to job.
4. MSE- M/S chemical/biological hazardous.
5. MSF- Ergonomics, psychosocial & organizational factors.
6. MSG- Safety guideline for confined space entry.
7. MSH- About the use and handling of hazardous materials.
8. IC- Industrial accident.
9. MSD- Physical hazards.
10. ZAG- Zero accident goal.

4.1. Proposed framework for vision zero

Figure 3 shows the proposed framework in that 47 factors are combined in 8 components based on the similarity of question and these factors are necessary for any company to avoid the accident and lost time injuries. These factors are necessary to know to avoid accidents in the industry.

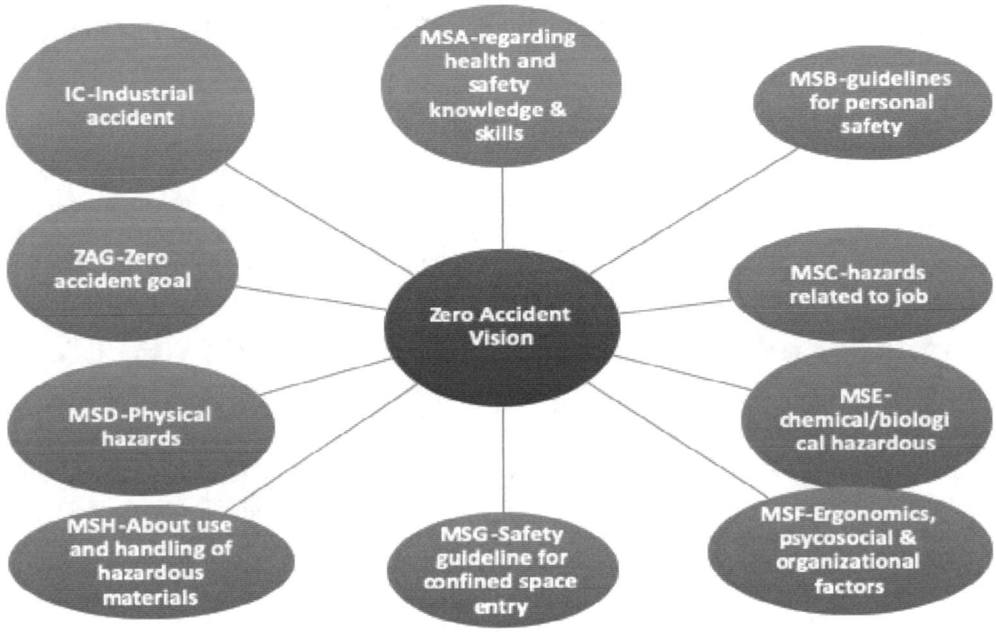

Fig. 3. Proposed Conceptual framework

4.1.1. MSA- M/S Regarding health and safety knowledge & skills

Griffin [21] have suggested that management values and knowledge & skill motivation are the key tools for the achievement of the zero accident vision. These tools determine the safety performance and categorize it in its components. He has also suggested that management values, safety inspection, personal training, and safety communication work as safety climate and these safety climates are caused by safety knowledge.

These safety climates lead the safety knowledge, safety motivation, motivation towards participation. The result gives a systematic framework for the investigation of the impact of worker's perception, behavior, and organizational safety outcomes. And this study also provides an idea of a core dimension of safety climates which are employee knowledge, skill, and motivation.

4.1.2. MSB- Guidelines for personal safety

Diaz and Neuhauser [13] has suggested the nascent nature of cultural assessment in health care, culture researchers lack consensus and clarity about domains important in a culture of safety; how to score and present improvements in culture over time; the relationship between culture and clinical and operational outcomes; and ultimately, how to package a tool kit to measure, score, and improve culture. He has suggested the teamwork climate and perception of management are the key enablers for personal safety.

4.1.3. MSC- M/S hazards related to job

Willquist & Torner [38] have classified the hazard related in the job are two types in which first is longterm sickness or lost-time injuries and the second is permanent disability of body parts. He had shown some statistics from the year 1996-1997 the risk of permanent disability of 5.57 per 1000 employees. The authors have concentrated on the improvement of the safety behavior of the company and employees and proposed a framework to evaluate or identify the hazards related to the job of the industry. The first stage of the analysis is the identification phase in which we defined what type of hazard is this and categorize it. The second phase is input to the analysis for the investigation on the base of data from the history, hazard identified through the method. And the third stage is analysis phase in which real analysis of the hazard is taking place, and this consist of the forward and backward method, morphological method and in the last stage which is the output stage in which we get the cause and consequences of hazard.

4.1.4. MSE- M/S chemical/biological hazards

Montano [29] has described the hazard related to the chemical or biological in the industry he admits that a substantial amount of disease or hazard is accounted for by these hazards, and these diseases are chronic illness, cancer, allergies, muskeleton disorder. He suggested that the identification of carcinogenic agents was based on the IARC classification. And the identification of the hazard which comes under the category of biological is put in ISCO category. According to him the biological hazard includes cellular or noncellular biological entities, which are capable of provoking infection or disease. Risk infection or biological disease is categorized into four groups. In the first group that is not likely to cause disease, and in the second group agents can cause disease, and in the third group the hazard cause which is severe and serious for the worker's health. This risk of chemical and biological hazard and work-related disease are highly concerned in the research. Major ISCO occupational groups were summarised following tasks. In the first task identification of occupational that is at higher risk, in second identification of common occupational and industrial hazard, In the third synthesis of some evidence regarding the societal cost and evidence, assessment and incidence or methodological problems.

4.1.5. MSF- Ergonomics, psychosocial & organizational factors

David [10] has discussed the identification of physical, psychological and individual occupational risk factors involved in the industry. The study is based on the level of a variety of factors that are responsible for these risks. He has suggested the assessment technique which are self-reports, observational method, direct method are involved. In self-reports data is collected from the workers in the form of interview questions or video recording to find the ergonomic and psychological factors, which are assessments of musculoskeletal symptoms , estimation of its magnitude and severity, frequency and estimation of time interval, mechanical exposure estimate, workers rating on load or estimation of related pain and discomfort, analysis of working condition and to find the higher risk group. And he had discussed some tools for the measurement of these risks like OWAS, RULA, REBA, and QEC, etc. for checking the whole body, upper body, and entire body assessment. He has suggested some advanced observational techniques like ROTA, TRAC, HARBO, PEO, PATH, etc. for the assessment of dynamic and static tasks, long-duration observation and various tasks performed in working condition.

4.1.6. MSG- Safety guideline for confined space entry

Burlet-Vienney et al. [7] discussed the safety precautions for working in the confined space these include the training of the workers and information available to them. The collection of the information and making some preventive measure for them, use of proper ventilation to maintain the proper atmospheric condition like oxygen, the management of the combustible dust present due to fire, monitoring of the toxic gases, proper idea of the rescue procedure for fast rescue, prohibition of the entry when there is some process involved, mandatory of using personal protective equipment on the work. He has visited the different organizations and observed the different type of confined space involved in that organization like crawl space, valve pit, access shaft, reservoir, technical room, tank truck, and palting bath etc. And the accidents occur in these areas are drowning, electrocution, hit by a falling object, fall from the height, engulfment of the poisoning substance, asphyxiation and caught or struck by machinery.

4.1.7. MSH- About use and handling of hazardous materials

Plummer et al. [32] have suggested that the use of hand gloves for hand protection, it is a necessary thing that is required to the workers who are handling the toxic and hazardous materials and cleaning the hazard on the sites. Furthermore, he has also suggested that there are types of chemicals are present in the industry and to tackle these chemicals various types of gloves are also required, compulsory the use of two pairs of gloves (one over the other). This double layer gloving technique is used to minimize the accident and increase safety in the industry. He has performed some physical activity on how to handle the hazardous materials because it is very prone to an accident in the industry and its proper permanent solution is required. He has concluded from the result that gloves type affects the timely completion of the job, and double gloving reduced the speed and dexterity of the worker during the performance of the task, he also suggests that error rates are more when we use double gloving.

4.1.8. MSD- Physical hazards

Li & Buckle [27] has discussed the work-related musculoskeletal disorders associated in the workers by different causes these termed as risk factors include physical workload factor, force, the posture of work, movement and vibration of the base or machinery, psychosocial stressors, and some individual factors are responsible for the physical hazard in the industry. For the measurement of these workload factors he normally asses it in the term of intensity, repetitiveness and time interval. For the solution, he has performed the task and with the help of RULA he has done the categorization of body posture, digital numbering of these body parts, marking of the posture angle, frequency occurrence force and muscle activities, record these and give the score to find the factors which are responsible for the physical hazard.

5. Conclusions

This research aims to attain a proper understanding of the ZAV and what are the enablers or factors necessary to implement it successfully in the industry. In the analysis, the questionnaire contains 47 factors and is divided into the components. From the result, we can see why these factors are necessary for the industry and based on the result we have developed a generalized framework for vision zero.

References

1. Abrahamsen, E.B., Asche, F., Milazzo, M.F., (2013), An evaluation of the effects on safety of using safety standards in major hazard industries, Safety Science, 59, 173-178.
2. Albert, A., Hallowell, M.R., Skaggs, M., Kleiner, B., (2017), Empirical measurement and improvement of hazard recognition skill, Safety Science, 93, 1-8.

3. Albery, S., Borys, D., Tepe, S., (2016), Advantages for risk assessment: Evaluating learnings from question sets inspired by the FRAM and the risk matrix in a manufacturing environment, Safety Science, 89, 180-189.
4. Allen, M.W., (1992), Communication and organizational commitment: Perceived organizational support as a mediating factor, Communication Quarterly, 40(4), 357-367.
5. Bahn, S., (2013), Workplace hazard identification and management: The case of an underground mining operation, Safety Science, 57, 129-137.
6. Burlet-Vienney, D., Chinniah, Y., Bahloul, A., Roberge, B., (2015), Occupational safety during interventions in confined spaces, Safety Science, 79, 19-28.
7. Burlet-Vienney, D., Chinniah, Y., Bahloul, A., Roberge, B., (2015), Design and application of a 5 step risk assessment tool for confined space entries, Safety Science, 80, 144-155.
8. Cudworth, A., (2009), The positive impact of communication on safety at shell, Strategic communication management, 14(1), 16.
9. Curcuruto, M., Conchie, S.M., Mariani, M.G., Violante, F.S., (2015), The role of prosocial and proactive safety behaviors in predicting safety performance, Safety Science, 80, 317-323.
10. David, G.C., (2005), Ergonomic methods for assessing exposure to risk factors for work-related musculoskeletal disorders, Occupational Medicine, 55(3), 190-199.
11. De Silva, K.G.V.K., Gunasekera, M.Y., De Alwis, A.A.P., (2017), Derivation of a societal risk acceptance criterion for major accident hazard installations in Sri Lanka, Process Safety and Environmental Protection, 111, 388-398.
12. Dekker, S.W., (2014), The bureaucratization of safety, Safety Science, 70, 348-357.
13. Diaz, M., Neuhauser, D., (2005), Pasteur and Parachutes: When Statistical Process Control is Better than a Randomized Controlled Trial, BMJ Quality & Safety, 14 (2), 140-143.
14. Drupsteen-Sint, L., (2014), Improving organisational safety through better learning from incidents and accidents, Copenhagen: Centre for Industiral Production; TNO.
15. Drupsteen, L., Wybo, J.L., (2015), Assessing propensity to learn from safety-related events, Safety Science, 71, 28-38.
16. Fera, M., Macchiaroli, R., (2010), Appraisal of a new risk assessment model for SME, Safety Science, 48(10), 1361-1368.
17. Garretson, I.C., Mani, M., Leong, S., Lyons, K.W., Haapala, K.R., (2016), Terminology to support manufacturing process characterization and assessment for sustainable production, Journal of Cleaner Production, 139, 986-1000.
18. Goerlandt, F., Khakzad, N., Reniers, G., (2017), Validity and validation of safety-related quantitative risk analysis: A review, Safety Science, 99, 127-139.
19. Gonzalez, A.A., Patroni, S.A., Vidal, J.G. (2015), Developing competencies in the process of hazard identification in an enterprise related to the field of logistic and food, Procedia Manufacturing, 3, 5052-5058.
20. Granerud, R.L., Rocha, R.S., (2011), Organisational learning and continuous improvement of health and safety in certified manufacturers, Safety Science, 49(7), 1030-1039.
21. Griffin, M.A., Neal, A., (2000), Perceptions of safety at work: a framework for linking safety climate to safety performance, knowledge, and motivation, Journal of occupational health psychology, 5(3), 347.
22. Keyton, J., (2011), Communication and organizational culture: A key to understanding work experiences, Sage.
23. Kines, P., Lappalainen, J., Mikkelsen, K.L., Olsen, E., Pousette, A., Tharaldsen, J., Tómasson, K., Törner, M., (2011), Nordic Safety Climate Questionnaire (NOSACQ-50): A new tool for diagnosing occupational safety climate, International Journal of Industrial Ergonomics, 41(6), 634-646.

24. Kontogiannis, T., Leva, M.C., Balfe, N., (2017), Total safety management: principles, processes and methods, Safety Science, 100, 128-142.
25. Langeland, T.A., (2009), Language and Change: An Inter-organisational Study of the Zero Vision in the Road Safety Campaign, PhD Thesis, University of Stavanger, Stavanger.
26. Legg, S.J., Olsen, K.B., Laird, I.S., Hasle, P., (2015), Managing safety in small and medium enterprises, 189-196.
27. Li, G., Buckle, P., (1999), Current techniques for assessing physical exposure to work-related musculoskeletal risks, with emphasis on posture-based methods, Ergonomics, 42(5), 674-695.
28. Li, H., Yang, X., Wang, F., Rose, T., Chan, G., Dong, S., (2016), Stochastic state sequence model to predict construction site safety states through real-time location systems, Safety Science, 84, 78-87.
29. Montano, D., (2014), Chemical and biological work-related risks across occupations in Europe: a review, Journal of Occupational Medicine and Toxicology, 9(1), 28.
30. Nascimento, F.A., Majumdar, A., Ochieng, W.Y., Schuster, W., Studic, M., (2016), Fundamentals of safety management: The offshore helicopter transportation system model, Safety Science, 85, 194-204.
31. Nyborg, W.L., (2000), Biological effects of ultrasound: development of safety guidelines: Part I: personal histories, Ultrasound in Medicine & Biology, 26(6), 911-964.
32. Plummer, R., Stobbe, T., Ronk, R., Myers, W., Kim, H., Jaraiedi, M., (1985), Manual dexterity evaluation of gloves used in handling hazardous materials, Proceedings of the Human Factors Society Annual Meeting, 29(8), 819-823.
33. Rollenhagen, C., (2010), Can focus on safety culture become an excuse for not rethinking design of technology?, Safety Science, 48(2), 268-278.
34. Rozenfeld, O., Sacks, R., Rosenfeld, Y., Baum, H., (2010), Construction job safety analysis, Safety Science, 48(4), 491-498.
35. Tchiehe, D.N., Gauthier, F., (2017), Classification of risk acceptability and risk tolerability factors in occupational health and safety, Safety Science, 92, 138-147.
36. Verma, A., Khan, S.D., Maiti, J., Krishna, O.B., (2014), Identifying patterns of safety related incidents in a steel plant using association rule mining of incident investigation reports, Safety Science, 70, 89-98.
37. Weick, K.E., Sutcliffe, K.M., (2001), Managing the unexpected, Vol. IX, San Francisco: Jossey-Bass.
38. Willquist, P., Törner, M., (2003), Identifying and analysing hazards in manufacturing industry- a review of selected methods and development of a framework for method applicability, International Journal of Industrial Ergonomics, 32(3), 165-180.
39. Xu, X., Yuan, Z., Gong, M., He, L., Wang, R., Wang, J., Yang, Q., Wang, S., (2017), Occupational hazards survey among coal workers using hand-held vibrating tools in a northern China coal mine, International Journal of Industrial Ergonomics, 62, 21-26.
40. Zhao, M., Liu, X., (2017), Reprint of: Regional risk assessment for urban major hazards based on GIS geoprocessing to improve public safety, Safety Science, 97, 112-119.
41. Zwetsloot, G.I.J.M., Kines, P., Ruotsala, R., Drupsteen, L., Bezemer, R., (2015), Success factors for the implementation of a Zero Accident Vision (ZAV) (No. TNO 2015 R11506), TNO.
42. Zwetsloot, G.I., Kines, P., Ruotsala, R., Drupsteen, L., Merivirta, M.L., Bezemer, R.A., (2017), The importance of commitment, communication, culture and learning for the implementation of the Zero Accident Vision in 27 companies in Europe, Safety Science, 96, 22-32.
43. Zwetsloot, G.I., Kines, P., Wybo, J.L., Ruotsala, R., Drupsteen, L., Bezemer, R.A., (2017), Zero Accident Vision based strategies in organisations: Innovative perspectives, Safety Science, 91, 260-268.

IOP Conf. Series: Materials Science and Engineering 1017 (2021) 012007 doi:10.1088/1757-899X/1017/1/012007

Parametric optimization of TIG welding of duplex stainless steel without filler rod by PCA method

S. Mondal[1], G. Nandi[2], P. K. Pal[2]
[1]Department of Mechanical Engineering, Haldia Institute of Technology, Haldia 721657, India
[2]Department of Mechanical Engineering, Jadavpur University, Kolkata-700032, India

E-mail: sandip_mondal2004@rediffmail.com

Abstract. The weld quality is mostly depended on the welding input factors. Welding quality is controlled by appropriately controlling the using process factors with sound knowledge base. The major influencing Tungsten Inert Gas (TIG) welding parameters are welding current, gas flow rate and speed of the welding. Controlling these parameters, the quality of the Duplex Stainless Steel (DSS) welding joints are improved. These welding process parameters are optimized to achieve the better quality of DSS welding joints. Here, the Principal Component Analysis (PCA) method and concept of Taguchi's signal to noise ratio (S/N) are used to optimize the above TIG welding process parameters of UNS 2205 DSS. The quality of the TIG welding of DSS has been evaluated in term of tensile test. Statistical tool like Analysis of Variance (ANOVA) is applied to evaluate the significance of the individual factors on the desired responses like ultimate tensile strength, yield strength etc. The optimal result is verified by the additional experiment. This result shows the application possibility of the PCA method for incessant development of welding quality of DSS in many fields, like chemical industries, oil & gas refineries, manufacturing industries etc.

1. Introduction

Duplex stainless steel ancestors have been enhanced through initiating nitrogen accompaniments for the period of last fifty decades. The exceptional blend of double phase configuration of ferrite with austenite has been seen in DSS. Austenite structure provides a unique arrangement of strength. Ferrite phase provides local corrosion resistance. DSS is being extremely used in oil and chemical industry. Several works has been reported for studying the mechanical behaviour of TIG welded DSS. The proper knowledgebase regarding TIG welding of duplex stainless steel is not sufficiently rich. More research is required to develop a sound knowledgebase of the various aspects related to duplex stainless steel welding.

Palani and Murugan [1] investigated the cladding effect on the weld bead geometry using different welding factors. Diaz et al. [2] showed the significant role of material characteristics of dissimilar stainless steels using finite element technique with the birth and death procedure. Zou et al. [3] investigated the effect of oxygen on crystallographic orientation in DSS weld. Magudeeswaran et al. [4] observed the phase structure of duplex stainless steel. They showed the equal amounts of bcc ferrite and fcc austenite phase structure of DSS material. They also observed that DSS had higher strength compare to single-phase austenitic stainless Steel. Lakshminarayanan et al. [5] researched on the mechanical and microstructural properties of AISI 409M ferritic stainless joints. Badjia et al. [6] observed the mixture of recrystallization texture and deformation in the material of single phase BCC and FCC structure. The crystallographic orientation of HAZ (heat affected zone) area was affected by the welding process, when the material was annealed at 1050° C. Juang and Tarng [7] determined the best TIG weld bead structure. They used Taguchi technique to analysis the weld pool structure using different welding process factors like gas flow rate, arc gap, speed and welding current. Tarng and Yang [8] optimized the weld bead geometry of GTAW process by Taguchi method. Tarng et al. [9] investigated on weld bead geometry of TIG welding of stainless steel. They used Taguchi method to determine the optimum welding process parameters. Tarng et al. [10] researched related to grey-based Taguchi technique to decide optimal process factors of submerged arc welding, concern of various weld properties. Yilmaz and Uzun [11] performed the destructive tests of welding joints created by GMAW and GTAW process on austenitic stainless steel. Rawlands et al. [12] showed that Taguchi technique was the most accepted method to solve the optimization difficulties within the area of manufacturing technology.

In the present work, some experiments and analyses have been completed regarding TIG welding of duplex stainless steel. Work material, considered in the present work is duplex stainless steel UNS 2205 (UNS means unified numbering system of North America). Nine butt-welded samples are made. Dimension of each sheet is 75mm x 50mm x 3mm. TIG welding is performed using varied input parameters. The parameters are current, gas flow rate and speed of welding. Taguchi's orthogonal array design is used to design the experiments by these controllable parameters. The levels of the parameters have been fixed based on trial runs and available literature. The effects of TIG welding process factors on ultimate tensile strength of these butt welding joints of UNS 2205 DSS have been experimented and optimized the above parameters of TIG welding using PCA method. Then, ANOVA test has been carried out to identify the significance of the individual factors on the desired responses like ultimate tensile strength, yield strength by Minitab 16.2.1.

2. Experimental work

2.1. Base metal

In the present work, duplex stainless steel (ASTM/UNS: 2205) is selected as the base material. The austenite and ferrite, both phase structures are observed in the microstructure of duplex stainless steel. After purchasing, the materials have been tested by Met-Lab Laboratory Services, recognised by Government, Public & Privet sector industries, at Mumbai-400004 to confirm about the composition of these materials. The testing results have been confirmed satisfactorily that the materials are UNS 2205 DSS. These materials are supplied by Aich Enterprise, Calcutta- 700141. The chemical composition of this base material is given in Table 1.

Table 1 Composition of UNS 2205 DSS

Composition of elements in %	C	Si	Mn	P	S	Cr	Mo	Ni	Al
	0.0210	0.2800	1.7200	0.0220	0.0140	22.6500	3.1800	4.7300	0.0100
	Co	**Cu**	**Nb**	**Ti**	**V**	**Pb**	**Fe**	**N**	
	0.0780	0.0090	0.0400	0.0080	0.0110	0.0030	67.123	0.1010	

2.2. Experimental plan

The TIG welding process factors for DSS joints are current, gas flow rate, and speed of welding. The range of these process factors are chosen based on the experimental trials. Three levels are taken for every parameter. Welding process factors and their levels are listed in Table 2. In this work, Taguchi orthogonal array design is most appropriate for process optimization, because it uses comparatively less number of observations to get optimum result. ANOVA will be carried out to identify the significance of the individual factors with the help of PCA method on the desired responses like ultimate tensile strength, percentage elongation etc. Taguchi Orthogonal Array (3^{rd}level design) is given in Table 3 below.

Table 2 Welding process factors with levels

Factor	Unit	Notation	Level		
			1	2	3
Welding current	Amp	C	80	85	90
Gas flow rate	l/min	F	7	7.5	8
Speed of welding	mm/s	S	2.3	2.8	3.5

Table 3 Taguchi's 3rd level orthogonal array design

Run	Welding current (C)	Gas Flow rate (F)	Welding Speed (S)
1	1	1	1
2	1	2	2
3	1	3	3
4	2	1	2
5	2	2	3
6	2	3	1
7	3	1	3
8	3	2	1
9	3	3	2

3. Experimental procedure

The Butt welding joints of UNS 2205 duplex stainless steel sheets have been completed by the IGBT digital welding inverter (400a, iii phase) of Electra engineering (India) Pvt. Limited. The photographic view of welding equipment is shown in Fig.1. In this welding joint, filler rod is not used. Edge preparation has not been done. Welding is performed by 2.5 mm diameter of non consumable electrode. Cup size is 3621.Tungsten shielding gas is argon used. Photographic view of welding sample no. 6 is shown in Fig. no.2.

Fig.1 The photographic view of welding equipment

Fig.2 Photographic view of welding sample no. 6

3.1. Testing of welds
After visual inspection of nine TIG welding specimens, the X-ray radiography tests are completed to determine the internal soundness of welds. Now, tensile test specimens of TIG welding samples have been prepared as per ASTM E8 standard [13] by WEDM machine to calculate yield strength, ultimate tensile strength etc.

3.2. X-ray radiography test
The X-ray radiography tests are completed for all the welding specimens. The X-ray radiography test of TIG welded DSS samples are performed at Inspection Survey & Surveillance (India) Pvt. Ltd. The results of X-ray radiography test are shown in Table 4. The X-ray radiography images of welding specimen no. 1 & 2 are shown in Fig. 3.

Table 4 Results of X-ray radiography test

Identification No.	Segment	Film Size	Technique	Thickness (mm)	Type of joint	Observation	Remarks
SMP-01	A	3″x6″	Single Wall Single Image	3.5	Butt	Little undercut	Acceptable
SMP-02	A	3″x6″	Single Wall Single Image	3.5	Butt	No significant defects	Acceptable
SMP-03	A	3″x6″	Single Wall Single Image	3.5	Butt	No significant defects	Acceptable
SMP-04	A	3″x6″	Single Wall Single Image	3.5	Butt	No significant defects	Acceptable
SMP-05	A	3″x6″	Single Wall Single Image	3.5	Butt	Capping undercut	Acceptable
SMP-06	A	3″x6″	Single Wall Single Image	3.5	Butt	No significant defects	Acceptable
SMP-07	A	3″x6″	Single Wall Single Image	3.5	Butt	No significant defects	Acceptable
SMP-08	A	3″x6″	Single Wall Single Image	3.5	Butt	No significant defects	Acceptable
SMP-09	A	3″x6″	Single Wall Single Image	3.5	Butt	No significant defects	Acceptable

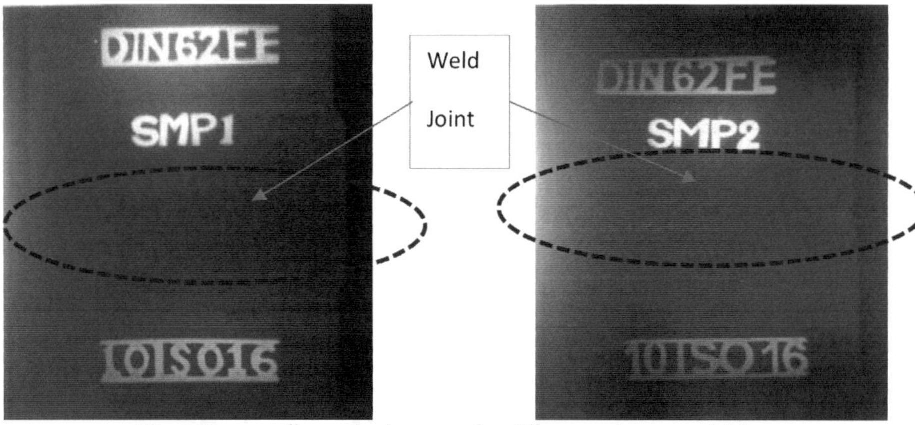

Fig. 3 X-ray radiography images of welding specimen no. 1&2

3.3. Tensile test and results

The tensile test samples have been made by Electronica Sprintcut-734 WEDM (Connected load 15 KVA, Input power supply 3 Phase, AC 415 V, 50 Hz) wire cutting machining. Photographic view of WEDM and tensile specimen preparation by WEDM are shown in Fig.4 (a) and 4(b).

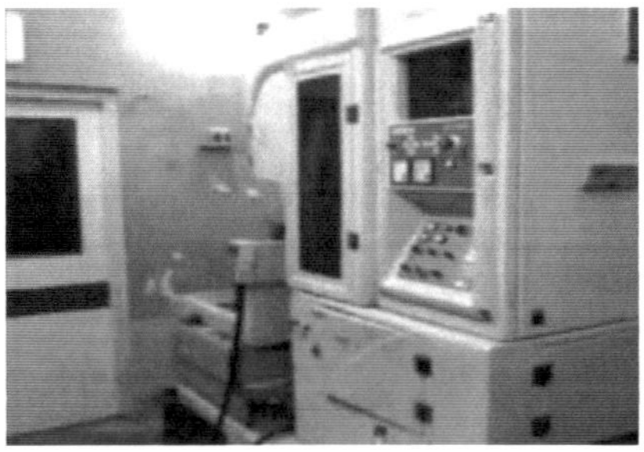

Fig.4 (a) Photographic view of WEDM

Fig.4 (b) Tensile specimen preparation by WEDM

As per requirement, the tensile properties i.e. ultimate tensile strength (UTS) and yield strength (YS) of the welding joints are determined. The photographic view of tensile test specimens is shown in Fig 5.

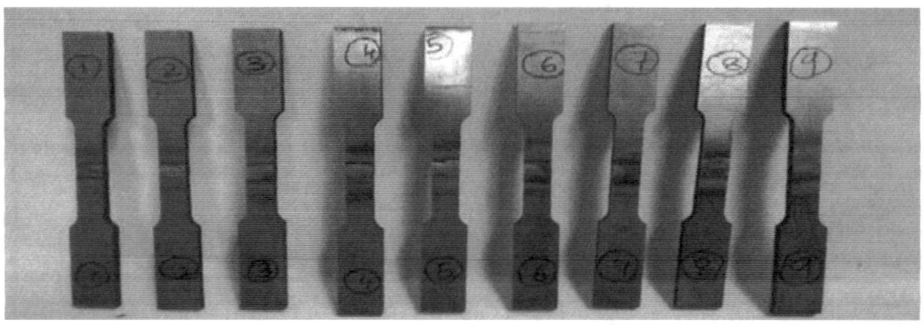

Fig.5 Photographic view of tensile test specimens

The tensile test is performed by tensile testing machine INSTRON made by Blue Star Engineering & Electronics Ltd., model no. : BSUT-60-JD-SERVO, serial no. : 2016/048. The maximum capacity of this machine is 600 KN. Photographic view of tensile testing machine is shown in Fig.7. Photographic view of tensile sample no.6 before test and after test is also shown in Fig. 6 (a) and (b).

(a) (b)

Fig.6 Photographic view of sample no. 6, (a) tensile sample before test (b) tensile sample after test

Data related to tensile strength are tabulated in Table 5. All these data are analysed by PCA method. The optimal parameter combination is also determined to achieve the desire quality of weld within experimental domain.

Table 5 Tensile test results

Sample No.	Yield Strength (MPa)	Ultimate Tensile Strength (MPa)
1	183.3	460
2	236.7	595
3	253.3	635
4	243.3	610
5	260	645
6	253.3	630
7	256.7	640
8	243.3	610
9	260	650

4. Optimization by using principal component analyses (PCA)

At first, the experimental data of yield strength and ultimate tensile strength are converted into S/N ratio for analysis using PCA method. S/N ratio of yield strength and ultimate tensile strength are shown in table no.6. According to this technique, two types of S/N ratios are categorised. One is lower the better (LB) and another is higher the better (HB). The S/N ratios of LB and HB are calculated with the help of equation 1 and 2 respectively. Then, investigational results (Table 5) are normalized. The main purpose is to maximize the experimental results. Here, Higher-the-Better (HB) criteria are considered. Results are normalized by the following equations:

(a) LB (lower the better)

$$\gamma_{ij} = -10\log[\frac{1}{n}\sum_{i=1}^{n} y_i^{-2}] \quad (1)$$

(b) HB (higher the better)

$$\gamma_{ij} = -10\log[\frac{1}{n}\sum_{i=1}^{n} y_i^{2}] \quad (2)$$

Where, γ_{ij} denotes the S/N ratios i^{th} experiment and j^{th} response calculated from observed values, y_i represents the experimentally observed value of the ith experiment and n=1 is the repeated number of every experiment of L9 orthogonal array. After normalization of results, verification is completed whether responses are correlated or not, by Minitab.

Then, the Eigen vector, Eigen value, accountability proportion (AP) and cumulative accountability proportion (CAP) are calculated. Eigen analysis of the correlation matrix is given in Table no.7.

Table 6 S/N ratio of Yield strength (YS) and ultimate tensile strength (UTS)

Sl. No.	SN-YS	SN-UTS
1	-45.2632	-53.2552
2	-47.484	-55.4903
3	-48.0727	-56.0555
4	-47.7228	-55.7066
5	-48.2995	-56.1912
6	-48.0727	-55.9868
7	-48.1885	-56.1236
8	-47.7228	-55.7066
9	-48.2995	-56.2583

Table 7 Eigen analysis of the correlation matrix

Variable	Eigen vectors	
	PC1	PC2
SN-YS	0.707	0.707
SN-UTS	0.707	-0.707

	(ξ_1)	(ξ_2)
Eigen value	1.9991	0.0009
AP	1.000	0.000
CAP	1.000	1.000

The equation MPI= ξ_1*1.000 + ξ_2*0.000 is used to calculate the Multi-response Performance Index (MPI). The MPI results are considered as the single objective function. These are optimised to get the optimum result. The factorial combination that maximized MPI is the optimal parametric arrangement. This parametric arrangement of process parameters is ensured the high tensile strength of the TIG weld joint of DSS. The principal components and calculated MPI are tabulated in Table 8. Optimal setting is determined by the plot (Fig. 7). The optimum arrangement becomes C1F1S1 that means welding current is 80 Amp, gas flow rate is 7 l/min and speed of welding is 2.3 mm/sec. The confirmatory test verifies this optimum result. The result of S/N ratio for MPI is 33.1149. In confirmatory experiment, this result becomes 34.0920. Therefore, this optimum welding process parameter setting has improved the welding quality.

Table 8 Principal components and calculated MPI

Sample no. Calculated MPI	SN-Yield Strength (MP_a)	SN-Ultimate Tensile Strength (MP_a)	
1	-45.2632	-53.2552	-45.2632
2	-47.484	-55.4903	-47.484
3	-48.0727	-56.0555	-48.0727
4	-47.7228	-55.7066	-47.7228
5	-48.2995	-56.1912	-48.2995
6	-48.0727	-55.9868	-48.0727
7	-48.1885	-56.1236	-48.1885
8	-47.7228	-55.7066	-47.7228
9	-48.2995	-56.2583	-48.2995

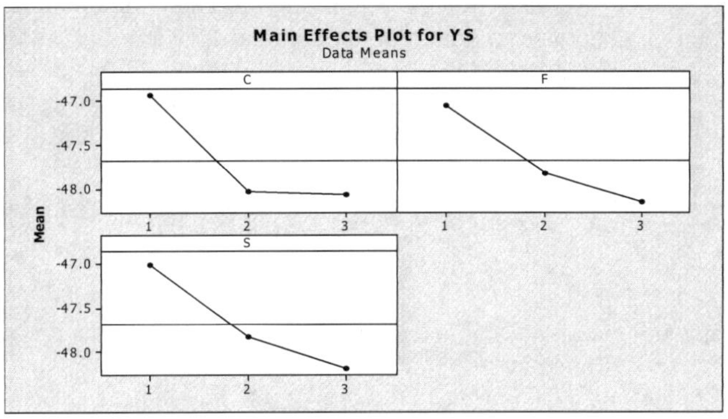

Fig.7 Main Effect plots of the MPI (S/N ratio)

5. Conclusions

Based on the result of present investigation and analysis of the experimental data of TIG welding of duplex stainless steel, the following conclusions come out:

i. In optimum weld process parametric combination, TIG welding of duple stainless steel provides satisfactory butt welding joints.
ii. X-ray radiography results show some welding defects under some parametric conditions (in correspondence with welded specimen nos. 1and 5). But, most of welding samples are defect free.
iii. The welding joints which are made by proper process parameters combination provide good ultimate tensile strength.
iv. Optimum parametric combination is successfully evaluated by the mean main effect plot. The optimum weld process parametric setting becomes C1F1S1 that means welding current is 80 Amp, gas flow rate is 7 l/min and speed of welding is 2.3 mm/sec.
v. Principal component analyses (PCA) and Taguchi method are mostly capable for satisfactory analysis and optimization of welding process parameters on the desired responses like ultimate tensile strength, yield strength.

References

[1] Palani P K and. Murugan N 2006 Sensitivity analysis for process parameters in cladding of stainless steel by flux cored arc welding Journal of Manufacturing Processes **8** 90-100

[2] Diaz J J D C, Rodriguez P M, Nieto P J G and Castro-Fresno D 2010 Comparative analysis of TIG welding distortions between austenitic and duplex stainless steels by FEM Journal of Applied Thermal Engineering **30** 2448-2459

[3] Zou Y, Ueji R and Fujii H 2014 Effect of oxygen on weld shape and crystallographic orientation of duplex stainless steel weld using advanced A-TIG (AA-TIG) welding method Journal of Materials Characterization **91** 42–49

[4] Magudeeswaran G, Nair S R, Sundar L and Harikannan N 2014 Optimization of process parameters of the activated tungsten inert gas welding for aspect ratio of UNS S32205 duplex stainless steel welds Journal of Defense Technology **10** 251-260

[5] Lakshminarayanan A K, Shanmugam K and Balasubramanian V 2009 Effect of Welding Processes on Tensile and Impact Properties, Hardness and Microstructure of AISI 409M Ferritic Stainless Joints Fabricated by Duplex Stainless Steel Filler Metal Journal of iron and steel research, international **16** 66-72

[6] Badjia R, Bacroix B and Bouabdallah M 2011 Texture, microstructure and anisotropic properties in annealed2205 duplex stainless steel welds Journal of Materials Characterization **62** 833–843

[7] Juang S C and Tarng Y S 2002 Process parameters selection for optimizing the weld pool geometry in the tungsten inert gas welding of stainless steel Journal of Materials Processing Technology **122** 33-37

[8] Tarng Y S and Yang W H 1998 Optimization of the weld bead geometry in gas tungsten arc welding by the Taguchi method International Journal of Advanced Manufacturing Technology **14** 549-554

[9] Tarng Y S, Yang W H and Juang S C 2000 The use of fuzzy logic in the Taguchi method for the optimization of the submerged arc welding process International Journal of Advanced Manufacturing Technology **16** 688-694

[10] Tarng Y S, Juang S C and Chang C H 2002 The use of grey-based Taguchi methods to determine submerged arc welding process parameters in hard facing Journal of Materials Processing Technology **128** 1-6

[11] Yilmaz R and Uzun H 2002 Mechanical properties of austenitic stainless steels welded by GMAW and GTAW Journal of Marmara for Pure and Applied Sciences **18** 97-113

[12] Rowlands H, Antony J and Knowles G 2000 An application of experimental design for process optimization The TQM Magazine **12** 78-83

[13] ASTM E8 2006 Standard Test Methods for Tension Testing of Metallic Materials ASTM International

IOP Conf. Series: Materials Science and Engineering 1017 (2021) 012008 doi:10.1088/1757-899X/1017/1/012008

Corrosion Resistant Superhydrophobic Aluminum Alloy: A Review

Bijayani Panda[1]

[1]Department of Mechanical Engineering, CMR Institute of Technology Bengaluru, India

Email: bijayani.p@cmrit.ac.in

Abstract. Aluminum alloys are widely used in aerospace and space applications due to their high strength to weight ratio. The most prevalent method to prevent corrosion of aluminum alloys is anodic oxidation or anodization treatment. The anodized layer is porous and hence there are chances of penetration of moisture and corrosion of underlying substrate underneath the anodized layer. Surface modification of the aluminum alloy by superhydrophobic coating is a very attractive method to improve the corrosion resistance of aluminum alloys. The inspiration for the fabrication of superhydrophobic coatings is obtained from various examples in nature such as strider's legs and lotus leaf. The fabrication of superhydrophobic coatings involves two important steps; generation of a rough surface and deposition of low surface energy material. The present paper reviews the various researches undertaken for the production of superhydrophobic surfaces on aluminum and aluminum alloys. Finally, the scope of future research activities in the fabrication of superhydrophobic coatings has been discussed in the paper.

1. Introduction

Aluminum alloys are abundantly used in the aviation industry. Aluminum is known to have good resistance to corrosion due to the formation of an adherent oxide coating on its surface. However, aluminum alloys are prone to corrosion in presence of aggressive environment [1]. Especially, in marine environments, aerosols are generated which contain small salt particles. These aerosols can be transported by wind and deposited on the metal surface that transform into salt solution droplet which can accelerate the corrosion process [2,3]. Earlier, chromate coatings were used which is now subjected to severe governmental regulations due to the carcinogenic nature of hexavalent chromium [4]. Aluminum alloys used in aerospace and space application are presently given anodization treatment to provide corrosion resistance. However, the porous oxide layer produced by anodization can easily allow for penetration of Cl^- ions in saline environment [5]. Hence, the anodic oxidation of aluminum alloys is not fully effective in aggressive environments.

There are two types of superhydrophobic surfaces. Highly adhesive surface allows the water droplets to stick to the surface even when the surface is turned upside down while low adhesive superhydrophobic surface have rolling angle less than 10° [6]. The second type of superhydrophobic surface is of great interest to researchers; recently, much attention is given to super-hydrophobic surfaces having water contact angle greater than 150° and water sliding angle less than 10° [7-10]. It is known that metals have a high surface energy such that a water droplet will be adsorbed on its surface and show a hydrophilic state (water contact angle < 90°). Superhydrophobic coatings change the wettability of the metal surface from hydrophilic to superhydrophobic, minimizing the contact area between the corrosive solution and the substrate [11]. In the presence of superhyodrophobic surface, the corrosion currents are shown to reduce drastically in corrosive environments [12]. Since superhydrophic surfaces possess excellent repulsion to water, they can be effective in preventing corrosion [2, 13,14].

2. Superhydrophobic Coatings: Inspiration and History

Nature has several non-wetting low adhesive surfaces like lotus leaves, water strider's legs, wheat leaves, duck feathers etc. [6, 15-17]. Especially, the 'lotus effect' [18] has generated great interest from researchers because of the application of super-hydrophobic surfaces in generating self-cleaning [19], anti-wetting [20], anti-icing [21], anti-corrosion [22], friction reducing [23] and antimicrobial [24] surfaces. The lotus leaves have self-cleaning tendency by rolling off adhering dust particles along with water droplets [25, 26]. The research on superhydrophobic coatings is inspired by this non-wetting phenomenon present in nature [2, 27-28].

The development of superhydrophobic coatings started around 200 years [15]. Thomas Young [29], first proposed the concept of surface wettability in 1805. The first superhydrophobic surface was produced by Ollivier in 1907; he could obtain a water contact angle of around 180° using soot, lycopodium powder and arsenic trioxide [30]. Langmuir [31] reported in 1920 that an organic compound absorbed as a monolayer, could change the wetting and frictional properties of solid surfaces. This finding motivated several researchers to work on chemical modification to control surface wettability. An optimum combination of roughness along with low surface energy is important for fabricating coatings possessing superhydrophobic property (Fig. 1) [25, 26, 32].

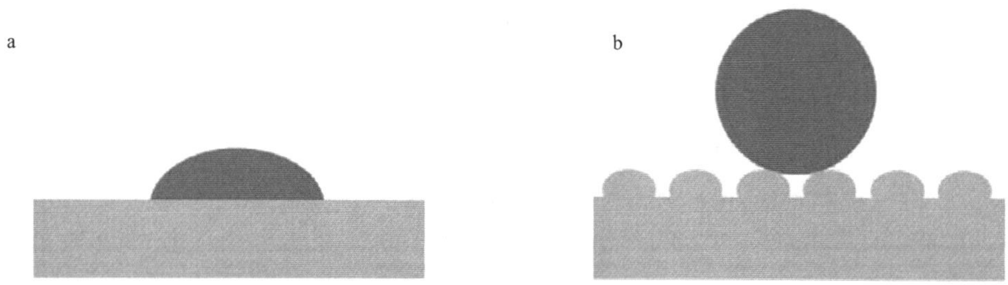

Fig.1 Schematic drawing showing the effect of (a) surface modification and (b) enhanced

Wenzel developed a relationship between the macroscopic surface roughness of a solid and the contact angle. He showed that surface roughness can enhance hydrophobicity [25, 33]. The model developed by Wenzel suggested that the surface area of a hydrophobic solid increases with the increase in roughness. According to Wenzel, the contact angle of water (θ) on a rough surface of any material is given by

$$\cos \theta = r.\cos \theta^* \quad (1)$$

where r is the roughness parameter (ratio of the real surface area to the projected surface area), θ* is the contact angle of water on the smooth surface made of the same material. Since r is greater than 1, a hydrophobic surface will become more superhydrophobic when it is made rough, according to the above equation.

Cassie and Baxter [25, 34] proposed that porous and rough surfaces have the capability to trap air between water and solid. This theory proposed by Cassie and Baxter is known as the composite wetting model. According to this model, a composite interface of solid and gas is formed between a water droplet and a superhydrophobic surface. The apparent contact angle (θ) of a water droplet on a solid superhydrobhobic surface is given by

$$\cos\theta = r_f f_{SL} \cos\theta_o - 1 + f_{SL} \quad (2)$$

where θ_o is Young's contact angle on an ideally flat surface, f_{SL} is the fraction of the wetted surface and r_f is the roughness of the wetted surface. From Eq. 2, it can be understood that a small value of f_{SL} would result in a high value θ on a rough hydrophobic surface. The two models discussed above are schematically shown in Fig. 2.

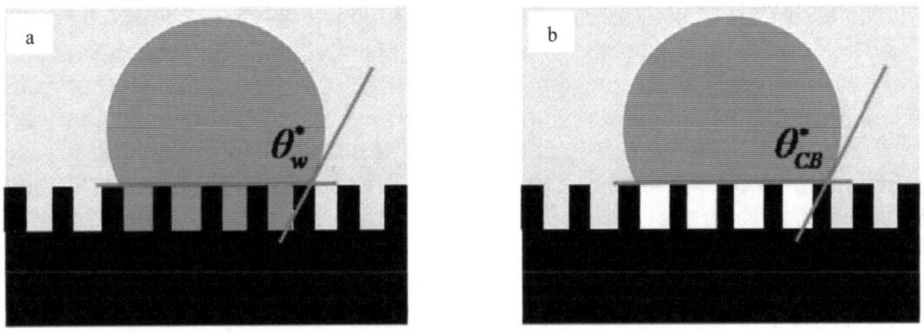

Fig. 2 Schematic representation of (a) Wenzel model and (b) Cassie-Baxter model

3. Fabrication of Superhydrophobic Coating on Aluminum/Aluminum Alloys

In case of lotus leaves, water droplet readily remains on the peak of the micrometer/nanometer scale rough structure on the leaf because air is trapped in the grooves present in between the rough structures [6, 35-37]. Similarly, the superhydrophobic coating is composed of hills and valleys; the air gets trapped in the valleys. Hence, it becomes difficult for the corroding anion (like Cl^-) to reach the bare metal surface [12]. Artificially, superhydrophobic surfaces can be fabricated in two steps: generation of a rough surface and the reduction of surface energy of the surface [6, 9, 38]. Methods such as anodic oxidation [39,40], electrochemical etching [41,42], chemical etching [43,44], laser etching [45], boiling water immersion [46], and coating [47,48] were developed to fabricate suitable rough structures required by superhydrophobic surfaces on aluminum substrates. The surface energy can be reduced by grafting a low surface energy material like triethoxyoctylsilane, flouroalkylsilane and fatty acids onto the roughened surface [6, 9]. Some reported research work carried out by various researchers is discussed in the following paragraphs.

Song et al. [6] produced a superhydrophobic coating on aluminum substrate (≥99.9 wt% purity) in two steps; the first step involved dipping the substrate in a solution of $CuCl_2$ for roughening the surface while the second step was to produce the desired low surface energy by dipping the roughened substrate in an ethanolic solution of flouroalkylsilane.

In one research, superhydrophobic coating (contact angle of around 160°) was fabricated on silicon substrate by repeatedly dipping in a saturated fatty acid solution (stearic acid) and consequent drying in air [26]. The repeated dipping and drying produced a natural double scaled rough coating of low surface energy. Hence, the low surface energy material used also increased the surface roughness of coating with each dip-dry cycle.

A solution containing NaCl, intended for electrochemical machining of the aluminium metal substrate, along with fluoroalkylsilane for creating a rough surface, was reported to produce a superhydrophobic surface having static water contact angle of 166° and a tilting angle of around 1° [42]. Guo et al. [49] produced superhydrophobic surfaces by etching aluminum alloy AA 2024 in an alkali solution followed by coating with low surface energy materials; perfluorononane(C_9F_{20}) and perfluoropolyether (PFPE).

They studied the effect of etching time and the concentration of NaOH on the surface roughness and its effect on the superhydrophobicity. At an optimum surface roughness of 2.7-5.8 μm, they could obtain superhydrophobicity on the aluminum alloy surface. This was possible due to the entrapment of air between microprotrusions and nanoparticles formed on the surface of the aluminum alloy.

Boiling water has been used for obtaining a rough surface on aluminium 6061 alloy followed by sputtering of polytetrafluoroethylene (PTFE) on the aluminum substrate surface to reduce the surface energy [46]. The treatment of aluminum surface with boiling water leads to the formation of boehmite ($Al_2O_3(H_2O)$) which is used in industries for corrosion protection. The consequent superhydrophobic coating obtained had a static water contact angle of around 164°. Zhang et al. [50] fabricated a simple superhydrophobic coating on aluminium alloy by immersing it in a solution containing hydrochloric acid and fatty acid molecules. Xie et al. [9] developed a two step method to fabricate superhydrophobic surfaces on aluminium (chemical composition: 4.5%Cu, 1.5%Mg, 0.6%Mn, 0.5%Si, 0.5%Fe, 0.3%Zn). They obtained this by treatment of the aluminum with boiling NaOH solution, followed by immersion in an ethanolic solution of lauric acid.

Song et al. [51] fabricated petal like structure, similar to lotus leaf on the surface of aluminum by a one step immersion process using perfluorodecyltriethoxysilane (PDES). Saleema et al. [38, 52] prepared superhydrophobic surface on aluminum alloys (AA 6061) by immersing the substrate in a solution comprising of sodium hydroxide and flouroalkyl-silane molecules. Ou et al. [53] fabricated a single step superhydrophobic coating on aluminium alloy AA 5083 by dipping the sample in a solution containing low surface energy perfluorooctyltrichlorosilane (PFOTS) molecules, ethanol and water. The treated alloys produced a rough surface covered by PFOTS molecules, showing a contact angle of 150° and sliding angle lower than 10°. Chen et al. [10] used chemical etching method using an alkali treatment in 1 mole /l NaOH aqueous solution, followed by immersing in a solution of hydrochloric acid and acetic acid to produce a superhydrophobic surface on aluminium (99.8 % purity). They could obtain a water contact angle of around 154° and water sliding angle of around 5° due to the presence of a combination of protuberances and micro-pores, which effectively trapped air. Moreover, as the etching time increased, the protuberances and micro-pores increased in number and decreased in size, leading to more effective air entrapment. However, beyond a certain etching time, the structure formed was destroyed due to excessive etching.

Electrochemical machining in $NaNO_3$ is found to help in dissolving grain boundaries and dislocations and produce a micro-nano rough surface [41]. Another work involves the development of superhydrophobic coating by brush-painting process [54]. This was achieved by using poly (furfuryl alcohol)/copper composite coatings followed by immersion in stearic acid. Superhydrophobic property was obtained due to the formation of copper stearate. The low surface energy of copper stearate has been reported elsewhere [55]. Immersion of aluminum alloy (chemical composition: 5.51 wt% Cu, 1.68 wt% Mg and balance Al) in a solution of stearic acid and N,N-dicyclohexylcarbodiimide in n-hexane for 24 h can produce the desired low surface energy after surface roughening in HNO_3/H_2O_2 solution [56].

4. Enhancement of Corrosion Resistance of Aluminum Alloys

Some published research articles have specifically focused on the corrosion resistance of superhydrophobic coatings. These are discussed in this section. One research mentions the use of a solution of myristic acid in ethanol and immersion of anodized aluminum in the solution for obtaining the super-hydrophobic coating [57]. They found that aluminum corrosion was largely inhibited due to the formation of superhydrophic film on the anodized aluminum surface.

Similarly, He et al. [5] fabricated superhydrophobic coating by melting myristic acid adsorbed onto anodized aluminium of 99.9 % purity. The anodization was carried out in 15 wt. % sulphuric acid. They reported that the surface treatment of anodized aluminum significantly improved the corrosion resistance of the substrate in seawater environment. The static contact angle in NaCl solution was found to be 154°. The surface roughness was increased by the anodization treatment. When the chemical modification was done by melting myristic acid, the acid filled up the nano-pores of the anodized surface.

Shi et al. [47] fabricated superhydrophobic surfaces on aluminum substrates (99.9 % purity) by acid etch followed by nano-silica particle deposition and then heat treatment. The nano-silica particles were deposited by immersing the substrate in an aqueous solution of nano-silica in tetramethylamonium hydroxide. The surface was then reacted with fluorosilane. They found improvement in the corrosion performance of the aluminum in NaCl environment.

Lu et al. [2] fabricated a superhydrophic surface on aluminium (99.9 % purity) by first producing an anodized layer and then subjecting the surface to a fluorination treatment. The anodization treatment was done using phosphoric acid in which aluminum was used as the working electrode and stainless steel was used as the counter electrode. The anodization treatment produced the desired surface roughness with a pillar-like or pore-like structure, depending on the time of anodization treatment given to the substrate. It was observed from Eq. 2 that the pillar like structure has lower f_{SL} than pore like structure and the pillar like structure results in a higher contact angle compared to pore like structure. The fluorination treatment was done by coating with silane-ethanol solution for 5 minutes at room temperature. The resultant coating was found to be highly adherent and scratch resistant. The deliquesced NaCl particle easily slipped off the superhydrophobic surface produced. The corrosion performance of the aluminum was largely improved in marine atmospheric condition. This was due to the presence of an air film between the saline solution and the superhydrophobic layer.

In another published work, anodization treatment was done followed by low temperature plasma treatment to produce the required combination of micro and nano structure [58]. Trichlorooctadecyl-silane was used to reduce the surface energy of the aluminum substrate. They could obtain a static water contact angle of around 158°.

It has been reported by Wang et al. [58] and Lu et al. [2] that anodization on aluminum alloys gives a suitable rough nano-pore structure film which was found to be beneficial in depositing the low surface energy materials for producing a superhydrophobic surface. Li et al [40] have shown that hard anodization of 2A12 aluminum alloy resulted in a composite structure made up of micro-scaled alumina columns and pores which results in an increase in water contact angle. The improved wetting property was reported to be due to decrease in surface energy caused. Moreover, the anodized layer not only provides a rough surface, but also an insulated layer with superior adherence to the substrate [2].

In general, the anodic as well as cathodic current densities decrease when an aluminum substrate is provided with a superhydrophobic coating. In addition, the corrosion potential also shifts to more noble potential [5, 44, 57]. The corrosion currents have been reported to reduce by around three orders of magnitude [57].

The superhydrophobic coatings on anodized aluminum substrate show a two time constant behavior under electrochemical impedance spectroscopy studies [5]. The two types of equivalent circuit models are shown in Fig. 3 (a) and (b). R_{ct} is the charge transfer resistance at the Al-coating interface, C_{dl} is the double layer capacitance corresponding to interaction at the substrate/coating interface and Rs is the solution. R_c correspond to resistance to the electrolyte in the pores on the coating or the resistance of ion conducting paths that develop in the coating. C_c is the capacitance of the surface coating [47]. The R_c of superhydrophobic film is found be higher when the anodized aluminum is treated with the organic coating. Higher R_{ct} values are ascribed to the improved anticorrosive property of the aluminum substrate.

Air is entrapped in the roughness grooves of the anodized layer due to which the penetration of aggressive species like Cl^- ions is largely reduced while the adsorbed organic coating gives the excellent anti corrosion properties. A Warburg impedance element can occasionally be observed, which indicates the penetration of water into the coating and the formation of a new electrolyte/Al interface under the coating [47].

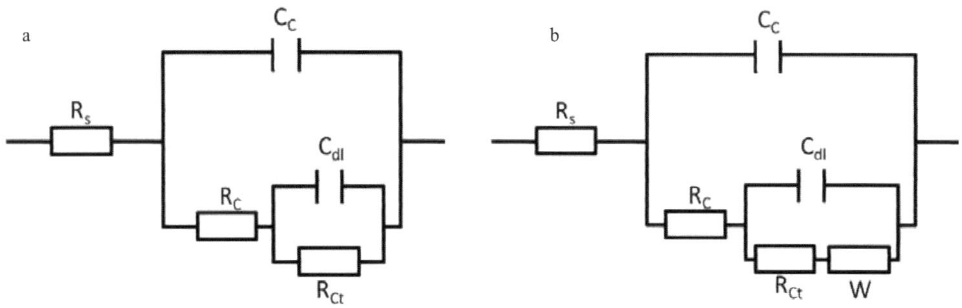

Fig. 3 Schematic of the equivalent circuit model for roughened and organic compounds

5. Future Prospects

All the published literature mentions superhydrobhobicity as the reason for improved corrosion resistance. However, what has largely been overlooked by most of the studies is that in a superhydrophobic surface, capillary condensation in groove of the film may induce penetration of the corrosive solution (like saline solution) into the film. This, in fact, has been reported by Wang et al. [13]. If the water or moisture is able to penetrate to reach the substrate under such condition, the corrosion of the substrate can easily take place. Hence, if additional metal ions are incorporated in the organic base, it can provide anodic/cathodic protection to the substrate.

Another important aspect which has been addressed to a lesser extent is the production of an optimum rough surface on the aluminum alloy. Very few of the published research have shown the effect of different surface roughness on the superhydrophobicity and the corrosion resistance of the alloy. The process parameters can play an important role in the roughness of the surface, which in turn will affect the superhydrophicity of the immersion coating as well as its adherence.

6. Conclusions

Aluminum alloys are usually prevented from corrosion by anodic oxidation or anodization treatment. However, the anodized layer is porous and there are chances of penetration of moisture and corrosion of underlying substrate underneath the anodized layer can take place. Surface modification of the aluminum alloy by superhydrophobic coating is a very attractive method to improve the corrosion resistance of aluminum alloys. The paper reviews the various methods by which fabrication of superhydrophobic coatings have been done by various researchers. The important steps involve the generation of a rough surface and deposition of low surface energy material. These have been done either in one or in multiple steps. The simplest method reported is by immersion as no special equipment is required and the process is relatively simple and can be carried out on an industrial scale. Anodization treatment of aluminum alloys seems to be a very versatile method for producing a combination of micro and nano scale rough structure. Future scope of work can be to improve the corrosion resistance of the substrate by anodic/cathodic protection by incorporating additional metal ions to the organic coating and the study of effect of surface roughness on the corrosion resistance of the substrate.

References

[1] Badawy W A, Al-Kharafi F M and El-Azab A S 1999 Corros. Sci. 41 709.
[2] Lu Z, Wang P and Zhang D 2015 *Corros. Sci.* **91** 287.
[3] Li S X and Hihara L H 2010 Corros. Eng. Sci. Technol. 45 49.
[4] EPA – Environmental Protection Agency, 1988.
[5] He T, Wang Y, Zhang Y, lv Q, Xu T and Liu T 2009 Corros. Sci. 51 1757.
[6] Song J, Xu W, Liu X, Lu Y, Wei Z and Wu L 2012 Chem. Eng. J. 211–212 143.
[7] Nakajima A, Hashimoto K and Watanabe T 2000 Langmuir 16 7044.
[8] Nakajima A, Fujishima A, Hashimoto K and Watanabe T 1999 Adv. Mater. 11 1365.
[9] Xie D and Li W 2011 *Appl. Surf. Sci.* **258** 1004.
[10] Chen Z, Guo Y and Fang S 2010 *Surf. Interface Anal.* **42** 1.
[11] Li C, Ma R, Du A, Fan Y, Zhao X and Cao X 2019 J. Alloys Compd. 786 272.
[12] Liu T, Chen S, Cheng S, Tian J, Chang X and Yin Y 2007 Electrochim. Acta 52 8003.
[13] Wang P, Zhang D, Qiu R, Wu J and Wan Y 2013 Corros. Sci. 69 23.
[14] Naing T H, Janudom S, Rachpech V, Mahathaninwong N and Thiwong S 2019 Mater. Res. Express 6 116415.
[15] Liu M, Wang S and Jiang L 2017 Nat. Rev. Mater. 2 17036 1.
[16] Fogg G E 1944 Nature 154 515.
[17] Cassie A B D and Baxter S 1945 Nature 155 21.
[18] Barthlott W and Neinhuis C 1997 *Planta* **202**1.
[19] Fürstner R, Barthlott W, Neinhuis C and Walzel P 2005 Langmuir 21 956.
[20] Srinivasan S, Choi W, Park K C, Chhatre S S, Cohen R E and McKinley G H 2013 Soft Matter 9 5691.
[21] Cao L, Jones A K, Sikka V K, Wu J and Gao D 2009 Langmuir 25 12444.
[22] Xu Q F and Wang J N 2009 New J. Chem. 33 734.
[23] Choi C H and Kim C J 2006 Phys. Rev. Lett. 96 066001.
[24] Wu R M, Jiang H Y, Liu W Y, Deng J, Yuan Z Q and Liu Q L 2012 Adv. Mat. Res. 583 346.
[25] Eduok U and Szpunar J 2018 Prog. Nat. Sci. 28 354.
[26] Kyle R, Joseph A and Chiara Neto A 2010 Aust. J. Chem. 63 525.
[27] Ahmad Z and Patel F 2012 Int. J. Corros. 2012 1.
[28] Mohamed A M A, Abdullah A M and Younan N A 2015 Arab. J. Chem. 8 749.
[29] Young T 1805 Phil. Trans. R. Soc. Lond. 95 65.
[30] Ollivier H 1907 J. Phys. Theor. Appl. 6 757.
[31] Langmuir I 1920 Trans. Faraday Soc. 15 62.
[32] Li J, Liu X, Ye Y, Zhou H 2011 J. Chen, Appl. Surf. Sci. 258 1772.
[33] Wenzel R N 1936 Ind. Eng. Chem. 28 988.
[34] Cassie A B D and Baxter S 1944 Trans. Faraday Soc. 40 546.
[35] Neinhuis C and Barthlott W 1997 Ann. Bot. 79 667.
[36] Barthlott W and Neinhuis C 1997 Planta 202 1.
[37] Feng L, Li S, Li Y, Li H, Zhang L, Zhai J, Song Y, Liu B, Jiang L and Zhu D 2002 Adv. Mater. 14 1857.

[38] Saleema N, Sarkar D K, Gallant D, Paynter R W and Chen X G 2011 ACS Appl. Mater. Interfaces 3 4775.
[39] Wu W C, Wang X L, Wang D A, Chen M, Zhou F, Liu W M and Xue Q J 2009 Chem. Commun.1043.
[40] Li J, Du F, Liu X L, Jiang Z H and Ren LQ 2011 J. Bionic Eng. 8 369.
[41] Xu W J, Song J L, Sun J, Dou Q L and Fan X J 2011 J. Mater. Sci. 46 5925.
[42] Song J L, Xu W J and Lu Y 2012 J. Mater. Sci. 47 162.
[43] Yin L, Wang Y Y, Ding J F, Wang Q J and Chen Q M 2012 Appl. Surf. Sci. 258 4063.
[44] Yin B, Fang L, Hu J, Tang A Q, He J and Mao J H 2012 Surf. Interface Anal. 44 439.
[45] Li X H, Xu Q, Li Z H and Huang W H 2011 International Conference on Electronics and Optoelectronics (ICEOE) 2 V2.
[46] Jafari R and Farzaneh M 2011 Appl. Phys. A 102 195.
[47] Shi X M, Nguyen T A, Suo Z Y, Wu J L, Gong J and Avci R 2012 Surf. Coat. Technol. 206 3700.
[48] Chakradhar R P S, Dinesh Kumar V, Shivakumara C, Rao J L and Basu B J 2012 Surf. Interface Anal. 44 412.
[49] Guo Z G, Zhou F, Hao J C and Liu W M J. 2006 Colloid Interface Sci. 303 298.
[50] Zhang Y, Wu J, Yu X and Wu H 2011 Appl. Surf. Sci. 257 7928.
[51] Song H and Shen X 2010 Surf. Interface Anal. 42 165.
[52] Saleema N, Sarkar D K, Paynter R W and Chen X G 2010 Acs. Appl. Mater. Inter. 2 2500.
[53] Ou J, Hu W, Xue M, Wang F and Li W 2013 Acs. Appl. Mater. Inter. 5 9867.
[54] Chen S, Guo B and Wu W 2011 Appl Phys A 105 861.
[55] Huang Y, Sarkar D K and Chen X 2010 Mater. Lett. 64 2722.
[56] Wang Q, Zhang B W, Qu M N, Zhang J Y and He D Y 2008 Appl. Surf. Sci. 254 2009.
[57] Yin Y, Liu T, Chen S, Liu T and Cheng S 2008 Appl. Surf. Sci. 255 2978.
[58] Wang H, Dai D and Wu X 2008 Appl. Surf. Sci. 254 5599.

Sustainable Construction Using EPS Beads in Light Weight Blocks to form Innovative Foam Concrete as a Green Building Material

A S Moon[1] and Dr. A Patel[1]

[1]Department of Civil Engineering, RKDF University, Bhopal – Madhya Pradesh, India

Email id: ash04moon@gmail.com

Abstract. We assume to build up another sort of basic light weight concrete by supplanting the coarse total with EPS dabs. Extended Polystyrene (EPS) this is light in weight and it has been utilized as in building applications. EPS is in granular structure and it is utilized in light weight cement to make the solid lighter from 950 kg/m³ to 1350kg/m³. In this paper we did test examination on the designing properties, as compressive quality, modulus of flexibility, drying shrinkage and creep, of polystyrene total cement fluctuating in density. Factors, for instance, water/solid extent, polystyrene/solid extent, solid substance, advancement, compaction, fire, resistance. At the completion of the examination, a philosophy for arranging EPS light weight solid mixes is plot.

1. Introduction

By and by Millions, Trillion of tons EPS squander is delivered which is destructive for condition hence to all the more likely use the EPS as a substitute of coarse totals. Most polystyrene items are presently not reused because of the absence of motivation to put resources into the compactors and calculated frameworks required for reusing. Because of low thickness of polystyrene froth or EPS dots, it isn't practical to gather. Generally, if a waste material experiences an underlying compaction measure. The material changes thickness from ordinarily 30 kg/m3 to 330 kg/m3 and it turns into a recyclable ware of high an incentive for creation of reused plastic pellets, protection sheets and different EPS materials for development applications; numerous makers can't deliver adequate piece due to assortment issues.

EPS dabs are utilized for bundling of products for simple taking care of and transportation. However, the removal of EPS dabs is turning into an issue for the garbage removal division. Since, EPS is lightweight, non-biodegradable, hydrophobic and misleadingly inactive in nature, and furthermore have great warm and sound protection; it tends to be utilized as a minimal effort trade of the coarse totals for the light weight concrete. Since, it a byproduct from the bundling business, using it in development of lightweight solid won't just lessen the cost however all decrease the weight on garbage removal divisions.

1.1 Definition: EPS or expanded polystyrene is a resolute cell plastic at first planned in Germany in 1950. It has been used in packaging courses of action since 1958. It is 98% air anyway the rest is delivered utilizing nearly nothing, round EPS spots - themselves made particularly of carbon and hydrogen.

Figure 1: EPS Beads

EPS dots can be used to make low thickness concretes required for building applications like cladding sheets, Partition dividers, composite ground surface structure and weight bearing strong squares. Decision of EPS touch all out was made prevalently in view of its low thickness, closed cell structure, hydrophobic and essentialness immersing characteristics. The EPS globules incorporated the mix contributes just to its low thickness.

2. History of Light Weight Concrete

Light weight cement concrete have been utilized in development since before the times of the Roman Empire. The most punctual sorts of lightweight cement were made by utilizing Grecian and Italians pumice as the lightweight total. Ordinary hydrated consumed lime was utilized as the cementations material in the blend. These early lightweight cements, by reason of the clearly frail materials, missed the mark in auxiliary execution of what we expect and accomplish today. They were, in any case, incredibly strong, and existing instances of these early lightweight cements are still to be found in different early structures of the Mediterranean region. The essential utilization of light weight concrete is to lessen the dead heap of the solid structure, which at that point permits the auxiliary architect to diminish the size of the section, balance and other burden bearing components.

2.1 Light Weight Concrete Using Eps

Lightweight concretes (LWCs) can be used in various improvement fields. It will in general be used for fixing wooden floors of old structures, passing on dividers of low warm conduction, associate decks, floating quay, etc. For the essential applications, the lightest possible material is used, i.e., regularly it has a specific gravity of 0.5, the quality being of less noteworthiness. Nevertheless, for some assistant applications, a compressive quality higher than 40MPa is a portion of the time essential, which drives the organizer to improve a material with a specific gravity close to 1.8. In such a case, lightweight sums, for instance, expanded glass or soil, partake in the resistance of the composite. The potential results offered by new cement based materials prescribe that it is possible to improve the compressive quality versus the specific gravity, or to show up at equivalent quality for lower express gravity. The purpose of this report is to achieve a mix plan for Lightweight EPS Concrete with thickness lesser than 1800kg/m^3 and enough high compressive quality so it will in general be used being developed explanation.

3. Literature Review

R. Gawale et al. 2016[5] performed explores and considered that thickness, compressive quality of the examples with EPS dots and the examples without EPS dots in the blend plan. Admixture Poly-carboxyl ate ether (PCE) is included blend configuration to build the functionality of the solid. The ends were discovered to be that quality of light weight solid utilizing EPS globules is low for lower thickness blend. This brought about augmentation of voids all through the example brought about by the Air entraining admixture. Along these lines the decline in compressive quality of the solid.

A Tayale et al. 2018[6] performed tests thought about the properties, for example, compressive quality and warmth protection, of EPS cement to the standard solid 3D square. The solid shapes comprises of 5%, 10%, 15%, 20%, 25%, and 30% EPS in blend plan M25. The blocks were tried 7 days, 14 days and 28 days of restoring. Creator presumed that the heaviness of the 3D shape was diminished by 63% when the EPS globules are included by 25%. During warming cycle, the EPS globules that came in direct contact with the warmed surface shrinked prompting arrangement of voids in the solid.

J Singh et al. 2017[7] performed attempts various things with M20 mix structure of coarse and fine absolute in degrees of 0%, 10%, 20%, 30%, 40% and half in the water solid extent of 0.50 was concentrated with silica rage as a midway replacement of cement. Maker reasons that the usefulness of mixes was seen to increase with increase in rate replacing of sums with polystyrene.

S M Abd et al. 2016 [8] examined the convenience and the relationship of the compressive quality, thickness assortment in the models for the 7 days and 28 days. The development in the EPS content in solid mixes reduces the compressive nature of concrete. Augmentation in the EPS content in solid mixes diminishes the thickness of concrete. Value increases with increase in EPS globules content

M D Karthick et al. [9] played out the trials on the solid blocks for usefulness and compressive quality, creator additionally played out the split tractable test. Creator infers that expansion in polystyrene volume, expands the voids when contrasted with the control blend. There is volume rot in the connection of compressive quality and volume portion of the EPS dabs. Creator additionally saw that decrease of the viable cross-area flexural tallness influenced by the EPS size part.

A Mandlik et al. 2013[10] played out the examinations solid blending in with and without the EPS dabs the cost viability is likewise concentrated by the creator , Author additionally infers that EPS concrete with no extraordinary holding specialist show great functionality and could undoubtedly be compacted and wrapped up. EPS can be utilized as a substitute for the coarse total and positive application is performed by the creator .EPS dabs can be an utilized as substitute in a non basic part.

Wenboshi et Al. 2015[11] performed tries and presumed that strength of the EPS concrete was acquired by making correlation between example when applying cyclic heap of 40KN, 50KN, and 60KN for 50000 or multiple times. Creator reasoned that hydroxyl-propyl cellulose is blended in the EPS concrete for improving the functionality of the grout and the impact of its blending proportion on the solid compressive quality is considered. To build the compressive quality the polymer emulsion is blending in the solid grout which bond different blends and connection between blending proportion and compressive quality is examined by creator.

4. Scope of The Study

An undertaking is made to take a gander at the it is lighter than the standard concrete with a dry thickness of 300 kg/m^3 up to1840 kg/m^3.The rule specialties of lightweight concrete are its low thickness and warm conductivity. The preliminary assessment mechanical properties of concrete made

IOP Conf. Series: Materials Science and Engineering 1017 (2021) 012009 doi:10.1088/1757-899X/1017/1/012009

with broadened polystyrene spots (EPB) as a fragmentary replacement of coarse and fine all out in degrees of 0%, 10%, 20%, 30%, 40% and half in the water solid extent of 0.50 was concentrated with silica fume as a deficient replacement of cement.

4.1 Puzzle Statement

Right now a huge number of huge amounts of waste polystyrene is delivered on the planet. This will at last reason contamination and is unsafe to the biological system. Public and worldwide natural guidelines have become more firm progressively which have made it costly to arrange. Thusly utilizing waste polystyrene in concrete production not just takes care of the issue of arranging this super light strong waste yet additionally helps safeguard common assets.

4.2 Purpose With the snappy unforeseen development and inventive augmentation, the need of substitutes for absolute in concrete has extended. Bit by bit new materials are being used as replacement of sums in strong turn of events, for instance, expanded glass, broadened polystyrene spots, etc. A result must be dictated by taking EPS touches and various tests have coordinated on it following 7 and 28 days independently to evaluate the properties of light weight strong squares. Lightweight concretes (LWCs) can be used in various advancement fields. EPS specks can be used to make low thickness concretes required for building applications like cladding sheets, Partition dividers, composite ground surface structure and weight bearing strong squares.

4.3 Objectives: This task targets accomplishing the accompanying:

- To use and supplant the assets to spare assets and condition.
- To research the physical properties of various materials to be utilized in work.
- Provide shield to the earth by using waste appropriately.
- Use of mechanical waste in a valuable way and give practical development material.

5. Methodology

To accomplish the goal of present examination, broad and thorough test program has been arranged. The whole examination has been ordered into different unmistakable periods of work for through and methodical methodology.

The materials utilized for planning concrete are chosen from those by the regular solid industry. Materials used for LWC using EPS spots are Crush sand stone, solid, Fly trash, EPS dabs and engineered admixtures. LWC can be planned and built utilizing an expansive scope of cementing materials, and this is basic for LWC to pick up prevalence.

Cement: The term concrete is commonly used to imply powdered materials which make strong concrete qualities when gotten together with water. These materials are all the more appropriately known as pressure driven concretes, Portland concrete being the most significant in development Cement is a fine grayish powder which, when blended in with water, frames a thick glue. 53 evaluation Ordinary Portland concrete adjusting to BIS 12269-1987 is utilized.

Fly ash: Fly debris or Pulverized fly debris is a development from the consuming of walloped coal accumulated by mechanical separators, from the fuel gases of warm plants. The synthesis changes with kind of fuel expended, load on the warmer and sort of separation. The fly flotsam and jetsam contain round shiny particles reaching out from 1 to 150 micron in estimation and besides experiences a 45-micron sifter. The mixture properties of fly flotsam and jetsam are referred to underneath satisfactory strong quality, sums should be hard and strong, freed from disastrous contaminating impacts, and falsely consistent. Sensitive and porous stone can confine quality and wear resistance, and a portion of the time it may moreover isolate during mixing and unfairly impact usefulness by growing the proportion of fines.

IOP Conf. Series: Materials Science and Engineering 1017 (2021) 012009 doi:10.1088/1757-899X/1017/1/012009

Crushed Sand Stone: Cementing sands reasonable for LWC are squashed sand, adjusted sands and Siliceous sand and calcareous sands can be utilized. The amount of fines under 0.125 mm is to be considered as powder. A base measure of fines (emerging from the fasteners and the sand) must be accomplished to dodge isolation.
Chemical admixture: Polycarboxylate ether (PCE) type is extremely effective dispersants for Calcium Aluminate concrete. They give glorious convenience to the material in the new state, and phenomenal physical properties in the cemented state.
Water: Water is utilized for blending and restoring according to IS 456:2000. From sturdiness thought water concrete proportion ought to be confined as in the event of typical cement and it ought to ideally be under 0.4 are tried for their significant properties before using them for making concrete.

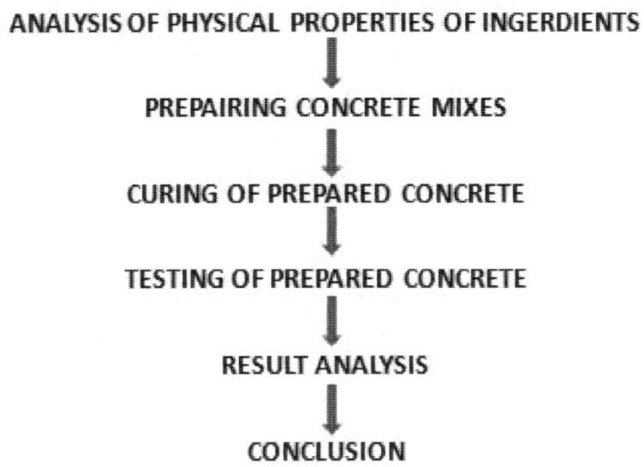

Figure 2: Path of Work

6. Procedure of EPS Light weight blocks

A. Materials and mix proportions

The materials utilized in this investigation were conventional Portland concrete adjusting to BS12: 1991, waterway sand with a fineness modulus of 2.85, squashed rock with a greatest size of 10 mm, accessible round EPS dots. The M20 evaluation of solid blend was utilized in this examination.

B. Properties of EPS Light weight blocks

Table 1: Properties of EPS Light weight blocks

Property Average value Density	13kg/m^3
Compressive strength	0.09MPa
Flexural strength	0.21MPa
Water absorption	4%by volume
Specific gravity of cement	3.15
Specific Gravity of fine aggregate	2.40
Specific Gravity of EPS beads	0.011
Fineness modulus of fine aggregate	3.00
Bulk Density of Fine Aggregate	1643 kg/m^3

C. Mixibility: A strategy like „sand-wrapping" was applied on the EPS specks. EPS spots were wetted from the outset with 30% of the mixing water and a short time later the remainder of the materials is incorporated. Mixing was continued until a uniform and streaming mix was gotten. The Cement to EPS extent was kept 1:1 by volume, the thickness of hardened concrete lessened 76, 57, 47, and 39%, separately, when diverged from control test, 1:1 extent of cement to EPS was picked and since EPS is hydrophobic, it has been represented that water maintenance is zero regardless, when the globules are immersed in water reliably for one month.

D. Casting and Curing: Various test examples EPS concrete was set up at various level of EPS dots (by vol. of coarse totals). The solid was intended for M25 blend plan according to the IS 10262: 1982. The 3D shape example of size 15x15x15 cm were set up at 5%, 10%, 15%, 20%, 25% and 30% of EPS (by vol. of coarse totals). Following 7, 14 and 28 days of restoring, they were tried for compressive quality.

7. Comparision Between Common Brick and Eps Concrete Blocks

1) The compressive quality of Light weight EPS solid squares is more than of mud block.
2) The water safe property of Light weight EPS solid squares is acceptable and it is valuable in development work.
3) The usefulness of Light weight EPS solid squares is acceptable as contrast with traditional blocks.
4) Rapid and moderately basic development is conceivable with the Light weight EPS blocks.
5) Economical as far as transportation as well as decrease in labor utilizing EPS blocks.
6) Significant decrease of in general weight results in sparing auxiliary edges, footing or heaps utilizing EPS blocks.

8. Applications of Eps Concrete Blocks

1) It is generally utilized as free fill protection in workmanship development where it improves imperviousness to fire, decreases clamor transmission, don't spoil and termite obstruction.
2) It is utilized underway of warmth protected light divider board.
3) Used in the scaffold to forestall freezing.
4) Environmentally nurseries are the structures which are less unsafe to condition. Green house are comprised of the materials which produces less CO2 and are reusable. To lessen the measure of plastic on the earth utilization of plastic as EPS dots is a smart thought for ecologically green tasks. It gives warm protection to building which forestalls the vitality misfortune in structures so it is valuable in earth green tasks.

Figure 3: EPS blocks in walls

Figure 4: EPS blocks in step

5) Light weight concrete utilized in development of steps, windows, garden dividers, and so forth.

6) As street bedding EPS concrete is utilized as street bedding in inflexible asphalts. It very well may be utilized in base course of unbending asphalt.

Figure 5: EPS blocks in street divider Figure 6: EPS blocks in Tiles

7) It is utilized in development of rooftops sections, little houses with load bearing dividers, and so forth. Since, the quality of L. W. C.

Figure 7: EPS concrete in rooftops sections Figure 8: EPS concrete in little houses

8) EPS concrete can be used in geo-stabilization projects to improve load bearing capacity of sub grade. It can be used to support pavements and foundations.
9) EPS concrete can be used as sub grade for railroad tracks.
10) EPS concrete can be used in Flooring.

9. Conclusion

• Starting finding have shown that the lightweight strong using EPS globules has an alluring solidarity to be an elective improvement material for the advancement of section divider, trail, railing divider, bed concrete
• The nature of light weight strong using EPS touches are low for lower thickness mix. This achieved expansion of voids all through the model achieved by the Air entraining admixture. In this manner the decrease in compressive nature of the strong.
• The substitution by utilizing EPS has indicated a positive application as a substitute material in building nonstructural individuals, and it likewise fills in as an answer for EPS removal.
It is inferred that supplanting of totals with EPS dots is better and efficient. By supplanting the totals concrete gets lighter than standard cement. Strength of solid better assessed. This investigation gives a plan to utilize a waste material in development work. This investigation constrains us to utilize other waste material and check for development work.

10. Limitations

1) The quality of lightweight solid utilizing EPS dabs are low for lower thickness blend.
2) Density is one of the significant boundaries which can control numerous physical properties in light weight cement and it is for the most part constrained by the sum.
3) The utilization of different strands, for example, coconut shell filaments, plastic filaments AQUA poly fibers, etc. can expand the heap conveying limit of cement in pressure and flexure.
4) It has been found from test information which shows that the compressive quality relies upon the incorporations size of EPS dabs, the littlest the size the most elevated the presentation.
5) Mixing time is longer than customary cement to guarantee appropriate blending.

References

[1]. IS 383:1970 "Specifications for coarse and fine aggregates from natural sources for concrete", *Bureau of Indian Standards*, New Delhi.
[2]. IS 456: 2000, "Code of practice for plain and reinforced concrete", *Bureau of Indian Standards*, New Delhi.
[3]. IS: 10262: 1982, "Recommended guidelines for concrete mix design" *Bureau of Indian Standards*, New Delhi.
[4]. IS 2386:1963 "Methods of tests for aggregates for concrete", *Bureau of Indian Standards*, New Delhi.
[5]. Roshan Gawale1, Shubham Mishra. "Light Weight Concrete by using EPS Beads". *International Conference on Emerging Trends in Engineering and Management Research* (ICETEMR-16), ISBN: 978-81-932074-7-5.
[6]. Ankur Tayal, Gaurav Gupta, Praveen Choudhary. "Lightweight Concrete using Recycled Expanded Polystyrene Beads" *International Research Journal of Engineering and Technology* (IRJET) **Volume: 05 Issue: 05** | May-2018.
[7]. Jaydeep Singh, Sangeeta Dhyani. "Light Weight Concrete using EPS" *International Journal Of Research in Technology and Management* ISSN 2454-6240, www.ijrtm.com, **Volume 3 Issue 3**, June 2017
[8]. Suhad M Abd, Dhamya Gh.,Maan Hattem1 and Dunya Khalil. "Effective Replacement of Fine Aggregates by Expanded Polystyrene Beads in Concrete", *International Journal of Engineering Research and Science and technology*. ISSN 2319-5991, www.ijerst.com, **Vol. 5, No. 3**, August 2016
[9]. M Dhivakar Karthick, S. Bhagavath Ramya. "Experimental study on strength behavior of concrete block with partial replacement of EPS beads", *International Journal of Engineering Research & Technology* (IJERT). ISSN: 2278-0181, ICITMSEE – 2018, **Volume 6, Issue 10**.
[10]. Abhijit Mandlik, Tarun Sarthak Sood, Shekhar Karade, Sangram Naik, Amruta Kulkarni. "Light-weight concrete using EPS," *International Journal of Science and Research* (IJSR). ISSN (Online): 2319-7064
[11] K. Krishna BhavaniSiram, K. Arjun Raj "Concrete + Green = Foam Concrete", *International Journal of Civil Engineering & Technology* (IJCIET), **Volume 4, Issue 4**, July – August 2013, pp. 179 - 184, ISSN Print:0976 – 6308, ISSN Online: 0976 – 6316.
[12] K. Krishna BhavaniSiram "Foam Concrete – The Present Generation's Building Solution", *International Journal of Civil,Structural,Environmental and Infrastructure Engineering Research and Development* (IJCSEIERD), **Volume 3, Issue 4**, October 2013, pp. 59 - 62, ISSN:2249 – 6866.
[13] Ali J. Hamad "Materials, Production, Properties and Application of Aerated Lightweight Concrete", *International Journal of Materials Science and Engineering* **Vol. 2, No. 2** December 2014.
[14] Dr. N. Arunachalam, V. Mahesh, P. Dileepkumar, and V. Sounder "Development Of Innovative Building Blocks (TNSCST- Sponsored Research Project)" *IOSR Journal of Mechanical and Civil Engineering* (IOSR-JMCE), e-ISSN: 2278 - 1684, p-ISSN: 2320 - 334X, pp 01-07.
[15] F. Ansari. Practical Implementation of Optical Fiber Sensors in Civil Structural Health Monitoring. *Journal of Intelligent Material Systems and Structures*, 18(8):879-889, 2007.
[16] Aasif Ahmad Ganie, Shree Prakash, Light Weight Concrete (Partial Replacement Of Coarse Aggregate Using Polystyrene Beads) International *Journal For Technological Research In Engineering*, ISSN (Online): 2347 – 4718, **Volume 5, Issue 11**, July-2018.

[17] Soni Kumari, Development of Light Weight Geoblocks for WALL Building Units using EPS Beads and Fly Ash. *International Journal of Engineering Research & Technology* (IJERT), ISSN: 2278-0181, **Vol. 6 Issue 04**, April-2017.

Acknowledgement

I take a great pleasure and immense pride in presenting our project study report on 'Sustainable Construction Using EPS Beads In Light Weight Blocks To Form Innovative Foam Concrete As A Green Building Material'.

I wish to express our gratitude heartfelt thanks to my project guide, Dr Arun Patel. Head of Civil Engineering Department, RKDF University, Bhopal for his encouragement, advice and guidance throughout the study. In addition to this the experiences and knowledge of my guide have been motivation during the period.

My special thanks to our Director, RKDF University, Bhopal for extending all the possible help and allow me to use all the resources that are available in the institute. Finally my gratitude to all our faculty members, supporting staffs and friends who help us directly and indirectly in our work.

IOP Conf. Series: Materials Science and Engineering 1017 (2021) 012010 doi:10.1088/1757-899X/1017/1/012010

Role of micro- factors on microstructure and on the tribological performance of HVOF coatings: A review

[1]Alok Vats, [2]Amar Patnaik, [2]M L Meena, [3]Dinesh Shringi

[1]Mechanical Engineering Department, MBM Engineering College, Jodhpur, India
[2]Mechanical Engineering Department, MNIT Jaipur, India
[3]Mechnaical Engineering Department, MBM Engineering College, Jodhpur, India

E-mail: alokvats1091@gmail.com

Abstract. Mechanical components, whether stationary or moving are subjected to wear and corrosion. A combined effect of these two is even more deleterious. Thermal spray technique proves to be an effective solution to address these issues. HVOF coatings owing to their inherent characteristics like high hardness, adhesion, density (porosity <1%) its provide enhanced mechanical and tribological performance and so are widely used across industries. Micro- factors like presence of pores, oxides and cracks, matrix and binder properties, addition of Cr, cohesive strength and boundary regions etc. do influence the coating behaviour, besides the deposition parameters. The objective of this paper is to review the effect of above mentioned factors on the wear and corrosive performance of the coatings for industrial and research viewpoint.

1. Introduction

Thermal spray comprises of number of coating processes. These processes are deployed to deposit both the metallic and non-metallic coatings. They are grouped into three classes which include spray using flame, electric arc and plasma arc techniques. The energy from these sources is used to transfer the heat to the coating material. Due to the effect of heat the material in powdered, wire and rod form gets converted into molten or a semi-molten phase. Acceleration of particles takes place by process gas through nozzle. The first particles that strike the surface form bonds with it. The subsequent particles led to an increase in thickness. As a result, a lamellar structure is obtained [1]. The schematic of HVOF equipment and process has been shown in Figure 1 while, coating features are shown in Figure 2. Better micro- mechanical properties make HVOF a better choice. These properties include density, bonding strength, surface hardness, fracture toughness, Coating Thickness, Residual Stress, Excellent Wear and Corrosion Resistance and Fine Surface Finish etc. [2]. Due to the enhancement of the micro- mechanical and wear behaviour, the HVOF coatings are used in marine, high temperature and excessive wear environments. As a result, the application areas of HVOF coatings are many and quite diversified. They include Aircraft industry, Power plants, Automobile and Transportation sector, Hydraulic machinery, Petroleum, Printing and Paper industry, Manufacturing, Metal Processing and Textile Machinery etc. [2]. The reason for enhanced features of HVOF coatings is their compact and hard microstructure with relatively less porosity. It is this microstructure which provides the HVOF coatings its well-known coating characteristics. The microstructure of the HVOF coatings not only depends on the deposition and geometrical parameters like Spray distance, Fuel flow, Oxygen, Nitrogen and Air flow rate at specified pressures, Nozzle type and geometry, Feedstock powder, powder feed rate and combustion chamber pressure, but also on micro- factors like presence of pores, oxides and cracks, matrix and binder properties, addition of Cr, cohesive strength and boundary regions etc. The microstructure of these coatings may also contain the precipitates, hard phases and salts like borides and carbides. The conformance of these phases is done by XRD, EDAX, while the surface characterization is done by SEM analysis. In this paper a review on the role of these micro-factors in influencing the wear and corrosion performance and material behaviour at large, has been done.

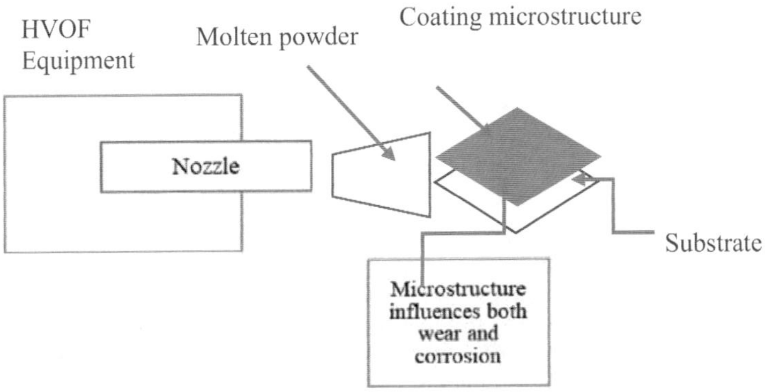

Figure 1. Schematic diagram showing HVOF coating process

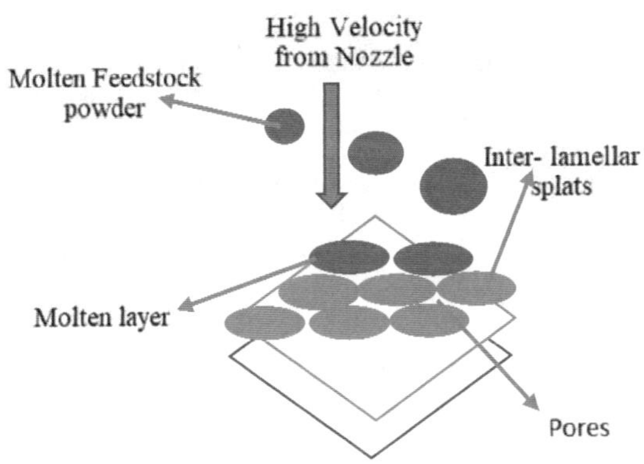

Figure 2. Schematic diagram showing HVOF coating features

2. Classification of Micro- Factors Influencing Coating Performance

The various micro- factors can be classified into three groups: Wear influencing factors, Corrosion influencing factors and Factors influencing both of them.

2.1 Corrosion influencing factors

2.1.1 Addition of Cr.

In the HVOF coatings, generally the inclusion of Chromium into WC-10Co-4Cr coating enhances the ability to resist corrosion. Because of the dissolution of W and Cr into Co binder localised corrosion can also occur. All this happens during the spraying process. So, different locations on a particular coating can give rise to different galvanic potentials and hence, corrode differently. It occured in the context of WC-17-Co coated samples, where the metal matrix's anodic potential was more than the WC particles (figure 3). The micro-galvanic potential between binder phase and hard WC particles resulted in corrosion of binder phase (Co) [3]. While, the insertion of Cr to WC-Co coatings could not uplift the erosion resistance. These coatings exhibited mainly ductile response to erosion at a velocity of 30 m/s. But the mode of erosion response changed to a composite mode at speeds higher than 30 m/s. This mode involved both the ductile and brittle modes [4]. Corrosion resistance can also be enhanced due to the compact micro-structure produced by the oxy- fuel process (HVOF). The presence of the Cr in the metal binder, as in the case of WC-10Co-4Cr, WC-40Cr_3C_2-25Ni-Cr, Cr_3C_2-25Ni-Cr and EHC improved performance in corrosive media. It is to be noted that the EHC coating also contained Cr. But micro-cracks in this coating had a deleterious effect on the corrosion performance [5].

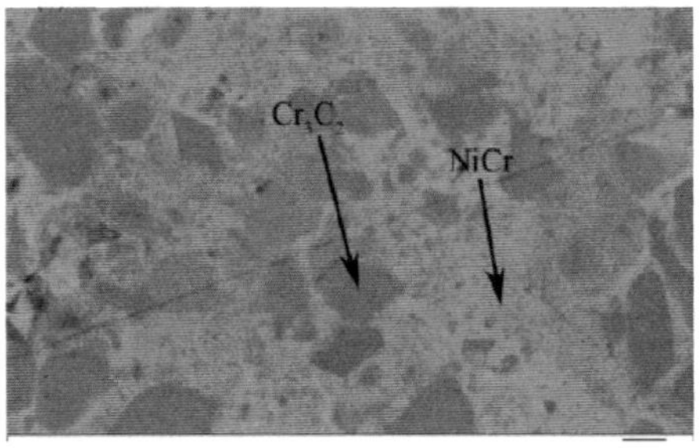

Figure 3. Diagram showing the Electron Micrograph of Cr_3C_2-25NiCr HVOF coating [3].

Figure 4. SEM micrograph of C21 ODS-NiCoCrAlY coating showing oxide layer [5].

2.1.2 Porosity, oxides and cracks.

Porosity influences corrosion greatly, as it occurred in the case of Cr_3C_2-25NiCr coating (Fig. 3). The coating had lower porosity and provided a superior resistance to corrosion than the WC-17Co and WC-10Co-4Cr coatings [3]. Similar situation also occurred with the M-6-(HV) coating during the cavitation-erosion testing. Crack started at the defects (pores, oxides, etc.) as the period of incubation was over [6]. Cracks in the wear scar of HVOF coating and the Ion implanted Ti-6Al-4V substrate caused the penetration of electrolytic media into the cracks. This sort of environment supported the formation of crevice corrosion [7]. Porosity also influenced the hardness level. This phenomena was particularly evident while comparing the CRCZ, CRCY and CRNY coatings [8]. The low porosity of oxy- fuel sprayed (HVOF) Al_2O_3–40%-TiO_2 and Cr_3C_2–20Ni-Cr coatings, led to dense micro-structure. These compact coatings also had nice adhesion to the base. As a result, bonding of coating with the substrate was also high. The oxide layer of Al_2O_3 and TiO_2 was observed in Al_2O_3–40%-TiO_2– coated samples. These coatings had the greatest corrosion resistance [9]. The number of holes in oxy-fuelled Cr_3C_2-25Ni-Cr coated samples (carbon steel substrate) were quite less in comparison to the plasma sprayed samples. The corrosion test was conducted in the sea water environment that is, 3.5% NaCl. It was found that the amount of pitting and crevice corrosion in the HVOF samples was less than plasma-coated ones [10] . The 250-micron WC–Co thick samples with NiCrAlY inter-coat had more number of pores. As a result, the electrolytic charged carriers penetrated into substrate by the way of these pores. It led to the delamination of coatings. The tendency to delaminate increased with increase in coating thickness's value. Consequently, the thick coating (450 microns) decreased the strength of bonding [11]. The corrosive properties of the Fe-based metallic (glass) coatings containing W, Mo, Ni and Co in definite amount were higher than that of the steel (1000 series) by nearly 5 times. Coatings having Co and Mo compositions had better resistance to corrosion, as compared to Ni and W coating. In this study, the oxide films acted as a barrier. The variation in the barrier film properties led to this difference [12]. In the Fe-based (amorphous) coatings the porosity level was lower in HVAF method (0.9 ± 0.1 vol%) as compared to HVOF ones (2.3 ± 0.3 vol%). It resulted in a better corrosion performance of the former. Also, the size of the pores was less in case of air- fuel (HVAF) coated samples. The large pore size in oxy- fuel (HVOF) coatings was caused by shrinkage. The particles in the complete molten state underwent shrinkage in oxy- fuelled coatings. Pores were created during this process. The velocity of impact and temperature, in combined form, enhanced the peening effect. It resulted in less number of pores [13]. Again, the lesser amounts of oxides in the HVAF sprayed FeCrMnSiNi and FeCrMnSiB coatings resulted in a greater inter-locking of particles. It resulted in higher hard-ness value of air-fuelled coatings compared to oxy-fuelled ones. The high speed particles in HVAF method generated a stronger peening effect. It increased the CR (Compressed residual stresses) and the surface hardness of the coating. There was lower splat's adherence in the areas of high oxide content and splashing droplets. Consequently, the weight loss was higher in these areas. Thus, the amount of oxide content of the HVAF coatings had a marked effect on the cavitation wear performance [4]. The MCls (Chloride salt of metal) were formed at 450 °C, during the study of corrosion resistance. There was a deposition of both HVOF and HVAF coatings. They had the approximate formulae of Cr_3C_2-25Ni-Cr (CC1), Cr_3C_2-50NiCr-Mo-Nb (CC625) and Cr_3C_2-37-WC-18Ni-Co-Cr (CW). The volatilisation of MCls was slow. The corrosion rate, at this stage was governed by the generation of solid oxo- metallates such as K_2CrO_4 [14]. The oxide scale in case of Cr_3C_2-20TiO_2 coated carbon steel after air oxidation of coated specimen was rich in Fe. But, due to absence of oxygen, there was no formation of Fe_2O_3. The absence of oxygen on the surface of this coating also indicated that oxidation was confined to the top most layer. This coating proved good at high temperature oxidation tests (figure 5) [15].

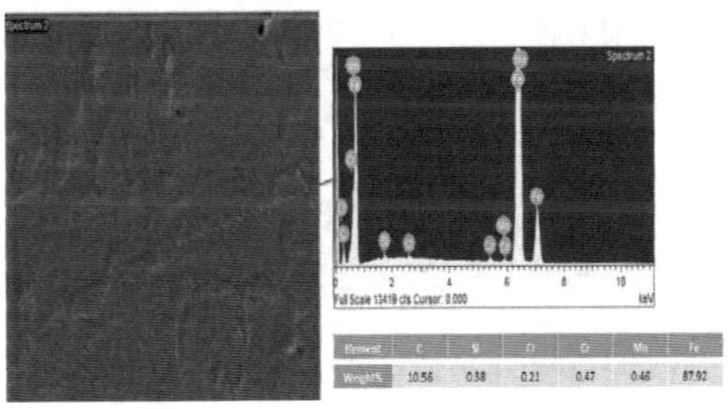

Figure 5. XRD of oxide scale in case of Cr_3C_2-20TiO_2 coated carbon steel [15].

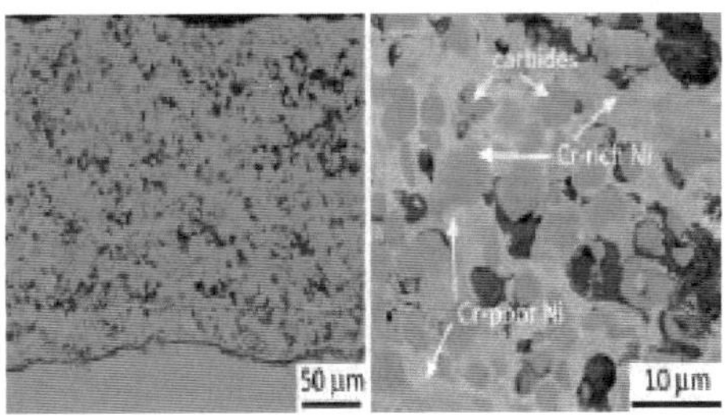

Figure 6. Schematic diagram showing HVOF coating features [20].

2.1.3 Diffusion paths.

Because of the lamellar structure, there were considerable amounts of diffusion paths available for the NiCoCrAlY coatings. Most of these paths passed through the grain boundaries between lamellae. By this, metallic ions (e.g., Ni- or Al-ion) diffused outwardly, towards the surface and formed a thicker oxide scale. In contrast, the nano-CeO_2/nano-Al_2O_3 ODS-NiCoCrAlY coating, had a uniform distribution and finer structure (of the oxide nanoparticles) in the lamellae and grain boundaries. Therefore, during exposure to high temperature oxidation test, the diffusion paths were reduced and therefore the outward diffusion of metallic ions became very limited [16]. The same is shown in figure 04.

2.1.4 The boundary regions and pre-existing defects.

It includes the inter particle boundary areas (high content of oxides, pores, and structural defects) and the inter splat boundary regions. They are the weak paths in the material. Removal of material takes place first at the boundary areas, as it occurred in [6]. It resulted in the erosion of the cavitation resistant Cavi-Tec particles. Similar situation occurred in the case of HVOF sprayed coating where delaminated areas were seen primarily at the inter-splat boundaries [17]. The F coating (highly brittle) was made of fine WC–10Co–4Cr particles. Micro-cracks easily initiated at the defective portions. These portions are the boundary regions between the layers and pores. The extension of micro-cracks along the boundaries of W_2C and W particles took place. It led to the formation of large scale pits. Delaminated wear areas were primarily observed on the F coating and was followed by micro-cutting. It indicated the highest magnitude of wear rate [18]. In the case of, WC-Co coatings, there was a loss of volume at the pores. Wear occurred primarily at lamellae, particularly at inter-faces. There existed a fairly strong co-relation b/w mechanical properties and the cavitation resistance [19]. The HVOF sprayed Cr_3C_2 - 174 -NiCr, Ni-Cr-B-Si and WC-Co-Cr coatings showed a lamellar microstructure. It also revealed flattened splats and pores present at inter-spat boundaries, as shown in figure 6 [20]. The pits were the key features in corroded samples of HVOF Fe-Cr-Mn-Co-Si coating. While larger pits were observed at the corroded surface of un-coated 304 SS [21]. A change in weight of the Fe-based (amorphous) HVOF coating in KCl media showed a great loss (between 96 to 168 h). The air-fuelled samples underwent less reduction in mass. The products of corrosive reaction in oxy-fuel coatings resulted in mass gain. While, the reduction of mass in air-fuelled samples was because of vaporization of MCls [13].

2.2 Wear influencing factors

2.2.1 Cohesive strength.

The mechanical properties like E (young's modulus) and fracture toughness are the properties which depend on cohesive strength at lamellar level. While hardness, relies on both inter- and intra-lamellar strength. Inter-lamellar cohesion plays a less important role in abrasion of oxy- fuel and air- fuel Cr_3C_2–Ni-Cr coatings [22]. Low porosity is also an indication of high cohesion, as in the case of Ni_3Ti and Ni_3Ti+(Cr_3C_2+20Ni-Cr) coatings (porosity 1 to 2%). The non- or partial melting of Ni_3Ti and Ni_3Ti +(Cr_3C_2+20Ni-Cr) particles led to less porous coatings. The oxy- fuel coatings had good cohesive properties. It was evident by less amount of porosity and larger deformation [23].

2.2.2 Matrix and Binder.

Presence and absence of binder also plays a significant role in determining wear performance. The Al_2O_3 (pure ceramic coating of highly brittle) coating on S355 possessed no ductile binder. It contributed to the enhanced mass and volume loss in erosion-corrosion conditions [24]. However, in the case of WC-12Co at 90 m/s, the generation of plastic grooves in the binder took place. It weakened the matrix mechanically and also resulted in un-supporting of WC grains. Consequently, the pull out of WC particles occured [25]. The metallic phase amount in the coating matrix is another key issue, as in the case of CRSF and MSN coatings. These coatings had a reduction in ceramic content, leading to a lower hardness than Cr_3C2-Ni-Cr and Cr_3C_2-CoNiCrAlY [8].

2.2.3 Precipitation hardening at high temperatures.

The precipitation of Cr_3Si (hard and stable) in the thermal-sprayed HEA (high entropy alloy) coating (800 °C temperature heat treatment) resulted in better wear resistance. Here, base material for comparison was bearing steel SUJ2 (12 m/mm^3 700 HV) [26].

2.2.4 Bonding strength.

includes the adhesion strength between coating and damaged work-piece as well as the coating cohesion strength. From dimensional recovery point of view, both of these are of great importance. Similar case occurred during the erosion wear of HVOF coatings at particle impact speeds of 3.61m/s. The angle of impact has little effect on coating deterioration. However, for speeds of the order of 9.33m/s, the highest values of coating wear were at 90° angle of impact [27]. An increase in the thickness of coating, impacted the bonding strength and tensile stresses [11]. The transverse cracking and slight porous nature justified the good adhesive features of oxy-fuelled coatings. In this study, metallic form of glass was sprayed on AISI 1035 steel. The coatings were 20 to 33 μm thick [12]. The bonding strength for plasma sprayed samples was 1500-2000 Psi which was lower than oxy-fuelled ones (2000-2700 Psi) [28]. In another case, the presence of NiCr bond coat provided good adherence with the substrate [29].

2.2.5 Hard and brittle phase.

The brittle phase such as W_2C was present in the oxy- fuelled WC coating which was nearly 360 μm in thickness. It led to an increase in the hardness from about 300 HV to 1120 HV. Consequently coatings' resistance was increased [30]. XRD analysis of WC–10Co–4Cr coated samples indicated the creation of compounds like WC & W_2C. Also, there was presence of Ni and Cr_7O_3 peaks in Ni–20Cr_2O_3 powder. Because of the formation of these hard phases the hardness of WC–10Co–4Cr coated surface was superior to Ni–20Cr_2O_3 coatings. In terms of wear Both coatings proved better than uncoated SS304 [31]. In order to characterize the performance of the coatings, the micro-hardness is considered as one of the most important property. On comparing with the base material, the hardness values of both Cr_3C_2-25ne of the important property NiCr and NiCrMoNb coatings enhanced by 173% and 10% respectively. The Cr_3C_2-25NiCr coating had higher hardness because of the presence of Cr_3C_2 hard phase in NiCr matrix [32]. Heat treatment was responsible for alteration of micro-structure of the coatings. In this treatment, amorphous phase was recrystallized, WC was decarburized and new eta phases were formed. Cracking took place in WC-12Co coatings owing to the excessive brittleness [33]. In the Ni- based coatings, like carbides and borides (hard phases) were embedded in the γ-Ni phases. The harder particles under gone less erosion. But the softer phase (γ-Ni) was displaced [34] .

2.3 Wear and corrosion influencing factors

2.3.1 Oxidation and decomposition.

The oxidized and decomposed forms of WC-Co-Cr coatings were more common in HVOF as compared to HVAF. As a result, a fine and uniform micro-structure was obtained for HVAF coating method. The HVAF coating underwent a small change in the composition of different phases. There changes were measured in relation to the feed-stock powder. There was the presence of W_2C in oxy-fuelled coating. The cohesion between the particles was low [35]. Due to the presence of carbide forming elements, there was a partial decomposition of MCs. It led to the formation of a wave like patterns [36]. In the case of oxy-fuelled and Air-fuelled Cr_3C_2-based coatings at high temperature (550°C), the main corrosion mechanism was electro-chemical oxidation of Cl^-. It led to the formation of volatile MCl. It paved the way for formation of voids. They created a number paths for corrosive media to attack the coating. The stability of HVAF sprayed Cr3C2-50Ni-Cr-Mo-Nb coating was the greatest. The amount of WC in air fuelled CW2 coatings was greater than the oxy-fuelled ones. WC had a higher reactivity at 550°C. Due to this, there was an enhancement in the rate of corrosion for air-fuelled coatings, as shown in fig. 7 [14]. The performance of HVOF sprayed WC–10Co4Cr cermet powder reinforced by Y (III) and Zr (IV) ceramics on X2CrNiMo-17-12-2 steel was studied. The erosion resistance of WC–10Co-4Cr coated material in fly-ash slurry media was increased to 1.45 times with the addition of ZrO_2. On the other hand, there was not much influence of addition of ZrO_2 (zirconia) for bottom-ash slurry. In the presence of 2H+ ions, the Zr^{2+} and $2O^{2-}$ ions were separated from each other. Owing to this oxidation of ZrO_2 phase the pits on coating surface were created. As a result, WC matrix was weakened [37]. The oxidation test was conducted in air at 900 °C, in the case of oxy-fuelled 718/Ni-Cr-Al-Y coating. This test led to the creation of different iron oxides like FeO, Fe_2O_3, and Fe_3O_4 on the cast iron substrate. The coating of the alloy undergone lower rate of

IOP Conf. Series: Materials Science and Engineering 1017 (2021) 012010 doi:10.1088/1757-899X/1017/1/012010

oxidation. The generation of oxides like $NiCr_2O_4$, Al_2O_3, Cr_2O_3 and TiO_2 led to high oxidation resistance [38]. The attack of corrosion started by actively oxidizing the binder, that is Co and Cr. It resulted in the establishment of micro-galvanic potential b/w the anodic binder and the cathodic phase. The binder phases corroded at the carbide/matrix interface. It was eventually preceded by the oxidation of dissolved W in the binder. The Formation of a passivation layer was composed of oxides of Cr, Co and tungsten [39].

2.3.2 Amorphous/ crystalline structure of coating.
Compared to AISI 1035 steel, of Fe-based (amorphous) coatings resulted in a 15-35 percent lower corrosion potential. Also, the coatings had 1.1-4.7 times greater polarisation resistance [12]. Meanwhile, the WC phase grew smaller in the HVOF spraying process of WC-10Co4Cr coatings. This phase was embedded in the completely melted Co/Cr phase, suggesting that the splats were in a half melted state. The size of the WC phase did not change in the LT-HVOF spraying process.

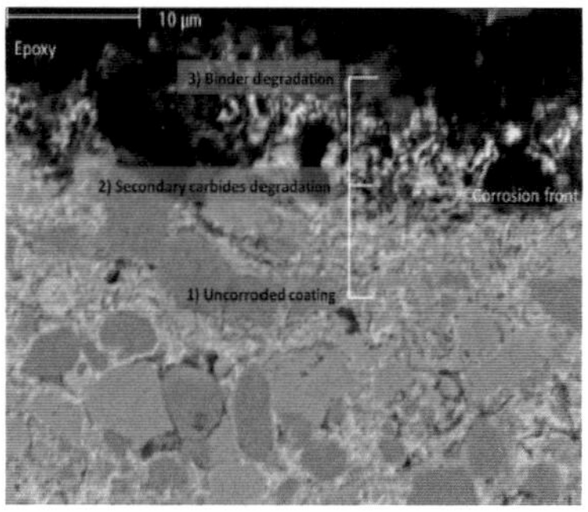

Figure 7. SEM showing details of corroded coating CC1.1 [14].

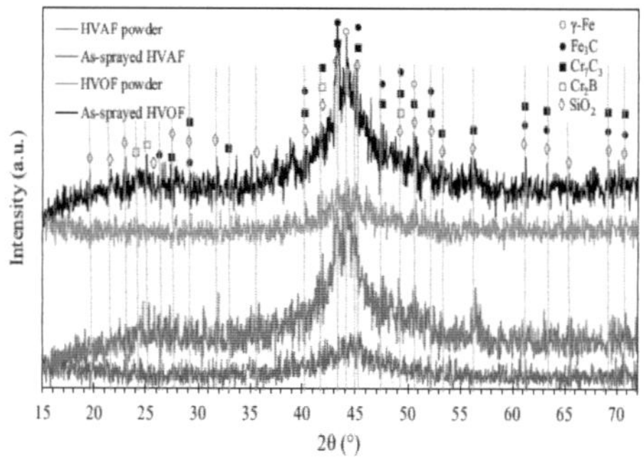

Figure 8. Peak analysis of the Fe-Cr-Ni--Mo-B-Si-C powder and coatings [13].

When deposition of splat was fast, the cracks were formed at the interface. It indicated that the splats were only slightly melted. The residual stress of the coating depended on a number of factors related to temperature and in flight features of particles. The building up of residual stress will create cracks [40]. The Nickel- based coatings containing calcium compounds were more wear resistant. Better hardness of unmelted cenospheres was the reason behind this [29]. The cobalt and chromium rich layer on oxy- fuelled WC-10Co-4Cr coating surface resulted in high wear. Also, binder was softer than the matrix. That is why, the binder itself was preferentially degraded [41]. While, in case of Fe-based (amorphous) coatings on the Mo-steel substrate, the crystalline compounds like solid solution of austenite, FeC_3, Cr_7C_3, SiO_2 as well as Cr_2B and Fe_2B were observed in the air- fuelled and oxy-fuelled coatings (fig. 8). Their formation was more in the HVOF coatings [13]. The HVAF coatings, for both FeCrMnSiNi and FeCrMnSiB alloys, had lower spraying temperatures, comparing to the oxy-fuelled coated ones [4]. As the time period for cavitation-erosion testing of oxy- fuelled Fe-based (amorphous/nano-crystalline) coated samples was increased, different wear regimes were observed. Material removal during the erosion process was classified into two types. They were the creation of the nano- crystallines and the spallation of the nano- crystallines. The amount of nano- crystallines increased significantly with testing time [42]. The HVOF sprayed Cr3C2-25 percent NiCr coating offered better resistance to corrosion than uncoated and Ni-20 percent Cr coatings. In both cases, the formation of small quantities of NiS and Cr_2S_3 and the lack of FeS showed that the Fe fid not diffuse externally. It resulted in a better resistance to corrosion. The alloying elements Ni, Cr, O, and C were present on the top of the scale [43]. In the oxy- fuelled, Fe-based coating, i.e. T800, a number of hard phases of carbides and borides were present. It was treated at a temperature greater than the temperature of the onset of crystallisation. T700 preserved the most of the amorphous structure even at elevated temperatures. While the Fe23 (C, B)6, Fe17Y2, Al_2O_3, Al13Fe4, and alpha-Fe crystalline phases appeared in large quantities for T800[44].

References

[1] T. S. Coatings, "Introduction to Thermal Spray Processing," 2004.
[2] S. Metco, "High Velocity Oxy-Fuel (HVOF) Solutions."
[3] S. Li *et al.*, "Corrosion behavior of HVOF sprayed hard face coatings in alkaline-sulfide solution," *Appl. Surf. Sci.*, vol. 416, pp. 69–77, 2017, doi: 10.1016/j.apsusc.2017.04.149.
[4] L. L. Silveira, A. G. M. Pukasiewicz, D. J. M. de Aguiar, A. J. Zara, and S. Björklund, "Study of the corrosion and cavitation resistance of HVOF and HVAF FeCrMnSiNi and FeCrMnSiB coatings," *Surf. Coatings Technol.*, vol. 374, no. June, pp. 910–922, 2019, doi: 10.1016/j.surfcoat.2019.06.076.
[5] Q. Wang, S. Luo, S. Wang, H. Wang, and C. S. Ramachandran, "Wear, erosion and corrosion resistance of HVOF-sprayed WC and Cr 3 C 2 based coatings for electrolytic hard chrome replacement," *Int. J. Refract. Met. Hard Mater.*, vol. 81, no. November 2018, pp. 242–252, 2019, doi: 10.1016/j.ijrmhm.2019.03.010.
[6] S. Lavigne, F. Pougoum, S. Savoie, L. Martinu, J. E. Klemberg-Sapieha, and R. Schulz, "Cavitation erosion behavior of HVOF CaviTec coatings," *Wear*, vol. 386–387, pp. 90–98, 2017, doi: 10.1016/j.wear.2017.06.003.
[7] V. Totolin, V. Pejaković, T. Csanyi, O. Hekele, M. Huber, and M. Rodríguez Ripoll, "Surface engineering of Ti6Al4V surfaces for enhanced tribocorrosion performance in artificial seawater," *Mater. Des.*, vol. 104, pp. 10–18, 2016, doi: 10.1016/j.matdes.2016.04.080.
[8] L. Baiamonte *et al.*, "Thermal sprayed coatings for hot corrosion protection of exhaust valves in naval diesel engines," *Surf. Coatings Technol.*, vol. 295, pp. 78–87, 2016, doi: 10.1016/j.surfcoat.2015.10.072.
[9] M. A. Zavareh, A. A. D. M. Sarhan, P. A. Zavareh, B. B. A. Razak, W. J. Basirun, and M. B. C. Ismail, "Development and protection evaluation of two new, advanced ceramic composite thermal spray coatings, Al_2O_3-40TiO_2 and Cr_3C_2-20NiCr on carbon steel petroleum oil piping," *Ceram. Int.*, vol. 42, no. 4, pp. 5203–5210, 2016, doi: 10.1016/j.ceramint.2015.12.044.
[10] H. Watson *et al.*, "Electrochemical corrosion behavior of carbon steel pipes coated with a protective ceramic layer using plasma and HVOF thermal spray techniques for oil and gas," *Ceramics International*, 2015, doi: 10.1016/j.ceramint.2015.10.134.
[11] K. Raghu Ram Mohan Reddy, N. Ramanaiah, and M. M. M. Sarcar, "Corrosion behavior of duplex coatings," *J. King Saud Univ. - Eng. Sci.*, vol. 28, no. 2, pp. 246–250, 2016, doi: 10.1016/j.jksues.2014.04.006.
[12] A. A. Burkov and P. G. Chigrin, "Effect of tungsten, molybdenum, nickel and cobalt on the corrosion and wear performance of Fe-based metallic glass coatings," *Surf. Coatings Technol.*, vol. 351, no. July, pp. 68–77, 2018, doi: 10.1016/j.surfcoat.2018.07.078.
[13] E. Sadeghi and S. Joshi, "Chlorine-induced high-temperature corrosion and erosion-corrosion of HVAF and HVOF-sprayed amorphous Fe-based coatings," vol. 371, no. January, pp. 20–35, 2019.
[14] D. Fantozzi, V. Matikainen, M. Uusitalo, H. Koivuluoto, and P. Vuoristo, "Chlorine induced high-temperature corrosion mechanisms in HVOF and HVAF sprayed Cr3C2-based hardmetal coatings," *Corros. Sci.*, vol. 160, no. February, 2019, doi: 10.1016/j.corsci.2019.108166.
[15] K. Premkumar, K. R. Balasubramanian, S. P. Sivapirakasam, and R. Shankar, "Oxidation resistance of Ni-Cr-TiO2 powder HVOF coating on carbon steel at elevated temperature,"

Mater. Today Proc., no. xxxx, pp. 1–7, 2020, doi: 10.1016/j.matpr.2020.01.206.

[16] F. Ghadami, A. Sabour Rouh Aghdam, A. Zakeri, B. Saeedi, and P. Tahvili, "Synergistic effect of CeO2 and Al2O3 nanoparticle dispersion on the oxidation behavior of MCrAlY coatings deposited by HVOF," *Ceram. Int.*, vol. 46, no. 4, pp. 4556–4567, 2020, doi: 10.1016/j.ceramint.2019.10.184.

[17] Š. Houdková, Z. Pala, E. Smazalová, M. Vostřák, and Z. Česánek, "Microstructure and sliding wear properties of HVOF sprayed, laser remelted and laser clad Stellite 6 coatings," *Surf. Coatings Technol.*, vol. 318, pp. 129–141, 2017, doi: 10.1016/j.surfcoat.2016.09.012.

[18] T. Gong *et al.*, "Influence of WC carbide particle size on the microstructure and abrasive wear behavior of WC-10Co-4Cr coatings for aircraft landing gear," *Wear*, vol. 362–363, pp. 135–145, 2016, doi: 10.1016/j.wear.2016.05.022.

[19] M. S. Lamana, A. G. M. Pukasiewicz, and S. Sampath, "Influence of cobalt content and HVOF deposition process on the cavitation erosion resistance of WC-Co coatings," *Wear*, vol. 398–399, no. December 2017, pp. 209–219, 2018, doi: 10.1016/j.wear.2017.12.009.

[20] S. Tkachenko *et al.*, "Wear of grinding rotors with thermally-sprayed coatings in a high-speed mill," *Wear*, vol. 412–413, pp. 49–59, 2018, doi: 10.1016/j.wear.2018.05.016.

[21] G. Y. Koga, "Corrosion and wear properties of FeCrMnCoSi HVOF coating," 2018, doi: 10.1016/j.surfcoat.2018.10.101.

[22] G. Bolelli *et al.*, "Sliding and abrasive wear behaviour of HVOF- and HVAF-sprayed Cr_3C_2-NiCr hardmetal coatings," *Wear*, vol. 358–359, pp. 32–50, 2016, doi: 10.1016/j.wear.2016.03.034.

[23] N. C. Reddy, B. S. A. Kumar, H. N. Reddappa, M. R. Ramesh, P. G. Koppad, and S. Kord, "HVOF sprayed Ni_3Ti and Ni_3Ti+(Cr_3C_2+20NiCr) coatings: Microstructure, microhardness and oxidation behaviour," *J. Alloys Compd.*, vol. 736, pp. 236–245, 2018, doi: 10.1016/j.jallcom.2017.11.131.

[24] T. Peat, A. M. Galloway, A. I. Toumpis, and D. Harvey, "Evaluation of the synergistic erosion-corrosion behaviour of HVOF thermal spray coatings," *Surf. Coatings Technol.*, vol. 299, pp. 37–48, 2016, doi: 10.1016/j.surfcoat.2016.04.072.

[25] N. Vashishtha, R. K. Khatirkar, and S. G. Sapate, "Tribological behaviour of HVOF sprayed WC-12Co, WC-10Co-4Cr and Cr_3C_2–25NiCr coatings," *Tribol. Int.*, vol. 105, pp. 55–68, 2017, doi: 10.1016/j.triboint.2016.09.025.

[26] W. L. Hsu, H. Murakami, J. W. Yeh, A. C. Yeh, and K. Shimoda, "On the study of thermal-sprayed Ni0.2Co0.6Fe0.2CrSi0.2AlTi0.2 HEA overlay coating," *Surf. Coatings Technol.*, vol. 316, pp. 71–74, 2017, doi: 10.1016/j.surfcoat.2017.02.073.

[27] M. A. R. Mojena, A. S. Roca, R. S. Zamora, M. S. Orozco, H. C. Fals, and C. R. C. Lima, "Neural network analysis for erosive wear of hard coatings deposited by thermal spray: Influence of microstructure and mechanical properties," *Wear*, vol. 376–377, pp. 557–565, 2017, doi: 10.1016/j.wear.2016.12.035.

[28] M. A. Zavareh, E. Doustmohammadi, A. A. D. M. Sarhan, R. Karimzadeh, P. Moozarm Nia, and R. S. Al/Kulpid Singh, "Comparative study on the corrosion and wear behavior of plasma-sprayed vs. high velocity oxygen fuel-sprayed Al8Si20BN ceramic coatings," *Ceram. Int.*, vol. 44, no. 11, pp. 12180–12193, 2018, doi: 10.1016/j.ceramint.2018.03.275.

[29] M. Doddamani, M. Mathapati, and M. R. Ramesh, "Microstructure and tribological behavior of plasma sprayed NiCrAlY/WC-Co/cenosphere/solid lubricants composite coatings," *Surf. Coatings Technol.*, vol. 354, no. September, pp. 92–100, 2018, doi: 10.1016/j.surfcoat.2018.09.018.

[30] C. Zheng *et al.*, "Experimental study on the erosion behavior of WC-based high-velocity oxygen-fuel spray coating," *Powder Technol.*, vol. 318, pp. 383–389, 2017, doi: 10.1016/j.powtec.2017.06.022.

[31] J. Singh, S. Kumar, and S. K. Mohapatra, "Tribological analysis of WC–10Co–4Cr and Ni–

$20Cr_2O_3$ coating on stainless steel 304," *Wear*, vol. 376–377, pp. 1105–1111, 2017, doi: 10.1016/j.wear.2017.01.032.
[32] V. Sreenivasulu and M. Manikandan, "High-temperature corrosion behaviour of air plasma sprayed Cr3C2-25NiCr and NiCrMoNb powder coating on alloy 80A at 900 °C," *Surf. Coatings Technol.*, vol. 337, no. January, pp. 250–259, 2018, doi: 10.1016/j.surfcoat.2018.01.011.
[33] N. Vashishtha, S. G. Sapate, P. Bagde, and A. B. Rathod, "Effect of heat treatment on friction and abrasive wear behaviour of WC-12Co and Cr3C2-25NiCr coatings," *Tribol. Int.*, vol. 118, pp. 381–399, 2018, doi: 10.1016/j.triboint.2017.10.017.
[34] M. E, H. X. Hu, X. M. Guo, and Y. G. Zheng, "Comparison of the cavitation erosion and slurry erosion behavior of cobalt-based and nickel-based coatings," *Wear*, vol. 428–429, no. September 2018, pp. 246–257, 2019, doi: 10.1016/j.wear.2019.03.022.
[35] R. E. Kumar, M. Kamaraj, S. Seetharamu, and A. K. S., "A pragmatic approach and quantitative assessment of silt erosion characteristics of HVOF and HVAF processed WC-CoCr coatings and 16Cr5Ni steel for hydro turbine applications," *Mater. Des.*, vol. 132, pp. 79–95, 2017, doi: 10.1016/j.matdes.2017.06.046.
[36] L. Janka, J. Norpoth, C. Tomastik, V. Matikainen, and P. Vuoristo, "Laser-induced precipitation of (W1- xCrx)2C: Mixed carbide phase for improved wear resistance of thermally sprayed hardmetal coatings," *Surf. Coatings Technol.*, vol. 337, pp. 177–185, 2018, doi: 10.1016/j.surfcoat.2017.12.061.
[37] J. Singh, S. Kumar, and S. K. Mohapatra, "Tribological performance of Yttrium (III) and Zirconium (IV) ceramics reinforced WC–10Co4Cr cermet powder HVOF thermally sprayed on X2CrNiMo-17-12-2 steel," *Ceram. Int.*, vol. 45, no. 17, pp. 23126–23142, 2019, doi: 10.1016/j.ceramint.2019.08.007.
[38] H. Vasudev, L. Thakur, A. Bansal, H. Singh, and S. Zafar, "High temperature oxidation and erosion behaviour of HVOF sprayed bi-layer Alloy-718/NiCrAlY coating," *Surf. Coatings Technol.*, vol. 362, no. January, pp. 366–380, 2019, doi: 10.1016/j.surfcoat.2019.02.012.
[39] J. A. Picas, M. Punset, E. Rupérez, S. Menargues, E. Martin, and M. T. Baile, "Corrosion mechanism of HVOF thermal sprayed WC-CoCr coatings in acidic chloride media," *Surf. Coatings Technol.*, vol. 371, pp. 378–388, 2019, doi: 10.1016/j.surfcoat.2018.10.025.
[40] T. Han, C. Deng, X. Zhang, and Q. Liu, "A model of splats deposition state and wear resistance of WC-10Co4Cr coating," *Ceram. Int.*, vol. 44, no. 4, pp. 4230–4236, 2018, doi: 10.1016/j.ceramint.2017.12.003.
[41] G. A. Ludwig, C. F. Malfatti, R. M. Schroeder, V. Z. Ferrari, and I. L. Muller, "WC10Co4Cr coatings deposited by HVOF on martensitic stainless steel for use in hydraulic turbines: Resistance to corrosion and slurry erosion," *Surf. Coatings Technol.*, vol. 377, no. January, 2019, doi: 10.1016/j.surfcoat.2019.124918.
[42] L. Qiao, Y. Wu, S. Hong, and J. Cheng, "Ultrasonic cavitation erosion mechanism and mathematical model of HVOF sprayed Fe-based amorphous / nanocrystalline coatings College of Mechanics and Materials , Hohai University , 8 Focheng West Road , Key Laboratory of Nuclear Materials and Safety Asses," *Ultrason. - Sonochemistry*, 2018, doi: 10.1016/j.ultsonch.2018.11.010.
[43] S. M. Muthu and M. Arivarasu, "Investigations of hot corrosion resistance of Hvof coated Fe based super alloy a-286 in simulated gas turbine environment," *Eng. Fail. Anal.*, p. 104224, 2019, doi: 10.1016/j.engfailanal.2019.104224.
[44] F. Huang *et al.*, "Effect of heat treatment on erosion- corrosion of Fe- based amorphous alloy coating under slurry impingement," 2019, doi: 10.1016/j.jallcom.2019.153132 .

Schematic review of plasma arc oxidation process for Mg Alloy Bio Implants.

Priya Jadhav[1], Arunkumar Bongale[1] and Satish Kumar[1]

[1]Department of Mechanical Engineering, Symbiosis Institute of Technology (SIT), Symbiosis International (Deemed University), Lavale, Pune - 412 115, Maharashtra State, India.

priya.jadhav@sitpune.edu.in

Abstract. The biomaterials are having diverse applications in temporary implants. The implant performance and compatibility inside the human body element are enormous factors of research. The implant material requires excellent mechanical properties that can endure corrosion, higher stress, load-bearing capacity. The compatibility with local tissues of the human body is a crucial point of tolerance. This article is based on a review of papers in the area of Magnesium alloy bioimplants. The purpose of the review is to understand different methods and recent developments for Mg alloy material in the domain. Different methods for producing coatings and the effect of the addition of nanoparticles are discussed.

1. Introduction

The biomaterial plays a vital role in the implants. They either support or replace human tissue. Hence they should intend the same properties as that of bone and tissue. In most of the implants, the function desired is to give stability at the point of contact between the bones in the process of healing or replace the joints permanently. The metals are widely used for biomaterial as it can be configured with the same characteristics of bone by improving its material characterization. The review only focuses on Mg alloy and its compatibility to be used as a biomaterial.

The mg alloy has mechanical strength required for implants with the ability to control the corrosion rate by alloying with different elements. To improve the corrosion resistance of magnesium implants different surface modification methods of coatings are used. The important aspect of bio implant is an antibacterial surface treatment and different methods to produce such surface coatings. The coating is nothing but a surface modification technique ether by improving properties if the existing one or applying different material on the base metal.

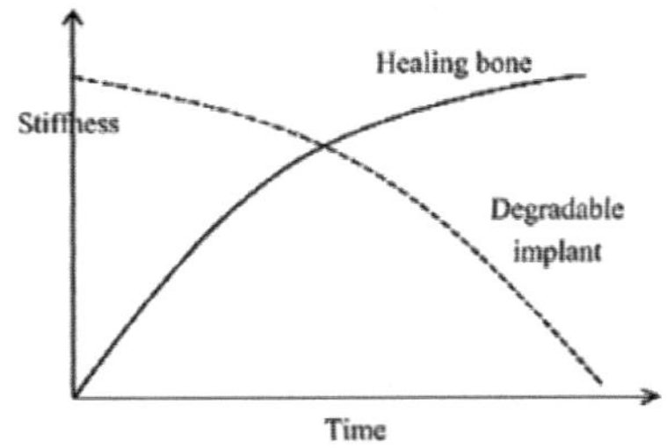

Figure 1 The gradual decrease in stiffness of biodegradable magnesium implant and concurrent healing zone [1]

Table 1 shows different materials used for bio-implants as per various locations in the human body. Apart from metals Polymers are also widely used for implants however they have some limitations like post-implantation recoiling which is difficult in the case of cardiovascular implants. Table shows different methods of producing coatings on the implants.

Table 1 Different metals used for bio implants

Material	Application	Reference
Ti alloy	Used for producing pins, bolts, screws	[1]
Mg alloys	Orthopaedic implants	[2]
Stainless steel	Long term/ Permanent implants	[3]
Co –Cr alloy	Total joint replacement	[4]

2. Required Material properties for orthopedic implants

The requirement of the orthopedic implant is high stability considering various forces inside the body. As the implant has to sustain at least up to the healing process, the biocompatibility is required to be higher. Hence the tissue friendly biomaterials are preferred for implants. However, these implants face degradation due to a longer presence in the bone and due to tissue environment. The characteristics desired are mechanical properties similar to bone, biocompatibility with surrounding tissues, uniform degradation behavior.

3. Alloying elements in implants

The addition of Copper enhances corrosion rate when used in NaCl environment. Still, the concentration of copper needs to be limited as the excess amount can cause deleterious biological effects such as neurodegenerative disease and specially cytotoxicity.It is added to improve the corrosion resistance of the implant. The Mn is preferred because of its essential bio element in the human body. It plays many roles within a cellular system such as different co-factors for many metalloenzymes. On the other hand excess use of the Mn leads to a neurological disorder like Almizer.Aluminum is considered as the material with good mechanical properties and corrosion resistance. Due to low weight and passivating element, it's known with good corrosion resistance. On the demerit side, the use is limited since in high doses it causes neurotoxicity which alters functioning due to the brain-blood barrier. The addition of Zinc increases yield strength in the base material. Magnesium alloys are preferred as implants for high Young's modulus 3-20 GPa which is similar to

Young's modulus of Bone 20 Gpa. When Mg combined with Zinc it works effectively against the hydrogen gas evaluation. Study shows that the Zn rich Mg alloy rarely form hydrogen gas during simulation in body fluids[8].It is less detrimental than other elements such as Mn, Al due to high absorbing ability in human body cell. On contrast the higher amount of Zn can damage to stomach lining cell[9]

Table 2 Different metals used in orthopedic implants using various techniques of the coating

Material	Technique for processing	Properties	Reference
Ti Alloy	Anodizing	High corrosion resistance	[1]
Stainless steel	Phase reversion	Superior corrosion resistance for wear and corrosion resistance for long stability of the implants	[5]
Co Cr Mn alloy	Heat treatment	High wear and corrosion resistance, biocompatibility	[6]
HA	Sol gel synthesis	Improved densification and biocompability	[6]
CaP	Sol gel combustion method	Good viability, integration with enamel	[7]

4. Review of literature

The research is carried on the different classical biomaterials as well as non-metallic Nano phase materials. The study is mainly carried out on different base metals , different types of implants, different coating methods, surface characterization , additives and suspended particles and addition of nano particles.Researchers studied the effect of the surrounding medium like aggressive ion, chloride, phosphate using Potetiometric polarization, impedance spectroscopy for various characteristics. From results, it has been proved that chloride ions can induce pitting corrosion. The presence of phosphate reduces corrosion. hydrogen carbonate ions increase the rate of corrosion but make the surface passivation.[10].Effect of the hydrothermal treatment with a different time in hours on MAO coated Mg alloy studied by the author [11]. Due to the hydrothermal treatment in Ca-EDTA solution and MAO coating, corrosion of the Mg alloy reduces boosting the healing process. Due to micro-arc oxidation, the layer of the magnesium alloy forms over the surface. This layer reacts with body fluid, prevents the peeling off due to the implant process. Hence the process is beneficial for orthopedic implants.

Song et. Al. [12]studied the compatibility of the Mg-Zn-Y-Nd-Zr alloys in a bone implant. Different in vitro tests were carried out to find the effect of the alloying on the implant. As the main objective is to find the cyctocompatibility of the implant, the samples were tested for cytotoxicity for degradation in implanting. The metallic irons released by alloy promote the ostogenetic differentiation.As the calcium content in the body can be improved with the help of the degradable implant material along with the corrosion-resistant material. Hence achieving biodegradable material with a positive effect in the area where the implant is placed can be made. The material of the implant is taken as Mg-0.5Ca-xY, with a variation of Y content, which results in better compatibility, refining of microstructure, and reduces the corrosion. SEM images show that the improvement in the microstructure is due to the cubic structure of the Yttrium in Mg alloy. The alloy of Mg, Ca, and Y formed using the powder metallurgy by heating all elements 710-720 0C. The researcher only discusses the formation of the alloying of Y in different compositions and improvement in structures with the help of SEM images [13].The biomaterial can also be enhanced by the plasma electrolyte oxidation method, where the coating formed on Mg alloy is corrosion resistant. The author discussed the experimental evaluations of such type e of coated material for in vivo testing results. As compared to uncoated material, the micro-arc oxidation coating optimizes the degradation rate and can be used as a safe biomaterial [15].Few Mg alloys enhancement can be done by adding nanoparticles like Graphene. The research work shows that adding Graphene strengthens the alloy and reduces the wear rate. These types of implant materials can be used for the longer sustaining requirement of the part inside the body. Also, it helped to reduce the degradation of the implant material [16].

Most of the bio-implants formed using the powder metallurgy with heat treatment. The Mg alloy, Mg-Nd-Zn-Zr, developed by the heat treatment and then immersed in NaCl solution for a longer time. The effect of the soaking for the different periods discussed for the corrosion behavior of the alloy metals. The material formed with LSPO (Long stacking period ordered) structure shows a slower corrosion rate [17].The alloy AZ31 and AZ91 of Mg was formed by a micro-arc oxidation method to study the coatings for anti-corrosive properties. From experimental results, AZ91 Mg alloy found to be more resistant to corrosion as compared to AZ31. The researcher also concluded that once the corrosion started on AZ 91 material, it spreads faster due to differences in morphologies of the two alloys [18].Gao et al. worked on the addition of the Graphene oxide coatings for the bio implant material. The tests taken in body fluid shows that GO (Graphene oxide) shows considerable improvement in the corrosion resistance when along with hydroxyapatite (HA). As per experiments conducted, this has been found that hybrid coating of the GO forms nucleation and crystallization of HA, creating a uniform coating[19].

The coating of the Graphene oxide layer on Ti alloy was studied by researched to get the enhancement of the properties of the TI alloy. From experimental results, it has been found that Graphene shows considerable improvement in the corrosion resistance [1].The idea of bio absorbable stents were also proposed in 2006. Due to permanent joint the patient has to face the in stent restenosis. It causes due to reaction between the blood and stent surface and irritation of vascular tissue. These stents are first step towards the biodegradable temporary stents. The material get dissolved in long term. The experimental details of the Mg stents are mentioned by author [20].The composition of the Mg alloy as biomaterial is studied by adding the alloying element such as Mg-Gd-Y, Mg-Gd-Zn. These materials were tested in saline environment in cast state. Due to alloying of these rare earth element the biomaterial shows uniform corrosion resistance and good corrosion resistance in case of Mg –Gd alloy[21].

Further to improve the orthopedic implants the addition of alloying element Y is studied. The material system of Mg-0.5Ca-xY were studied for different mechanical testing like indentation micro hardness, micro scratch and stiffness. Due to alloying the friction coefficient get reduced resulting in improvement in the mechanical properties. The researcher also proposed further scope of experimentation by heat treatment of the implant [2].The orthopaedic biomaterial FeMnSi were studied for electrochemical behaviour of the coating of HA-ZrO_2 by pulsed laser deposition technique. The SEM testing was also carried out to find out the morphological properties. The SEM shows degradation of material during the polarization test. With heat treatment the phase change of the

IOP Conf. Series: Materials Science and Engineering 1017 (2021) 012011 doi:10.1088/1757-899X/1017/1/012011

alloying material was achieved to improve the characterization of the biomaterial like Mg-2Zn-0.5Nd-0.5Zr alloy. The increase in the solution temperature up to 500 ° C changes the phase increase grain size 4-5μm. Due to homogeneous material structure the material shows uniform degradation behaviour [22].Furthermore, the study on the addition of Gd in Mg alloy was done by Yao et.al. by varying percentage content of Gd in Mg-2 Zn-0.5 Zr-xGd for the study of microstructure and biocorrosion behavior. Gd is proven to be material adding good corrosion resistance. Hence the different proportion of material was tested for improvement in mechanical properties and anticorrosive performance. A particular weight percentage the nanoscale rod-like phases of (Mg Zn) 3Gd found that give better ultimate strength and yield strength. Also, the significant corrosion resistance with the uniform rate of corrosion was observed [14].

The addition of nanomaterial graphene on Ti alloy is also considered as the potential area. Instead of adding Graphene particle in suspended form Asgar et al. studied the effect of the formation of the graphene oxide layer on the base metal. The material characterization is analyzed using IR ,X ray diffraction. It was observed that at a temperature around 200 OC , the reduction in GO coating occurs. This showed good corrosion resistance on potentiometer at the reduced contact angle[1].Adhesion of coating is also an important factor in coating deposition. An increase in adhesion is effective in the chemical bonding and biodegradation behavior of the implant. The process of improvement in adhesion is called as surface activation. This is achieved by pretreatment methods. In the case of magnesium the increase in surface adhesion results in a reduction of the release of Mg ions. Surface treatment of CoCrMo also results in the release of harmful ions to the body in vivo testing. For Titanium alloy the surface treatment method improves the formation of the oxide layer, adhesion, and bonding of hydroxyapatite coating [8].

The mg oxides are stronger ions formed during the formation of coating which enhances the mechanical properties. To increase its effect the metal phenolic networks are formed by deposition of nano-sized metal oxides using a chemical reaction between the metal substrate and phenolic ions. The researcher studied the formation of magnesium oxide films on mg based substrate using, MG phenolic network. The results showed that the concentration of Mg alloy is three times higher than the uncoated samples. In vitro tests showed that osteocompability significantly changes due to the increase in the Mg ions in the coating. The coating also reduces the formation of gas cavities around the implantation area increasing biocompatibility of the material [23].

The synthesis of the double-layered fluorine-doped hydroxyapatite or polycaprolactone coating on Mg-2Zn-3Ce was done by electro deposition and a dip-coating method. The result showed the uniform layer of nanostructured. This composite coating gives significant protection for Mg against corrosion. The immersion test carried out in body fluid for 10 days showed higher compressive strength improving the degradation resistance in body fluid. The bond strength also improved [24].The addition of nanoparticles in the alloying base material requires consideration of the success of implant, antibacterial activity, cell adhesion. Hence the nanomaterial design should be cell adhesive and wettability. The existing implants are costly and require a special facility to produce the material eg. 3D printing. Also, the implants' life can be improved along with reducing the pain and inflammation at the location in the body. The production of smart material also in the stage of development[25].Different review of the mg alloy as the orthopaedic implants are done by researchers to identify potential area on improvement for the Mg alloy[17], [26]–[28]. Researchers developed different experimental and numerical methods of improving characterization and surface topology [29],[30].With experimentation the efforts were also made to conduct the finite element analysis of implants in body fluid. This helps to determine the various stresses implant has to withstand and loading conditions [31].Few researchers also worked on the method of producing coatings by plasma arc electrolytic oxidation on Mg alloy[32]. The different parameters like electrolyte compositions, different input requirement of current, voltage and power supply. Also depending upon the composition of the electrolyte used and additives in different composition , the material characterization changes.[11], [18], [32]–[34].The oxidation coatings were developed by researchers to optimize characteristics of the surface modification .[35][36] .These coatings were analyzed using SEM analysis , XRD analysis to identify the distribution of alloying element and uniform coating distribution.[36],[37],[11], [33], [38]. Another type of biological testing is immersion testing in which samples are immersed in solution similar as body fluid [39]This helps to study the effect of the environment similar to body fluid on coating degradation. As the implant has to withstand up to healing of the joint it should not corrode easily up to specific time of healing. This gives the corrosion resistance of the material[40]. As study conducted by researcher for AZ31 mg alloy [33], AZ91D [30], MG-6Zn-Ca[38] which shows improvement in degradation behaviour due to alloying. The additives were added in base Mg alloy to strengthen the mechanical characteristics. The material like graphene was tested on Mg alloy as it forms the oxide layer on the coating as it promotes crystallization and nuclear growth of layer. Due to the mineralization of hydroxyapatite in modified SBF forms the dense structure of HA/GO hybrid coating. This promotes the corrosion resistance in the m- SBF[43].

5. Conclusion-

As per the literature review on the use of nanoparticles in the producing biomaterials, the few nanoparticles show significant improvement in the mechanical properties. The use of nanomaterial is increasing for the biomaterial due to properties enhancement. Different methods of producing coating also help to improve the properties regarding the application of biomaterial as per application location. Still the use of nanoparticles needs to be analyzed further to investigate the effect on the tissues of the body. When the implant is exposed to the body for a longer period, the nanoparticle has to be biocompatible enough to remain unaffected on tissues and the surrounding muscles of the body. Hence work can be done in finding out the biocompatibility of the materials like fullerene and graphene with alloys of Mg and Ti.Along with this, the material can also be tested for producing an oxide layer on the base material for improvement in corrosion resistance, wear-resistance, and other mechanical properties.To define the optimum value of the coating layer the parameters like electrode type, the distance between anode and cathode, nanoparticles, electrolytes, current and voltage requirements can be configured for a different set of combinations. This may help in producing a better oxide layer.

The main advantage of the magnesium and its alloy, if employed as an orthopedic implant, is its biocompatibility and the biodegradability properties, which ultimately can be used to avoid the need for a second surgery for the affected patients after implantation. But, the main drawback of these alloys is the rapid and rigorous corrosion, which is an inherent property of the magnesium and its alloys when they are in contact with the fluids with chloride content [16]. The human body fluids, such as blood plasma, consists of chlorides in it [17]. The magnesium orthopedic implants, when getting in contact with the bloodstream, they react with the chlorides and result in an alkaline environment and hydrogen liberation around the region. This will ultimately result in delayed healing of the tissues around the region and may lead to poisoning of the tissues also [18]. Concerning the mechanical integrity of the orthopedic implants, the corrosion also is responsible for the deterioration of mechanical properties of the implant leading to fracture and failure of the same before the complete healing has been achievedAlso, a comparison of Ti and Mg alloy at optimum configuration can be identified before in vivo testing of biomaterial

References

[1] H. Asgar *et al.*, "Functionalized graphene oxide coating on Ti6Al4V alloy for improved biocompatibility and corrosion resistance," *Mater. Sci. Eng. C*, vol. 94, pp. 920–928, 2018.
[2] C. Munteanu, Ștefan Lupescu, B. Istrate, V. I. Antoniac, M. Benchea, and A. Savin, "Some

Tribological Aspects of Mg-0.5Ca-xY Biodegradable Materials," *Key Eng. Mater.*, vol. 782, pp. 136–141, Oct. 2018.

[3] R. Müller *et al.*, "Surface engineering of stainless steel materials by covalent collagen immobilization to improve implant biocompatibility," *Biomaterials*, vol. 26, no. 34, pp. 6962–6972, Dec. 2005.

[4] M. Niinomi, "Metallic biomaterials," *Journal of Artificial Organs*, vol. 11, no. 3. pp. 105–110, Sep-2008.

[5] R. D. K. Misra, C. Nune, T. C. Pesacreta, M. C. Somani, and L. P. Karjalainen, "Understanding the impact of grain structure in austenitic stainless steel from a nanograined regime to a coarse-grained regime on osteoblast functions using a novel metal deformation-annealing sequence," *Acta Biomater.*, vol. 9, no. 4, pp. 6245–6258, 2013.

[6] K. Yamanaka, M. Mori, and A. Chiba, "Nanoarchitectured Co-Cr-Mo orthopedic implant alloys: Nitrogen-enhanced nanostructural evolution and its effect on phase stability," *Acta Biomater.*, vol. 9, no. 4, pp. 6259–6267, 2013.

[7] R. Tang and Y. Cai, "Calcium Phosphate Nanoparticles in Biomineralization and Biomaterials," *Nanotechnologies Life Sci.*, 2010.

[8] V. Huynh, N. K. Ngo, and T. D. Golden, "Surface Activation and Pretreatments for Biocompatible Metals and Alloys Used in Biomedical Applications," *Int. J. Biomater.*, vol. 2019, 2019.

[9] D. Persaud-Sharma and A. Mcgoron, "Biodegradable magnesium alloys: A review of material development and applications," *J. Biomim. Biomater. Tissue Eng.*, vol. 12, no. 1, pp. 25–39, 2012.

[10] Y. Xin, K. Huo, H. Tao, G. Tang, and P. K. Chu, "Influence of aggressive ions on the degradation behavior of biomedical magnesium alloy in physiological environment," *Acta Biomater.*, vol. 4, no. 6, pp. 2008–2015, Nov. 2008.

[11] S. Y. Kim, Y. K. Kim, M. H. Ryu, T. S. Bae, and M. H. Lee, "Corrosion resistance and bioactivity enhancement of MAO coated Mg alloy depending on the time of hydrothermal treatment in Ca-EDTA solution," *Sci. Rep.*, vol. 7, no. 1, pp. 1–11, 2017.

[12] X. Song *et al.*, "Investigation on the in vitro cytocompatibility of Mg-Zn-Y-Nd-Zr alloys as degradable orthopaedic implant materials," *J. Mater. Sci. Mater. Med.*, vol. 29, no. 4, Apr. 2018.

[13] B. Istrate, C. Munteanu, Ştefan Lupescu, V. I. Antoniac, and E. Sindilar, "Structural Characterization of Mg-0.5Ca-xY Biodegradable Alloys," *Key Eng. Mater.*, vol. 782, pp. 129–135, Oct. 2018.

[14] H. Yao, J. Wen, Y. Xiong, Y. Liu, Y. Lu, and W. Cao, "Microstructures, mechanical properties and corrosion behavior of as-cast Mg-2.0Zn-0.5Zr-xGd (wt %) biodegradable alloys," *Materials (Basel).*, vol. 11, no. 9, Aug. 2018.

[15] M. H. Song, W. J. Yoo, T. J. Cho, Y. K. Park, W. J. Lee, and I. H. Choi, "In vivo response of growth plate to biodegradable Mg-Ca-Zn alloys depending on the surface modification," *Int. J. Mol. Sci.*, vol. 20, no. 15, Aug. 2019.

[16] A. A. Hamzah, R. S. Selvarajan, and B. Y. Majlis, "Graphene for biomedical applications: A review," *Sains Malaysiana*, vol. 46, no. 7, pp. 1125–1139, 2017.

[17] G. Barati Darband, M. Aliofkhazraei, P. Hamghalam, and N. Valizade, "Plasma electrolytic oxidation of magnesium and its alloys: Mechanism, properties and applications," *J. Magnes. Alloy.*, vol. 5, no. 1, pp. 74–132, 2017.

[18] C. Liu, J. Yuan, H. Li, and B. Jiang, "Role of substrates in the corrosion behaviors of micro-arc oxidation coatings on magnesium alloys," *Metals (Basel).*, vol. 9, no. 10, 2019.

[19] F. Gao *et al.*, "Biomimetic synthesis and characterization of hydroxyapatite/graphene oxide hybrid coating on Mg alloy with enhanced corrosion resistance," *Mater. Lett.*, vol. 138, pp. 25–28, Jan. 2015.

[20] P. Erne, M. Schier, and T. J. Resink, "The road to bioabsorbable stents: Reaching clinical reality?," *CardioVascular and Interventional Radiology*, vol. 29, no. 1. pp. 11–16, Feb-2006.
[21] J. Kubásek, D. Vojtěch, and J. Čapek, "Properties of biodegradable alloys usable for medical purposes," 2012.
[22] J. Chen, S. Wei, L. Tan, and K. Yang, "Effects of solution treatment on mechanical properties and degradation of Mg-2Zn-0.5Nd-0.5Zr alloy," *Mater. Technol.*, vol. 34, no. 10, pp. 592–601, Aug. 2019.
[23] M. Asgari *et al.*, "Mg-Phenolic Network Strategy for Enhancing Corrosion Resistance and Osteocompatibility of Degradable Magnesium Alloys," *ACS Omega*, vol. 4, no. 26, pp. 21931–21944, 2019.
[24] H. R. Bakhsheshi-Rad, E. Hamzah, M. Kasiri-Asgarani, S. Jabbarzare, N. Iqbal, and M. R. Abdul Kadir, "Deposition of nanostructured fluorine-doped hydroxyapatite-polycaprolactone duplex coating to enhance the mechanical properties and corrosion resistance of Mg alloy for biomedical applications," *Mater. Sci. Eng. C*, vol. 60, pp. 526–537, Mar. 2016.
[25] S. Kumar, M. Nehra, D. Kedia, N. Dilbaghi, K. Tankeshwar, and K. H. Kim, "Nanotechnology-based biomaterials for orthopaedic applications: Recent advances and future prospects," *Materials Science and Engineering C*, vol. 106. Elsevier Ltd, 01-Jan-2020.
[26] S. Kamrani and C. Fleck, "Biodegradable magnesium alloys as temporary orthopaedic implants: a review," *BioMetals*, vol. 32, no. 2, pp. 185–193, Apr. 2019.
[27] S. Prasadh, V. Ratheesh, V. Manakari, G. Parande, M. Gupta, and R. Wong, "The potential of Magnesium based Materials in Mandibular reconstruction," *Metals (Basel).*, vol. 9, no. 3, pp. 1–20, 2019.
[28] C. Liu, Z. Ren, Y. Xu, S. Pang, X. Zhao, and Y. Zhao, "Biodegradable Magnesium Alloys Developed as Bone Repair Materials: A Review," *Scanning*, vol. 2018, 2018.
[29] A. Dehghanghadikolaei, H. Ibrahim, A. Amerinatanzi, and M. Elahinia, *Biodegradable magnesium alloys*, 2nd ed. Elsevier Ltd., 2019.
[30] Y. W. Song, D. Y. Shan, and E. H. Han, "Electrodeposition of hydroxyapatite coating on AZ91D magnesium alloy for biomaterial application," *Mater. Lett.*, vol. 62, no. 17–18, pp. 3276–3279, Jun. 2008.
[31] R. Jiang, D. J. Bull, D. Proprentner, B. Shollock, and P. A. S. Reed, "Effects of oxygen-related damage on dwell-fatigue crack propagation in a P/M Ni-based superalloy: From 2D to 3D assessment," *Int. J. Fatigue*, vol. 99, pp. 175–186, 2017.
[32] S. R. Paital and N. B. Dahotre, "Calcium phosphate coatings for bio-implant applications: Materials, performance factors, and methodologies," *Materials Science and Engineering R: Reports*, vol. 66, no. 1–3. pp. 1–70, 14-Aug-2009.
[33] P. Shi, B. Niu, S. E, Y. Chen, and Q. Li, "Preparation and characterization of PLA coating and PLA/MAO composite coatings on AZ31 magnesium alloy for improvement of corrosion resistance," *Surf. Coatings Technol.*, vol. 262, pp. 26–32, Jan. 2015.
[34] L. H. Li *et al.*, "Deposition of microarc oxidation-polycaprolactone duplex coating to improve the corrosion resistance of magnesium for biodegradable implants," *Thin Solid Films*, vol. 562, pp. 561–567, Jul. 2014.
[35] L. Ren, T. Wang, Z. Chen, Y. Li, and L. Qian, "Self-lubricating PEO–PTFE composite coating on titanium," *Metals (Basel).*, vol. 9, no. 2, pp. 1–11, 2019.
[36] K. Bai *et al.*, "Fabrication of chitosan/magnesium phosphate composite coating and the in vitro degradation properties of coated magnesium alloy," *Mater. Lett.*, vol. 73, pp. 59–61, Apr. 2012.
[37] G. Liu, J. Wang, K. Bian, and P. Zhu, "Preparation and characterization of nanostructured dibasic calcium phosphate coating on magnesium alloy wire," *Mater. Lett.*, vol. 209, pp. 323–326, Dec. 2017.
[38] S. Gaur, R. K. Singh Raman, and A. S. Khanna, "In vitro investigation of biodegradable polymeric coating for corrosion resistance of Mg-6Zn-Ca alloy in simulated body fluid,"

Mater. Sci. Eng. C, vol. 42, pp. 91–101, Sep. 2014.
[39] N. Bellotti, B. del Amo, and R. Romagnoli, "Progress in Organic Coatings," *Tara Tann. a Nat. Prod. with antifouling Coat. Appl.*, 2012.
[40] P. Tian, D. Xu, and X. Liu, "Mussel-inspired functionalization of PEO/PCL composite coating on a biodegradable AZ31 magnesium alloy," *Colloids Surfaces B Biointerfaces*, vol. 141, pp. 327–337, May 2016.

IOP Conf. Series: Materials Science and Engineering 1017 (2021) 012012 doi:10.1088/1757-899X/1017/1/012012

Influence of machining parameters on taper square areas during slant type taper profiling using wire electric discharge machining

I.V.Manoj[1], Narendranath S[1]

[1]Department of Mechanical Engineering, National Institute of Technology Karnataka, Surathkal 575025, India.

E-mail:vishalmanojvs@gmail.com

Abstract. The oblique/ tapered machining is a unique method in wire electric discharge machining to get taper profiles difficult to machine materials. These profiles have many applications like dies, nozzles, inserts, cutting tools and other components. In the present study, a simple square profile is machined using a slant type fixture to achieve a tapered profile made of Hastelloy X. A simple square is machined at different slant angles to get a 0°, 15° and 30° tapered profiles. This paper aims to study the effects of the machining parameters like wire guide distance, wire offset, corner dwell time and cutting speed override on the area of machined profiles. Both scanning electron microscope and coordinate measuring machine were used to measure the areas of 1mm and 5mm. It is observed that as the wire guide distance and cutting speed override increases the area decreases. Whereas in case of wire offset, as offset increases the areas also increase and corner dwell time doesn't affect on the area. As the angle increased the area also increased leading to bigger taper profiles.

1. Introduction

Wire electric discharge machining (WEDM) is an electrothermal process where the sparks are generated between the workpiece and electrode due to high potential difference. It is a non-contact machining process where all the conductive and hard materials can be precisely machined [1-3]. This forms a versatile way for machining for different materials like nickel, titanium alloys, composites etc. which can be used in many for complex machining [4-6]. The tapered cutting is unique machining used in dies, forming cutters, splicing moulds, vanes and blades etc. [7, 8]. During traditional taper cutting, there are many disadvantages like wire guide wear, friction, wire breakage, angular inaccuracies, poor surface due to insufficient flushing etc. have been observed during taper machining of materials [9-11]. Many kinds of research have been made to improve the material, mechanism and machining ways. Manoj et al. [12] have highlighted the effect of taper angles on different response parameters which was produced with a unique fixture. Naveed et al. [13] reported optimization of machining complex profile which involves curved features and inclined surfaces. It was observed that there was a 33.3% reduction in angular error, a 14.3% reduction in radial error, a 12% increase in cutting speed and a 14.4% reduction in surface roughness of workpiece.

Sharma et al. [14] have machined a complex profile slot where the surface and metallurgical effects of machining parameters were highlighted. Firouzabadi et al. [15] have reported convex, concave corner radii error in corner profiling for different corner angles and achieved a reduction of 17% in residual material thickness in convex curvatures. Werner [16] explored a method in machining curvilinear profiles on wire-electric discharge machines. The optimal machining parameters were found by modern CAD/CAM systems and tool travel study. Bisaria H and Shandilya P [17] has claimed that at lower pulse parameters, the corner error for 60°, 90°, and 120° profile was decreased by 43.38%,

31.12%, and 29.04%, respectively compared to high parameters. Mouralova et al. [18] explored the profile and area parameters of WEDMed aluminium. They have identified optimal parameters giving best surface quality and least kerf width. Wang et al. [19] machined micro gear profile with 50 μm wire and less than 1.5 μm machining error with an excellent surface (Ra= 0.9 μm). Mandal et al. [20] have employed wire offset value for making the wire travel in a proper path and it was also used to compensate the overcut value. Chaubey and Jain [21] have machined bevel and helical gears and showed that WEDM is the most economic and sustainable alternative. Dodun et al. [22] made an attempt to improve geometrical accuracy in corner machining of thin parts. It was concluded that the thickness and profile angle of the workpiece also plays a vital role in machining accuracy.
From the literature, it can be observed that profile accuracy was very important while component manufacturing. In the present study, a unique slant fixture was employed which avoids the disadvantages in conventional taper operation in WEDM. Many disadvantages like bending of wire, angular inaccuracies, wire break, guide wear etc. occurs in conventional machining of tapered profiles. The effect of different input parameters on a square profile was studied at different taper profiles at 0°, 15° and 30° angles. A simple square profile was used in machining of Hastelloy X using WEDM. The input parameters such as wire guide distance (WDG), corner dwell time (CDT), wire offset (WO) and cutting speed override (CSO) were used.

2. Workpiece and procedure

Hastelloy X is a nickel-based superalloy that provides a good balance of high- temperature strength, oxidation resistance, and fabricability. It has wide applications as transition ducts, combustor cans and liners, injector nozzles, flame holders, exhaust struts and many hot gas path components. The composition of the as-received alloy is shown as in table 1 which was tested and verified with the aid of optical emission spectrometry (SpectraMax 130779).

Table 1. Chemical composition of Workpiece.

%	C%	Si%	Mn%	P%	S%	Cr%	Mo%	Fe%	Co%	W%	Ni%
Composition	0.06	0.21	0.65	0.027	0.01	20.65	8.24	18.05	0.69	0.29	50.88

2.1. Experimental Procedure

The profile to be machined has to be programmed in the WEDM process, the wire traces in the workspace where the workpiece was clamped. The sparks from the wire melt the workpiece forming a profile. The square profiles of different dimensions namely 1mm, 3mm and 5mm were machined for the investigation as shown in fig. 1. The programming was performed using computer numerically controlled software called ELCAM. The Zinc coated copper wire of diameter 0.25mm was used as an electrode that traces the profile. The deionized water was employed as dielectric fluid throughout the study. The machining was performed with the aid of slant fixture in order to achieve a tapered profile. The fig. 1 (a) shows the workpiece rotated at 0° and 30° slant angles attached to fixture. This whole setup with Hastelloy X workpiece was fastened to the WEDM table as shown in fig. 1 (b).

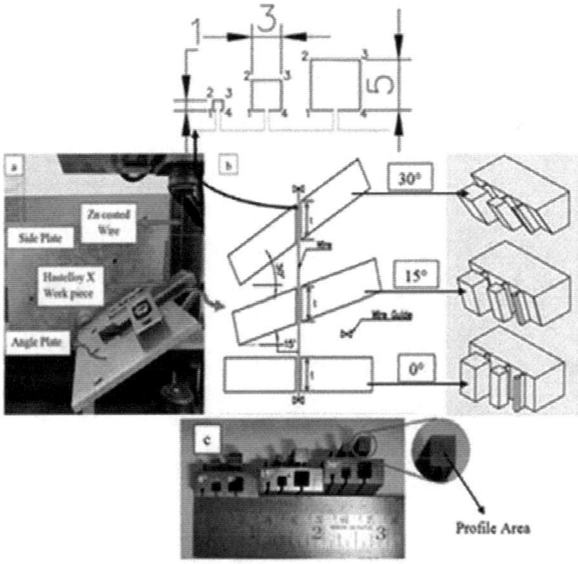

Figure 1. (a) Workpiece at different slant angle by slant fixture (b) Slant Fixture (c) Area of the profile.

3. Results And Discussion

3.1. Variation of areas at different parameters

The parameters were chosen based on the initial experiments [23, 24]. As there were 4 input parameters and from the initial experiments, the interactions were very low. So Taguchi's L_{16} orthogonal array was used for the investigation. The 1mm square was measured using the Image J software form the SEM images obtained from 'JEO JSM-6368OLA'. The 5mm squares were calculated from coordinate measuring machine (CMM) 'TESA VISIO 200'. Table 2 shows various profiles areas machined at different slant angles. The 'Minitab' software was employed to get ANOVA and main effects plots.

Table 2. Slant Areas at Different Parameters for Different Angles.

Sl. No.	WDG (mm)	CDT (s)	WO (μm)	CSO (%)	Area in mm² 1mm	5mm
			0°			
1	40	0	0	31	0.773	23.086
2	40	33	40	54	0.786	22.969
3	40	66	80	77	0.986	23.112
4	40	99	120	100	1.059	24.131
5	50	0	40	77	0.690	22.323
6	50	33	0	100	0.664	21.914
7	50	66	120	31	1.033	24.284
8	50	99	80	54	0.765	23.432
9	60	0	80	100	0.701	23.032
10	60	33	120	77	0.822	23.633
11	60	66	0	54	0.588	21.826
12	60	99	40	31	0.806	22.425

13	70	0	120	54	0.938	23.392
14	70	33	80	31	0.865	22.766
15	70	66	40	100	0.529	22.13
16	70	99	0	77	0.460	21.754
			15°			
17	75	0	0	31	0.953	25.486
18	75	33	40	54	0.957	25.369
19	75	66	80	77	1.166	25.612
20	75	99	120	100	1.239	26.431
21	85	0	40	77	0.855	24.723
22	85	33	0	100	0.833	24.213
23	85	66	120	31	1.198	26.584
24	85	99	80	54	0.926	25.832
25	95	0	80	100	0.881	25.332
26	95	33	120	77	1.006	26.033
27	95	66	0	54	0.751	24.126
28	95	99	40	31	0.986	24.825
29	105	0	120	54	1.100	25.792
30	105	33	80	31	1.049	25.166
31	105	66	40	100	0.709	24.330
32	105	99	0	77	0.640	24.154
			30°			
33	100	0	0	31	1.253	25.946
34	100	33	40	54	1.247	26.205
35	100	66	80	77	1.476	26.760
36	100	99	120	100	1.539	27.869
37	110	0	40	77	1.215	25.582
38	110	33	0	100	1.143	24.845
39	110	66	120	31	1.538	27.656
40	110	99	80	54	1.316	26.384
41	120	0	80	100	1.181	26.209
42	120	33	120	77	1.276	26.863
43	120	66	0	54	1.091	24.728
44	120	99	40	31	1.286	25.563
45	130	0	120	54	1.440	26.433
46	130	33	80	31	1.309	26.381
47	130	66	40	100	1.009	24.915
48	130	99	0	77	0.950	24.195

3.2. ANOVA table and Main Effects plot for areas at different angles

The ANOVA and main effect plots were formulated as shown in table 2. Fig. 2, 3 and 4 show the main effect plots for different dimensions and at various slant angles. Table 3 shows the ANOVA analysis of areas with percentage contribution. From the ANOVA and main effects plot, the WO is the most significant and highest contributing factors compared to all the input parameters on the area. It was noticed that as the WO increases the area of the square profile also increases due to the increase in offset. Similar effects were also observed by Mandal et al. [20] and Kanlayasiri and Jattakul [25]. The ANOVA table also shows the contributing factors. It can be noticed that the error percentage of contribution was very low (0-3%). The WDG was the next highest contributing parameter followed by the CSO parameter on the area at different slant angles. The CDT parameter was the least influential and contributing factor for different slant areas [23].

Table 3. Slant Areas at different parameters for different angles.

Sl. No.	Factor	DF	Area 1mm² Sum of squares	Area 1mm² % Contribution	Area 5mm² Sum of squares	Area 5mm² % Contribution
			0°			
1	WDG	3	0.0960	21.27	1.4734	16.01
2	CDT	3	0.0004	0.10	0.0569	0.62
3	WO	3	0.2683	59.78	6.8701	74.67
4	CSO	3	0.0458	10.20	0.4186	4.55
5	Error	3	0.0387	8.65	0.3818	4.15
			15°			
1	WDG	3	0.0961	21.39	1.6550	17.46
2	CDT	3	0.0005	0.13	0.0844	0.89
3	WO	3	0.2685	59.74	6.9577	73.41
4	CSO	3	0.0473	10.52	0.4618	4.87
5	Error	3	0.0370	8.22	0.3184	3.36
			30°			
1	WDG	3	0.1006	22.40	3.150	19.86
2	CDT	3	0.0029	0.65	0.012	0.08
3	WO	3	0.2664	59.33	11.887	74.94
4	CSO	3	0.0407	9.06	0.692	4.36
5	Error	3	0.0384	8.55	0.121	0.76

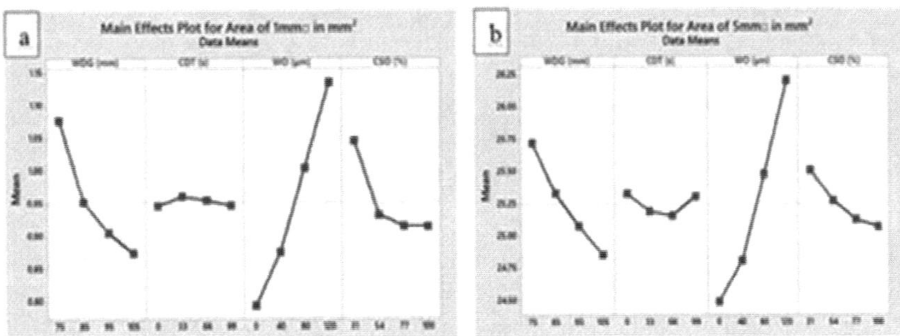

Figure 2. Effects plots of areas (a) 1mm and (b) 5mm at 0°.

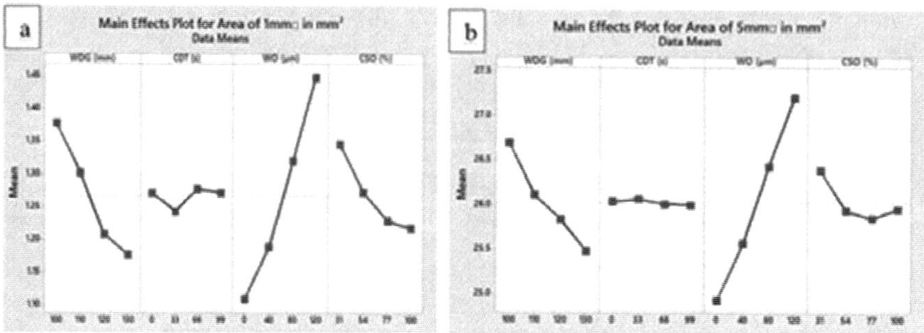

Figure 3. Effects plots of areas (a) 1mm and (b) 5mm at 15°.

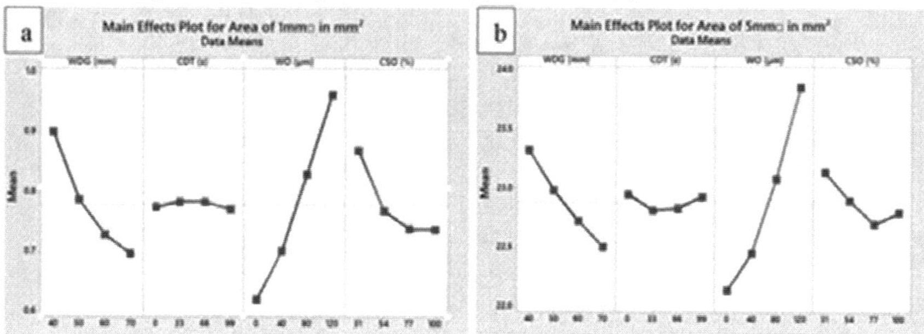

Figure 4. Effects plots of areas (a) 1mm and (b) 5mm at 30°.

3.3. Effect of WO on area

The WO parameter is the most significant and contributing factor (59.33-74.94%) influencing the area during machining. It can be observed from the main effects plot that as WO parameter increases the area also escalates. This phenomenon is noticed due to the increase in offset distance during profiling for all dimensions and at all the slant angles. This leads to wire machining higher circumference than the programmed profile [25]. Fig. 2, fig. 3 and fig. 4 shows a clear increase in the area of the profile offset distance increases (WO parameter increases). There was small variation noticed at 40μm. These variations in the area were observed due to the vibrations in the wire during profiling [26]. As the slant angle increases, workpiece thickness also increases which increases the wire vibrations [24]. This leads to variations in machining the area of the taper profile.

3.4. Effect of WDG on area

From the ANOVA table and main effect plot, the WDG parameter is the next contributing factor affecting the area. As the WDG parameter increases the wire lag also increases due to the flexible nature of wire [25]. Therefore as the WDG increases the area reduces. From fig. 2 (b) and fig. 3 (b) it can be observed as the WDG increases the areas reduces. There were variations noticed in fig. 2 (a) at 50mm, fig. 3 (a) at 85mm and fig. 4(a), (b) at 120mm. These variations were due to the phenomenon of the increase in vibrations due to the change in wire length and flushing [24, 27].

3.5. Effect of CSO on areas

It can be noticed that as the cutting speed override increases the area decreases. The cutting speed override controls the cutting speed of the wire without varying the parameters during machining. At higher cutting speed the corner error increases and there is a decrease in areas [28]. The cutting speed after a certain limit was limited to influence on the profile which after remains constant [29]. A similar trend can be identified in fig. 2, 3 and 4 main plots. There was a slight increase from 77% to 100% in fig 2(b) and 4(b) due to the variation in vibrational forces due to sparking at different cutting parameters [27, 30]. These vibration forces cause irregular random vibrations which affect the areas while profiling. It also indicates after a certain speed the areas will not further decrease.

3.6. Variation of areas at different angles

It can be observed from the fig.5 that as the slant angle increases, the area increases for both 1mm and 5mm square profiles. This observed due to slant angles provided by the fixture while machining. As the slant angle increases as shown in fig. by tilting the slant fixture, the material available for machining also increases. Although the profile traced by the wire remains the same due to the slant/tilting of workpiece increases the area of profile [23, 31].

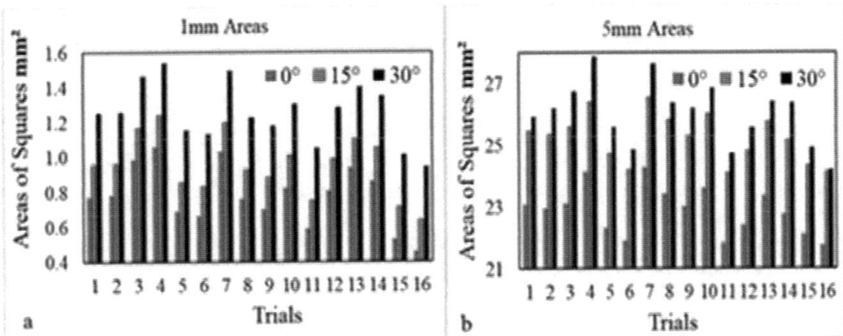

Figure 5. Variation of (a) 1mm and (b) 5mm square areas

4. Conclusion

From the above experiments carried out with slant angle, wire guide distance, corner dwell time, wire offset and cutting speed override as input parameter and area as a response parameter. The future scope of this research can be by knowing the overcut and effects of input parameters. WO parameter can be optimized for different profiles and can be machined with required tolerances without any finishing operation. The following conclusions can be drawn.

1. During slant type taper profiling, the WO parameter was the most significant parameter having contributing factor ranging from 59.33-74.94% and CDT parameter was the most insignificant parameter.
2. There was a decrease ranging from 4.54-18.89% in the area as WGD increased at all the angles. After 77% of CSO parameter, the area remains nearly constant indicating cutting speed doesn't affect the area.
3. As the angle was increased from 0° to 30°, there was a maximum increase in the area ranging from 15.87-85.45%% in square profiles.

5. References

[1] Pramanik A 2014 Problems and solutions in machining of titanium alloys *Int. J. Adv. Manuf. Tech.* **70** 919–928.
[2] Joy R, Manoj I V, Narendranath S 2020 Investigation of cutting speed, recast layer and micro-hardness in angular machining using slant type taper fixture by WEDM of Hastelloy X *Mater. Today.* **27** 1943-1946
[3] Maurya R, Porwal R K 2020 EDM of Hastelloy – An overview *Mater. Today.* **26** 311-315.
[4] Abyar H, Abdullah A, Akbarzadeh A 2019 Prediction Algorithm for WEDM Arced Path Errors Based on Spark Variable Gap and Nonuniform Spark Distribution Models *J. Manuf. Sci. Eng.* **141** 011011

[5] Bisaria H, Shandilya P 2019 Experimental investigation on wire electric discharge machining (WEDM) of Nimonic C-263 superalloy *Mat. Manuf. Process.* **34** 83-92.
[6] Maurya R, Porwal R K, Kumar V 2018 Experimental investigation & modelling of wire EDM process during machining of Nicrofer 5716 *Mater. Today.* **232** 164–177.
[7] Manikandan N, Arulkirubakaran D, Palanisamy D, Raju R 2019 Influence of wire-EDM textured conventional tungsten carbide inserts in machining of aerospace materials (Ti–6Al–4V alloy) *Mat. Manuf. Process.* **34** 103-111.
[8] Maher I, Ahmed A. D. Sarhan, Hamdi M. 2015 Review of improvements in wire electrode properties for longer working time and utilization in wire EDM machining. *Int. J. Adv. Manuf. Technol.* **76** 329–351
[9] Kinoshita N, Fukui M, Fujii T 1987 Study on Wire-EDM: Accuracy in Taper-Cut *CIRP Annals* **36** 119-122.
[10] Plaza S, Ortega N, Sanchez JA, Pombo I, Mendikute A 2009 Original models for the prediction of angular error in wire-EDM taper-cutting *Int. J. Adv. Manuf. Technol.* **44** 529–538.
[11] Yan H, Liu Z, Li L, Li C, He X 2017 Large taper mechanism of HS-WEDM *Int. J. Adv. Manuf. Technol.* **90** 2969–2977.
[12] Manoj I V, Joy R, Narendranath S 2020 Investigation on the Effect of Variation in Cutting Speeds and Angle of Cut during Slant Type Taper Cutting in WEDM of Hastelloy X *Arab. J. Sci. Eng.* **45** 641-651.
[13] Naveed R, Mufti N A, Ishfaq K, Ahmed N, Khan S A 2019 Complex taper profile machining of WC-Co composite using wire electric discharge process: analysis of geometrical accuracy, cutting rate, and surface quality *Int. J. Adv. Manuf. Technol.* **105** 411-423.
[14] Sharma P, Chakradhar D, Narendranath S 2018 Analysis and Optimization of WEDM Performance Characteristics of Inconel 706 for Aerospace Application. *Silicon* **10** 921–930.
[15] Firouzabadi H A, Parvizian J, Abdullah A, Tehrani A F 2015 Analysis of Residual Material and Machining Error on Straight and Corner Curved Paths in Roughing of WEDM *Int. J. Adv. Manuf. Technol.* **76** 447-459.
[16] Werner A 2016 Method for enhanced accuracy in machining curvilinear profiles on wire-cut electrical discharge machines *Precis.Eng.* **44** 75-80.
[17] Bisaria H, Shandilya P 2019 Processing of curved profiles on Ni-rich nickel–titanium shape memory alloy by WEDM *Mater. Manuf. Process.* **34** 1333-1341.
[18] Mouralov K, Kovar J, Klakurkova L, Bednar J, Benes L, Zahradnicek R 2018 Analysis of surface morphology and topography of pure aluminium machined using WEDM. *Measurement* **114** 169-176.
[19] Wang Y, Chen X, Wang Z L, Donga S 2018 Fabrication of micro gear with intact tooth profile by micro wire electrical discharge machining *J. Mat. Process. Tech.* **252** 137-147.
[20] Mandal A, Dixit A R, Das A K, Mandal N 2016 Modeling and Optimization of Machining Nimonic C-263 Superalloy using Multicut Strategy in WEDM *Mat. Manuf. Process.* **7** 860-868.
[21] Chaubey S K, Jain N K 2018 Investigations on surface quality of WEDM-manufactured meso bevel and helical gears. *Mat. Manuf. Process.* **14** 1568-1577.
[22] Dodun O, Gonçalves-Coelho A M, Slătineanu L, Nagîţ G 2009 Using wire electrical discharge machining for improved corner cutting accuracy of thin parts *Int. J. Adv. Manuf. Technol.* **41** 858–86,.
[23] Manoj I V, Narendranath S 2020 Variation and artificial neural network prediction of profile areas during slant type taper profiling of triangle at different machining parameters on Hastelloy X by wire electric discharge machining. *P. I. Mech. Eng. E-J. Pro.* **234** 673–683.
[24] Manoj I V, Joy R, Narendranath S 2020 Investigation on the Effect of Variation in Cutting Speeds and Angle of Cut during Slant Type Taper Cutting in WEDM of Hastelloy X *Arab. J. Sci. Eng.* **45** 641-651.

[25] Kanlayasiri K, Jattakul P 2013 Simultaneous optimization of dimensional accuracy and surface roughness for finishing cut of wire-EDMed K460 tool steel *Precis. Eng.* **37** 556-561.

[26] Habib S, Okada A 2016 Experimental investigation on wire vibration during fine wire electrical discharge machining process Int. J. Adv. Manuf. Technol. **84** 2265-2276.

[27] Habib S 2017 Optimization of machining parameters and wire vibration in wire electrical discharge machining process *Mech. Adv. Mater. Mod. Process.* **3** 1-9.

[28] Selvakumar G, Kuttalingam KGT, Prakash SR 2018 Investigation on machining and surface characteristics of AA5083 for cryogenic applications by adopting trim cut in WEDM *J. Braz. Soc. Mech. Sci. & Eng.* **4** 1-8.

[29] Sanchez J A, Rodil J L, Herrero A, Lacallea L N L, Lamiki A 2007 On the influence of cutting speed limitation on the accuracy of wire-EDM corner-cutting J. Mater. Process Technol. **182** 574-579.

[30] Sarkar S, Sekh M, Mitra S, Bhattacharyya B 2011 A novel method of determination of wire lag for enhanced profile accuracy in WEDM *Precis. Eng.* **35** 339-347.

[31] Manoj I V, Joy R, Narendranath S 2019 Investigation of machining parameters on corner accuracies for slant type taper triangle shaped profiles using WEDM on Hastelloy X *IOP Conf. Ser.: Mater. Sci. Eng.* **591** 1-11.

An Insight: Machining of Titanium Alloys & Associated Tool Wear

Mangal Singh Sisodiya[1], **Vivek Bajpai**[1]

[1]Department of Mechanical Engineering, Indian Institute of Technology (ISM) Dhanbad, India

Email ID: mangal_284@yahoo.com

Abstract. Titanium alloy has some important applications in the aerospace industries along with other sectors, but its machinability efficiently is still a topic of intent research. Several factors work together to hinder its effective machining by various cutting tools. Although work has been done to some extent by some well cutting tools such as carbide tools however, very little work has been stated and established by using CBN tools or cryogenically treated CBN tools. The review has been done elaboratively to bring out various aspects of titanium alloy and its machining by various conventional cutting tools along with the CBN cutting tool.

1. Introduction

Titanium in its alloy form is primarily used in aerospace industries mainly because of three factors which are high specific resistance, fracture resistance and an exceptional corrosion resistance. They are also being used in various other industrial and commercial areas including paper and pulp, food processing, etc. They are indeed well-known engineering materials available in various alloys and forms. However, although being used extensively, they are quite expensive when compared to other metals due to problems during fabrication & melting, difficult melting, etc E.O. Ezugwu et al. (1995) [1] stated that titanium and its alloys are not so good due to certain inherent properties as follows:

- Titanium is quite reactive and tends to weld to the cutting tool leading to tool failure.
- It has low thermal conductivity which indirectly affects tool life
- It's high strength at higher temperatures and low elastic modulus further makes it difficult to machine.

All these inferences have been further supported in the works of H. Hong et al. (1993) [2]. Such poor machinability have led big companies to invest in large amounts in developing techniques so that machining cost is reduced. [1]

2. Allotropes of Titanium

R.M. Duncan et al. [3] stated that Titanium undergoes allotropic transformation at 882°C from low temperature hexagonal α-phase to high temperature BCC β-phase. The alloying elements changes the transformation temperature [4]. Elements which help to raise the transformation temperature are specifically the α-stabilizers and, aluminium which is an important α-strengthening element at normal and temperatures till 550°C [5]. On the other hand, elements which decrease the transformation temperature are the β-stabilisers which includes β-isomorphous and β-eutectoid [5]. The work further states that β-isomorphous alloying substances are readily soluble in β-titanium but the β-eutectoid compounds are soluble to a certain extent in β-titanium.

3. Inherent Unfavourable properties of titanium

Though researchers and scholars have helped to evolve the machining of this element and still a lot of efficient work remains, the main hurdle being faced is due to certain inherent properties which are enlisted and described as follows:

- High Cutting Temperature: It is specified in the work of W. Konig (1979) [6] that 80% of the total heat generated during machining of titanium is taken up by the tool due to poor thermal conductivity of titanium and hence, it is well known by now that very high cutting temperatures are created while machining this hard material and that too, close to the cutting edge of tool which is a major cause of tool wear.
- High Cutting Pressure: The cutting forces while machining titanium are almost same as when machining steel [7]. W. Konig (1979) [6] has stated that there exists high stresses in tools when machining titanium than while machining nickle-based alloy and three to four times of that while machining steel Ck 53N. This may be attributed to the small chip-tool contact area on the rake face of the tool [8] and due to the resistance offered by titanium alloy to deformation at higher temperatures which reduces at temperatures above 800°C [9].
- Low modulus of elasticity & consequent springback: A Pramanik (2014) [10] states that the low modulus of elasticity leads to bouncy conditions in titanium, i.e. to say that there is more strain and hence, deformation in titanium under a certain degree of force.
- Hardening due to Diffusion and Plastic Deformation: G. He (1985) [11] brought out the fact that when the temperature of cutting zone attains 600-700°C and crosses it, diffusion of oxygen and nitrogen molecules from the air takes place which eventually hardens the surface layer of the titanium workpiece. M. Donachie Jr. (1982) [12] brought out the fact that titanium retains its strength at even higher temperatures unlike most other metals. It is also recollected that both the phenomenon of diffusion and plastic deformation are responsible for hardening of titanium while machining it however, plastic deformation is the more pronounced factor [13,14].
- Mechanism of Chip formation: A Hosseini et al. (2014) [15] says that titanium and its alloys probably has some unique combination of mechanical and metallurgical properties that actually make them extremely difficult to machine than their counterparts. Keeping in mind all the above shortcomings that might pop up while machining titanium and its alloy, a good cutting tool, therefore, must entail the following [16,17]:

4. Machining of Titanium Alloys by Various Tools

4.1. Machining of Titanium alloys with HSS tools:

Inspite of good toughness of HSS tools, the softening takes place at around 600°C and hence cannot be used for applications that has working temperature above 500°C [18,19]. It has been effectively brought out that HSS tools are not suitable for machining titanium and its alloys when the cutting speed exceeds 30 m/min [20,21]. However, some highly alloyed grades and certain general-purpose grades can definitely be used to machine titanium, but care has to be taken that the cutting speed does not exceed 30 m/min [22]. In case of machining of titanium alloys, the degradation of the tool takes place due to a lot of factors, however, it is predominantly due to the plastic deformation that takes place [23]. Further, the crater wear that takes places due to the high temperature that is produced by chips rubbing on the rake surface of the tool is also another major factor for tool deterioration [24]. Both these factors makes HSS tools unfit for machining titanium and its alloys.

4.2. Machining of Titanium alloys with carbide tools:

One of the most preferred carbide for machining of Titanium and its alloys is the straight grade cemented carbide (WC-Co) [25]. Cutting speeds greater than 60 m/min is not preferred while machining with cemented carbide tools [26]. Higher speeds, i.e. greater than 60 m/min, leads to plastic deformation due to intense heat generation [27]. Apart from the diffusion wear, another type of wear that usually takes place in carbide tools is the adhesion wear and temperatures above 740°C and contact pressure of 0.23 GPa [28]. The adhesion wear increases as the tool material is progressively and repeatedly damaged.

Higher crater wear has been noticed in tools with coarser grains and higher flank wear has been noticed in tools with smaller grains [29]. The coating layer is removed by either abrasion wear or by the chemical reaction between the titanium workpiece and the coating such as titanium nitride and titanium carbonitride [30].

4.3. Machining of Titanium using Ceramic Tools:

A Chakraborty et al (2000) [31] reveals that ceramic tools have a higher compressive strength, hot hardness, better chemical inertness and oxidation resistance than its counterparts. However, they are subjected to mechanical and thermal shocks and they easily break away during interrupted machining due to being brittle and lacking toughness.

Kejia zhung et al 2014 [32] observed notch wear in ceramic tools mainly occur due to incidence of a hard film in workpiece surface.

Z. Yin et al 2015 [33] found that to ensure quality of machining surface and efficiency, special super-hard materials advance ceramics cutting tools have been progressively working for machining of difficult-to-cut materials via good wear resistance, high compressive strength and excellent chemical stability.

4.4. Machining of Titanium using PCD Tools:

Observations have stated that PCD tools has a much lower wear rate when utilised for machining of titanium due to the formation of titanium carbide as a protective coating on the rake face of tool [34]. They even show acceptable performance when utilised for machining of Ti-6Al-4V which is one of the common alloys of titanium [35]. Although it has an acceptable performance however, the high costs involved hinders its much widespread utilisation.

E.O. Ezugwu et al 2003 observed that special materials of polycrystalline diamond (PCD) and polycrystalline cubic boron nitride (pcBN) are used to maintain a high hardness at elevated temperatures [36]

4.5. Machining of Titanium by CBN Tools:

CBN is the second highest hard material, has a high melting point, high hot hardness at elevated temperatures and has a high oxidation resistance with stability upto 2000°C [37]. F Klocke (2011) [37] further specifies that the grains of CBN are extremely small (1-50μm) and are sintered together by binder under high temperature and pressure and are called polycrystalline cubic boron nitride or PCBN. Both CBN and PCBN tools are used to machine hard-to-cut materials like forged steels, alloy steels, etc [38]. It was reported that the combined effect of elevated temperature and compressive stress may reach at a stage that the tool is not being able to bear and hence, deformation and consequent wear starts [39]. Although CBN and PCBN tools can be utilised for higher cutting speed applications like 150 m/min for Ti-6Al-4V [40] but they are hardly used because of their expensive nature when compared to carbide tools and therefore, are not efficient economically.

5. Cryogenic Treatment

Cryogenics refers to the science of extremely low temperatures, be it lower than 120 K / -153°C [38] or 100 K /-173°C [41]. Cryogens are generally available as liquid gases such as liquid nitrogen, oxygen, helium, argon, etc. [42]. However, in case of machining, sometimes temperatures above 100 K / -173°C is also considered as cryogenics as in the case of cryogenic machining using solid and/or liquid carbon dioxide [43,44,45] and sometimes even to the extent of using chilled air at very low temperatures [46,47]. In case of cryogenic machining, the cryogenic component is introduced in the cutting operation in the form of cutting fluid to change the properties of the materials or to dissipate the heat generated due to cutting at the cutting zone [48]. Cryogenic processing or in other words cryo-processing is part of the heat treatment cycle as an extended process which eventually enhances the tool material properties [49]. Many advantages of such cold processing has been known for ages for example Swiss watchmakers used to leave watch components in the cold of the Alps for several months in order to improve upon the wear resistance of the parts [50,51]. For steel cutting tools, the austenite retained in any cutting tool can severely lead to tool wear and the cutting edge getting chipped off [52].

However, such a generic process cannot be recommended for all tools and it is necessary to study every tool individually for cutting process parameters which include hardness, toughness and wear resistance of tool [53]. It is previously noted that changes in tool material properties happens when the tool is being cooled down to cryogenic temperature and further, heated upto the room temperature [54].

It has been reported that cryogenic processing indeed increases the wear resistance of tool by up to 126% if we use HSS tools to machine grey cast iron or to machine carbon steel [55,56]. Firouzdor et al. (2008) [56] even studied the effect of tempering after doing the cryogenic process and came up with the conclusion that the tool life in this case increases by 49% when compared to tools which were non-tempered after the cryogenic process. Leskovsek et al. (2006) [57] studied the effect of tempering temperature on the wear resistance of tools. It was highlighted that the tool which was cryogenically treated and was tempered at 600°C performed a little better than tools which were only tempered at the same temperature. Molinari et al. (2001) [58] undertook a study and gave that the specimen which was un-tempered after doing the cryogenic treatment showed the most promising results in the sense of highest wear resistance. It is indeed a true fact that cryo-processing indeed improves tool materials and there are several theories given by several researchers for the same but none of them could explain the underlying reason [59]. The first such explanation for the improved properties was the transformation of retained austenite into martensite allotrope which is much harder [60]. Another explanation brings out that the cryogenic treatment done enhances more of carbide particles in the martensite matrix and further, refines the microstructure and provides a more uniform carbide distribution. These carbides were classified into primary carbides, large secondary carbides and small secondary carbides by Das et al. (2007) [61]. So far as titanium alloys are concerned, S. Y. Hong et al. (1999) [62] suggested that it will be beneficial to freeze the tool rather than cooling or freezing the workpiece but keeping in view the workpiece titanium, it was also put forward that it will be beneficial to cool the workpiece as well as the tool considering the reduction in chemical reactivity that will be caused between the two due to the treatment as well as freezing the tool enhances the tool properties, provides stability to cutting tool material and reduces friction between the tool and the workpiece at the cutting zone.

Although extensive work has been carried out already on cryogenic machining using cryogenic substance as a coolant but comparatively less work has been carried out on cryo-processing with no concrete explanations for the improvement in properties via cryo-freezing of tool has been provided yet.

6. Tool Wear

Researchers have actively studied about the machinability of Ti-6Al-4V and have concluded that PCBN (Polycrystalline Cubic Boron Nitride) and PCD (Polycrystalline Diamond) tools may be used to machine Ti-6Al-4V effectively [63]. PCBN allows for hard turning of workpiece materials for which the hardness is 58-62 HRC [64]. K. Bouacha et al. (2014) [65] optimised the various parameters of hard turning and the responses that were recorded was tool wear, surface roughness, cutting forces and metal volume removed. The most important factors which attributed to the surface roughness was that of feed rate and cutting speed.

Y. Jia et al. (2018) [66] machined Ti-6Al-4V by PCBN Tools (Polycrystalline Cubic Boron Nitride) manufactured by DI Corporation in a dry turning operation. CBN insert's performance depends largely also on the type and content of binder material. The most significant reason for such type of tool wear is the dissolution diffusion of binder material into the chip [67]. However, the Binderless CBN inserts (BCBN) are free from above effect and exhibit superior characteristics as well. A. R. Zareena et al. (2005) [68] machined titanium alloy using CBN, BCBN and PCD tools using high pressure lubricant and concluded that BCBN tools are free from cobalt binder and hence, there is no diffusion dissolution taking place and therefore, the tool life is more, surface finish is better and cutting forces are also less and hence, can be used effectively to machine titanium alloys both from an economic and functional point of view.

7. Conclusion

The previous works were reviewed extensively in connection to titanium alloy grade-5 turning using cryogenically treated as well as untreated CBN inserts although much needs to be still explored in this regard. There seemed to be a huge gap in context of variation of cryogenic treatment duration and treatment cycles with the study of its effects on tool wear on machining. Further, there seemed to be totally minute exploration on using solid CBN or binder less CBN for titanium machining and studying the wear on tool correspondingly. Hence, we propose to take up the variation of cryogenic duration and cycles on tool wear as a future work and dwell into the effects it produces.

References

[1] E.O. Ezugwu, Z.M. Wang, "Titanium Alloys and their machinability- A review", Journal of Materials Processing Technology, 68 (3), 262-274, 1997

[2] H. Hong, A.T. Riga, J.M. Cahoon, C.G. Scott, Wear 162, 34-39, 1993

[3] R.M. Duncan, P.A. Blenkinsop, R.E. Goosey, IMI Titanium, P.O. Box 216, Witton, Birmingham B6 7BA, UK, 63-87

[4] D. Eylon, S. Fulishiro, P.J. Postans, F.H. Froes, J. Met. 36 (11), 55-62, 1984

[5] IMI Titanium, Commercially Pure Titanium, Kynoch Press, Birmingham, 1978

[6] W. Konig, Proc. 47th Meeting of AGARD Structural and Materials Panel, Florence, Sept. 1978, AGARD, CP256, London, 1.1-1.10, 1979

[7] N. Zlatin, Modern Math. Shop 42 (12), 139-144, 1970

[8] D.C. Kirk, Tools and dies for industry, Proc. Conf. 76/77. Met. Soc., London, 77- 98, 1976-7

[9] B.B. Johnson, Tips on Milling Titanium--and Tools to Do the Job

[10] A Pramanik, Problems and solutions in machining of titanium alloys, Int J Adv Manuf Technol 70(5–8), 919–928, 2014

[11] G He, Zhang Y, Experimental investigations of the surface integrity of broached titanium alloy, CIRP Ann-Manuf Technol 34(1), 491–494, 1985

[12] M J Donachie, Introduction to titanium and titanium alloys, American Society for Metals, Titanium and Titanium Alloys Source Book, 7, 1982

[13] X Yang, C R Liu, Machining titanium and its alloys, Mach Sci Technol 3(1), 107–139, 1999

[14] H Child, A Dalton, ISI special report 94, London, 1968

[15] A Hosseini, H Kishawy, Cutting tool materials and tool wear, in Machining of Titanium Alloys, Davim, J. Ed. Springer: Berlin, Heidelberg, 31-56, 2014

[16] A Machado, J Wallbank, Machining of titanium and its alloys—a review, Proc Inst Mech Eng B J Eng Manuf 204(1), 53–60, 1990

[17] H J Siekmann, How to machine titanium, Tool Eng 34(1), 78–82, 1955

[18] B Kramer,On tool materials for high speed machining, J Eng Ind 109(2),87–91, 1987

[19] E Ezugwu, J Bonney, Y Yamane, An overview of the machinability of aeroengine Alloys, J Mater Process Technol 134(2), 233–253, 2003

[20] M Rahman, Z-G Wang, Y-S Wong, A review on high-speed machining of titanium alloys, JSME Int J, Series C 49(1), 11–20, 2006

[21] J Pérez, J Llorente, J Sánchez, Advanced cutting conditions for the milling of aeronautical alloys, J Mater Process Technol 100(1), 1–11, 2000

[22] P.D. Hartung, B Kramer, B V Turkovich, Tool wear in titanium machining, CIRP Annal Manuf Technol 31(1), 75–80, 1982

[23] A Machado, J Wallbank, Machining of titanium and its alloys—a review, Proc Inst Mech Eng B J Eng Manuf 204(1), 53–60, 1990

[24] E. M.Trent, P. K. Wright, Metal cutting. Butterworth-Heinemann, Boston, 2000

[25] C Che-Haron, Tool life and surface integrity in turning titanium alloy, J Mater Process Technol 118(1), 231–237, 2001

[26] N Narutaki et al, Study on machining of titanium alloys, CIRP Annal Manuf Technol 32(1), 65–69, 1983

[27] P Dearnley, A Grearson, Evaluation of principal wear mechanisms of cemented carbides and ceramics used for machining titanium alloy IMI 318. Mater Sci Technol, 2(1), 47–58, 1986

[28] M I Sadik, B Lindström, The effect of restricted contact length on tool performance, J Mater Process Technol 48(1), 275–282, 1995

[29] R Edwards, R Edwards, Cutting tools, Institute of Materials, London, 1993

[30] M Rahman, Y S Wong, A R Zareena, Machinability of titanium alloys. JSME Int J Series C 46, 107–115, 2003

[31] A Chakraborty, K Ray, S Bhaduri, Comparative wear behavior of ceramic and carbide tools during high speed machining of steel. Mater Manuf Process 15(2), 269–300, 2000

[32] Kejia Zhuang, Dahu Zhu, Xiaoming Zhang, , Han Ding "Notch Wear Prediction Model In Turning Of Inconel 718 With Ceramic Tools Considering The Influence Of Work Hardened Layer" Wear 313, 63-74, 2014

[33] Z. Yin, C. Huang, J. Yuan, B. Zou, H. Liu, H. Zhu, Cutting performance and life prediction of an Al2O3/TiC micro–nano-composite ceramic tool when machining austenitic stainless steel, Ceram. Int. 41 (2015) 7059–7065.

[34] F Nabhani, Machining of aerospace titanium alloys, Robot Comput Integr Manuf 17(1), 99–106, 2001

[35] E Kuljanic et al, Milling titanium compressor blades with PCD cutter, CIRP Annal Manuf Technol 47(1), 61–64, 1998

[36] Z-C Lin, D-Y Chen, A study of cutting with a CBN tool, J Mater Process Technol 49(1),149–164, 1995

[37] E.O. Ezugwu, J. Bonney, Y. Yamane, An overview of the machinability of aeroengine alloys, J. Mater. Process. Technol. 134 (2003) 233–253.

[38] Z Zoya, R Krishnamurthy, The performance of CBN tools in the machining of titanium alloys, J Mater Process Technol 100(1), 80–86, 2000

[39] E Ezugwu et al, Evaluation of the performance of CBN tools when turning Ti–6Al–4V alloy with high pressure coolant supplies, Int J Mach Tools Manuf 45(9), 1009–1014, 2005

[40] K D Timmerhaus, R. P. Reed, Cryogenic engineering: fifty years of progress. New York: Springer Verlag, 2007

[41] I J Karassik, J P Messina, P Cooper, C C Heald, Pump handbook, 3rd edition, New York: McGraw-Hill, 2001

[42] A Shokrani , V Dhokia , P M Escalona & S T Newman, State-of-the-art cryogenic machining and processing, International Journal of Computer Integrated Manufacturing, 26(7), 616-648, DOI: 10.1080/0951192X.2012.749531, 2013

[43] L D Chiffre, J L Adreason, S Laderberg, I B Thesken, Performance testing of cryogenic CO2 as cutting fluid in parting/grooving and threading austenitic stainless steel, CIRP Annals – Manufacturing Technology, 56 (1), 101–104, 2007

[44] E Abele, B Schramm, Using PCD for machining CGI with a CO2 coolant system, Production Engineering, 2 (2), 165–169, 2008

[45] C Machai, D Biermann, Machining of β -titanium-alloy Ti–10V–2Fe–3Al under cryogenic conditions:Cooling with carbon dioxide snow. Journal of Materials Processing Technology, 211, 1175–1183, 2011

[46] H Tsai, H Hocheng, Investigation of the transient thermal deflection and stresses of the workpiece in surface grinding with the application of a cryogenic magnetic chuck. Journal of Materials Processing Technology, 79 (1–3), 177–184, 1998

[47] B Yalcin, A Ozgur, M Koru, The effects of various cooling strategies on surface roughness and tool wear during soft materials milling. Materials & Design, 30 (3), 896–899, 2009

[48] A Shokrani, V Dhokia, S T Newman, Environmentally conscious machining of difficult-to-machine materials with regard to cutting fluids. International Journal of Machine Tools and Manufacture, 57, 83–101, 2012

[49] F J Da Silva, S D Franco, A R Machado, E O Ezugwu, A M Souza Jr, Performance of cryogenically treated HSS tools, Wear, 261 (5–6), 674–685, 2006

[50] W S Busch, Introduction to cryogenics, Society of Manufacturing Engineers, Chicago, Technical Paper CM69- 252, 1969

[51] D J Kamody, Cryogenics process update, Advanced Materials & Processes, 155 (6), H67–H69, 1999

[52] D Harris, Another big step forward in modern metalworking, Society of Manufacturing Engineers, TECH Paper No. AD70-734, 4 P, 1970

[53] S S Gill, H Singh, R Singh, J Singh, Cryoprocessing of cutting tool materials – a review, The International Journal of Advanced Manufacturing Technology, 48 (1–4), 175–192, 2009

[54] S Kalia, Cryogenic processing: A study of materials at low temperatures. Journal of Low Temperature Physics, 158 (5–6), 934–945, 2009

[55] B R Ramji, H N N Murthy, M Krishna, M J Raghu, Performance study of cryogenically treated HSS drills in drilling gray cast iron using orthogonal array technique. Research Journal of Applied Sciences, Engineering and Technology, 2 (5), 487–491, 2010

[56] V Firouzdor, E Nejati, F Khomamizadeh, Effect of deep cryogenic treatment on wear resistance and tool life of M2 HSS drill. Journal of Materials Processing Technology, 206 (1–3), 467–472, 2008

[57] V Leskovsek, M Kalin, J Vizintin, Influence of deep-cryogenic treatment on wear resistance of vacuum heat-treated HSS. Vacuum, 80 (6), 507–518, 2006

[58] A Molinari, M Pellizzari, S Gialanella, G Straffelini, K H Stiasny, Effect of deep cryogenic treatment on the mechanical properties of tool steels. Journal of Materials Processing Technology, 118 (1–3), 350–355, 2001

[59] S S Gill, J Singh, R Singh, H Singh, Metallurgical principles of cryogenically treated tool steels – a review on the current state of science. The International Journal of Advanced Manufacturing Technology, 54, 59–82, 2011

[60] M Preciado, P M Bravo, J M Alegre, Effect of low temperature tempering prior cryogenic treatment on carburized steels. Journal of Materials Processing Technology, 176 (1–3), 41–44, 2006

[61] D Das, A K Dutta, V Toppo, K K Ray, Effect of deep cryogenic treatment on the carbide precipitation and tribological behavior of D2 steel. Materials and Manufacturing Processes, 22 (4), 474–480, 2007

[62] S Y Hong, Z Zhao, Thermal aspects, material considerations and cooling strategies in cryogenic machining, Clean Technologies and Environmental Policy, 1(2), 107–116, 1999

[63] A J D Oliveira, A E Diniz, D J Ursolino, Hard turning in continuous and interrupted cut with PCBN and whisker-reinforced cutting tools, Journal of Materials Processing Technology,209 (12-13), 2009

[64] R Shetty,T K Jose,G D Revankar, S S Rao, S D Shetty, Surface Roughness Analysis during Turning of Ti-6Al-4V under Near Dry Machining using Statistical Tool, Internationa Journal of Current Engineering and Technology, Vol. 4(3), 2014

[65] K Bouacha, M A Yallese, S Khamel, S Belhadi, International Journal of Refractory Metals and Hard Materials, Vol. 45, 160-178, 2014

[66] Y Jia, L Zhu, Technics Research on Polycrystalline Cubic Boron Nitride Cutting Tools Dry Turning Ti-6AL- 4V Alloy Based on Orthogonal Experimental Design, MATEC Web of Conferences 142, 03002, 2018

[67] H Sumiya, S Uesaka, S Satoh, Mechanical Properties of High Purity Polycrystalline CBN Synthesized by Direct Conversion Sintering Method, J. Mater. Sci., 35, 1181–1186, 2000

[68] A R Zareena, M Rahman, Y S Wong, Binderless CBN Tools, a Breakthrough for Machining Titanium Alloys, Journal of Manufacturing Science and Engineering, Vol. 127, 277-279, 2005

IOP Conf. Series: Materials Science and Engineering 1017 (2021) 012014 doi:10.1088/1757-899X/1017/1/012014

Optimization of Abrasive Wear Behaviour of Cenosphere Filled Basalt-Epoxy Composites Using Taguchi Approach

Vikas sharma[1], M L Meena[1], A K Chaudhary[21*], and Sundeep Kumar[3]

[1]Mechanical Engineering Department, Malaviya National Institute of Technology, Jaipur, 302017, India

[2]Department of Industrial and Production Engineering, College of Technology, GBPUAT, Pantnagar, 263145, India

[3]Department of Technical Education, Jaipur Government of Rajasthan, 302017, India

E-mail address: aruncdme@gmail.com

Abstract. The current research work investigates the three-body abrasive wear analysis of cenosphere particulate filled chopped basalt fiber (40 wt.% constant) reinforced epoxy composites for developing light weight, low cost, and sustainable composite. The effect of cenosphere particulate on the composite was investigated. The composite laminates are fabricated by traditional open mold casting technique. The cenosphere particulate is mixed (0-15 wt.%, step of 5 wt.%) in the composites. Abrasive wear process input parameters like: filler loading, normal load, sliding distance, and abrasive size are optimize by using Taguchi design of experiments. From the results, it is clearly seen that 10 wt.% cenosphere filled basalt-epoxy composite (EBC 10) shows lowest abrasive wear rate in a steady state abrasive wear condition (with a variation in sliding distance and normal load). Taguchi experimental design technique is an effective optimization technique and gives a better outcome.

1. Introduction

Researchers are mostly used either synthetic or natural fiber as reinforcement in the fiber reinforced polymer based composites (FRCs). In the present researchers, mineral fiber has also been used by researchers in fiber reinforced polymer based composites for developing light weight, low cost, and sustainable composite. In the category of mineral fibers, basalt fiber is one of reinforcement (mineral fiber) having excellent strength and modulus. Basalt fiber is produced volcano (lava erupted from a volcano) [1-3].

For improving the strength, and wear resistance of fiber reinforced polymer composites, researchers are used ceramic fillers like: silicon oxide (SiO_2), aluminum oxide (Al_2O_3), and magnesium oxide (MgO) etc. In the current research, solid industrial waste material like: Fly ash, Cenosphere, Red mud, Marble powder, Granite dust, etc. has also been used by researchers as filler material in fiber reinforced polymer based composites for developing better wear and friction resistant composite. Several research scholars worked to improve the wear and friction properties of composites by reinforcing with waste particulate materials [4-6].

The hollow microsphere fly ash particles are termed as cenosphere [7]. For reference purpose some important research work are here. Alkadasi et al. [8] investigates the mechanical characteristics of fly ash filled polybutadiene rubber based composite and reported that hardness, tensile strength, and tensile modulus improved with fly ash content. Sunil et al. [9] studied the tribological behaviour of submicron size cenosphere filled vinyl ester based composite and they found that cenosphere filler content improved the wear resistance of the composite. Chand et al. [10] investigates the tribological characteristics of cenosphere filled High Density Polyethylene (HDPE) based composite and they reported that tribological performance of composite has been improved with cenosphere filler. Ray et al. [11] fabricated the fly ash filled (40 wt.% and 50 wt.%) vinylester based composites and result shows that composite with fly ash filler shows superior wear resistance than neat (without fly ash) composite. Suresha et al. [12] investigates the wear performance of cenosphere filled glass-epoxy composites and they found that sliding distance and normal load were most significant factors for wear loss.

In the present research work, fabricated the cenosphere (0, 5, 10, 15 wt.%) filled basalt fiber (40 wt.% constant) reinforced epoxy based composites via open mold casting technique for investigating the three-body abrasive wear behaviour of the composites and optimizing the process parameters by using Taguchi design of experiments.

2. Materials and Method

2.1 Required Materials and Fabrication of Composites

The details of raw materials and its absolute properties are listed in Table 1, while Figure 1 shows the formulation of composites and fabrication procedure by Open mold casting technique.

Table 1. Properties of materials used in analysis (supplier's data).

		Properties			
		Density (g/cc)	Tensile strength (MPa)	Young's modulus (GPa)	Thermal conductivity (W/mK)
Materials	Epoxy Resin	1.19	52	3.2	0.35
	Basalt Fiber (BF)	2.66	4500	87	0.031
	Cenosphere (Grade 'VA LD 300')	0.85	-	-	-

IOP Conf. Series: Materials Science and Engineering 1017 (2021) 012014 doi:10.1088/1757-899X/1017/1/012014

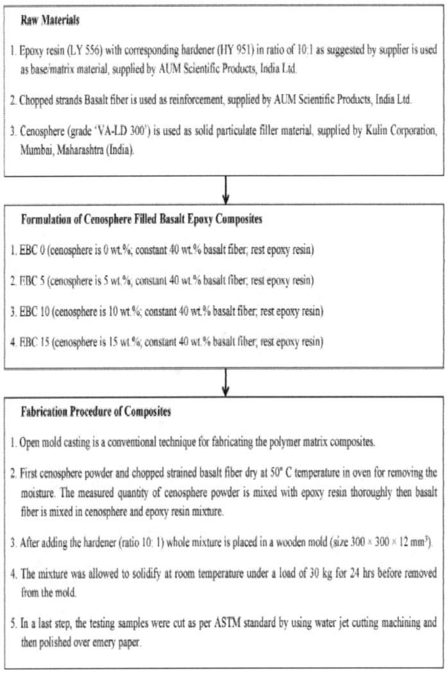

Figure 1. Raw materials, formulation, and fabrication procedure.

2.2 Three-Body Abrasive Wear Test Rig

The abrasive wear test (three-body) of fabricated composite samples is conducting on dry sand abrasion test rig. Model TR-50; supplied by MAGNUM Bangalore, INDIA; ASTM G65 standard. Figure 2 shows the illustration of abrasion test rig. Table 2 listed the control parameters and their levels for experiment. Composite sample dimension (75 × 25 × (10-12) mm^3) in shape of rectangular is used for experiment. The sand particles are strike among the sample and rubber wheel through the nozzle at the feed rate of 255 $\pm$ 5 g/min^2. The weight of composite samples before experiment and after experiment is measured using precision electronic balance and then weight loss (initial weight – final weight) during the experiment is calculated. Finally, abrasive wear rate is calculated by using following Eq.

$$\text{Ws} = \frac{\Delta V}{\rho \times F_n \times S_s} \ (\text{mm}^3/\text{N-m}) \qquad (1)$$

Where,
W_s = Abrasive wear rate (mm^3/N-m).
ρ = Actual density of fabricated composite samples.
F_n = Normal load.
S_s = sliding distance.

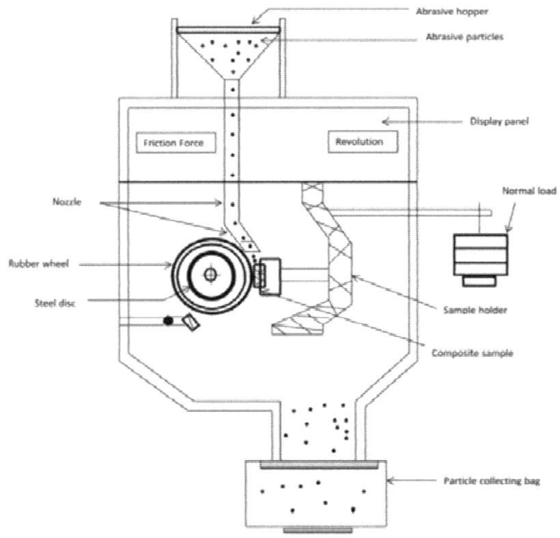

Figure 2. Three-body abrasion (dry sand) test rig.

Table 2. Dry sand abrasion test rig control parameter and their levels.

Control factors	Levels				Units
	I	II	III	IV	
A: Filler loading	0	5	10	15	wt.%
B: Normal load	20	40	60	80	N
C: Sliding distance	500	1000	1500	2000	m
D: Abrasive size	100	200	300	400	μm

2.3 Taguchi's Design of Experiment

Taguchi's design of experimental technique is better alternate of conventional full factorial technique for finding the optimal results by reducing the number of experimental runs. Taguchi's design of experimental technique helps to find out the optimal parameter. Table 2 shows the selected input parameters and their levels for analysis using Taguchi's method. Taguchi analyses find out the input parameter, that is highly influence the response performance (abrasive wear rate). The experimental runs are simulated by software Minitab 17. There are four input parameters and four levels (see Table 2), so L_{16} (4^4 = 16) orthogonal array is selected for the analysis. For any component minimum abrasive wear rate (response/output) is better, so smaller-the-better type signal-to-noise ratio (S/N ratio) has selected for analysis [13].

$$\frac{S}{N} = -10 \log \frac{1}{n} \sum Y^2 \quad (2)$$

Where, n is the Number of observations and Y is the Observed data.
After the Taguchi analysis of observed data, the optimal process parameters are finding with the help of Analysis of Variance (ANOVA) tool. The influence of each process parameter and their optimal level on the output characteristics (Abrasive wear rate) is estimated.

3. Results and Discussion

3.1 Steady State Abrasive Wear Test for Prepared Composites

3.1.1 Abrasive Wear Rate of Composites as a Function of Sliding Distance

The steady state abrasive wear rate of unfilled (0 wt.%, virgin composite EBC 0) and cenosphere filled (5, 10, and 15 wt.%) basalt-epoxy composites are analyzed by dry sand three-body abrasive wear analysis. The abrasive wear rate of unfilled and cenosphere filler basalt-epoxy composite samples with respect to sliding distance (500, 1000, 1500, and 2000 m) are depicted in Figure 3. Other parameters like sliding velocity (200 rpm), normal load (40 N), and abrasive size (300 μm) keeping as constant during the experiment.
The influence of sliding distance on abrasive wear rate of fabricated composites is depicted in Figure 3. From the figure it is clearly seen that the abrasive wear rate tends to decreased with sliding distance from 500 to 2000 m. Composite with 10 wt.% cenosphere (EBC 10) shows minimum abrasive wear rate: from 1.16×10^{-2} mm^3/N-m (for 500 m sliding distance) to 6.56×10^{-3} mm^3/N-m (for 2000 m sliding distance). Result shows that composite with 10 wt.% cenosphere (EGC 10) shows lowest abrasive wear rate than other composites. The reason for decreasing wear rate with filler can be, the filler (cenosphere) creates a hard surface of composite. At the lower sliding distance during abrasion, the sand particles are in contact with epoxy matrix and the epoxy has lower hardness than sand particles, so at lower sliding distance the ratio of material removal is high. After that, at higher sliding distance, the sand particles are in contact with fiber and filler (cenosphere) and fiber/filler act as a barrier to the process of abrasion. But, composite with 15 wt.% cenosphere filler does not show any improvement (in decreasing trend) in wear rate. So, from the analysis it is clear that cenosphere filler up to certain percentage, improves the wear rate (in decremented rate) of composites across the sliding distance range. Similar results were also stated by Arivalagan et al. [14] in their study of with and without fly ash filled carbon fabric reinforced epoxy composite.

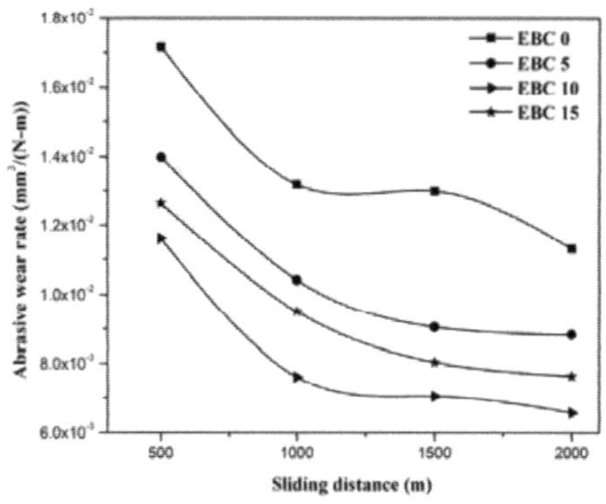

Figure 3. Effect of sliding distance on Abrasive wear rate.

3.1.2 Abrasive Wear Rate of Composites as a Function of Normal Load

Figure 4 represents the steady state abrasive wear rate of unfilled (0 wt.%, virgin composite EBC 0) and cenosphere filled (5, 10, and 15 wt.%) basalt-epoxy composites. The abrasive wear rate is analyzed with respect to normal load (20, 40, 60, and 80 N) keeping other parameters as constant: sliding distance (1000 m), sliding velocity (200 RPM), and abrasive size (300 μm).

The wear rate of composite samples with respect to normal load (20, 40, 60, and 80 N) is depicted in Fig. 4. Result shows that wear rate of composite samples shows improvement (in decremented order) with increase in normal load. Also shows that, addition of cenosphere filler, improves the wear resistance of the composites. Basalt-epoxy composite with 10 wt.% cenosphere filler (EBC 10) shows lowest wear rate from 1.59×10^{-2} mm^3/N-m (at 20 N normal load) to 1.23×10^{-2} mm^3/N-m (at 80 N normal load) and the unfilled cenosphere (EBC 0) composite shows highest wear rate from 2.44×10^{-2} mm^3/N-m (at 20 N normal load) to 1.77×10^{-2} mm^3/N-m (at 80 N normal load). The hardness of the composite surface improves due to addition of cenosphere filler, and due to filler particles, adhesion between reinforcement and matrix becomes superior that's prevents pre mature matrix failure. But higher percentage of cenosphere filler in composite (EBC 15) does not show any improvement (in decremented rate) in wear rate because higher percentage of filler creates more voids in composite, and due to voids, the bonding between matrix and reinforcement becomes poor. The order of wear rate for entire range of normal load is EBC 10 << EBC 15 << EBC 5 < EBC 0. Related results were also stated by Patnaik et al. [15] in their study of mechanical and abrasive wear characteristics of particulate filled glass-epoxy composites.

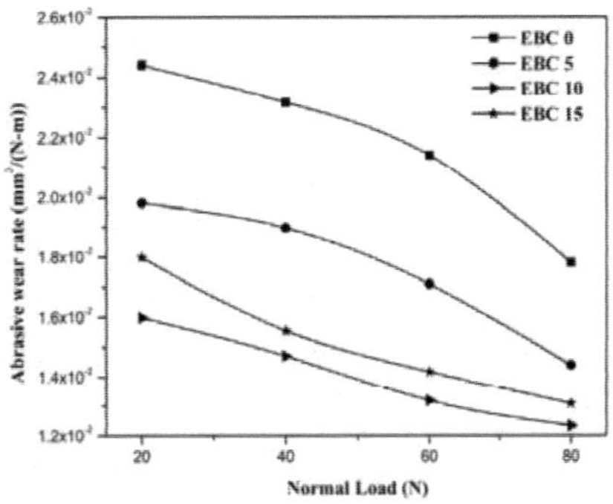

Figure 4. Effect of normal load on abrasive wear rate.

3.2 Taguchi's Design of Experiments and ANOVA Analysis

L_{16} Taguchi's orthogonal array is used to conduct experimental runs. The experimentally evaluated abrasive wear rate and corresponding Signal to Noise (S/N) ratio value of fabricated composites are represents in Table 3. Figure 5 shows the main effect plot of S/N ratio.

The mean value of S/N ratio for all experiment runs is 42.19 dB (see Fig. 5) and the control parameters like A: Filler loading (Level 3, S/N = 45.6174), B: Normal load (Level 1, S/N = 46.1859), C: sliding distance (Level 3, S/N = 46.6616), and D: abrasive size (Level 2, S/N = 44.8407) are designate minimum abrasive wear rate. 10 wt.% filler loading (A_3), 20 N normal load (B_1), 1500 m sliding distance (C_3), and 200 μm abrasive size (D_2) are the optimum control parameter combinations for minimum abrasive wear rate (see Fig. 5). Taguchi's analysis also gives the ranking order; it is depicted in Table 4. As per Table 4 the ranking order of significant input parameters is: C: Sliding distance, B: Normal load, A: Filler loading, and D: Abrasive size.

Further, the percentage contribution of individual control parameter and their order of significance on the response (Abrasive Wear Rate) of composites are analyzed with Analysis-of-variance (ANOVA) tool. The ANOVA analysis is performed at 5% significance level and denoted by 'p'. ANOVA analysis and p-value of each parameter depicted in Table 5. Lower value of 'p' for a parameter indicates, higher will be its contribution on the total variation of abrasive wear rate [16]. The order of significant control parameter and its percentage contribution (P) for abrasive wear performance of composite is: Sliding distance (p = 0.016 and P = 50.99%), Normal load (p = 0.043 and P = 24.51%), Filler loading (p = 0.089 and P = 14.07%), and Abrasive size (p = 0.172 and P = 8.03%). So, from the ANOVA analysis it is clearly seen that, sliding distance, normal load, and filler loading are most significant parameters for abrasive wear rate, respectively. The abrasive size is least significant parameter for abrasive wear rate.

Table 3. Taguchi's L_{16} orthogonal array based experimental design.

Exp. No.	Filler Loading (wt.-%)	Normal Load (N)	Sliding distance (m)	Abrasive Size (μm)	Abrasive Wear Rate (mm^3/N-m)	S/N Ratio (dB)
1	0	20	500	100	0.0107580	39.3654
2	0	40	1000	200	0.0100679	39.9412
3	0	60	1500	300	0.0061733	44.1897
4	0	80	2000	400	0.0062460	44.0880
5	5	20	1000	300	0.0103009	39.7425
6	5	40	500	400	0.0250950	32.0083
7	5	60	2000	100	0.0079359	42.0081
8	5	80	1500	200	0.0041982	47.5387
9	10	20	1500	400	0.0019356	54.2637
10	10	40	2000	300	0.0067632	43.3970
11	10	60	500	200	0.0094290	40.5107
12	10	80	1000	100	0.0060967	44.2981
13	15	20	2000	200	0.0027002	51.3721
14	15	40	1500	100	0.0092744	40.6543
15	15	60	1000	400	0.0168799	35.4526
16	15	80	500	300	0.0155896	36.1433

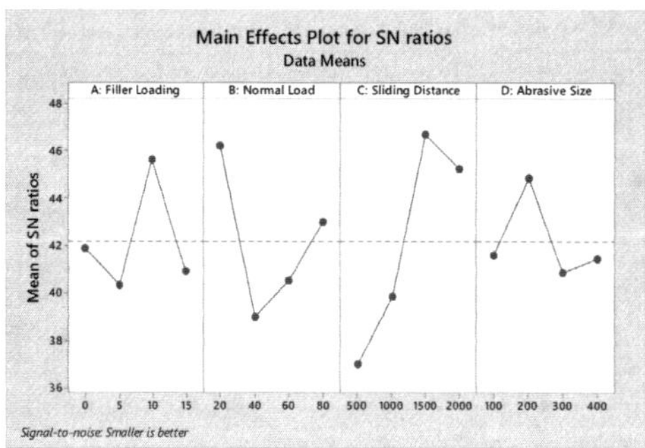

Figure 5. Effect of input parameters on abrasive wear rate.

Table 4. Input control parameters ranking order of significance.

Level	Filler Loading (wt.-%)	Normal Load (N)	Sliding Distance (m)	Abrasive Size (μm)
1	41.90	46.19	37.01	41.58
2	40.32	39.00	39.86	44.84
3	45.62	40.54	46.66	40.87
4	40.91	43.02	45.22	41.45
Delta	5.29	7.19	9.65	3.97
Rank	3	2	1	4

Table 5. ANOVA results for abrasive wear rate.

Input control factors	DF	Adj. SS	Adj. MS	F-value	p-value	*P* (%)
A: Filler Loading	3	67.85	22.618	5.93	0.089	14.07
B: Normal Load	3	118.19	39.397	10.32	0.043	24.51
C: Sliding Distance	3	245.81	81.938	21.47	0.016	50.99
D: Abrasive Size	3	38.75	12.916	3.38	0.172	8.03
Error	3	11.45	3.817			
Total	15	482.06				

4. Conclusions

In this study, basalt fiber reinforced epoxy matrix composites with or without cenosphere particulates are fabricated by open mold casting technique. Abrasive wear rate under steady state condition as a function of sliding distance and normal load increased shows improvement (in decreasing trend). 10 wt.% cenosphere filled basalt-epoxy composite (EBC 10) shows minimum abrasive wear rate. Taguchi experimental analysis was done with four input control parameters and their four levels. From the Taguchi analysis parameter sliding distance was most significant for abrasive wear rate followed by normal load, filler loading, and abrasive size.

Acknowledgement

The author acknowledges the Mechancial engineering Department of Malaviya National Institute of Technology, Jaipur for the financial support. The author also expresses their sincere gratitude to Advance Research Lab for Tribology and Material Research Centre of the institute for experimentation and cauterization work.

References

[1] Choudhary M, Sharma A, Dwivedi M, and Patnaik A, 2019 *Fibers and Polymers* **20** 823.

[2] Panwar M J, patniak A, and Nagar R, 2017 *Polym. Compos.*, **38** 736.

[3] Ahmetli G, kocak N, Dag M and Kurbanli R, 2012 *Polym., Compos.*, **3** 1455.

[4] Kulkarni S M and Kishore, 2001 *Polymers and polymer composites*, **9** 25.

[5] Barkoula N M and Karger-Kocsis J, 2002 *Journal of reinforced plastics and composites*, **21** 1377.

[6] Patnaik A, Satapathy A, Mahapatra S S and Dash R R, 2008 *Journal of reinforced Plastics and composites*, **27** 1093.

[7] Dadkar N, Tomar B S, Satapathy B K, 2009 *Materials & Design,* **30(10)** 4369.

[8] Alkadasi N A N, Hundiwale D G, Kapadi U R, 2003 *J Appl Polym Sci*, **19** 1322.

[9] Sunil T and Chauhan S R, 2014 *J Compos Mater*, **18** 2831.

[10] Chand N, Sharma P and Fahim M, 2010 *Mater Sci Eng*, A **527** 5873.

[11] Dipa R and Gnanamoorthy R, 2007 *J Reinf Plast Compos*, **26** 5.

[12] Suresha B, Chandramohan G, Siddaramiah, Jayaraju T, 2008 *Poly Compos*, **29** 307.

[13] Amuthakkanna P, Manikandan V, Jappes J T W, Uthayakumar M, 2013 *Mater. Phys. Mech.*, **16** 107.

[14] Chen P, Li J, Zhang J, 2018 *J. Wuhan Univ. Technol. Mater. Sci. Ed.*, **33** 139.

[15] Patniak A, Satapathy A and Biswas S, 2010 Malays. Polym. J., **5** 37.

[16] Choudhary M, Singh T, Sharma A, Diwedi M and Patniak A, 2019 Mater. Res. Exp., **6** 105702.

IOP Conf. Series: Materials Science and Engineering 1017 (2021) 012015 doi:10.1088/1757-899X/1017/1/012015

Influence of deposition parameters on Tribological Performance of HVOF Coating: A review

[1]Alok Vats, [1]Arun Kumar, [2]Amar Patnaik, [2] M L Meena,

[1]Mechanical Engineering Department, MBM Engineering College, Jodhpur, India
[2]Mechanical Engineering Department, MNIT Jaipur, India

E-mail: alokvats1091@gmail.com

Abstract. Any mechanical component stationary or in relative motion is exposed to severe surface phenomena like corrosion, wear, abrasion etc. To address these issues various surface treatment processes are in industrial practice. In recent times, coatings techniques have been identified as cost effective alternative to other surface modification techniques. Thermal spray technique proves to be the prominent method to among a large variety of processes available. HVOF method owing to its capability to provide wear and corrosion resistant coatings with characteristics like high hardness, adhesion, density (porosity <1%) is widely accepted across industries. Process parameters seems to have significant influence on coating properties like porosity level, oxide content, micro hardness and tribological performance of coating. This paper intends to investigate the effect of above mentioned process deposition parameters on physical, mechanical, wear, erosion and corrosion properties of coating.

1. Introduction

A particular mechanical component can be subjected to different sorts of operating environments. Corrosion, wear, oxidation, abrasion, adhesion and radiation are regular phenomena that these components are regularly subjected to. The extent of deformation and wear depend on the surrounding media. When two or more parts are having a contact, the properties of both parts influence wear. Wear, friction and corrosion are surface phenomena and cause great damage to the parts. All industries including automobile, aerospace and biomedical receive the highest impact of these phenomena.

The wear and friction lead to a loss of about 2-4% of the GDP of an industrialized nation. Losses due to friction account for nearly 23% of the total worldwide consumption of energy [1]. The estimation of the severity of friction can be judged from the fact that near about 33.3% of fuel is wasted to limit friction in automobiles. A reduction in losses on account of friction can lead to savings upto 576,000 million Euros and 385,000 million litre of fuel. There is also a reduction of 960 million tons of CO_2 emission from this [2]. The advent of corrosion can cause more severe failures than that by friction alone. The total monetary value of corrosion is near about 3.4% of world's GDP [3].

The primary requirement of surface engineering is modification of surface properties through various processes e.g. Atomic Deposition, Particulate Deposition and Bulk Coatings. The characteristics of HVOF coatings are quiet superior. They include good bonding strength, dense and optimized structure and less number of pores. These coatings also possess great surface hardness. There are various spray methods e.g. by flame spraying, High Velocity Oxy-Fuel (HVOF) and plasma spray. Main aim of all these methods is to develop wear resistant and corrosion tolerant coatings.

There are various ways through which different class of materials like metals, ceramics, cermet and polymer materials can be deposited using HVOF method. The materials can take the powdered form, wired or rod form. They are inserted into a gun (torch). Inside the torch the material is heated near or slightly above the melting point. Material is converted to molten or semi-molten droplets. They are made to accelerate by the carrier gas. Their high velocity projects them upon the substrate. Upon striking, the droplets get converted into lamellar (thin) morphology. The heated droplets adhere to the surface. They keep on overlapping one over the other. They get interlocked during the process of solidification. A number of passes are required to produce a thick coating [4]. Table 1 shows the process attributes of various thermal spray methods.

The HVOF method is of great importance among these techniques. HVOF equipment of modern state of the art have gas speeds ranging up to 2000 m/s. The high speeds and medium temperatures allow very dense coatings to be developed with exceptional wear and corrosion behavior and excellent bond strengths [5]. Coatings sprayed by HVOF have found broad application in the maritime, aviation, automobile sectors and associated fields.

Thermal spraying with HVOF is a technique that melts and propels the powder material at high speed. It makes use of the oxygen and mixtures of the fuel gas, towards the substrate. Gases like H_2, C_3H_8, CH_4 propylene, acetylene, SPRAL 29 kerosene, ethylene, crylene, and LPG etc. helps in burning of fuel. The HVOF system is composed of a spray gun and three units. There are various units to feed powder, to measure and control the flow of the gases and to supply air and carrier gas. The powder feeder is made of a hopper, one vibrator for air, meter to monitor feed rate and controller. The material to be coated is injected from the feed unit via carrier gas and finally to the gun, where combustion takes place. By measuring feeding rate, the quantity of powder necessary for coating can be controlled. Based on the melt temperature and feed rate of the fuel, the powder material reaches the flame in the combustion zone. At this stage, it could be in molten or may be in semi-molten form. The HVOF cycle has a flame temperature of about 3000°C [6]. At supersonic velocities, the molten or semi-molten particles are then pushed from nozzle of gun onto the target / substrate. The splat wise deposition takes place at the substrate. Basic principle of Gas fueled HVOF technique is shown in figure 1.

There is a high speed and high impact force of the accelerated particles of powder. The coatings developed by the HVOF technique have less number of pores. The oxy-fuel coatings have a higher bond strength than other methods of spraying by plasma, flame and electric arc [6]. The typical features of HVOF coatings like inter-lamellar splats, porosity, semi- molten top layer due to high velocity semi- molten particles are shown in figure 2.

The process parameters like spraying distance, fuel and oxygen flow rate, rate at which powder is fed, particle speed, flame temperature have significant influence on coating properties. The most important coating characteristics include porosity level, oxide content, micro hardness, bond strength and tribological performance. The intent of this paper is to review the influence of process deposition attributes on physical, mechanical, wear, erosion and corrosion properties of thermally sprayed coating.

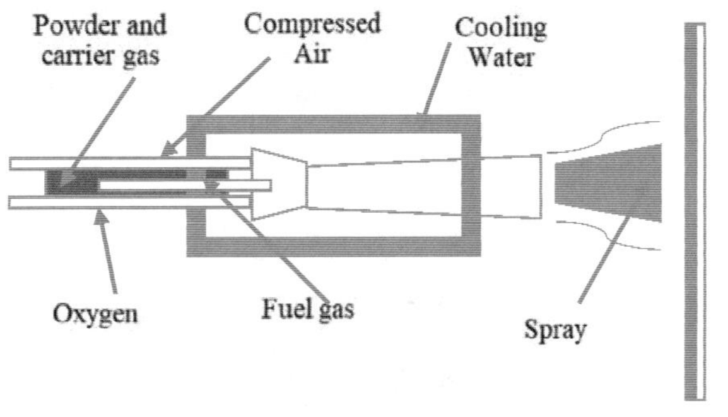

Figure 1. Gas fueled HVOF Thermal Spray Gun

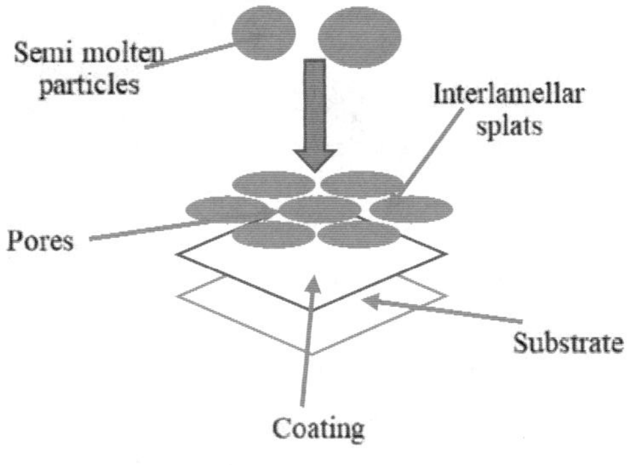

Figure 2. Schematic showing the typical features of HVOF coatings

2. Physical and mechanical properties of Coatings

The deformation was more pronounced in the liquid state as compared to the molten state. The coatings had a compact (dense) structure. The hardness of coating was because of high particle K.E. during HVOF technique. As a result, the droplets reacted with oxygen less frequently. All this happened during the flight time of the particles and their flattening. The diffusion mechanism allowed the process of oxidation to take place slowly. The high K.E. of the particles (plastic condition) allowed flattening by process of deformation. It resulted in the formation of dense and pore-free coating. The amount of oxygen in the coatings was low.

Micro size and nano size coatings of Tungsten Carbide-Cobalt (WC-12%Co) were thermally sprayed on carbon steel using both plasma and HVOF spray. The fuel rate was taken as 0.0063 L/S. The powder feed rate for the particles of micro and nano size was taken as 77 and 90 g/min respectively. Spray distance for HVOF process was taken as 381mm while it was 127 mm for plasma spray [7]. For HVOF micro, the coatings were designated as H1 and for H2 as HVOF nano. All these parameters resulted in very dense HVOF coatings (porosity $< 1.3\%$) and a very low (less than 1.69%) amount of oxide. These features are evident in figure 3.

It was observed that coating hardness produced by micro-sized powder spraying by HVOF when averaged was found to be 1066 HV (H1). While, for the nano-sized powder coatings produced by (H2) sprayed by HVOF method it was found to be 1367 HV. It was due to distribution of the nano-sized powders over large area. The finer distribution of nanostructured carbides in the matrix was also responsible for greater hardness of H2 coating.

Lekatou et al. [8] conducted a similar analysis. They studied the traditional and nano WC 12%Co HVOF coatings deposited on Al7075 T6. They investigated that the nano-sized powder coatings had considerably lower porosity. Their mechanical properties like micro-hardness and fracture- toughness were also better. But the degree of decarburization was higher, as compared to their traditional counterparts.

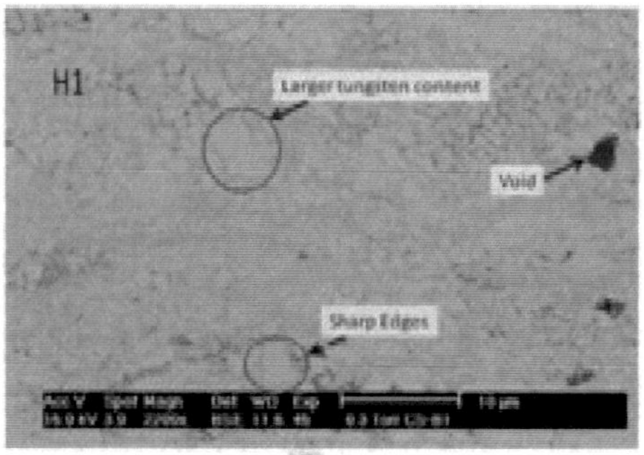

Figure 3. Coatings of H1 showing voids and sharp edges in cross section mode [7]

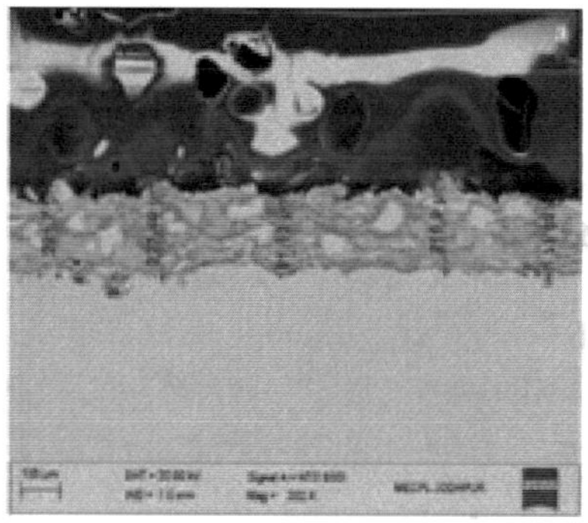

Figure 4. SEM micrograph of Ni_3Ti coating showing cross sectional thickness and lamellar microstructure [10].

Researchers recorded Higher hardness and less porosity [9], in the three coatings (Cr_3C_2-self fusing alloy, Cr_3C_2-Ni-Cr-Al-Y, Cr_3C_2-Ni-Cr) deposited by HVOF spray. The deposition of Cr_3C_2-Ni-Cr-Al-Y coatings took place by Atmospheric Plasma Spray (APS), on nimonic discs. They discovered that APS deposited CRNY coatings were more porous (6-12%) than the coatings sprayed by HVOF (3-8%). The deposition parameters were oxygen flow (800, 870, 940 slpm), kerosene flow rate (20.8, 22.7, 24.6 litres per hour) and deposition distance (320, 350, 380 mm). They resulted in a higher hardness of HVOF coatings (1100-1300 HV0.1) compared to the plasma sprayed coatings (700-800 HV 0.1).

Katranidis et al. [10] investigated the effect of variable parameters e.g. spraying distance, angle of spray and traverse speed of gun on the complex geometry. There was a rotation of samples in front of the spraying gun. Results of WC-17 percent Co powders on steel substrates through HVOF revealed that in the short spray distance, micro-mechanical properties like compressive residual stresses, micro-hardness and porosity were higher. The particles appeared to hit at lower temperatures and elevated velocity. It resulted in high transfer of peening stress. Porosity value was unaffected when it was spray from very long distances from the gun at different angles. It was because peening stresses in the microstructure of the coating became less important. Gun traverse velocity did not impact the levels of porosity.

The production of Ni3Ti and Ni3Ti+(Cr3C2 + 20Ni-Cr) coatings by HVOF technique on AISI 420 stainless steel showed better mechanical properties. MDN 420 and ASTM B3265 titanium alloy (Ti-15) acted as the substrate. The microstructure of the coatings was concentrated. The layers reflecting the laminar structure [11] are shown in Figure 4. The coating micro-hardness was found to be higher than that of the substrate. It was due to the greater density, lower porosity and high micro-cohesion between the slosh of coating materials. Low porosity (1-3%) indicated that the HVOF technique produced sufficient deformation of the particles. They travelled at high velocity and had high temperature. It caused high cohesion of individual splats [11].

It was planned to increase the solid particle erosion resistance of WC 10Co 4Cr ceramic powder [12]. In this study, the influence of ceramic-reinforced (Yttrium and Zirconium) coating powders was investigated. The coatings were manufactured by HVOF technique on SS X2-Cr-Ni-Mo-17-12-2. The fuel flow rate of LPG was 70 slpm. With the aid of the lab scale pot tester, tribological properties were tested. Coal ashes were used as the erodent (i.e., fly ash and bottom ash). In comparison with traditional WC-10Co4Cr cermet. Yttria enhanced WC-10Co4Cr coating showed superior tribological and mechanical properties like low porosity and high micro-hardness. In fly ash slurry, the zirconia reinforced WC-10Co4Cr coating had demonstrated excellent wear resistance.

3. Wear Resistance Properties

HVOF is among the versatile techniques of deposition of coating material. This method increases wear resistance. This is a process with no material constraint and has the ability to deposit coatings ranging from few micrometres to tens of millimetres. Wide variety of shapes and sizes are possible with HVOF. The carbide ceramic coatings HVOF are commonly used against tribo-corrosion in the petroleum industries. One such carbide coating (Cr3C2-NiCr) performed by the HVOF method on carbon steel [13] showed better tribological efficiency than the uncoated sample. A number of coating parameters such as the feed rate of powder (50-60 g / min), the flow rate of fuel (20-30 l / hr) and the spray distance (200 mm) were used. They played a major role in achieving a compact microstructure [13].

In industries such as mining and hydropower production, the coatings such as iron aluminides have proven to be protective for steel components. Amiriyan et al. [14] performed the HVOF sprayed Fe3Al / TiC coating wear test. They found that the mechanical and tribological performance of the alumina counter body coatings were improved. The spike in wear resistance of counter body coatings was because of the improvement in hardness. The development of TiC particle (in Fe_3Al matrix) assisted in this process. Sliding and abrasive wear tests conducted by Bolelli et al. [15] on HVOF as well as HVAF-sprayed Cr_3C_2 25 wt. % Nickel Chromium coatings (hard metallic). They used two different particle sizes (fine and coarse). Their study showed that the fine feedstock powder provided coatings having higher wear resistance than the coatings produced by coarse feedstock powder. It was because of greater cohesion between matrix and carbide within the lamellae.

The surface of the substrate (Ti6Al4V) was treated with W-DLC, HVOF and Ion-Implantation techniques. The researchers used the coatings of CrN with a topcoat (C: H: W). The Cr3C2 was embedded in the Ni / Cr matrix and Low Pressure Nitriding [16] was used to improve the tribo-performance of Ti6Al4V in seawater. The Ti6Al4V substrate coated with W-DLC coating showed the minimum friction. The stability of friction was due to the formation of a transfer layer of less shear strength at interface. The coating of W-DLC and HVOF showed better tribo-corrosion efficiency under Open Circuit Potential (OCP) conditions. But the Ion implantation and the untreated Ti6Al4V substrate did not perform that well. It can be attributed to their enhanced friction and shallow depth of wear scars (Figure 5).

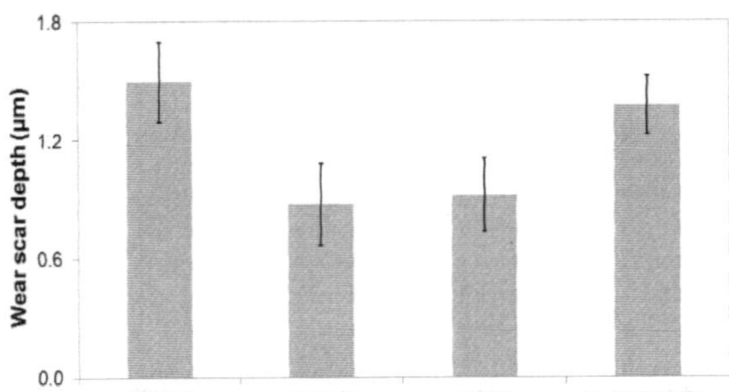

Figure 5. Depth of wear scar on target surface treatments [11].

Carbon steel is another most used material in oil and gas industries. It is exposed to regular tribo-corrosion [12]. Tribological and mechanical properties of HVOF sprayed layers on carbon steel were investigated in [17]. The coatings had the approximate formula Al_2O_3-40TiO_2 and Cr_3C_2-20NiCr. Both coatings were better in terms of thickness, porosity and substrate bonding. The highest wear resistance was that of the Cr_3C_2–20NiCr coating deposited by HVOF. The sample wear mechanism was based on the properties of the transferred layers. A variation in loads resulted in a change in the properties of these layers.

The effect of particle size on coating wear efficiency has been investigated [18]. The compact WC-10Co-4Cr coatings were deposited with average particle sizes of 0.6, 1.2 and 6.1 μm. All these three types of WC particles were deposited on steel substrates (ASTM 1045) for aircraft landing gear [18]. HVOF with process parameters such as kerosene (0.37 litre/ min), powder feed speed (85 gram/min) and stand-off distance (380 milli-meter) was used in this study. The low size of the particles resulted in low porosity. The best abrasive wear resistance was demonstrated by 1.2 μm WC particles coating. The attributed like plasticity and homogeneity of coatings was found to be increased with the increment in the WC particle size. The micro-cutting wear mechanism was found to be primary wear mechanism. Researchers [13] investigated erosion and abrasion wear performance of HVOF deposited WC coatings. The coatings had an approximate chemical composition of WC12Co, WC10Co4Cr and Cr_3C_2 25NiCr. The substrate was 316 grade stainless steel. The erosion (solid particle) and abrasion testing were performed. The Cr_3C_2-25NiCr coating showed the maximum wear rate both tests (Figure 6). Cr_3C_2-25NiCr coating had the lowest coefficient of friction. The COF increased initially as surface asperities were ploughed and localized contact stress was found to be increased. Formation of tribo-oxide film resulted in lower COF (Figure 7).

Effect of particle velocity on coating performance was observed by Kumar et al. [14]. They used WC-CoCr hard coatings of composites on the steels used for hydro turbines using two

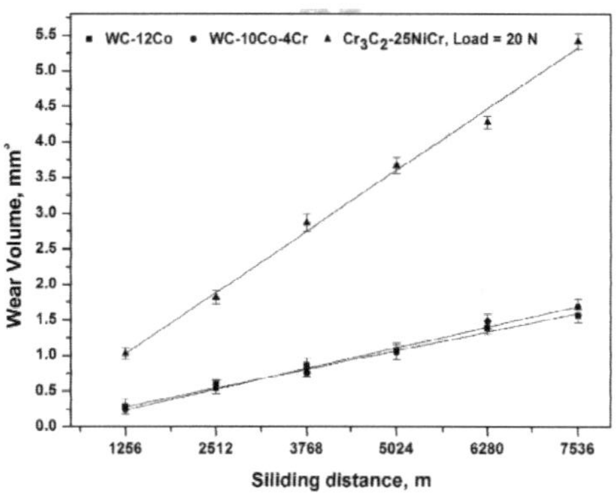

Figure 6. Variation of abrasive wear volume with sliding distance at load of 20 N [13].

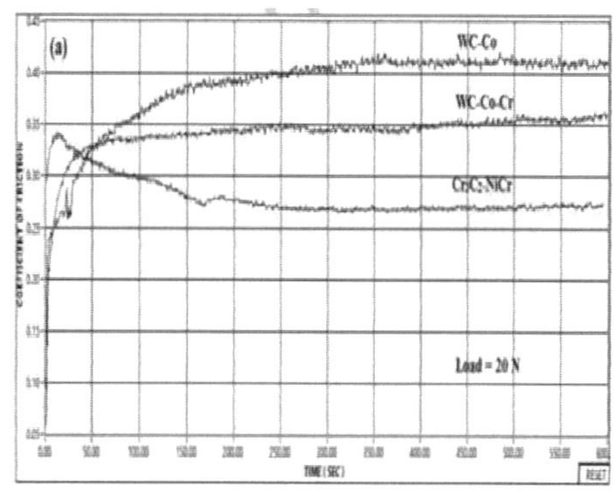

Figure 7. Variation in COF for at a load of 20 N [13].

different thermal spray techniques HVAF and HVOF. The parameters like spray particle velocities for HVAF were 1010 meter/s, 960 meter/s and 895 meter/s. HVOF used the particle speed of 680 meter/s. WC–CoCr coating sprayed at higher velocity by HVAF method exhibited comparatively better erosion resistance than others. It was true for all impact angles as well as jet velocity. HVAF samples performed better than HVOF coated ones. It was due to the better mechanical properties like adhesion, high density, increased hardness. The resistance to erosion was also found to be due to these attributes [14].

Another protective coating, $Al_8Si_{20}BN$ was deposited via HVOF and plasma spray. The carbon steel substrate was investigated by researchers in [15] with respect to coating properties. The characterization of coatings was done to know about phase composition, micro-structure, micro-hardness and bond strength. Response to wear was investigated by applying 5N, 10N, 15N and 20N loads on pin-on-disk machine. After which a comparison of the results was done. As per the findings, the durability of HVOF-coated models was greater than the plasma-coated samples. All these tests were performed under different loadings in the same environment [15].

In another study, WC 10Co4Cr powder, Cr_3C_2 25NiCr powder and carbide powder WC 40Cr_3C_2 25NiCr, was deposited by HVOF method. The process parameters like Fuel flow rate 23.8 litre/h, standoff distance 380 mm and Feed rate of 70, 45 and 55 g/min were used by the researchers. They exhibited better mechanical and wear properties [16].

The process parameters were kept nearly constant at the values 0.46 and 0.38 l/min (Fuel flow rate), 45 g/min (Powder feed rate) and 356 mm (Standoff distance). The effect of coating powder material on the performance of the WC based HVOF coatings was assessed. Song et al. [17] reported the better performance of HVOF sprayed WC Co Cr coating under load of 96, 240 and 318 N. WC-Cr_xC_y-Ni coatings could not perform that well. The wear performance of the coating was determined by phase composition of phases and relative proportion of hardening carbides. The properties of binder and the interaction of carbide formation and matrix in WC-Cr_xC_y-Ni powder were responsible for its wear resistance.

WC coatings proved beneficial in conditions where cracking of coating is a normal occurrence. Tribological performances of such HVOF coated WC-12Co on steel sample were examined. The process parameters like Kerosene flow rate (22.7 l/h), powder feeding rate (65 gram/min) and spray distance (380 mm) were used. The researchers Tao Luo et al. [18] carried out rubber wheel abrasion test under dry condition and ball on disc wear tests in sliding module. The results showed that the coating exhibited poor wear resistance in abrasion due to crack propagation. However, the mass loss that occurred at the cracks was in the initial stages. As the time passed by, both the cracked and non-cracked HVOF coatings showed a similar wear rate. It was due to the ability of the cracks to store wear debris. The debris caused an abrasion of the surface of HVOF coating during sliding.

Optimum wear resistance and mechanical properties were obtained at different set of spray parameters. Co-Mo-Cr-Si super alloy coatings, prepared on Mar-M247 (Ni-based super alloy) by HVOF showed better fretting wear resistance. The coatings showed a dense (porosity $< 0.8\%$) behavior. The process parameters like feed rate of powder (6.5 RPM), flow rate of fuel (6.0 GPH), O_2 flow rate (1950 SCFH) and spray distance (380 mm) were used. However, the effect of spraying parameters was less pronounced on hardness and wear tolerance of coatings [19].

4. Erosion and Corrosion Behaviour of Coatings

Effect of acid concentration on corrosion performance of HVOF deposited alumina coating was investigated. The AISI 4340 tool steel was used as substrate. Parameters like flow rate of fuel, feed rate of powder and spray distance were taken as 280 ml/min, 70 g/min and 50.8 cm respectively. The quantitative analysis of coated sample done by immersing in various media. They included HCl environment with concentrations of 5%, 10% and 15 vol.%. After 3 hours, it was found that the increment in acid concentration enhanced the corrosion rate. As a result, highly porous and deteriorated surfaces were obtained [20].

Aluminium and nickel industries face problems in pumps while transporting caustic liquor. Lima et al [21] used AISI 1020 carbon steel as substrate. Three powder materials of Cr_3C_2–25NiCr, WC–17Co and WC–12Co were used as feedstock in the HVOF method. Substrate was tested for slurry erosion testing. The erosive particles (400 gram/l) of Fe_2O_3 (HRC 40 to 50 and avg. dia. of 600 micron) were deployed as the erosive agent. Powder feed rate and Spray distance was taken in the range of 38-76 g/min and 250-350 mm respectively for three coatings. Impact angle was taken as 90°. WC coated samples showed the highest micro hardness values. The WC–12Co coated layer (Figure 8) showed the highest wear under experiment conditions. The wear grew up with time.

WC–17Co coatings performed better in erosion test. It was owing to a unique combination of high micro hardness and low porosity. Poor corrosion behavior of Cr_3C_2–25Ni-Cr coating was due to high roughness and porous nature. The destruction of the passivated layer lowered the corrosion performance [21].

Research work conducted by peat et al. [22] [23] explored the tribo-corrosion behavior of HVOF coating. WC-Co-Cr, Cr_3C_2-Ni-Cr and Al_2O_3 surface coatings were formed on S355 steel specimen. The test was conducted under three conditions. They were erosive liquid impingement, slurry-erosion and dry jet-erosion. Fuel flow rate was taken as 681.5, 0.455 and 732 (l/min) for WC-Co-Cr, Cr_3C_2-Ni-Cr and Alumina (Al_2O_3) coatings respectively. Spraying distance was 229, 355 and 178 mm for above coatings, respectively. It was found that Al_2O_3 coatings provided comparably poor protection under E-C (erosive corrosive) medium. Despite Cr_3C_2-NiCr provided great corrosion resistance. But the coating failed poorly due to erosive wear under liquid impingement conditions. It led to highest loss of material out of all the investigated coatings. This wear behavior continued in erosion under dry as well as slurry conditions.

In another study, the Corrosion behavior of HVOF deposited WC-17Co, WC-10Co-4Cr and Cr_3C_2-25Ni-Cr coatings in solution of alkaline sulfide having S^{2-}, 0.2 mililitre/L, and pH value of 10 was evaluated. 18 days of immersion testing was conducted. The examination of corrosion rate showed that Cr_3C_2-Ni-Cr coating had less porosity. It was the reason for the best resistance to corrosion. In solutions of alkaline-sulphide (pH=10), the number of pores, formation of passive film and setting up of micro-galvanic potential between hard phase and binder phase played a great role in determining the corrosion behavior of HVOF coatings [24].

The prediction of erosion wear rate using an Artificial Neural Network (ANN) was done for hard coatings. The layers of coatings were deposited by different spray techniques including thermal. The division of coating was done in three groups of WC, CrC and metallic alloy coatings. Different microscopic characteristics that helped in controlling the mechanical properties and the tribo-performance of the 3 classes were analyzed. The results showed that the velocity of impact of the particles had a higher effect on the erosion behavior. But the angle of impact did not effect the erosion resistance much [25].

Zheng et al. [26] investigated commercial WC–10Co–4Cr coating deposited by using HVOF on the AISI 4140 substrate. ASTM G-73 standards were followed while conducting erosion testing. They revealed that resistance to erosive wear of the coating samples was remarkably improved. The wear performance was compared to that of un-coated material. At low angle of impact, the main erosive wear mechanism was micro-cutting as well as horizontal cracks failure (Figure 9). In a similar study [27], there was an analysis of erosive wear performance of WC– 10Co–4Cr and Ni–20Cr_2O_3 coated SS 304 substrate samples. It was observed that WC–10Co–4Cr coated samples gave a better performance than Ni–20Cr_2O_3 [27].

The CaviTec powder in the atomized state was used to perform the HVOF coating. In this study, the C-E (cavitation erosion) testing of HVOF coatings was done. By milling the feed stock powder prior to deposition, the performance of the coatings was enhanced. A great improvement in the wear characteristics was obtained optimizing the deposition parameters. Damaged surface of Cavi-Tec coatings indicated that the defect sites acted as point of initiation for material loss and degradation of surface. Pores and the boundaries at the inter-splat regions were included in the defects (Figure 10) [28]. In a similar effort to reduce the wear of grinding rotors in high speed mill, WC-CoCr coating was deposited by HVOF technique. The process parameters like Ethylene (C_2H_4) (60 slpm) and spray distance (200 mm) were used. When tested for dry erosion, the coating had lowest weight loss than APS-sprayed Cr_3C_2 -25NiCr and NiCrBSi coatings [29].

The slurry erosion-corrosion behavior of HVOF deposited WC-10Co-4Cr coatings (cermet) and the base SS 1Cr-18Ni-9Ti was examined by Hong et al. [30]. Spray of the feed stock powder took place at parameters like flow rate of kerosene (0.38 L/min), flow rate of O_2 (897 L/min) and nozzle spraying distance (300 mm). There was a rig with rotating disc apparatus. The rig also had circulating system facility. It was used to perform the coating. Slurry of water and NaCl (3.5 wt.%) were used in

the test. The results revealed higher slurry wear E-C (erosive-corrosive) resistance was found in case of WC-10Co-4Cr coating in comparison to the SS substrate in both media.
Effect of porosity in relation to corrosion resistance was investigated in an another study [31]. The corrosion andand erosion resistance (in the slurry environment) of HVOF deposited WC-10Co-4Cr coated samples was examined. The parameters like Feed rate (40 gm/min), rate of fuel flow (60 ltr/min) and spraying distance (200 mm) were used. The AISI 410 stainless steel (application in hydraulic turbines) was used as substrate. The Coatings had better wear performance during the slurry erosion test as compared to the uncoated steel substrate. In another study, it was found that the corrosion resistance was improved because of lower porosity. Presence of 5% TiO_2 in HVOF sprayed 80Ni-15Cr_2O_3 and 80Ni-15Al_2O_3 coating led to an increment in the resistance to erosive wear of mild steel pipeline. The process parameters like Kerosene flow rate (70 /min) and Spray distance (140 mm) were deployed. The application of this steel material is in mining and chemical sector. The prime purpose transportation of solids. Experimental results indicated that rotational speed played an important role in erosive wear of samples in given media, time duration of test and nature of eroding particles. Coating with composition of 80Ni-15Cr_2O_3–5TiO_2 performed better than 80Ni-15Al_2O_3–5TiO_2 coating [32].

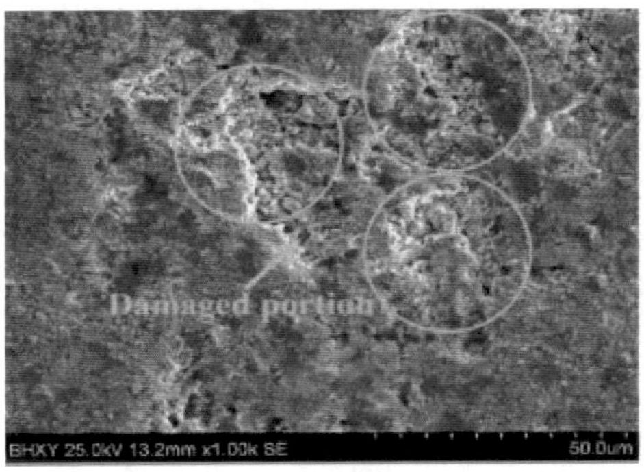

Figure 8. Eroded morphology of WC coating at low Impact angle (15°) [26].

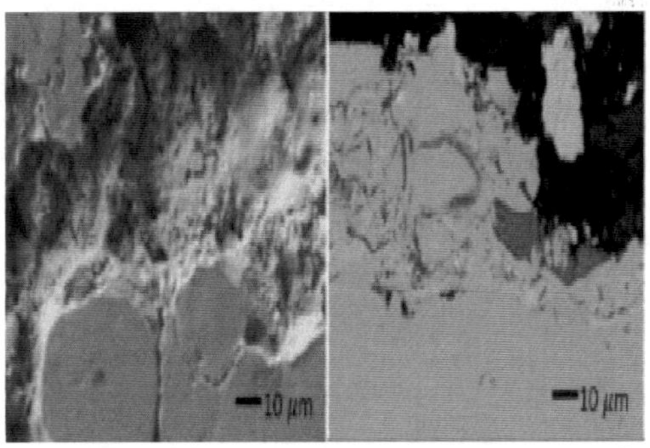

Figure 9. Cross sectional SEM image of coating after 2h of cavitation-erosion [28].

5. Conclusion
Wear resistance of HVOF coatings is better than other methods, except HVAF. They also have an excellent corrosion resistance. Their use is recommended for high temperature application also.

References
[1] K. Holmberg and A. Erdemir, "Influence of tribology on global energy consumption , costs and emissions," vol. 5, no. 3, pp. 263–284, 2017, doi: https://doi.org/10.1007/s40544-017-0183-5.
[2] K. Holmberg, P. Andersson, and A. Erdemir, "Tribology International Global energy consumption due to friction in passenger cars," *Tribiology Int.*, vol. 47, pp. 221–234, 2012, doi: 10.1016/j.triboint.2011.11.022.
[3] M. G. Gerhardus Koch, Jeff Varney, Neil Thompson, Oliver Moghissi and J. Payer, "International Measures of Prevention , Application , and Economics of Corrosion Technologies Study," 2016.
[4] M. R. Dorfman, "Thermal Spray Coatings," *Handb. Environ. Degrad. Mater.*, vol. 5, pp. 569–596, 2012, doi: 10.1016/B978-1-4377-3455-3.00019-5.
[5] M. Parco, L. Zhao, J. Zwick, K. Bobzin, and E. Lugscheider, "Investigation of HVOF spraying on magnesium alloys," *Surf. Coatings Technol.*, vol. 201, pp. 3269–3274, 2006, doi:

10.1016/j.surfcoat.2006.06.047.
[6] T. S. Sidhu, S. Prakash, and R. D. Agrawal, "State of the art of HVOF coating investigations - A review," *Mar. Technol. Soc. J.*, vol. 39, no. 2, pp. 53–64, 2005, doi: 10.4031/002533205787443908.
[7] S. Al-Mutairi, M. S. J. Hashmi, B. S. Yilbas, and J. Stokes, "Microstructural characterization of HVOF/plasma thermal spray of micro/nano WC-12%Co powders," *Surf. Coatings Technol.*, vol. 264, pp. 175–186, 2015, doi: 10.1016/j.surfcoat.2014.12.050.
[8] A. Lekatou, D. Sioulas, A. E. Karantzalis, and D. Grimanelis, "A comparative study on the microstructure and surface property evaluation of coatings produced from nanostructured and conventional WC-Co powders HVOF-sprayed on Al7075," *Surf. Coatings Technol.*, vol. 276, pp. 539–556, 2015, doi: 10.1016/j.surfcoat.2015.06.017.
[9] L. Baiamonte *et al.*, "Thermal sprayed coatings for hot corrosion protection of exhaust valves in naval diesel engines," *Surf. Coatings Technol.*, vol. 295, pp. 78–87, 2016, doi: 10.1016/j.surfcoat.2015.10.072.
[10] N. C. Reddy, B. S. A. Kumar, H. N. Reddappa, M. R. Ramesh, P. G. Koppad, and S. Kord, "HVOF sprayed Ni_3Ti and Ni_3Ti+(Cr_3C_2+20NiCr) coatings: Microstructure, microhardness and oxidation behaviour," *J. Alloys Compd.*, vol. 736, pp. 236–245, 2018, doi: 10.1016/j.jallcom.2017.11.131.
[11] V. Totolin, V. Pejaković, T. Csanyi, O. Hekele, M. Huber, and M. Rodríguez Ripoll, "Surface engineering of Ti6Al4V surfaces for enhanced tribocorrosion performance in artificial seawater," *Mater. Des.*, vol. 104, pp. 10–18, 2016, doi: 10.1016/j.matdes.2016.04.080.
[12] M. A. Zavareh, A. A. D. M. Sarhan, P. A. Zavareh, B. B. A. Razak, W. J. Basirun, and M. B. C. Ismail, "Development and protection evaluation of two new, advanced ceramic composite thermal spray coatings, Al2O3-40TiO2 and Cr3C2-20NiCr on carbon steel petroleum oil piping," *Ceram. Int.*, vol. 42, no. 4, pp. 5203–5210, 2016, doi: 10.1016/j.ceramint.2015.12.044.
[13] N. Vashishtha, R. K. Khatirkar, and S. G. Sapate, "Tribological behaviour of HVOF sprayed WC-12Co, WC-10Co-4Cr and Cr_3C_2–25NiCr coatings," *Tribol. Int.*, vol. 105, pp. 55–68, 2017, doi: 10.1016/j.triboint.2016.09.025.
[14] R. E. Kumar, M. Kamaraj, S. Seetharamu, and A. K. S., "A pragmatic approach and quantitative assessment of silt erosion characteristics of HVOF and HVAF processed WC-CoCr coatings and 16Cr5Ni steel for hydro turbine applications," *Mater. Des.*, vol. 132, pp. 79–95, 2017, doi: 10.1016/j.matdes.2017.06.046.
[15] M. A. Zavareh, E. Doustmohammadi, A. A. D. M. Sarhan, R. Karimzadeh, P. Moozarm Nia, and R. S. Al/Kulpid Singh, "Comparative study on the corrosion and wear behavior of plasma-sprayed vs. high velocity oxygen fuel-sprayed Al8Si20BN ceramic coatings," *Ceram. Int.*, vol. 44, no. 11, pp. 12180–12193, 2018, doi: 10.1016/j.ceramint.2018.03.275.
[16] Q. Wang, S. Luo, S. Wang, H. Wang, and C. S. Ramachandran, "Wear, erosion and corrosion resistance of HVOF-sprayed WC and Cr 3 C 2 based coatings for electrolytic hard chrome replacement," *Int. J. Refract. Met. Hard Mater.*, vol. 81, no. November 2018, pp. 242–252, 2019, doi: 10.1016/j.ijrmhm.2019.03.010.
[17] B. Song, J. W. Murray, R. G. Wellman, Z. Pala, and T. Hussain, "Dry sliding wear behaviour of HVOF thermal sprayed WC-Co-Cr and WC-CrxCy-Ni coatings," *Wear*, vol. 442–443, pp. 1–10, 2020, doi: https://doi.org/10.1016/j.wear.2019.203114.
[18] C.-J. L. Luo, Xiao-Tao, Gregory M. Smith, Yan Wang, Edward Gildersleeve, Sanjay Sampath, "Cracking induced tribological behavior changes for the HVOF WC-12Co cermet coatings," *Ceram. Int.*, vol. 45, no. 4, pp. 4718–4728, 2019, doi: https://doi.org/10.1016/j.ceramint.2018.11.164.
[19] W. jin Yang, L. Zou, X. ying Cao, J. hua Liu, D. jun Li, and Z. bing Cai, "Fretting wear properties of HVOF-sprayed CoMoCrSi coatings with different spraying parameters," *Surf. Coatings Technol.*, vol. 358, no. August 2018, pp. 994–1005, 2019, doi:

10.1016/j.surfcoat.2018.12.039.

[20] A. Khosravifard, E. Salahinejad, A. H. Yaghtin, A. Araghi, and A. Akhbarizadeh, "Tribochemical behavior of alumina coatings deposited by high-velocity oxy fuel spraying," *Ceram. Int.*, vol. 41, no. 4, pp. 5713–5720, 2015, doi: 10.1016/j.ceramint.2015.01.002.

[21] C. R. C. Lima *et al.*, "Developing alternative coatings for repair and restoration of pumps for caustic liquor transportation in the aluminum and nickel industry," *Surf. Coatings Technol.*, vol. 268, pp. 123–133, 2015, doi: 10.1016/j.surfcoat.2014.08.010.

[22] T. Peat, A. M. Galloway, A. I. Toumpis, and D. Harvey, "Evaluation of the synergistic erosion-corrosion behaviour of HVOF thermal spray coatings," *Surf. Coatings Technol.*, vol. 299, pp. 37–48, 2016, doi: 10.1016/j.surfcoat.2016.04.072.

[23] T. Peat, A. Galloway, A. Toumpis, D. Harvey, and W. H. Yang, "Performance evaluation of HVOF deposited cermet coatings under dry and slurry erosion," *Surf. Coatings Technol.*, vol. 300, pp. 118–127, 2016, doi: 10.1016/j.surfcoat.2016.05.039.

[24] S. Li *et al.*, "Corrosion behavior of HVOF sprayed hard face coatings in alkaline-sulfide solution," *Appl. Surf. Sci.*, vol. 416, pp. 69–77, 2017, doi: 10.1016/j.apsusc.2017.04.149.

[25] M. A. R. Mojena, A. S. Roca, R. S. Zamora, M. S. Orozco, H. C. Fals, and C. R. C. Lima, "Neural network analysis for erosive wear of hard coatings deposited by thermal spray: Influence of microstructure and mechanical properties," *Wear*, vol. 376–377, pp. 557–565, 2017, doi: 10.1016/j.wear.2016.12.035.

[26] C. Zheng *et al.*, "Experimental study on the erosion behavior of WC-based high-velocity oxygen-fuel spray coating," *Powder Technol.*, vol. 318, pp. 383–389, 2017, doi: 10.1016/j.powtec.2017.06.022.

[27] J. Singh, S. Kumar, and S. K. Mohapatra, "Tribological analysis of WC–10Co–4Cr and Ni–20Cr2O3 coating on stainless steel 304," *Wear*, vol. 376–377, pp. 1105–1111, 2017, doi: 10.1016/j.wear.2017.01.032.

[28] S. Lavigne, F. Pougoum, S. Savoie, L. Martinu, J. E. Klemberg-Sapieha, and R. Schulz, "Cavitation erosion behavior of HVOF CaviTec coatings," *Wear*, vol. 386–387, pp. 90–98, 2017, doi: 10.1016/j.wear.2017.06.003.

[29] S. Tkachenko *et al.*, "Wear of grinding rotors with thermally-sprayed coatings in a high-speed mill," *Wear*, vol. 412–413, pp. 49–59, 2018, doi: 10.1016/j.wear.2018.05.016.

[30] S. Hong, Y. Wu, W. Gao, J. Zhang, Y. Zheng, and Y. Zheng, "Slurry erosion-corrosion resistance and microbial corrosion electrochemical characteristics of HVOF sprayed WC-10Co-4Cr coating for offshore hydraulic machinery," *Int. J. Refract. Met. Hard Mater.*, vol. 74, pp. 7–13, 2018, doi: 10.1016/j.ijrmhm.2018.02.019.

[31] G. A. Ludwig, C. F. Malfatti, R. M. Schroeder, V. Z. Ferrari, and I. L. Muller, "WC10Co4Cr coatings deposited by HVOF on martensitic stainless steel for use in hydraulic turbines: Resistance to corrosion and slurry erosion," *Surf. Coatings Technol.*, vol. 377, no. January, 2019, doi: 10.1016/j.surfcoat.2019.124918.

[32] V. Singh, S. Kumar, and D. Ratha, "Synergistic effect of the addition of TiO 2 feedstock on solid particle erosion of Ni/Al_2O_3 and Ni/Cr_2O_3 coatings," *Wear*, vol. 426–427, no. January, pp. 250–257, 2019, doi: 10.1016/j.wear.2019.01.030.

IOP Conf. Series: Materials Science and Engineering 1017 (2021) 012016 doi:10.1088/1757-899X/1017/1/012016

Analysis of a Structure for Efficient Energy Utilization Using Design Builder

Bharath B[1], Abdus Sami[1], Kartikeya Rastogi[1]

[1]Department of Civil Engineering, JSS Academy of Technical Education Noida, India

Email id: bharath6148@gmail.com

Abstract. A major concern for the Indian Construction industry is its negligence of energy efficiency aspects of the various building projects being undertaken nowadays. Moreover, it is a concern for the citizens as well. Heat island and climate change phenomenon has led to a massive increment in the ambient temperatures, intensified reliance on energy resources, created discomfort conditions. By 2030, India's dependence on energy imports is expected to exceed 53% of the country's total energy consumption [5]. As growth is inevitable, a multifaceted and scalable solution is needed to temper the environmental impacts of growing cities. The growth in the cities leads to hot urban environments. The temperature variation between urban areas and adjacent rural areas is called the "heat island phenomenon". It is found that the heat island phenomenon triples the peak electricity demand, doubles the building`s cooling load, and reduces the mechanical cooling system`s coefficient of performance by 25% [5]. To ease our lives in such a climate we need to invent the cooling techniques and those too are energy and cost-efficient. Many of the environmentalists and architects are putting forward their contributions toward it, suggesting solutions to make use of more and more sustainable systems of energy not like those being practiced in earlier times like courtyards and cavities but newer techniques which merges out with the present form of architecture, few of which are namely, the green roof techniques, terracotta pot, outdoor cooling systems. In this paper, we have analysed the Academic Block 4 building of our college and estimated the size of an outdoor cooling plant based on weather data, occupancy, and constructional details of the building such as heat absorbed by various facades due to internal and solar gains, etc

1. INTRODUCTION

Analyzing the building from a structural point of view and ensuring the safety of various joints, beams, columns, slabs is an important task in construction but, these days due to rising environmental atrocities and phenomena like heat island and climate change analyzing the buildings from an energy efficiency point of view is also very important. It ensures less amount of energy consumption during the lifecycle of the building and fewer emissions consequently. Energy efficiency analysis is also important in getting LEED (Leadership in Energy and Environmental Design) and GRIHA (Green Rating for Integrated Habitat Assessment).

Considering India especially, which has a hot climate for major duration of year and keeping in view the kind of architecture being practised here one can easily predict that there is a great demand for passive or energy efficient cooling techniques. Design Builder is an Energy Plus based software tool used for energy, carbon, lighting and comfort measurement and control. Design Builder combines the fast three-dimensional building modelling with dynamic energy simulations along with its unique simple use. It is considered a novel software tool to make and evaluate building designs. Energy Plus

is a modern building performance simulation program that combines the best features from BLAST and DOE-2 along with some new capabilities [2].
We first created a model geometry of the building under analysis i.e. Department of Civil Engineering of JSS Academy of Technical Education. Weather data depending upon the location of the building was assigned to the structure for realistic analysis. The dimensions of the various facades, constructional details such as wall thickness, type of bricks, mortar, paints used, and windows openings were properly set up in the model in the most realistic way for obtaining accurate results from the simulation.

For creating the model geometry, we used Design Builder and for running the simulations and estimating the size of the cooling plant we used Energy Plus as the simulation engine. Design Builder allows complex buildings to be modelled in a simple fast way even by non-expert users. Design Builder is the first and most comprehensive program that creates a graphical interface to an Energy Plus dynamic thermal simulation engine. This graphical interface makes the design of the buildings, their energy performance, and CFD simulations allow them to be displayed in 3D to provide support for examination. Design Builder uses the latest version of the Energy Plus simulation engine for calculating the energy performance of buildings. The resulting data can be filtered as desired and presented in graphs or can be exported in tabular format for use in other applications.

METHODOLOGY:

Before starting to make the model geometry we had to set the site location so that weather data such as outside dry bulb temperature, humidity, solar gains, etc. which would be useful in the energy efficiency analysis can be assigned to our model.
Design Builder provides inbuilt weather data files that can be imported into the model from the libraries or these files can be downloaded from the internet as well. In this case, we have used an inbuilt weather data file of New Delhi – Safdarjung as this place is close to the building, we have analysed and has similar weather, similar sunrise and sunset timings, similar day and night temperatures for accurate results.
Using the different tools available for creating the geometry of the building we created a 3D model of Academic Block 4 which would be used for simulations further.

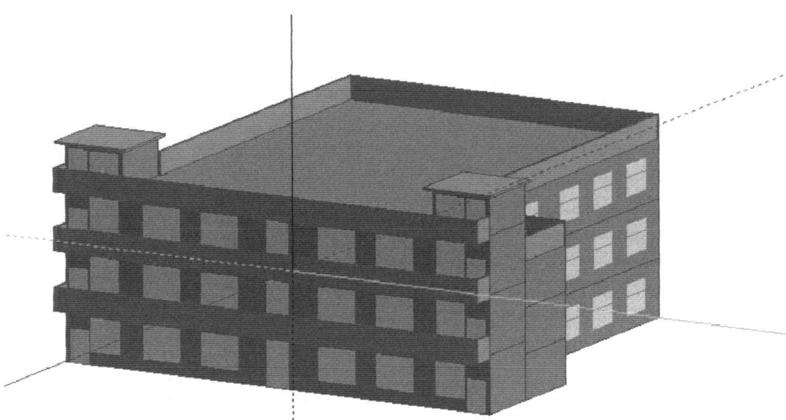

Figure 1: Model Geometry - Academic Block 4 building

Then in the next step, the constructional details of the various building components were customized and assigned respectively to the components. For example, in the picture below we can see the constructional details of the external walls of the AB4 building showing wall thickness, plaster, and paint details.

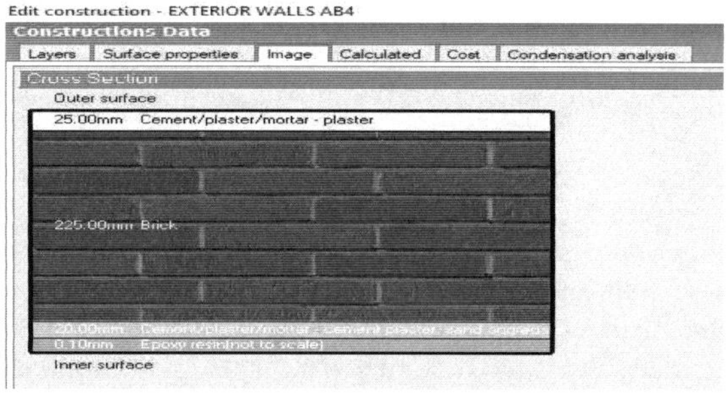

Figure 2: Constructional details of exterior walls

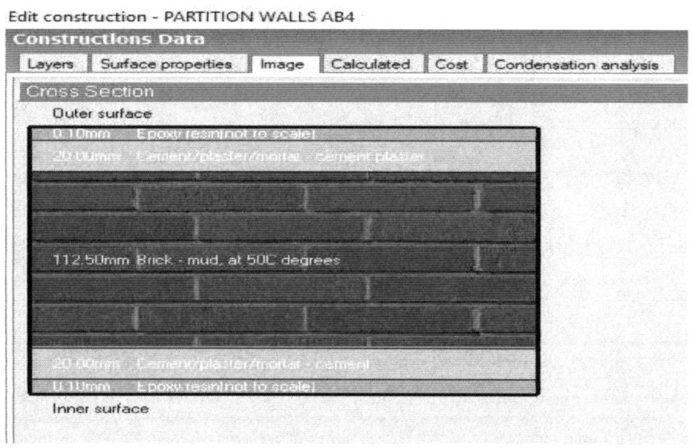

Figure 3: Constructional details of partition walls.

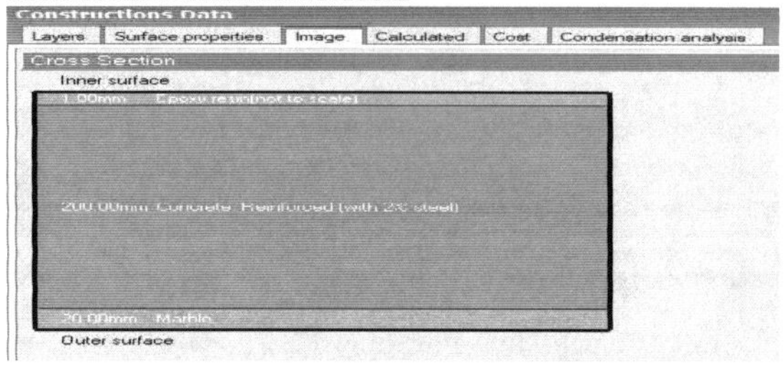

Figure 4: Floor slab details.

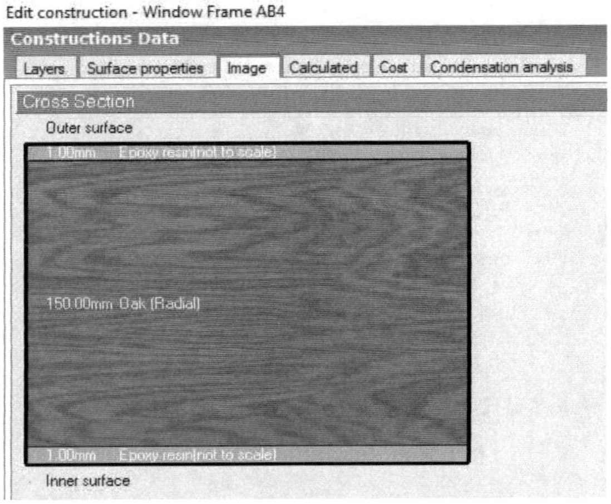

Figure 5: Details of a window, door frames

Different building materials have different tendencies to absorb, emit, or retain heat. How well a building material resists or conducts heat is judged based on R-Value and U-Value respectively. R-Values are additive for layers of materials. The higher the R-Value the better is the thermal performance of the material. Similarly, low U-Values indicate high levels of insulation. Analysing the building materials based on these values is a useful way of predicting the composite behaviour of an entire building element rather than relying on the properties of individual materials.

Based on the materials used the R-Value of external walls was found to be 0.582 meter squared kelvin per watt and the U-Value for the same was 1.718 watt per meter squared kelvin. The U, R-Values of internal partition walls were found to be 2.143 and 0.467.

Similarly, the details of windows such as window to wall percentage, the height of windows from the ground, type of glass used in windowpanes and its thickness, opening and closing schedules of windows, etc were set up:

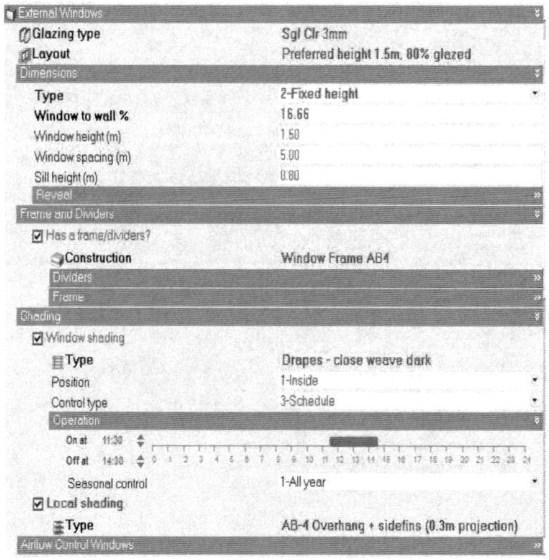

Figure 6: Details of window openings

The occupancy density and the number of days per week the building would be occupied is also set up by the 'Activity Tab'. Metabolic activities also affect the heat gain in a building due to the heat released in the form of carbon dioxide in the process of metabolism taking place within the occupants of a building.

Other important parameters that can be set up using this tab are the environmental controls such as the heating and cooling set-point temperatures and gains through office equipment and computers which play a major role in the design of a cooling plant and analyzing the heat gains because they release heat.

This completes the data entry portion for our model. After this step, we ran simulations and cooling design calculations for sizing the cooling plant. We have carried out the calculations for summer design day which is supposed to be the hottest day of the year so that results automatically fit in and be valid for other days of the year.

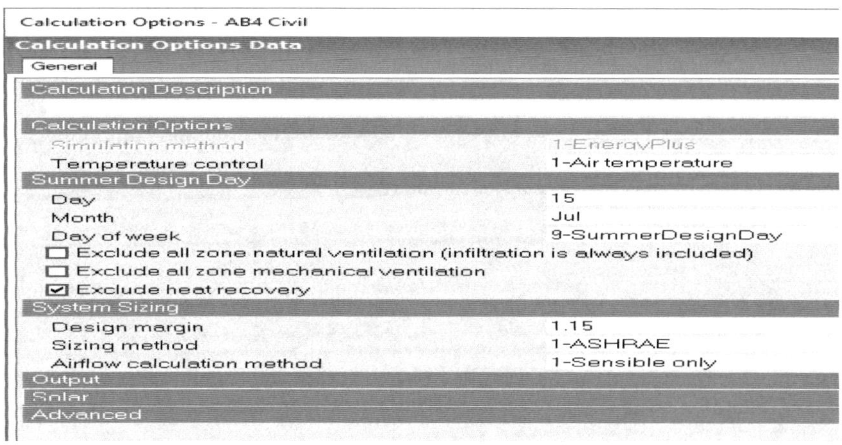

Figure 7: Inputs for obtaining cooling plant size.

IOP Conf. Series: Materials Science and Engineering 1017 (2021) 012016 doi:10.1088/1757-899X/1017/1/012016

RESULT AND CONCLUSION:

Table 1: Cooling Design Calculations for 15 July (hottest day of the year for our site location)

Zone	Design Capacity (kW)	Air Temp (°C)	Time of Max Cooling	Max Op Temp in Day (°C)	Outside Dry-Bulb Temperature
Ground Floor: Surveying X Geology Lab	24.23	23	Jul 16:00	27.1	41.6
Ground Floor: CAD Lab	25.35	23	Jul 10:00	26	38.6
Ground Floor: BMC and Transportation Lab	24.18	23	Jul 16:00	27.2	41.6
Ground Floor: Geotechnical and Environment Lab	22.58	23	Jul 10:00	26	38.6
First Floor: Seminar Hall	42.8	24	Jul 15:30	29	41.9
First Floor: Computer Desks	9.4	24	Jul 14:00	28.1	42.2
First Floor: Senior Professor Office	7.5	24	Jul 09:00	29.1	36.9
First Floor: HOD Office	7.56	24	Jul 09:00	29.2	36.9
First Floor: Department Library	18.72	24	Jul 14:00	28.2	42.2
First Floor: Presentation Room	14.64	24	Jul 09:00	29.3	36.9
First Floor: Staff Room	33.51	24	Jul15:30	29.1	41.9

Second Floor: Room 204	17.92	23	Jul 14:00	29.0	42.2
Second Floor: Room 202	21.84	23	Jul 09:00	29.5	36.9
Second Floor: Room 206	23.72	23	Jul 16:00	30.1	41.6
Second Floor: Room 205	23.68	23	Jul 16:00	30.0	41.6
Second Floor: Room 203	17.74	23	Jul 14:00	28.9	42.2
Second Floor: Room 201	21.89	23	Jul 09:00	29.5	36.9
Overall	357.26				

The final rendered image of the model created with the help of Design-Builder is shown below:

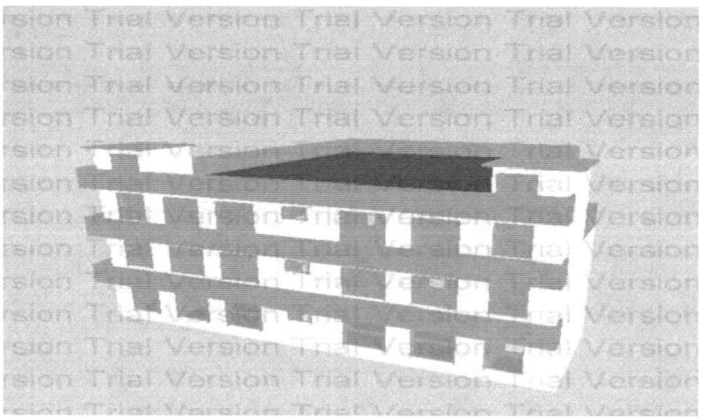

Figure 8: Final rendered image of AB 4

After carrying out the cooling design calculation for the design summer day i.e. 15 July the size of the cooling plant was estimated to be 357.26 kW or 102 tonnes. Keeping in mind the relation:

1 tonne = 3.504 kW

A design margin of 15% is included in this result keeping in mind extra occupancy or extremely hot weather conditions which may develop accidentally.

These energy calculations were done by the EnergyPlus simulation engine incorporated in the software.

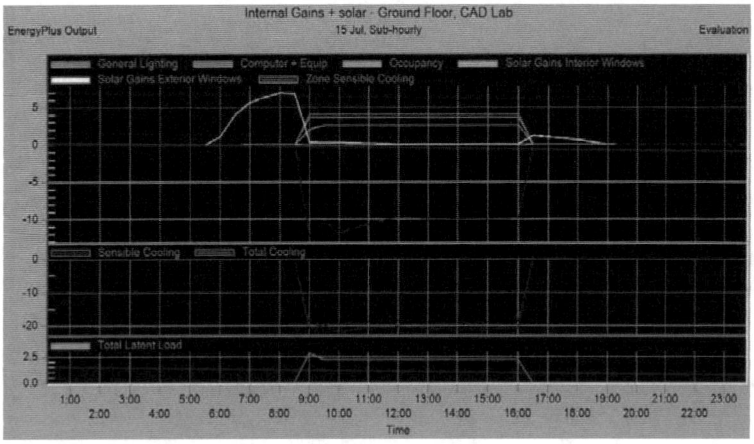

Figure 9: Graph of Variations of various gains and cooling with time.

This graph shows the variation of solar gains and internal gains to time for the Ground Floor CAD Lab. The yellow line represents the solar gains through the exterior windows. We can see how it starts from a point of time when the sun rises and continues to go up till the cooling plant is turned on as per the schedule. As the zone sensible cooling i.e., the blue line increases the solar gains decrease which can be seen by the downward slope of the yellow line. The heat gains due to various types of equipment such as computer systems, fans, etc. can be seen to remain at a fairly constant value throughout. These loads ultimately go down along with the gains due to occupancy when the building starts getting evacuated by the occupants after 4 o'clock in the evening.

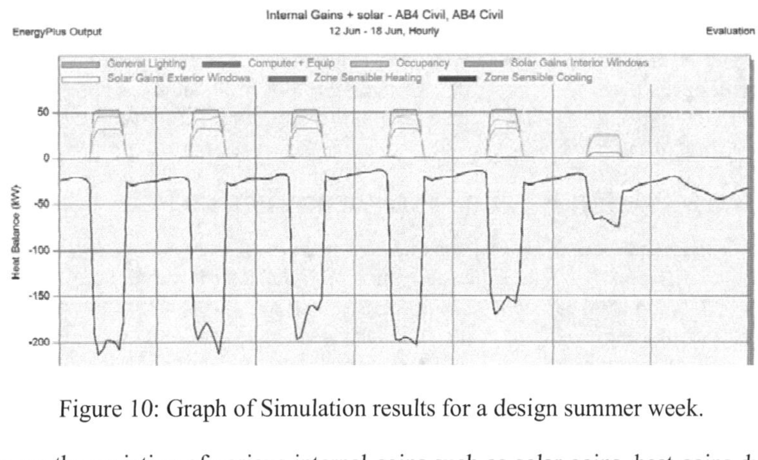

Figure 10: Graph of Simulation results for a design summer week.

Here, we can see the variation of various internal gains such as solar gains, heat gains due to various appliances such as fans, bulbs and tube lights, computer systems, and other laboratory machines such as electric oven and other testing machines for one week. It is evident from the graphical data how the gains remain fairly constant or tend to follow the same pattern for five days of the week i.e. two peaks of solar gains in the morning and evening respectively whereas lesser gains are seen while on the sixth

day as the building is occupied or is in use for a lesser time duration the graph's peak at lower levels. Other gains such as lighting and equipment rise steadily during the start of the day as these things are turned on and attain a constant level and then again go down as they are turned off in the evening. On Saturday lesser occupancy is seen due to half-day hence a reduction in gains is seen along with the cooling loads. On Sunday only solar gains can be seen as the building remains unoccupied and computers and other appliances are not in use for the whole day.It is important to note that the line representing zone sensible heating remains flat throughout because we have not provided a heating plant since it was not necessary as per the weather conditions and requirements.
The above graph is for design summer weak which means the results would not exceed these values for any other week during the rest of the year.
Hence, reliable data is generated using the simulation techniques. Also, by analyzing the different gains through the simulation results we can determine the areas where which we can tackle to reduce the gains and make our cooling or heating plant extra energy and fuel-efficient.

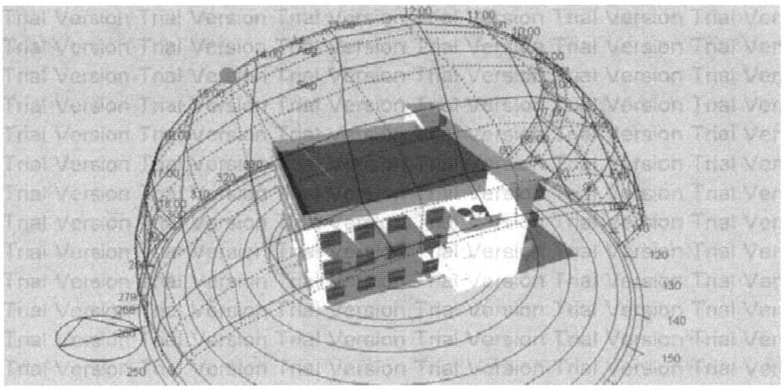

Figure 11: Sun-path diagram for a random instance.

The above sun path and shadow diagram is for 1 June 2:30 pm. We can see the shadows generated due to the position of the sun and also analyze through these diagrams which portion or façade of the building experiences solar gains at what time and for what duration of time.Various such diagrams can be generated with the help of the software corresponding to different dates and times of all the months in a year for an effective analysis. These diagrams help in designing the overhangs and side fins for the external windows of different facades.

USEFUL CHANGES WHICH MAY LEAD TO ENERGY EFFICIENCY:

With the help of these sun-path diagrams for different months of the year we can see that the major solar gains are experienced by the front, top, and rear façade of the building. In the previous simulations and cooling design calculations and investigations, we have found out that the major gain in the building is solar gain. Hence, to reduce the solar gains and consequently reduce the size of the cooling plant for better economics we would suggest that external blinds be used on these walls or window glazing be reduced from single to double glazing of vacuum type.

Double-glazed windows have two glass panes set parallel to each other in such a way that there is a little space or air column between them which acts as an insulating layer. Double glazed windows are essentially two panes of glass with a vacuum between them. The trapped air creates insulation that prevents heat loss or gains through the window because the air temperature cannot penetrate the vacuum and therefore bounces back. From our analysis, we could conclude that Double glazed windows cut down heat gains/loss by 30% in comparison to single glazed windows. Standard B rated double glazing saved four to five thousand rupees per year in cooling costs.

It was also noticed that providing a green roof instead of providing only the normal concrete slab also improves the energy efficiency of the building to a noticeable extent. Green roofs improve the microclimate by cooling and humidifying surrounding air. This cooling effect significantly increases the performance of air conditioning systems and reduces carbon emissions. It also binds dust and toxic particles such as smog; it also helps in rainwater retention. It protects the roof membrane and its water-proofing layer. While using a green roof in the analysis it was seen that temperature variations were reduced upto 4-5 degrees Celsius. Approximately 100 tonnes/year of carbon emission was reduced. A Significant fall in air conditioning loads in summers was seen but not much difference in monsoons was observed in cooling loads.

The analysis of shadows generated on various facades with the help of sun path diagrams helps us in deciding the size of overhangs and side fins of windows in such a way that for most of the time the facades remain in shade and consequently lesser heat gains leading to lesser consumption of energy can be observed ultimately increasing the efficiency of the air conditioning system of the building.

CONCLUSION:

From the results it is evident that it is extremely necessary to realize the importance of energy prospects of a building rather than only concentrating on and ensuring the structural safety and serviceability. It is necessary to state the useful role played by software in making our analysis easier and faster during calculations. Learning software like this always helps in the long run in terms of carrier growth and making effective analysis for educational purposes as well. We conclude that we have learned how to analyze the simulation results and developed a better understanding of it during the course of our work. Through our investigations and studies, we have designed the capacity of a fully functional cooling plant for the Academic Block 4. Also, our analysis has led us to a place from where we have suggested what useful changes in the construction of the building can be made to make it more energy efficient i.e., use less amount of energy for the same or a better output. Also, having faced the wrath of itchy and tiresome summers while attending lectures in the building due to the absence of cooling systems, we realize the necessity of conditioning and energy needs. This project made it possible for us to give back something valuable to our department and faculties and also provide for a comfortable environment by showing that an energy and cost efficient cooling system is possible to be installed in the Department of Civil Engineering, Academic Block 4.

REFERENCES:

1.Bernardo H, Quintal E and Oliveira F. 2017. Using a Calibrated Building Energy Simulation Model to Study The Effects Of Improving The Ventilation In A School. *Energy Procedia***113** (2017) 151 – 57.

2. Lawrie L K, Dr. Fredrick C. Winkelmann, Dr. C. O. Pedersen April 2000 EnergyPlus: energy simulation program. *ASHRAE Journal* 49-56.

3. Zhao J, Lam K P and Ydstie B E 2013 EnergyPlus model based predictive control using MATLAB, Simulink and MLE+. *Proceedings of BS-2013 13th Conference of International Building Performance Simulation Assoc*. 2466-73. (France : Chambery, France)

4. Rallapalli H S. 2010 A comparison of EnergyPlus and eQuest whole building energy simulation results for a medium sized office building. (U.S.A.: Arizona State University).

5. Aamir M. 2018 Energy Efficiency Analysis of Green Roofs.(New Delhi: Jamia Millia Islamia.).

6.Marjanovic V, Bogdanovic G, Petrovic N and Bojic M. May 2014. Optimization of low-rise building's geometrical form in design builder 433-38*8th International Quality Conference Centre for Quality*(Serbia:University of Kragujevac, Serbia)

7.Korondi P and Banoczi E.November 2014*Environmental Engineering and Management Journal***13** (11) 2743-2748

8.Baharvand M, Ahmad M H, Safikhani T andRoshida Binti Abdul Majid. 2013. Design Builder Verification and Validation for Indoor Natural Ventilation. *Journal of Basic and Applied Scientific Research.* J Basic Appl. Sci. Res., **3(4)** 182-89

9.Yua S, Cui Y, Xu X and Feng G 2015 Impact of civil envelope on energy consumption based on EnergyPlus. Procedia Engineering **121** 1528-34 (China: Shenyang Jianzhu University, Shenyang, China)

10.Wasilowski H A, Reinhart C F 2009 Modelling an Existing Building in Design Builder Energy Plus: Custom vs Default Inputs.*Eleventh International IBPSA Conference Glasgow, Scotland.* 1252-59

11.Mahdavi A. and Martens B. 2014Using a data driven approach to support the design of energy efficient buildings.(Singapore: National University of Singapore)

12. Beidi Li 2017 Use of building energy simulation software in early stage of design process (Sweden: KTH Royal Institute of Technology Stockholm, Sweden).

13. Daniel C and Leake J M 2011 Energy efficient design using building information modelling and energy simulation. University of Illinois

Development of an Internet of Things (IoT) based Lockout/Tagout (LOTO) device for Accident Prevention in Manufacturing Industries

S Kumar[1*] and S M Tauseef[1]

[1]Department of Safety, Health and Environment, University of Petroleum and Energy Studies, Dehradun, Uttarakhand – 248007, India

E-mail: sameercar@gmail.com

Abstract. Despite all the technological advances, safety remains the biggest challenge in Industries. Each year, thousands of electrocutions and hundreds of machinery-related accidental deaths are happening. A number of these accidents involve inadequate, unauthorised or improper use of hazardous energy control procedures, also known as lockout/tagout (LOTO) procedures. Although the organizations like Occupational Safety and Health Administration (OSHA), USA, The National Institute for Occupational Safety and Health (NIOSH) and others have defined a very detailed LOTO procedure, however, these LOTO procedures have many manual activities. With the emergence of IoT (Internet of Things) these manual activities can be combined with automated activities thereby enhancing safety. IoT based devices can be implemented to ensure authorized access to hazardous areas around operating machinery, improve documentation and monitoring of service during repair and/or maintenance that requires shutdown of the machinery, and prevent unexpected, unauthorised start up, or movement during maintenance of machinery or related activities. Acknowledging the importance of IoT enabled safety devices NIOSH is also exploring the possibility of implementing IoT based technologies to provide intelligent machine monitoring as part of a comprehensive LOTO program. This paper presents the development of an IoT based LOTO device, its implementation and effectiveness in improving safety and accident prevention. It is found that the IoT based LOTO device is effective in accident prevention (from 47 to 0 incidents) and its performance (from 697 to 349 Seconds) is reliable.

1. Introduction

Manufacturing industries has many equipment which require maintenance, repair and service to be done by an individual or a group of persons to make sure that no untoward incident happens during the operation of such equipment. These persons are trained to perform the job with safety and efficiently known as competent person. The competent person is required to follow a defined hazardous energy control program or LOTO procedure wherein they isolate the hazardous energy source and puts a lock so that no one accidently energize the equipment while it is under repair or maintenance. Different type of hazardous energy like electrical, hydraulics, chemical, pneumatic, potential etc. are used in machines in engineering industries.

1.1. Historical data on Accidental Deaths

Many agencies collect data for different type of accidents, worldwide. National Crime Record Bureau (NCRB) under Ministry of Home Affairs, Government of India too has started collecting and publishing data from the year 1967. As per NCRB Data for the year 2019, India had 411,824 accidental deaths in 2018 [1] and 421,104 in the year 2019 [2]. There were 13,432 deaths due to electrocution and 1001 machinery-related accidental deaths in the year 2019. Similar analysis of accident data available from the Mine Safety and Health Administration (MSHA) was conducted by Ruff et. al. [3] from the NIOSH. Based on the analysis of the accident data Ruff concluded that one-third of these accidents involved improper LOTO procedures as a contributing factor [4]. The MSHA has acknowledged this problem in their report RIN 1219-AB 91. It is reported that 76% fatalities are involved working near or around belt conveyors, where the mining workers become entangled in belt drives, belt rollers, and discharge points. Factors that were identified as responsible for these accidents were insufficient, inadequate or missing guards and number of crossovers in different strategic locations, and/or inappropriate LOTO procedures. It was concluded in the report that interlock systems that ensure that machine guards are properly secured in place; and/or ensure machines are properly locked out and tagged out during maintenance would reduce fatalities [5].

In order to ensure proper implementation of LOTO a number of researchers have explored various solutions like worker-worn electric field sensor [6], potential application of IoT technologies based intelligent machine monitoring systems for improved worker safety and design of safety related parts of control systems in machines [7] However, the solutions proposed by these researchers are limited to electric field sensors, machine monitoring systems and design of safety related parts of control systems in machines. IoT based LOTO device is not in their scope[8]. This paper presents the design and development of an IoT based LOTO device to enhance safety of LOTO procedures. Data is collected and analysed to test device performance and its effectiveness in improving safety.

2. Research Methodology for Development of IoT based LOTO device

The development of the IoT based LOTO device was done in 3 phases. Selection of the hazardous machines and hazardous energy were done in Phase-1, study of the existing conventional LOTO device and procedure were done in Phase-2 and device with different control technologies were developed in Phase-3. These trials and experiments were planned and carried out at JCB India Limited, Ballabgarh, India Plant.

2.1. Phase-1: Selection of machine and hazardous energy for IoT based LOTO

In this paper, data on safety incidents related to one of the Machine was collected. There were 30 Near Miss and 17 Minor Injury cases in the period of May 2011 to April 2018. Further analysis of these 47 incidents revealed that in 10 cases LOTO procedures were not implemented properly. This Computer Numerical Control (CNC) machine is used to do machining of Gear Box Parts and uses various hazardous energies[9]. In order to work a CNC machine, the machine operator loads the parts into a machine, starts the machine and unloads the parts after machining[10]. Machine operator calls maintenance person for any kind of support whenever the need arises. Maintenance and Service person also apply LOTO during scheduled preventive maintenance and repair work. There are 52 activities identified on this CNC Machine which requires LOTO implementation and procedure adherence. In these 52 activities 67% are electrical, 13 % are Hydraulic, 4% Mechanical and 4% Pneumatics[11]. The machine has 16 different types of LOTO devices in which majority 6 are circuit breaker (CB) LOTO device. Out of these 6 CB LOTO devices, there are 4 Miniature Circuit Breakers (MCB) of different ratings, which requires 4 different types of MCB LOTO devices. The above study and data related to CNC Machine reveals complexity of different LOTO activities and devices. Furthermore, Specification and Type of MCB LOTO devices depend on specifications and type of MCBs[12]. Based on the safety incidents and type of hazardous energy, we have selected to develop an IoT based LOTO device for controlling electrical energy on CNC A-81 Makino machine.

2.2. Phase-2: Study and Analysis of conventional LOTO device and procedure for CNC machine

Operation and Maintenance Persons follows LOTO procedures identified and defined by OSHA [13] for maintenance, repair or servicing of machines or equipment [14] as shown in 'Figure 1'. Data collected on CNC machines from May 2011-June 2018 reveals that LOTO procedures were not adhered mainly in steps of notifying the employees, attaching LOTO device and verifying LOTO.

- Notifying employees: In this step employees working on the machine and or near to the machine was not notified of the operation/maintenance person and safety incident had happened as the persons were not known about the hazardous machines available on the machine.

- Attaching the lockout device: In this step the conventional LOTO devices like padlock were not attached as per procedure and maintenance person was working on the machine thinking that the work will take less time and he will do it without applying LOTO.
- Verifying lockout: In this step maintenance person forgets to verify the lockout for isolation of hazardous energy[15].

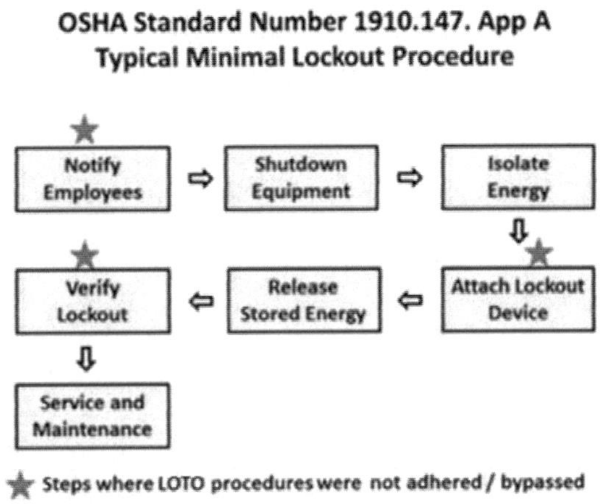

Figure 1. OSHA –Standard Number 1910.147 Typical LOTO Procedures

2.2.1 Limitations and Challenges of Conventional LOTO device and Procedure

- A maintenance / service person has sole responsibility and decision making of putting LOTO device and physical padlock, he/she can bypass and lock or unlock the LOTO device and padlock[16].
- Manager and Supervisor need to visit, see and verify the physical condition of actual padlock at the circuit breaker.
- LOTO procedure takes time (697 seconds as per data collected) and delays the actual trouble shooting and fault-finding process.

2.2.2 Leveraging IoT based LOTO device and procedure to overcome challenges

IoT based devices are being used in industries[17] and IoT based LOTO device can be helpful to Operation and Maintenance Persons in adherence of LOTO procedures identified and defined by OSHA.

2.2.3 Proposed Benefits and advantages of IoT based LOTO Device and Procedure

- The IoT based LOTO device will improve LOTO procedure adherence and reduction in safety incidents.
- It may help to reduce LOTO implementation time and improves productivity.
- Person cannot bypass the LOTO procedure as any action by him will send an automatic text message to his supervisor and manager.
- Any person other than authorized cannot lock/unlock the device or break the IoT based LOTO device, as any action by him/her will send a text message automatically to concerned manager, supervisor and electrician.

2.3. Phase-3 Development of IoT based LOTO device for CNC machine

The proposed device consists of Arduino Mega 2560, Global System for Mobile Communications (GSM) Module, Relay, Transistors, Liquid Crystal Displays (LCD), Transformers, Servo Motor and Mechanical device for carrying out IoT based LOTO Procedure in case of an electrical fault occurs or need arises[18] as shown in 'Figure 2'. IoT based communication has also been set up between different person like electrician, supervisor and managers based on their roles, responsibilities and approval hierarchy. As soon the fault occurs, complete system consisting of hazardous energy (in our case electrical energy is hazardous energy) gets locked and real-time message reaches the supervisor and manager. Since system is locked, it becomes safe to carry out repair work. After completion of the work, electrician makes a request to his supervisor to unlock the system from his IoT based device (in our case mobile phone), supervisor unlocks the IoT based LOTO and now electrician can physically switch on the hazardous energy. Focusing on standardization, ease of understanding and implementation, the localization of the IoT based device is done by using standard devices like Arduino and GSM modules which has advantage of being low cost, easily available and widely used over other industrial device like Programmable Logic Controllers (PLC). The device uses the state-of-the-art technique in IoT based ON-OFF switching for LOTO.

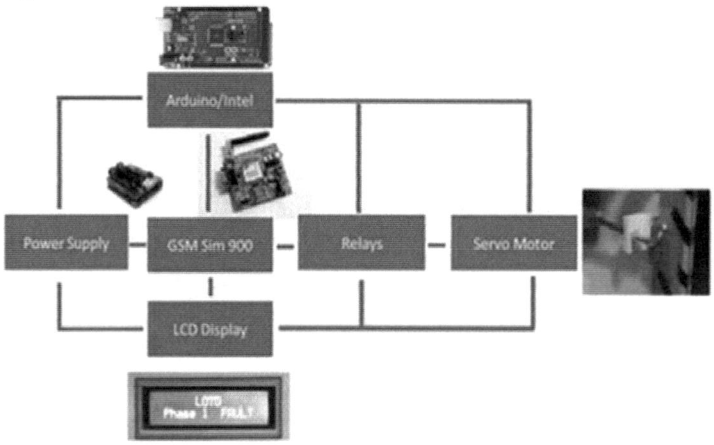

Figure 2. Schematic Block Diagram for developing IoT based LOTO device

3. Trial of IoT based LOTO Device for Performance, Reliability and Effectiveness

Several tests were performed on the developed IoT based LOTO device as shown in 'Figure 3' to find the effectiveness, performance, and to understand its effect on improving safety and accident prevention.

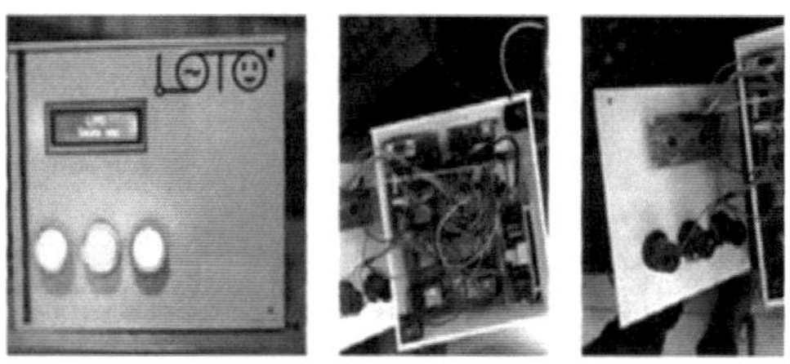

Figure 3. IoT based LOTO device

3.1 Plan for Experiment and Data Collection for IoT based device performance

The most important factor of the IoT based LOTO device is actuation time. The SMS must reach within standard/defined time and actuation must happen within standard/defined time

- Location of IoT based LOTO device
- Network Operators

- Network Signal Strength (as shown in table 1) -Being IoT based device, network strength is a very important factor [19].
- The signal is usually measured in dBm which is the power ratio in decibels of the radio power per milli-watt, a signal of -60dBm is nearly perfect and at -112dBm the call drops. Above -87dBm android mobile phone shows full 4 bars of signal.
- Arbitrary Strength Unit (ASU) is maps to (ASU - 141) ≤ dBm < (ASU - 140) [20]and ASU is an integer value proportional to the received signal strength measured by the mobile phone.
- Voltage Level
- Actuation Time- The time was noted down as start time and finish time and later difference were calculated

Table 1. Various factors of an experiment of an IoT based LOTO device

Network Strength- Number of Tower	Network Strength-dBm	Network Strength – ASU	Practical Impact on Communication
1	-107 to -113	27 to 33	Very Poor- Call Drop
2	-103 to -106	34 to 37	Poor
3	-101 to -102	38 to 39	Average
4	-91 to -100	40 to 49	Good
5	-51 to -90	50 to 89	Excellent

3.2 Data Collection and Analysis of IoT based LOTO device

Data collection was done in April 2018 for IoT based LOTO device on various parameters.

3.2.1 Parameters recorded

- Location, Date, Start Time, Finish Time, Total Actuation Time in Seconds,
- Voltage Level in Volts
- Name of Network Operator, Signal Strength in dBm, Signal Strength in ASU
- Signal Strength in terms of the Number of Bars displayed on the phone screen

3.2.2 Descriptive Statistics of actuation time

Table 2 shows the Actuation time in Seconds (difference of start and finish time) for 1080 readings.

Table. 2. Descriptive Statistics of 1080 readings for actuation time of developed device

Descriptive Statistics	*Op-A_Loc-SF*	*Op-A_Loc-OD*	*Op-A_Loc-Base*	*Op-B_Loc-SF*	*Op-B_Loc-OD*	*Op-B_Loc-Base*	*Op-C_Loc-SF*	*Op-C_Loc-OD*	*Op-C_Loc-Base*
	Condition-1	*Condition-2*	*Condition-3*	*Condition-4*	*Condition-5*	*Condition-6*	*Condition-7*	*Condition-8*	*Condition-9*
Mean	22.60	22.44	25.86	24.18	23.23	23.46	22.40	22.98	24.23
Standard Error	0.06	0.05	0.04	0.07	0.09	0.11	0.09	0.03	0.06
Median	22.50	22.00	26.00	24.00	23.00	23.00	22.00	23.00	24.00
Mode	22.00	22.00	26.00	24.00	23.00	23.00	22.00	23.00	24.00
Standard Deviation	0.67	0.58	0.45	0.72	1.03	1.15	1.02	0.34	0.68
Sample Variance	0.44	0.33	0.21	0.52	1.07	1.33	1.05	0.12	0.46
Kurtosis	-0.60	1.86	4.35	8.43	4.96	3.04	6.48	13.79	5.55
Skewness	0.67	1.17	-0.02	2.64	2.15	1.72	2.66	1.00	2.30

Range	2.00	3.00	3.00	4.00	5.00	6.00	5.00	3.00	4.00
Minimum	22.00	22.00	25.00	23.00	22.00	22.00	22.00	22.00	23.00
Maximum	24.00	25.00	28.00	27.00	27.00	28.00	27.00	25.00	27.00
Sum	2712.00	2693.00	3103.00	2901.00	2788.00	2815.00	2688.00	2758.00	2907.00
Count	120	120	120	120	120	120	120	120	120
Confidence Level (95.0%)	0.12	0.10	0.08	0.13	0.19	0.21	0.19	0.06	0.12

3.2.3 Regression Analysis of Trial Data

Regression Analysis were done for all the trials in different test conditions and depicted in table 3. Actuation time (y) of IoT based LOTO device has a strong positive correlation with the signal strength (x).

Table 3. Regression Analysis

Trial Condition	Location of IoT device	Network Operator	Regression Equation
1	Shop-floor or Indoor	A	$y = -1.4034x + 29.412$ $R^2 = 0.2918$
2	Outdoor	A	$y = -1.4034x + 29.412$ $R^2 = 0.2918$
3	Basement	A	$y = -0.9462x + 29.777$ $R^2 = 0.6769$
4	Shop-floor or Indoor	B	$y = -1.2414x + 28.902$ $R^2 = 0.7687$
5	Outdoor	B	$y = -1.8415x + 32.103$ $R^2 = 0.8504$
6	Basement	B	$y = -0.6933x + 26.208$ $R^2 = 0.3346$
7	Shop-floor or Indoor	C	$y = -1.5813x + 30.017$ $R^2 = 0.8008$
8	Outdoor	C	$y = -0.299x + 24.451$ $R^2 = 0.0896$
9	Basement	C	$y = -1.0402x + 29.201$ $R^2 = 0.874$

4. Analysis of Safety Incidents on a Machine before and after implementation of IoT based LOTO device

Data on various safety incidents and machine stoppages was collected from May 2011 to April 2018 on a Makino CNC Machine In May 2018, we applied IoT based LOTO device on this machine and collected data again from May 2018 to April 2019 to gauge the effectiveness of IoT based LOTO devices in enhancing safety.

4.1 Time Period of data collection

- May 2011 to April 2018 the machine was secured for maintenance and other related activities with the Conventional LOTO device.
- May 2018 to April 2019 the machine was secured for maintenance and other related activities with the IoT based LOTO device.

4.2 Nature of Accident or Safety Incident

- Minor injuries are injuries which do not involve medical treatment and requires only first aid treatment [21].
- Major injuries are non-fatal but severe injuries. They are defined by: Part of the body injured, Nature of injury, Incident type and Duration of medical leave. Examples include: Blindness, Amputation, Crushing, Deafness, Paralysis, fractures and dislocations: hand, head, back, abdomen and chest, neck, hip and pelvis, Exposure to electric current, drowning, Hypothermia etc.
- Near Miss- a person is just saved from the accident.

4.3 Descriptions of the Problems like Breakdown, Maintenance or repair work on the Machine and LOTO process adherence.

- Descriptions of the problems where LOTO applied and processes were adhered
- Descriptions of the problems where LOTO processes were not adhered

4.4 Descriptions of the Hazardous Energies and LOTO process adherence

There are many hazardous energies available on the machine, in case of any problem the person needs to perform LOTO based on the nature of the problem and applicable hazardous energies. The person needs to perform his work based on the standardized operating procedure.

4.5 Details of the Accidents or Safety Incidents

All the safety incidents, whether minor injuries, major injuries, near-misses and accidents were recorded and analysed for their possible root causes. Corrective and Preventive actions are also taken based on the incident analysis[21].

4.6 Details of the hazardous energies to be controlled on the machine for various repair, service and or maintenance work

The machine has many hazardous energies. LOTO needs to be applied to control these hazardous energies. The t data collected on the machine from May 2011-April 2019 with various hazardous energies like Electrical -98, Pneumatics-55, Hydraulics-41, Gravity and Mechanical -16 were observed with the service/repair problem.

4.7 Details of the Safety Incidents and LOTO procedure adherence

LOTO must be applied on the machine while working for service, repair or maintenance. There were safety incidents on the machine, like minor injuries, major injuries and near-misses during repair, maintenance or service of the machine. Data were collected from May 2011-April 2019 for all the cases where LOTO procedures were followed and LOTO were applied. The data were also taken in the cases where LOTO were not applied. Table 4 shows the data collection sheet of safety incidents. Table 5 shows the summary of safety incidents.

Table 4. Data Collection Sheet of Safety Incidents

Parameter No.	Description of Parameters in data collection sheet	Details filled in the data collection sheet
1	Serial Number	1
2	Date	23 May 2011
3	Month	01 May 2011
4	Description of Machine Problem where Service, Repair, Maintenance required	Door Opening Problem-Improper, Jerks are observed at ends.
5	Type of Hazardous Energy-1 to be controlled	Electrical

6	Type of Hazardous Energy-2 to be controlled	Gravity
7	No of times Conventional LOTO applied	1
8	No of Times IoT based LOTO applied	0
9	No of Times LOTO not adhered	1
10	Time Taken in LOTO Implementation (Minutes)	12
11	No. of Safety Incidents (Injuries, Near-Miss & Accidents)	1
12	Description of Safety Incident-1	Near Miss- Finger Pinching Right Hand First Finger
13	Possible Causes of Safety Incident-1	• Pnuematic LOTO Not followed, • Trapped air in the system was not released, LOTO procedure were not followed.
14	Corrective and Preventive Action for Safety Incident-1	• LOTO Adherence Audit • LOTO Retraining

Table 5. Summary of Safety Incidents from May 2011-April-2019

Time Period of the incident	No. of Safety Incident	No of Times Conventional LOTO applied	No of Times LOTO procedure not adhered	Time taken in LOTO Implementation (Minutes)
May 2011-April 2012	8	29	8	346
May 2012-April 2013	9	27	3	308
May 2013-April 2014	7	29	3	346
May 2014-April 2015	6	25	4	301
May 2015-April 2016	5	28	2	315
May 2016-April 2017	7	21	2	229
May 2017-April 2018	5	21	3	246
May 2018-April 2019	0	23	0	134

5. Results and Discussion

5.1 Interpretation of data analysis

5.1.1 Mean actuation time

Mean Actuation time for all the 9 test conditions was 23.49 Seconds. It means that a LOTO device can be designed considering the mean actuation time of 23.49 seconds. This actuation time should be considered while making SOP (Standard Operating Procedure) of LOTO Program.

5.1.2 Location of the LOTO device and network strength

For Network Operator A, it was found that the developed LOTO device has less actuation time & standard deviation at the Shop floor (mean 22.60 seconds and std. dev. 0.67 seconds) and Basement locations (mean 25.86 seconds and std. dev. 0.45 seconds). It means that Operator A can be used for the IoT based LOTO devices for equipment at Shop floor and Basement Location. Operator C can be used for outdoor locations. Network booster can be used to boost the network strength for operator B and C for improved signal strength.

5.2 Result of experiment on IoT based LOTO device

These results are very encouraging and shows reliability of the developed IoT based LOTO device. We could not find any malfunction in 1080 readings and IoT based LOTO device works best when network signal strength is good (-51 to -100 dBm).

5.3 Reduction in number of safety accidents or incidents on machine before and after IoT based LOTO device implementation.

The data collected (table 6, 'Figure 4') clearly shows that the number of safety incidents reduced from "47" to "0". The IoT based device is also able to reduce (table 7) finger (from 10 to 0), hand (from 5 to 0) and arm (from 2 to 0) related minor injuries.

Table 6. Types of Safety Incidents occurs

Type of Safety Incident	Before IoT based LOTO	After IoT based LOTO
	No. of Incidents from May 2011-April 2018	No. of Incidents from May 2018-April 2019
Near Miss	30	0
Minor Injury	17	0

Table 7. Body parts injured (minor)

Type of Safety Incident	Before IoT based LOTO	After IoT based LOTO
	No. of Incidents from May 2011-April 2018	No. of Incidents from May 2018-April 2019
Fingers	10	0
Hand	5	0
Arm	2	0

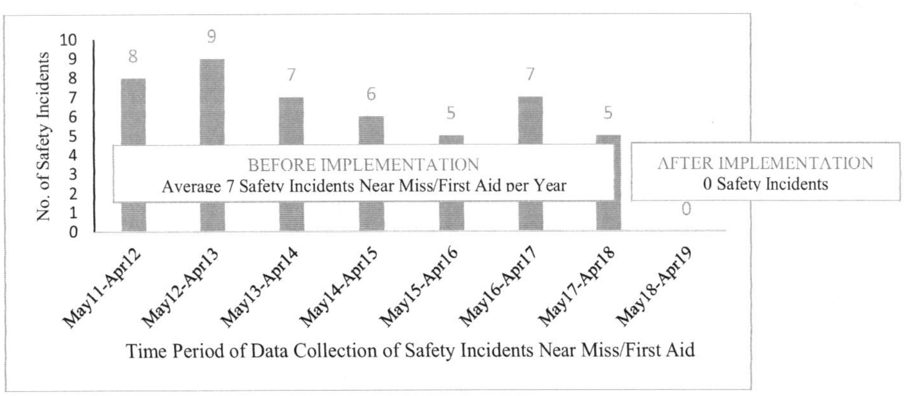

Figure 4. Number of safety incidents before and after implementation of IoT based LOTO device.

5.4 Reduction in LOTO implementation time on machine before and after IoT based LOTO device implementation.

The data collected from the machine before (May 2011-April 2018) and after (May 2018-April 2019) implementation of IoT based LOTO device on Machine with various hazardous energy (Table 8) clearly shows that average LOTO implementation time (seconds) is reduced by almost 50% from 697 to 349 on the machine. Lesser LOTO implementation time improves the availability of the machine for more productive work.

Table 8. Reduction in LOTO implementation time

LOTO Implementation Time	*Before IoT based LOTO*	*After IoT based LOTO*
	Data Collected from May 2011-April 2018	*Data Collected from May 2018-April 2019*
Number of Times LOTO Applied	180 Times	23 Times
Total Time Taken in LOTO Implementation (Minutes)	2091 Minutes	134 Minutes
Average Time Taken in LOTO Implementation (Minutes)	11.6 Minutes	5.8 Minutes
Average Time Taken in LOTO Implementation (Seconds)	697 Seconds	349 Seconds

6. Conclusion and Future Scope

A reliable and robust IoT based LOTO device has been developed for automating the LOTO procedure for 230 V AC electrical devices. The device mainly consists of a Power Supply, Microcontroller, GSM SIM Module, Relays, Servo Motors and Switches. The device can be programmed using open source like Arduino IDE. The device gives additional safety to the person as this device gets auto locked (based on the pre-programmed flow) and sends text messages to concerned person (based on the pre-programmed approval authority) for example- LOTO device gets locked and relevant electrician, supervisor and manager gets automated text message. It is different in a way that the electrician cannot bypass the LOTO procedure as any unauthorized action by him/her will send an automatic text messages to his/her supervisor and manager. It also makes sure only authorized person can perform LOTO. It also communicates to every authorized person in case of weak network signal or any tampering with the device.

- The IoT based LOTO device is 99.7 % faster than the conventional LOTO device with an average implementation time of 349 seconds compared to 697 seconds of conventional LOTO device. It helps electrician in troubleshooting, repair, maintenance or servicing work. It also helps in digital record keeping as data from microcontrollers/sensors can be recorded and verified.
- The IoT based LOTO device has resulted into reduction in the number of accidents and safety incidents from 7 to 0 incidents per year, it also helped to reduce LOTO implementation time from 697 seconds to 349 seconds.
- Reduction in accidents and incidents gives safer work environment and lesser LOTO implementation time improves productivity.
- IoT based LOTO device developed by us is simple, real time, digital, tested for functioning and endurance and verified with respect to present conventional LOTO devices which are customized, non-standard, un-reliable and manual.
- The present device has been made mainly for controlling the 230V AC hazardous energy at circuit breakers. Further improvements can be made on the device for other hazardous energies, faster actuation speed and better performance under low network strength conditions. A mobile app (an application) can also be developed for Android/iOS/other systems. The researcher is planning to apply for patenting the IoT based LOTO device.

References

[1] National Crime Record Bureau NCRB 2018 Snapshots 2018 - Accidental Deaths & Suicides in India 2018,

[2] National Crime Record Bureau (NCRB) 2019 Accidental Deaths and Suicides in India 2019

[3] Ruff T, Coleman P and Martini L 2011 Machine-related injuries in the US mining industry and priorities for safety research, *International Journal of Injury Control and Safety Promotion* **18** 11–20

[4] MSHA 2018 Safety improvement technologies for mobile equipment at surface mines, and for belt conveyors at surface and under- ground mines. *Federal Register 83 FR 29716* (to be codified at 30 CFR 56 and 75) **83** 29716–29719

[5] Kowalski-Trakofler K and Barrett E 2007 Reducing non-contact electric arc injuries: An investigation of behavioural and organizational issues *Journal of Safety Research* **38** 597–608

6] Zeng S, Powers J R and Newbraugh B H 2010 Effectiveness of a worker-worn electric-field sensor to detect power-line proximity and electrical-contact *Journal of Safety Research* **41**, 229–239,

[7] Porras-Vázquez A and Romero-Pérez J A 2018 A new methodology for facilitating the design of safety-related parts of control systems in machines according to ISO 13849:2006 standard, *Reliability and Engineering Safety.* **174** 60–70

[8] Grazia M, Angelo P, Francesca M and Setola R 2020 Integrating IoT technologies for an intelligent safety management in the process industry, *Procedia Manufacturing* **42** 511–515,

[9] Lotti G, Villani V, Battilani N and Fantuzzi C 2019 Machines interaction for CNC machines *IFAC Papers Online* **52** 31–36

[10] Barbosa M, Silva F, Pimentel C, and Gouveia R M, 2018 A Novel Concept of CNC Machining Center Automatic Feeder, *Procedia Manufacturing* **17** 952–959

[11] Kumar S and Tauseef S M 2018 Analysis of various lockout tagout (LOTO) devices used in industrial safety *International Journal of Engineering and Technology* **7** 1329.

[12] Brady Worldwide Inc. 2020 Your experts for safety, facility and equipment identification *Product Catalog,* 180–247.

[13] Occupational Safety and Health Administration Standard Interpretations OSHA 2010 *Electrical Codes, Standards, Recommended Practices and Regulations* 147 487–491

[14] Alonzo R J 2010 Developing electrical safe work practices *Electrical Codes, Standards Recommended Practices and Regulations* 183–212

[15] Emami-Mehrgani B, Nadeau S and Kenné J P 2018 Optimal lockout/tagout, preventive maintenance, human error and production policies of manufacturing systems with passive redundancy, *Journal of Quality in Maintenance Engineering* ,**20** 453–470

[16] Poisson P and Chinniah Y 2015 Observation and analysis of 57 lockout procedures applied to machinery in 8 sawmills *Safety Science* **72** 160–171

[17] Sandro N and Peter S 2020 Internet of things (IoT) Opportunities, issues and challenges towards a smart and sustainable future *Journal of Cleaner Production* **274** 3-4

[18] Kumar N S, Vuayalakshmi B, Prarthana R J and Shankar A 2016 IOT based smart garbage alert system using Arduino UNO, *IEEE Region 10 Annual. International Conference Proceedings/TENCON,* 1028–1034

[19] Meixner B, Kleinrouweler J W, and Cesar P 2018 4G/LTE channel quality reference signal trace data set, *Proceedings of 9th ACM Multimedia System Conference* 387–392.

[20] Wildpackets Inc 2002 Converting signal strength percentage to dBm values www.wildpackets.com 802 1-11

[21] Northwood J M, Sygnatur E F and Windau J A 2012 Updated BLS occupational injury and illness classification system *Monthly Labor Review* **135** 19–28

IOP Conf. Series: Materials Science and Engineering 1017 (2021) 012018 doi:10.1088/1757-899X/1017/1/012018

Fracture analysis of edge cracked FGM plate under compressive load along crack

Kundan Mishra[1], Achchhe Lal[1,] and B. M. Sutaria[1]

[1]Department of Mechanical Engineering, SVNIT, India

Email: mishrakp18@gmail.com

Abstract. Fracture analysis of FGM plate with edge crack under compressive point load along crack is carried out by using extended finite element method (XFEM). In this present work a FGM plate where material properties vary along width with edge crack is considered for the study. The mathematical formulation is carried out in MATLAB and the stress distribution and mixed mode stress intensity factors (SIFs) are obtained for different position of point load. The validation study is carried out for developed MATLAB code with previously published research paper and then fracture analysis of FGM plate with crack under compressive point load along crack is investigated. The present study is carried out to understand the fracture resistance of edge crack FGM plate when the action of load is along crack.

1. Introduction

The functionally graded materials (FGMs) are the types of composite materials where material properties can be varies as per the requirement in required direction. This is firstly utilized by Japanese scientist as thermal resisting material for different applications like aircraft and other engineering applications. During different manufacturing processes and operations, the materials my contain various discontinuities like voids, inclusions and cracks and the presence of these discontinuities strength of material reduces drastically and the stress induces near to the tip of crack is very high. Hence, it is necessary to understand the fractural behavior of material for the reliability and safety of materials.

Many researchers have started using XFEM to solve fracture problem instead of finite element method (FEM), as in FEM every time mesh refinement is required as crack propagates. Where as in XFEM based analysis enrichment functions are utilized near crack tip and crack face, and due to this remeshing is not required. In this direction Belytschko and Black [1] presented crack growth analysis by utilizing FEM with minimal remeshing. This work is carried out by using enrichment functions and the results obtained from 2-D cracked problem shows good accuracy. Duax et al [2] modelled geometries with different discontinuities like geometry with multiple crack, branched crack and crack emerging from hole by using XFEM. In this present work after solving number of problems on material with different discontinuities it is confirmed that this method is very much robust and accurate for solving fracture analysis. Ebrahimi et al [3] investigated the XFEM based fracture analysis of orthotropic material with crack by utilizing Heviside and asymptotic function. The mixed mode SIFs are determined by utilizing interaction integral for calculating J integral and the accuracy and effectiveness of this present method is also checked. Kim and Paulino [4] evaluated the mixed mode SIFs of FGM material with crack by interaction integral method. In this present work mixed mode SIFs are calculated for variety of problem on plate under different loadings. Kunaporn et al [5] presented the analysis of thin walled composite beam with crack along length of the beam analytically.

In this research paper response of damaged beam in terms of safety index and probability of failure is investigated. Sharma [6] presented the XFEM based fracture analysis of 2-D plate with different discontinuities. In this research work it is observed that the SIFs of one crack is greatly affected when it is interacted with other cracks and holes. It is also observed that presence of holes with crack is more severe than presence of inclusion with crack. Lal et al [7-8] presented stochastic mixed mode SIFs of edge cracked composite beam and centre cracked composite plate under tensile, shear and combined loadings by XFEM. In this research work various probabilistic approach are utilized to see the severity of various fracture parameter. Khatri and Lal [9] investigated fracture analysis of emanating crack from hole in isotropic plate by XFEM under in plane loadings. Here, it is observed that reliability of the structure can be enhanced by controlling various fracture parameters. Ahmed [10] presented XFEM based modeling of 1D and 2D domain with crack. In this present work the result is also compared with the analytical and experimental work. Kumar [11] and Oliver [12] presented different types of fracture related problems by theoretical and computational approach respectively.

From the previously presented research papers it is observed that many research articles are available on fracture analysis of FGM plate under different loadings. It is also observed that very less work has been done on the fracture analysis of FGM plate under point compressive load along crack direction.

In the present work, fracture analysis of edge cracked FGM plate under point compressive load is carried out. The mixed mode SIFs, stress distribution and deformation plots are presented for edge cracked FGM plate. The effect of position of point load on cracked FGM plate is also investigated. The mathematical formulation for the present work is carried out in MATLAB.

2. Problem formulation

The displacement field vector in XFEM framework is presented as.

$$u^h(x) = \sum_{i=1}^{n} N_i(x)\overline{u_i} + \sum_{i=1}^{n_c} N_i(x)H(x)a_i + \sum_{i=1}^{n_{t1}} N_i(x)\sum_{\alpha=1}^{4} \Phi_\alpha^1(x)b_i^{\alpha 1} + \sum_{i=1}^{n_{t2}} N_i(x)\sum_{\alpha=1}^{4} \Phi_\alpha^2(x)b_i^{\alpha 2} \quad (1)$$

where $\overline{u_i}$, a_i, $b_i^{\alpha 1}$ and $b_i^{\alpha 2}$ are the conventional degrees of freedom (dofs), and added dofs for crack face and crack tip. A FGM body with crack is shown whose area is denoted by Ω and its outer boundary Γ as in Fig. 1

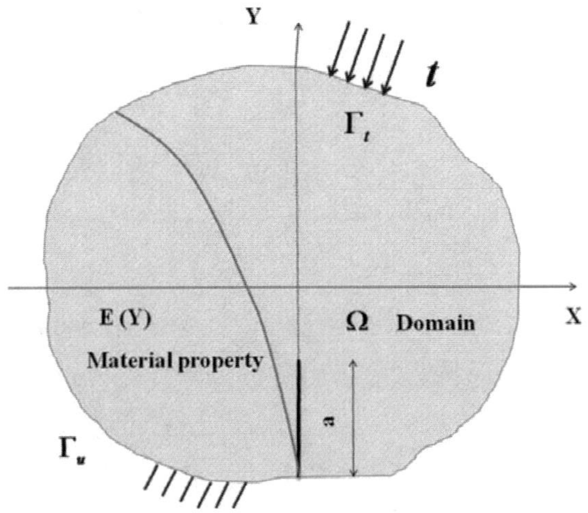

Fig.1 An arbitrary body with crack, under compressive loading

The total dofs is represented as

$$\text{dofs} = \text{size}(m) + \text{size}(m_c) + \text{size}(m_t) + \text{size}(m_h) \text{ Where, } m_t = m_{t1} + m_{t2} \tag{2}$$

Here, m , m_c and m_t are the nodes for FEM, enrichment for crack face and for crack tip respectively.,

$H(x)$ is the Heaviside function for the enrichment of the crack, and the value is +1 or -1 and Ψ is the Asymptotic functions for crack tip.

$$\Psi_\alpha^{t1} = \sqrt{r}\sin\left(\frac{\theta}{2}\right), \Psi_\alpha^{t2} = \sqrt{r}\cos\left(\frac{\theta}{2}\right), \Psi_\alpha^{t3} = \sqrt{r}\sin\theta\cos\left(\frac{\theta}{2}\right), \Psi_\alpha^{t4} = \sqrt{r}\sin\theta\cos\left(\frac{\theta}{2}\right) \tag{2a}$$

The MMSIF and J-integral can be represented as,

$$J = \frac{K_I^2 + K_{II}^2}{E_e} \text{ Where, } E_e = E \text{ for plane stress and } E_e = \frac{E}{1-\upsilon^2} \text{ for plane strain condition} \tag{3}$$

Equation (4) represents the the J integral for the body with crack, and Equation 5 represents the summation o two states.

$$J = \int_\Gamma \left(W\delta_{1j} - \sigma_{ij}\frac{\partial u_i}{\partial x_1} \right) n_j d\Gamma \tag{4}$$

$$J^{(1+2)} = \int_\Gamma \left[\frac{1}{2}\left(\sigma_{ij}^{(1)} + \sigma_{ij}^{(2)}\right)\left(\varepsilon_{ij}^{(1)} + \varepsilon_{ij}^{(2)}\right)\delta_{1j} - \left(\sigma_{ij}^{(1)} + \sigma_{ij}^{(2)}\right)\frac{\partial(u_i^{(1)} + u_i^{(2)})}{\partial x_i} \right] n_j d\Gamma \tag{5}$$

On further solving, get

$$J^{(1+2)} = J^{(1)} + J^{(2)} + \frac{2}{E_{eff}}(K_I^{(1)}K_I^{(2)} + K_{II}^{(1)}K_{II}^{(2)}) \tag{6}$$

Now, from equation (3) and (6) get,

$$I^{(1,2)} = \frac{2}{E_{eff}}(K_I^{(1)}K_I^{(2)} + K_{II}^{(1)}K_{II}^{(2)}) \quad (7)$$

Where, $I^{(1,\text{Mode I})}$ and $I^{(1,\text{Mode II})}$ are interaction integrals

The SIF in XFEM for two states $K_I^{(1)}$ and $K_{II}^{(1)}$ are evaluating by putting $K_I^{(2)}=1$ and $K_{II}^{(2)}=0$ and $K_I^{(2)}=0$ and $K_{II}^{(2)}=1$ in Equation. 7, we get

$$K_I^{(1)} = \frac{M^{(1,Mode\ I)}E_{eff}}{2} \quad \text{and} \quad K_{II}^{(1)} = \frac{M^{(1,Mode\ II)}E_{eff}}{2} \quad (8)$$

The above mathematical formulation is done in MATLAB by utilizing XFEM which is very much efficient for the fracture analysis. In this present approach enrichment functions are used to take care of mesh refinement near crack tip and crack face.

3. Results and Discussions

In FGM material properties can be varied as per the requirement. The validation study is carried out to check the effectiveness of the present method. Table 1 represents the mixed mode SIFs of edge cracked FGM plate under uniform tensile load. The present method is based on XFEM and the results of the present method are very much close to the previously published result as represented in Table 1. In this validation study material properties vary along x direction and loading is perpendicular to the crack. In this validation study dimension of plate is considered as (1×8) unit and material properties of FGM plate vary along W=1 unit. The variation of first mode SIF K_I with respect to a/W ratio for different value of modulus ratio (***E_2/E_1***) **is** presented in Table. 1. The present results obtained from XFEM is validated from the results of previous published research paper [4].

Table 1 Variation of first mode SIF for edge cracked FGM plate under tensile loading .

Methods	E_2/E_1	a/W				
		0.2	**0.3**	**0.4**	**0.5**	**0.6**
Kim and Paulino [4]	0.1	1.284	1.847	2.554	3.496	4.962
	0.2	1.390	1.831	2.431	3.292	4.669
	1	1.358	1.658	2.110	2.822	4.030
	5	1.132	1.370	1.794	2.366	3.448
	10	1.003	1.228	1.588	2.175	3.212
Present simulation XFEM	0.1	1.3191	1.8537	2.4601	3.2401	4.7243
	0.2	1.3970	1.8423	2.4415	3.2104	4.4662
	1	1.3698	1.7240	2.3459	3.0556	4.1763
	5	1.2581	1.4243	2.0927	2.6398	3.4428
	10	1.1775	1.3079	1.9040	2.3285	2.9254

The above mathematical formulation which is done in MATLAB is utilized for further study. In this present work material properties vary in Y direction and loading is along the crack. Fig. 2 (a-b) represent the dimension and loading of edge cracked FGM plate and enrichment for crack face ad crack tip.

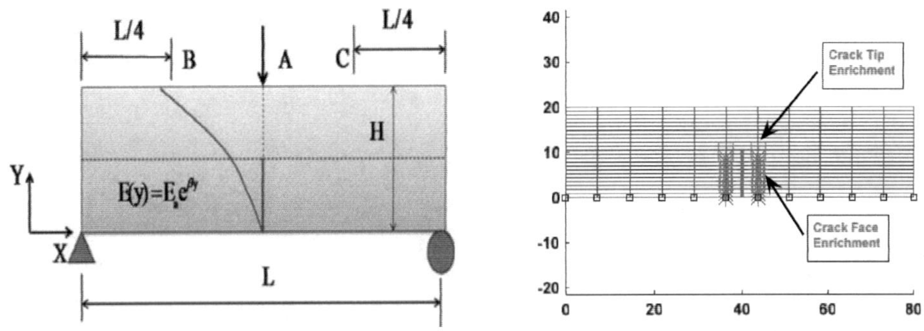

(a) (b)

Fig. 2 (a) Geometry of FGM plate with loadings (b) Crack enrichment

In FGM plate material property varies in such a way that lower side of FGM plate is 100% alloy where as upper side is 100% ceramic. Dimension of FGM plate is L=80 mm and H= 20 mm. The material properties of and E_a= 70GPa, υ_a=0.33, E_c=300 GPa and υ_c= 0.21 represent elastic modulus of alloy, poisson's ratio of alloy, elastic modulus of ceramic and poisson's ration of ceramic. The applied point load is 100 MPa. The elastic modulus of FGM plate is represented as

$$E(y) = E_a e^{\beta y} \text{, where } \beta = \frac{1}{H}\ln\left(\frac{E_c}{E_a}\right) \quad (9)$$

The volume fraction of ceramic (V_c) and alloy (V_a) in FGM plate along y direction are represented as

$$V_c = \frac{E(y) - E_a}{E_c - E_a} = \frac{E_a e^{\beta y} - E_a}{E_c - E_a} \quad (10a)$$

$$V_a = 1 - V_c(y) \quad (10b)$$

The poisson's ratio of FGM plate along x direction is represented as

$$\upsilon(y) = \frac{\upsilon_a V_a(y) E_c + \upsilon_c V_c(y) E_a}{V_a(y) E_c + V_c(y) E_a} \quad (11)$$

FGM plate is modelled and the Elastic Modulus, Volume fraction and Poisson's ration is plotted along x direction as shown in Fig. 3 and Fig. 4 respectively.

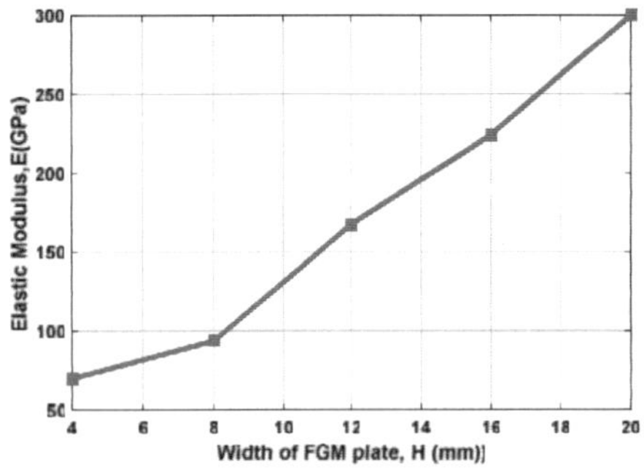

Fig. 3 Variation of Elastic Modulus along y direction

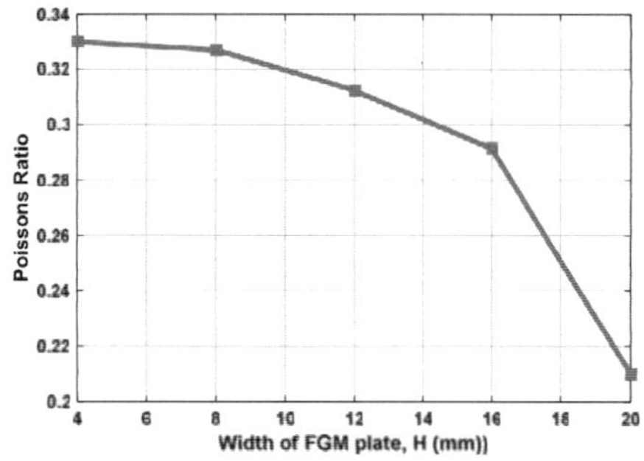

Fig. 4 Variation of Poisson's ration along x direction

Table (2-3) and represent the variation of mixed mode SIFs with respect to crack length for position of point load as shown in Fig. 2(a) for isotropic and FGM plate. The elastic modulus and poisson's ratio of Isotropic plate are 70 GPa and 0.33 respectively. The dimension and loading of Isotropic plate are considered as same as for FGM plate.

Table 3 represents the variation of mixed mode SIFs with respect to crack length for position of point load as shown in Fig. 2(a). From this present study it is observed that as crack length increases magnitude of MMSIF also increases. The maximum value of K_I and K_{II} are observed when loading is at centre of the plate, where as minimum value of K_I and K_{II} are observed when loading is at right of the plate. The maximum reduction of K_I and K_{II} for loading at B as compared to loading at A are 37.79 % and 58.57% at crack length 15 mm and 10 mm respectively, whereas the maximum reduction of K_I and K_{II} for loading at C as compared to loading at A are 75.83 % and 15.95% at crack length 15 mm and 10 mm respectively. It is observed that for same crack length and loading FGM plate has less value of first mode SIF (K_I) than Isotropic plate because in FGM elastic modulus increases along crack length direction.

Table 2 Normalized mixed mode SIFs of edge crack Isotropic plate

Crack	Loading at A		Loading at B		Loading at C	
length	K_I	K_{II}	K_I	K_{II}	K_I	K_{II}
10	2.0073	0.0814	1.2788	-0.1428	0.8484	0.0941
11	2.1035	0.0836	1.3340	-0.1462	0.8853	0.0961
12	2.2603	0.0912	1.4249	-0.1531	0.9473	0.1032
13	2.4165	0.0914	1.5171	-0.1599	1.0084	0.1036
14	2.6320	0.1087	1.6387	-0.1679	1.0842	0.1140
15	2.7456	0.1083	1.7256	-0.1778	1.1430	0.1137

Table 3 Normalized mixed mode SIFs of edge crack FGM plate

Crack	Loading at A		Loading at B		Loading at C	
length	K_I	K_{II}	K_I	K_{II}	K_I	K_{II}
10	2.0047	0.0815	1.2771	-0.1433	0.8473	0.0945
11	2.1012	0.0837	1.3323	-0.1469	0.8843	0.0965
12	2.2564	0.0911	1.4223	-0.1540	0.9456	0.1035
13	2.4103	0.0916	1.5130	-0.1606	1.0058	0.1041
14	2.5993	0.1089	1.6215	-0.1689	1.0801	0.1145
15	2.7309	0.1082	1.6987	-0.1788	1.1312	0.1141

Fig. 5(a) represents the deformation plot of cracked FGM plate under point load at A for crack length 10 mm. In Fig. 5(b-d) represent the stress distribution plot of FGM plate with edge crack under point load of 100 MPa at A, B and C respectively for crack length 10 mm. From the above distribution it is observed that maximum stress is observed when point load is at centre (A), whereas minimum stress is observed when point load is at right (C).

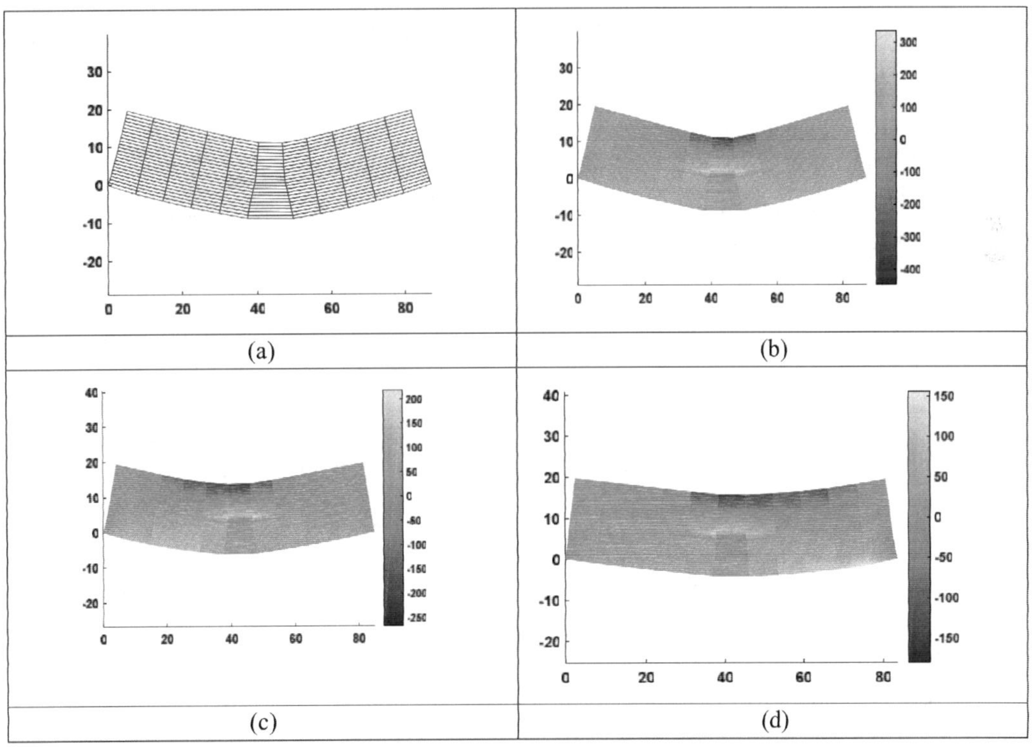

Fig. 5 (a) Deformed shape for load at A (b) Stress distribution for load at A (c) Stress distribution for load at B (d) Stress distribution for load at C

4. Conclusions

Mixed mode SIFs, deformation and stress plots are investigated for FGM plate with edge crack under under point load at different locations. The XFEM based mathematical formulation is done in MATLAB. The observations from the present study are given as

- From this present study it is observed that as crack length increases magnitude of MMSIF also increases.
- The maximum value of K_I and K_{II} are observed when loading is at centre of the plate, where as minimum value of K_I and K_{II} are observed when loading is at right of the plate.
- The maximum stress is observed when point load is at centre (A), whereas minimum stress is observed when point load is at right (C).
- In FGM material fracture resistance can be improved as compared to Isotropic material.

References

[1] T. Belytschko and T. Black, Elastic crack growth in finite elements with minimal remeshing, International Journal for Numerical Methods in Engineering 45(5) (1999), 601–620.

[2] C. Duax, N. Moes, J. Dolbow, N. Sukumar and T. Belytschko, Arbitrary branched and intersecting cracks with the extended finite element method, International Journal for Numerical Methods in Engineering 48(12) (2000), 1741–1760.

[3] S.H Ebrahimi, S. Mohammadi and A. Asadpoure, An extended finite element (XFEM) approach for crack analysis in composite media, International Journal of Civil Engineering 6(3) (2008), 198–207.

[4] J.H. Kim, and G.H Paulino, Simulation Finite element evaluation of mixed mode stress intensity factors in functionally graded materials, *Int. J. Numer. Methods Eng* 53 (2002), 1903-1935.

[5] C. Kunaporn, M.P. Singh, M.J. Patil and R.K. Kapania, Analysis of thin walled beam with crack of random location and size, AIAA Journal 50(6) (2012), 1265–1280.

[6] K. Sharma, Crack Interaction Studies Using XFEM Technique, Journal of Solid Mechanics 6 (2014) 410-421

[7] A. Lal, S.B Mulani and R.K. Kapania, Stochastic Fracture Response and Crack Growth Analysis of Laminated Composite Edge Crack Beams Using Extended Finite Element Method , International Journal of Applied Mechanics, 9(4) (2017) 1750061 (33 pages)

[8] A. Lal, S.P. Palekar, S.B. Mulani and R.K. Kapania, Stochastic extended finite element implementation for fracture analysis of laminated composite plate with a central crack, Aerospace Science and Technology, 60(2017), 131-151.

[9] K. Khatri and A. Lal, Stochastic XFEM fracture and crack propagation behavior of an isotropic plate with hole emanating radial cracks subjected to various in-plane loadings, Mechanics of Advanced Materials and Structures, 25(9) (2018), 732-755.

[10] A. Ahmed, eXtended finite element method (XFEM)-modelling arbitrary discontinuities and failure analysis, 2009.

[11] P. Kumar, Elements of fracture mechanics, McGraw hill education (India) private limited, 91, 2009.

[12] J. Oliver, A.E. Huespe, E. Samaniego and E.W.V Chaves, On strategies for tracking strong discontinuities in computational failure mechanics, Fifth World Congress on Computational Mechanics July 7–12Vienna, Austria., 2002.

Numerical study of heat transfer enhancement through a circular tube fitted with and without rectangular cut twisted tape insert

M U Hakim[1], M W Hridoy[2], M I Islam[1] and M S Rayhan[3]

[1]Department of Mechanical Engineering, Khulna University of Engineering & Technology, Khulna-9203, Bangladesh
[2]Department of Mechatronics and Industrial Engineering, Chittagong University of Engineering & Technology, Chittagong-4349, Bangladesh
[3]Dhaka University of Engineering & Technology, Dhaka-1700, Bangladesh

Email: masbah.067@gmail.com

Abstract. Numerical simulations have been carried out to investigate the heat transfer enhancement in a tube fitted with a rectangular cut twisted tape insert and compare the data with a smooth tube. For this purpose, CFD (Computational Fluid Dynamics) was utilized to simulate the three-dimensional model with appropriate boundary conditions. The numerical results show that the flow field can be adjusted and the thickness of boundary layer can be decreased by inserting the rectangular cut twisted tape insert so that the heat transfer can be enhanced in the tube. Water was used as the working fluid. The results obtained show an increase in Nusselt number as a result of increasing heat transfer rate for the tube provided with rectangular cut twisted tapes than the plain tube. The pressure drop across the tube wall for twisted tape with the rectangular cut is higher than the plain tube. This increase in the rate of heat transfer is due to the introduction of secondary swirl flow (due to twisted tape) along with the primary flow.

1. Introduction

Heat exchanger refers to the system that is used for heat transfer between two or more fluids. The sole purpose of heat transfer enhancement is to increase the effectiveness of the heat exchanger. The general mechanism of heat transfer enhancement involves induced swirl flow which potentially promotes fluid mixing resulting in a thinner boundary layer and a higher convective heat transfer rate [1]. The process can be both active and passive. In the active method, desired flow modification is achieved via an external power source. On the other hand, passive techniques include the incorporation of inserts, rough surfaces etc. for the geometrical modification of the flow channel to bring out the desired change in the flow without any external power source. In recent times, there has been a good number of researches performed on the performance augmentation of the heat exchanger. The works include both active and passive heat transfer enhancement techniques.

Megerlin et al. [2] used spiral brush inserts in short channels under high heat flux condition and obtained turbulent heat transfer coefficient enhancement up to 8.5 times. Different types of "Everter" inserts and several disk inserts were investigated by Maezawa and Lock [3]. Although the local heat transfer coefficient increased a significant amount for air flowing in an electrically heated tube, no marked improvement was observed beyond 10 diameters downstream of the insert. Also, a doubling of average heat transfer coefficient was recorded. But, on the negative side, friction factors increased up to 3.5 times. Eiasma-ard et al. [4] observed a 57% augmentation on the heat transfer rate with the combination of wavy-surfaced tube and helical tape compared to wavy-surfaced tube alone. On another investigation [5], Eiasma et al. reported a comparatively better performance in backward flow than the forward one in case of louvered strips. Bhuiya et al. [6] compared various helical tape inserts with different helix angles fitted in a tube. The experiment concluded that, under the same operating condition, decreasing of helix angles increase the Nusselt number, friction factors and thermal enhancement efficiency. Using a double-helical tape insert with 9° helix angle resulted in the maximum thermal efficiency of 215°. Zhang et al. [9] performed a three-dimensional numerical study on heat transfer in a tube with helical screw tape inserts of four different widths and found an enhancement of heat transfer up to 351% although the friction factor also increased to 1020%. Saeedinia et al. [10] found an increase in heat transfer of 45% on average in a tube with a wire coil with the highest wire diameter although the pressure dropped by about 63%. García et al. [11] compared the pressure drop and heat transfer performance between three types of enhancement methods involving corrugated tubes, wire coils and dimple tubes and concluded that, for Reynolds number between 200 and 2000, wire coils should be used whereas dimpled and corrugated tubes are suitable for Reynolds number above 2000.

In general, a twisted tape enhances the heat transfer coefficient by inducing swirl flow and thus promoting fluid mixing between core and wall regions. As a result, the thermal boundary layer is disturbed, causing the enhancement of heat transfer in the existing system. Murugesan et al. [12] observed an increased heat transfer rate with increasing depth ratio and decreasing width ratio in case of V-cut twisted tapes. Murugesan et al., on another investigation [13], compared the effectiveness of classical and square-cut twisted tapes and found out square-cut one to be better in augmenting heat transfer performance. Thianpong et al. [14] reported a 108% increased heat transfer in a tube with perforated twisted tape inserts than the plain tube.

Researches have also been done numerically to optimize the performance of twisted tape inserts. Zhang and Cheng [15] compared numerically the heat transfer performance between classical and edge-fold twisted tape inserts. Although the pressure dropped by around 74% in edge fold twisted tape inserts, 9.2% heat transfer enhancement occurred. Rahimi et al. [16] compared the performance of four different types of twisted tapes- classical, perforated, notched and jagged. The result suggested that jagged twisted tapes offer the best thermal performance. Cui et al. [17] in their three-dimensional numerical model, investigated the impact of gap width and operating parameters on thermal hydrodynamic performance.

On all the above researches it has been found that, although introducing tape inserts improve the heat transfer performance, but it is achieved by a penalty of increased friction loss. So, there has to be a compromise between these two to get the optimum performance. In this paper, heat transfer performance has been studied numerically between circular tubes fitted with and without rectangular cut twisted tape insert. This paper offers an optimization between heat transfer coefficient and friction factor based on the Reynolds number. For the numerical simulation, Ansys fluent 16.1 student version has been used.

2. Description of the computational model

This work aims to compare two circular tubes with and without rectangular cut twisted tape inserts in terms of heat transfer rate, convective heat transfer coefficient, Nusselt number and friction factor by simulation. For this purpose, a model is designed and analysed in Ansys fluent. Pressure based solver is used to solve the problem. For turbulence modelling, k-ε turbulence model with standard wall function has been opted. SIMPLE (semi-implicit method for pressure linked equations) algorithm has been opted for the steady-state numerical analysis to solve the pressure and velocity field for the computational flow domain. Convergence criteria have been set as 10-5 for all the equations.

IOP Conf. Series: Materials Science and Engineering 1017 (2021) 012019 doi:10.1088/1757-899X/1017/1/012019

(a) (b)

Figure 1. (a) 3D view of the tube; (b) design of rectangular cut twisted tape insert.

2.1. Physical modelling

The circular tube with and without a rectangular cut twisted tape insert has been investigated in the current study. Water as the working fluid entered the tube at an inlet temperature and flowed through the test section. The physical parameters of the tube are shown in Table 1. The properties of water are shown in Table 2. Thermal conductivity of the material of the tube is 202.4 W/m-K. The CAD design of the model is presented in figure 1. The numerical study was carried out at twist ratios (y/w) of 5.25. The ratio of twist length (y) to the twisted tape width at the large end (w) is referred to as the twist ratio. Twist angle is the angle between the reference plane before and after twisting. Some assumptions were made during the simulation process such as:

- The physical properties of water were remained constant and were evaluated at the bulk temperature.
- Steady and incompressible flow.
- The temperature of the tube wall was constant.

Table 1. Physical parameters of the tube.

Parameters	Value
Tube inside diameter	26.6 mm
Tube outside diameter	30 mm
Tube test length	900 mm
Outside surface area	0.0848 m^2
Inside surface area	0.0752 m^2
Cross-sectional area	0.000556 m^2

Table 2. Properties of water.

Properties	Value
Viscosity	0.00087 kg/m-s
Density	998.2 kg/m^3
Thermal conductivity	0.6 W/m-K
Specific heat	4182 J/kg-K

2.2. Mesh generation and grid independence

A grid independence study has been performed for the Nusselt number for different number of mesh (Table 3) to check the computational accuracy. The results are taken for Re= 2722 using a circular tube equipped with rectangular-cut twisted tape insert. It is found that the difference in Nusselt number between mesh number 495713 and 961055 is 0.5%. Therefore, mesh number 495713 is adopted for this numerical analysis.

Table 3. Mesh independence test.

Mesh number	Nu
136581	19.28
259416	21.54
495713	24.49
961055	24.62

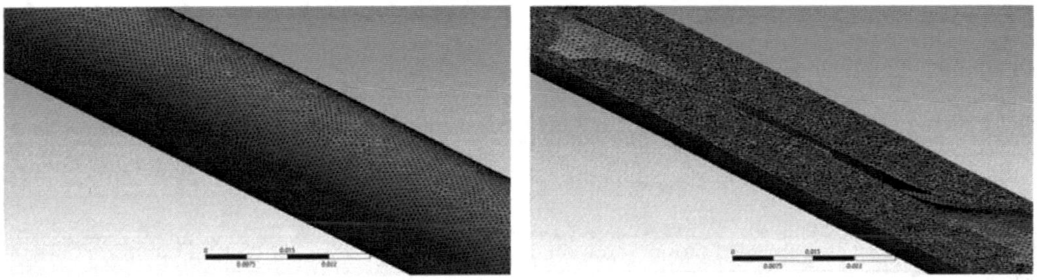

(a) (b)

Figure 2. (a) Meshing of the tube; (b) Meshing of the fluid domain.

2.3. Boundary conditions

The numerical simulation has been carried out opting finite volume method. As the boundary conditions, water with 303K temperature has been taken at the inlet. Also, seven different velocities have been considered ranging from 0.089 m/s to 0.161 m/s at the inlet. The velocities were uniform in each of the cases. No-slip boundary condition has been considered at the wall. Pressure outlet conditions have been applied at the exit.

2.4. Governing equations

Continuity equation: $\frac{\partial}{\partial x_i}(\rho u_i) = 0$ (1)

Momentum equation:

$$\frac{\partial}{\partial x}(\rho u_i u_j) = -\frac{\partial p}{\partial x_i} + \frac{\partial}{\partial x_j}[(\mu + \mu_t)\left(\frac{\partial u_i}{\partial x_j} + \frac{\partial u_j}{\partial x_i}\right) - \frac{2}{3}((\mu + \mu_t)\frac{\partial u_i}{\partial x_i}\delta_{ij} - \frac{2}{3}\rho k \delta_{ij}] + F \quad (2)$$

Energy equation: $\rho C_p u_i \frac{\partial T}{\partial x_i} = (k + k_T)\frac{\partial}{\partial x_i}(\frac{\partial T}{\partial x_i})$ (3)

Reynolds number: $R_e = \frac{\rho v L}{\mu}$ (4)

Nusselt number: $Nu_d = 0.023 Re_d^{0.8} pr^{0.4}$ (5)

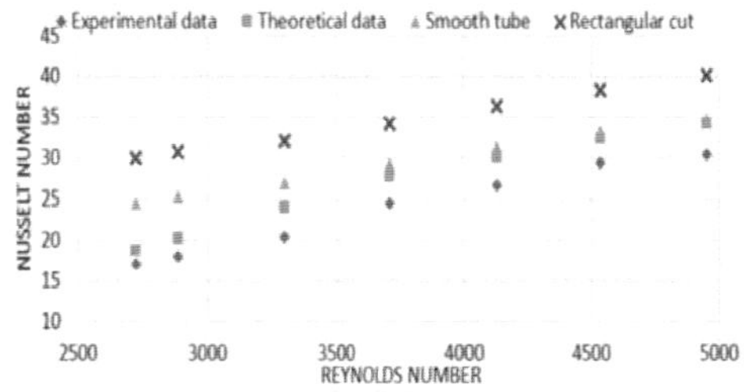

(a)

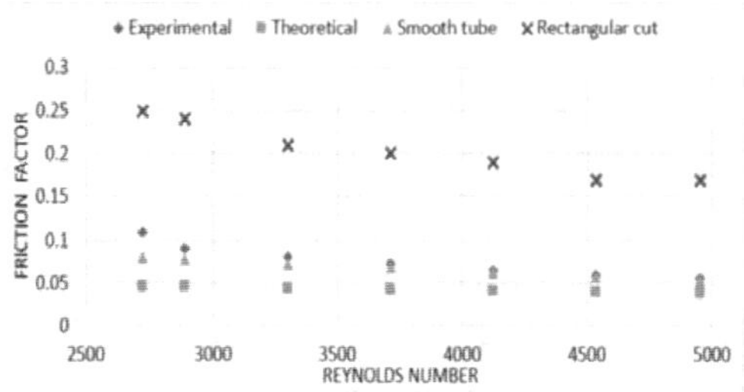

(b)

Figure 3. (a) Nusselt number validation for smooth tube; (b) friction factor validation for smooth tube.

2.5. Model validation

The numerical computation was first performed in a particular case of a smooth tube to validate the CFD simulation results with predicted theoretical data obtained by Gnielinski correlation [18] for Nusselt number and Petukhov [19] for friction factor. The predicted theoretical values of Nusselt number and friction factor are demonstrated in figure 3(a). The present results reasonably agree well with the available correlations with little variation for Nusselt number and friction factor. The friction factor is decreasing as the Reynold number increases. Friction factor for experimental, theoretical, smooth tube and rectangular cut twisted tape inserts are shown in the following figure 3(b) .

3. Results and discussion

The current study involves the using of rectangular cut twisted tape insert to augment the heat transfer inside a tube. The results obtained from the simulation clearly aligns with our proposal. Several parameters have been checked with varying Reynolds number. Although, heat transfer rate increases with increasing Re, the increase in friction factor is of concern.

Figure 4(a) shows the variation in Nusselt number with Reynolds number. Nusselt number is the ratio of convective to conductive heat transfer in the fluid boundary. The lowest value of Nusselt number recorded is 24.49 for Re=2722.029 for smooth tube. The value increased steadily with increasing Re and the maximum recorded value for smooth tube is 34.7 for Re=4949.144. For the tube with twisted tape insert however, the value is quite large. Here, the minimum recorded value of Nu is 29.95 which steadily increased to a maximum of 40.05 for the raising of Re from 2722.029 to 4949.144.

Figure 4(b) represents the variation of friction factor with Reynolds number. Friction factor vs Re depicts a declining curve. The maximum friction factor is found in case of the tube with twisted tape insert and the value is 0.25 for Re= 2722.029 which declines markedly as the Re increases. A total decrease of 0.08 is observed for increasing the Reynolds number to 4949.144. Friction factor curve for the smooth tube also shows a declining trend. The declination is however, less significant in this case. Friction factor decreases from 0.08 to 0.053 with the increase in Reynolds number from 2722.029 to 4949.144.

Figure 5 shows the relation between the theoretical and simulation friction factor for the smooth tube. Friction factor recorded from the simulation is higher than the theoretical friction factor. And the difference increases with increasing value. Maximum theoretical friction factor obtained is 0.047, whereas simulation results show the maximum value to be 0.08. The minimum value is 0.038 and 0.053 for theory and simulation respectively.

Thermal performance factor (TPF) represents the effectiveness of a heat transfer enhancement method and refers to the ratio of augmented heat transfer to the increased friction. Figure 6 shows that thermal performance factor is higher at lower Reynolds number and the maximum value is 0.837 for Re=2722.029. TPF decreases drastically with Reynolds number and it falls to 0.782 at Re=4949.144. The result suggests that, with increasing Re, the increase in friction factor is much higher compared to the corresponding increase in the heat transfer rate enhancement. As a result, despite the increase in heat transfer, overall effectiveness of the system is actually decreasing.

Temperature distribution inside the tube with insert is shown in figure 7. It is seen that the temperature is higher near the wall of the tube and as we go inside, it falls. The lowest temperature region is the region around the insert. Maximum recorded temperature for v=0.089 m/s is 351.7K whereas for v=0.161 m/s the maximum temperature is 348.7K. Also, it should be noted that the temperature difference between the regions is higher at low velocity, about 48.5K at 0.089 m/s and the difference lowers to 45.6K at 0.161 m/s. This seems logical as we know by now, increased velocity promotes increased mixing and decreases the difference in temperature between regions.

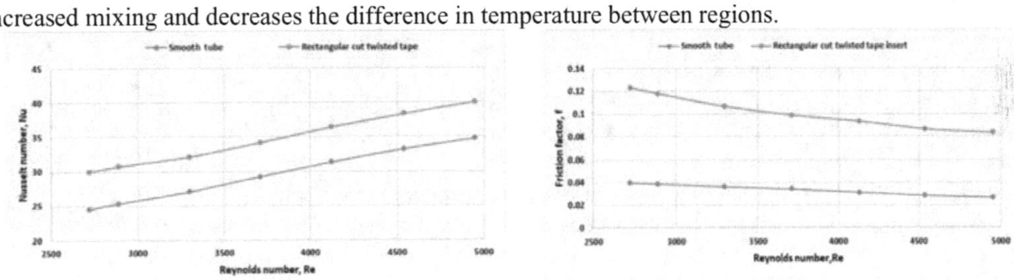

(a) (b)

Figure 4. (a) Variation of Nusselt number with Reynolds number; (b)Variation of friction factor with Reynolds number.

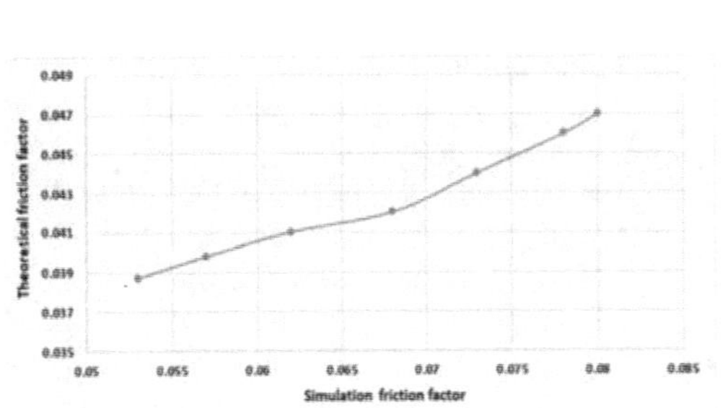

Figure 5. Comparison between simulated and theoretical friction factor (Smooth tube).

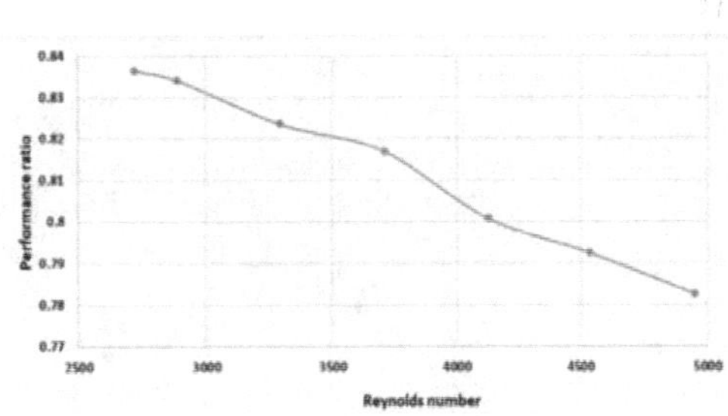

Figure 6. Thermal performance factor for rectangular cut twisted tape inserts.

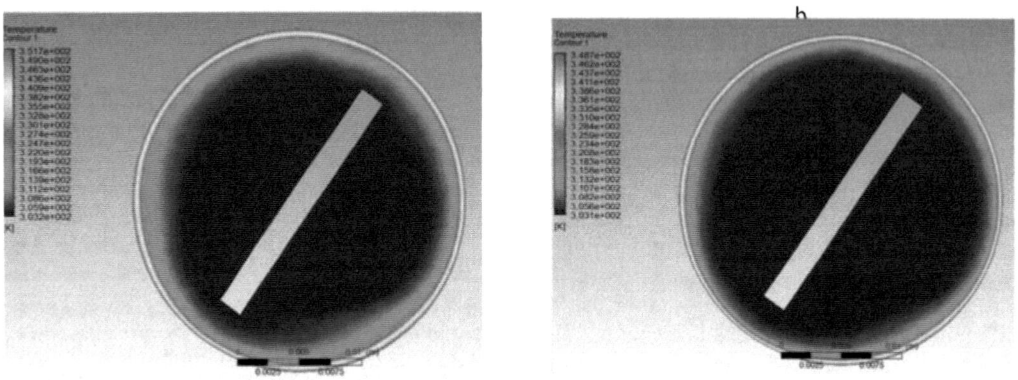

(a) (b)

Figure 7. (a) Temperature contour for tube with insert (at 0.089 m/s); (b) Temperature contour for tube with insert (at 0.161 m/s).

Turbulence kinetic energy and its distribution also depend largely on velocity. From Fig. 8 the difference can be observed for the tube with insert. From Fig. 8(a), at low velocity (0.089 m/s), the turbulence kinetic energy is much lower, ranging from $7.1*10^{-11}$-$1.967*10^{-6}$. Maximum energy region

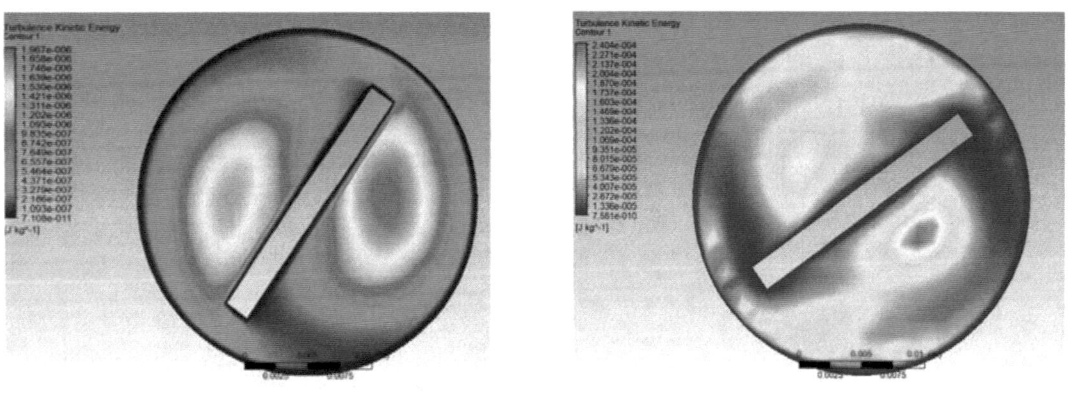

(a) (b)

Figure 8. (a) Turbulence kinetic energy for tube with insert (at 0.089 m/s); (b) Turbulence kinetic energy for tube with insert (at 0.161 m/s).

is observed around the insert and there is a notable variation in energy between different regions. In contrast, Fig. 8(b) shows a more uniform energy distribution in the tube at velocity of 0.161 m/s. And also, the overall turbulence kinetic energy ranges from $7.561*10^{-10}$-$2.404*10^{-4}$, which is much higher when compared to Fig. 7(a).

4. Conclusion

Heat transfer and friction factor characteristics of a tube fitted with rectangular cut twisted tape insert and without insert are numerically studied and validated by two performance criterions Gnielinski and Petukhov. Thus, the computational results obtained indicate that the rate of heat transfer and the pressure drop along tube fitted with rectangular cut twisted tape is higher than the plain tube. The insertion of the twisted tape in the flow passage results in the creation of swirl along with the normal flow. This swirl created increased the turbulence, flow length of the fluid flow, time of exposure of the fluid to the heat flux, resulting in augmented rate of heat transfer. As the Reynolds number increases, the swirling disturbance generated by rectangular cut twisted tape intensifies, thus causing higher temperature and velocity gradient near the tube wall. Therefore, both the heat transfer coefficient and the friction factor increase than the plain tube. The following conclusions can be drawn from the results of the present study.

- Heat transfer rates were increased by an average of 1.16 times as compared to the plain tube.
- Convective heat transfer coefficients were increased by an average up to 1.178 times as compared to the plain tube.
- Nusselt numbers were increased by an average up to 1.18 times as compared to the plain tube.
- Friction factors were increased by an average of 2.59 times as compared to the plain tube.

Appendix A. Nomenclature

ρ	Density (kg/m^3)
μ	Dynamic viscosity (Pa s)
k	Thermal Conductivity (W/m-K)
C_p	Specific heat (J/kg-K)
T	Temperature (K)
α	Thermal diffusivity (m^2/s)
v	Inlet velocity (m/s)
L	Length of tube (m)
w	Twisted tape width at large end
Re	Reynolds number
Pr	Prandtl number
Nu	Nusselt number

References

[1] M.M.K. Bhuiya, M.S.U. Chowdhury, J.U. Ahamed, M.J.H. Khan, M.A.R. Sarkar, M.A. Kalam, H.H. Masjuki, M. Shahabuddin. "Heat transfer performance for turbulent flow through a tube using double helical tape insert". *International Communications in Heat and Mass Transfer* **39** (2012) 818–825.

[2] F.F. Megerlin, R.W. Murphy, A.E. Bergles, Augmentation of heat transfer in tubes by use of mesh and brush inserts, *ASME J. Heat Transfer* **96** (1974) 145–151.

[3] S. Maezawa, G.S.H. Lock, *Proc. Sixth Int. Heat Transfer Conference (Toronto),* 1978

[4] Eiamsa-ard, S. and Promvonge, P. (2005) "Enhancement of heat transfer in a tube fitted with regularly-spaced helical tape swirl generators" *Solar Energy*.

[5] S. Eiamsa-ard, S. Pethkool, C. Thianpong, and P. Promvonge, "Turbulent flow heat transfer and pressure loss in a double pipe heat exchanger with louvered strip inserts," *International Communications in Heat and Mass Transfer*, vol. **35**, no. 2, pp. 120–129, 2008.

[6] M.M.K.Bhuiya, M. S. U. Chowdhury, J. U. Ahamed et al., "Heat transfer performance for turbulent flow through a tube using double helical tape inserts," *International Communications in Heat and Mass Transfer*, vol. **39**, no. 6, pp. 818–825, 2012.

[7] X. Zhang, Z. Liu, and W. Liu, "Numerical studies on heat transfer and friction factor characteristics of a tube fitted with helical screw-tape without core-rod inserts," *International Journal of Heat and Mass Transfer*, vol. **60**, no. 1, pp. 490–498, 2013.

[8] M. Saeedinia, M. A. Akhavan-Behabadi, and M. Nasr, "Experimental study on heat transfer and pressure drop of nanofluid flow in a horizontal coiled wire inserted tube under constant heat flux," *ExperimentalThermal and Fluid Science*, vol. **36**, pp. 158–168, 2012.

[9] A. García, J. P. Solano, P. G. Vicente, and A. Viedma, "The influence of artificial roughness shape on heat transfer enhancement: Corrugated tubes, dimpled tubes and wire coils," *Applied Thermal Engineering*, vol. **35**, no. 1, pp. 196–201, 2012.

[10] Murugesan P, Mayilsamy K, Suresh S. Turbulent heat transfer and pressure drop in tube fitted with square-cut twisted tape. *China J. Chem. Eng.* 2010; **18**:609e17.

[11] Murugesan P, Mayilsamy K, Suresh S, Srinivasan PSS. Heat transfer and pressure drop characteristics in a circular tube fitted with and without V-cut twisted tape insert. *International Communication of Heat Mass Transfer* 2011; **38**:329e34.

[12] Thianpong C, Eiamsa-ard P, Promvonge P, Eiamsa-ard S. Effect of perforated twisted-tapes with parallel wings on heat transfer enhancement in a heat exchanger tube. *Energy Procedia* 2012; **14**:1117e23.

[13] Zhang CY, Cheng TM. Three-dimensional numerical simulation of thermal-hydraulic performance of a circular tube with edge fold-twisted tape inserts. *J Hydrodynamic* 2010; **22**:662e70.

[14] Rahimi M, Shabanian SR, Alsairafi AA. Experimental and CFD studies on heat transfer and friction factor characteristics of a tube equipped with modified twisted tape inserts. *Chem. Eng. Process Process Intensifier* 2009; **48**:762e70.

[15] Y.Cui, M.Tian, L. Zhang, G. Li, and J. Zhu, "Three dimensional numerical simulation of convection-condensation of vapor with high concentration air in tube with inserts," *Chinese Journal of Chemical Engineering*, vol. **20**, no. 4, pp. 686–692, 2012.

[16] Gnielinski V. New equations for heat and mass transfer in turbulent pipe and channel flow. *Int. Chem. Eng.* 1976;**16(2)**:359-68.

[17] Petukhov BS. Heat transfer and friction in turbulent pipe flow with variable physical properties. *Advances in heat transfer*. 1970 Jan 1;**6**(503):i565.

Cumulative post-buckling buckling behavior of smart sandwich structure

Achchhe Lal [1] **and Kanif Markad**[1]

[1]Department of Mechanical Engineering, SVNIT, India

E-Mail: kmarkad13@gmail.com

Abstract: Cumulative post-buckling buckling behavior of smart sandwich structure has been presented with the combination of shape memory alloy (SMA) and shape memory polymer (SMP) in sandwich composite structure under uniform temperature distribution. The analysis is based on higher order shear deformation theory (HSDT) with von Karman nonlinearity by finite element method (FEM). Evaluation of nondimensional critical buckling temperature (NCBT) has been executed under varying plate thickness ratio. Thermal buckling performance of SMA reinforced SMP laminated composites under dynamic temperature condition with different strain rate and variation in SMA volume fraction is also presented first time to understand the cumulative post-buckling behavior.

1. Introduction

Smart materials have the characteristics to memorize a stimuli such as temperature, electric or magnetic field and correspondingly change its shape, natural frequency, stiffness, buckling, hardness and other mechanical properties. Shape memory alloys (SMA) are one of such class of smart material, which has the capability to regain its original shape/form from a deformation, when heated at an elevated temperature [1]. As compared to other smart materials, SMA has the highest recovery strain field and flexibility, due to its distinct thermo-mechanical, shape memory and pseudo-elastic properties towards the deviations in load, temperature and stresses [2]. In the past few decades, some alternatives are being explored to overcome limitations of SMA such as expense and poor fatigue. Among the various possible solutions Shape Memory Polymer (SMP) is a viable option and the further work is concentrated on it.

Buckling and post-buckling analysis of composite structures were taken into consideration by researchers, which affects stability of the structures in thermal environment [3, 4]. Embedding SMA, such as Nitinol wires, in the structure has satisfactorily enhanced the thermal buckling resistance, particularly in composites and sandwich panel subjected to thermal environment [5, 6]. Asadi et al. [7, 8] has gained analytical results for the vibration and thermal stability of hybrid composite structures strengthened with SMA fibers using FOBT technique for analytical solutions. When SMA wire reinforced composites are subjected to high temperature surplus strains are recovered, which induces internal tensions. This process create tension in structure and improves its thermal buckling and post buckling resistivity and provide hindrance to the resonance in the structure. There are few articles that deals with thermal buckling of composites or sandwich panels, which provides an opportunity to the authors to analyse the phenomenon in the current work. Analysing the phenomenon of SMP and SMA, it is noticed that the no one were combinedly analysed the behaviour of SMA embedded SMPC and this is the reason behind the current work.

2. General Mathematical formulation

Using FEM implementation in MATLAB, the post-buckling analysis of sandwich smart composite plate is carried out by HSDT adopting von-Karman kinematics in present study. Nine noded model which is having seven degrees of freedom (DOF) using C^0 continuity for the study of post-buckling behaviour of the structure.

2.1 Shape memory polymer matrix

In present study, facesheet and bottom sheet utilizes the shape memory polymer matrix having dynamic mechanical properties governed by T_g of polymer. A mathematical model is shown in Eq. (1) had been suggested to compute values of storage modulus $E_m(T)$ of the polymers at a wide span of temperatures across T_g with a constant frequency (Mahieux and Reifsnider [9]).

$$E_m(T) = (E_1 - E_2).\exp\left(-\left(\frac{T}{T_\beta}\right)^{m_1}\right) + (E_2 - E_3).\exp\left(-\left(\frac{T}{T_g}\right)^{m_2}\right) + E_3.\exp\left(-\left(\frac{T}{T_f}\right)^{m_3}\right) \quad (1)$$

Temperature dependent Poisson's ratio (μ) can be calculated through phase transition model for SMP material (Qi et al. [11]).

$$\mu = \mu_g v_g + \mu_r(1 - v_g) \quad (2)$$

$$v_g = 1 - \frac{1}{1 + exp\,[-(T - T_m)/Z]} \quad (3)$$

The properties of composite material are estimated through the theory of volume averaging (Shen et al. [13]). Following equations are utilized as per the theory to calculate numerous material properties.

$$E_{c1} = E_{f1}v_f + E_m v_m \quad (4)$$

$$E_{c2} = (1 - C)E_{c2}^1 + CE_{c2}^2 \quad (5)$$

$$\mu_{c21} = (1 - C)\mu_{c21}^1 + C\mu_{c21}^2 \quad (6)$$

$$\mu_{c12} = \mu_{c21}\frac{E_{c2}}{E_{c1}} \quad (7)$$

$$G_{c12} = (1 - C)G_{c12}^1 + CG_{c12}^2 \quad (8)$$

Where E_{c1} and E_{c2} are modulus of shape memory polymer composite (SMPC) along longitudinal and transverse directions respectively. μ_{c12} and μ_{c21} indicate Poisson's ratio of composites along 1-2 and 2-1 plane respectively. The superscript 1 and 2 indicate values of constants in series and parallel respectively. Elastic modulus of SMP has been presented in Fig. 1 for the range of temperature from 273 K to 373 K. T_g of SMP has been investigated to be 305 K (Gu et al. [10]).

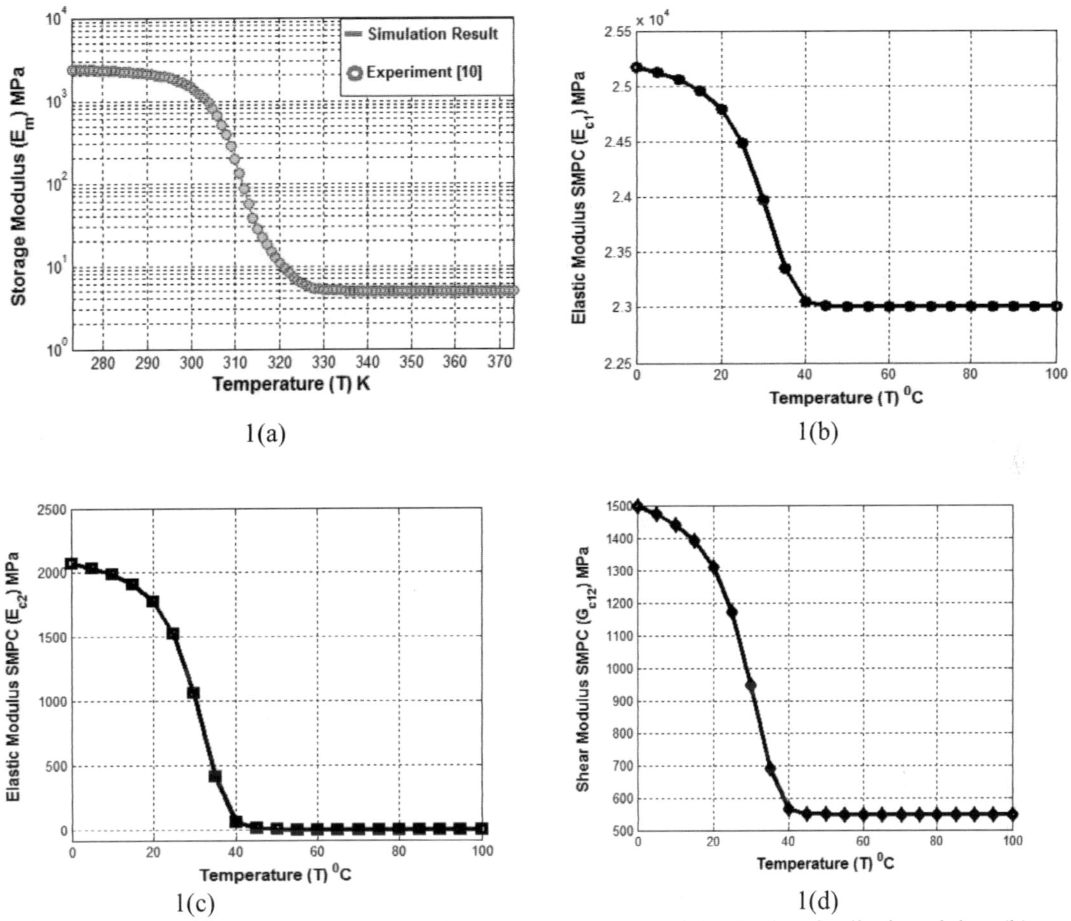

Fig. 1. Influence of variation of temperature on (a) storage modulus (a) longitudinal modulus, (b) transverse modulus (c) shear modulus of SMPC.

2.2 Displacement field equation

In present analysis sandwich smart composite plate is utilized whose length '*a*', width '*b*' and '*h*' is the thickness which is constant throughout. For the FEM analysis, following displacement field equation is get used.

$$U = u_0(x,y) + g_1(z)\Omega_x(x,y) + g_2(z)\zeta_x(x,y)$$
$$V = v_0(x,y) + g_1(z)\Omega_y(x,y) + g_2(z)\zeta_y(x,y) \quad (9)$$
$$W = w_0(x,y)$$

In Eq. (9) U, V and W are the displacements along x, y and z direction respectively at any point on the plate. u_0, v_0, w_0 are the midplane displacements, Ω_x, Ω_y are the rotations at z=0.

2.3 Strain and displacement relation

The strain vectors can be represented as,

$$\{\varepsilon\} = \{\varepsilon_x \varepsilon_y \ \varepsilon_z \ \gamma_{xy} \gamma_{xz} \gamma_{yz}\}^T = \{\partial u / \partial x \ \ \partial v / \partial x \ \ 0 \ \ \partial u / \partial x + \partial v / \partial y \ \ \partial u / \partial z + \partial w / \partial x \ \ \partial v / \partial z + \partial w / \partial y\}^T$$

$$\{\varepsilon\} = \{\varepsilon_l\} - \{\varepsilon_{nl}\} - \{\varepsilon_t\} \tag{10}$$

$$\{\varepsilon_t\} = \{\alpha_1 \Delta T \ \ \alpha_2 \Delta T \ \ 0 \ \ 0 \ \ 0\}$$

Displacement vector can be written as,

$$q = \begin{pmatrix} u & v & w & \zeta_y & \zeta_x & \Omega_y & \Omega_x \end{pmatrix}$$

The linear and nonlinear strain vector can be written as, $\{\varepsilon_l\} = [L]\{q\}$ and $\{\varepsilon_{nl}\} = \frac{1}{2}[A]\{\phi\}$ as per Lal et al. [14].

2.4 Strain and stress relationship

The relation between stress and strain for SMA inserted composite of kth layer with reference to reference coordinate system by following relation, Kumar and Singh [12]

$$\{\sigma\}^k = [\bar{Q}]^k \{\bar{\varepsilon}\}^k + \{\sigma_r\}^k V_s - \left([\bar{Q}]_m \{\alpha\}_m V_m\right)^k \Delta T \tag{11}$$

2.5 Finite element modeling

Displacement vector and field vector written as follows, Lal et al. [14]

$$\{q\}^e = \sum_{i=1}^{NN} N_i^e \{q_i\}^e, \ x = \sum_{i=1}^{NN} N_i \{x_i\}, \ y = \sum_{i=1}^{NN} N_i \{y_i\} \tag{12}$$

2.6 Strain energy in composite plate

Total strain energy in the sandwich smart composite plate is represented as,

$$\Pi_1 = \frac{1}{2} \sum_{i=1}^{NE} \left[q^{e^T} (k_l + k_{nl}) q^e \right] = \{q\}^T [K_l + K_{nl}] \{q\} \tag{13}$$

NE and *i* are the number of elements and elemental, respectively and *q* is the global displacement vector. Where, K_l and K_{nl} are the linear and nonlinear stiffness matrix as per Lal et al. [14]

2.7 Work done and strain energy due to thermal loading

Due to thermal load application, the work potential stored can be given as,

$$M_T = \sum_{k=1}^{NL} \int_{z_{k-1}}^{z_k} [\bar{Q}]\{\varepsilon_t\} dz; \quad M_A = \sum_{k=1}^{NL} \int_{z_{k-1}}^{z_k} [\bar{Q}]\{\varepsilon_t\} z dz; \quad M_P = \sum_{k=1}^{NL} \int_{z_{k-1}}^{z_k} [\bar{Q}]\{\varepsilon_t\} z^3 dz \tag{14}$$

$$N_O^T = M_T + M_A + M_P$$

Where N_O^T is the thermal compressive load acting over the sandwich smart composite plate.

2.8 Governing equation

The governing equation for the sandwich composite plate can be acquired by minimizing the total potential energy with respect to generalized displacements ($\partial(\Pi_1 - \Pi_2) / \partial q = 0$),

$$[K]\{q\} = \lambda[K_g]\{q\}$$

$$\text{Where, } [K] = [K_l] + [K_{nl}] \text{ with } [K_{nl}] = [K_{nl1}] + [K_{nl2}] + [K_{nl3}] \quad (15)$$

In the expression λ is the critical buckling temperature.

3. Results and discussion

In this study, various boundary conditions (BC) are applied which are directed as,
$SSSS: u = v = w = \Omega_x = \zeta_x = 0 \; at \; y = 0 \; and \; b; \; u = v = w = \Omega_y = \zeta_y = 0 \; at \; x = 0 \; and \; a$
$CCCC: u = v = w = \Omega_x = \zeta_x = \Omega_y = \zeta_y = 0 \; at \; y = 0 \; and \; b; \; u = v = w = \Omega_x = \zeta_x = \Omega_y = \zeta_y = 0 \; at \; x = 0 \; and \; a$
Convergence study has been done in Figure 2 to select the number of elements (neL) to calculate the results without tempering the results. From the figure it is clear that result converging from 3x3 mesh size, so 16 elements are selected for further study.

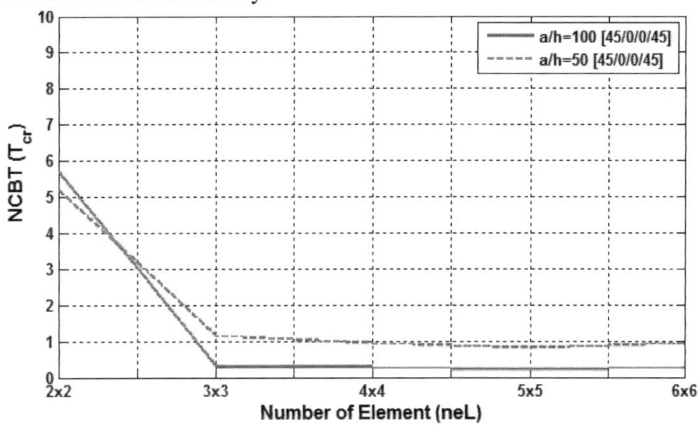

Fig. 2. Convergence study of FE solution for various mesh size

To validate the finite element computer program, problems that have analytical solutions are solved. The accuracy of the present HSDT model is verified by validating the results with [5 and 12] by obtaining nondimensional critical buckling temperature (NCBT) as, $T_{cr} = \Delta T \alpha_2$ as shown in Figure 3. Figure 4 shows NCBT variation with the SMA volume fraction and its validation with the Kumar and Singh [12] performed by layerwise theory. The result indicate the same nature of variation with the available results and also shows the effect of utilization of smart composite sandwich plate under clamped boundary condition. For the study material properties are considered, which is stated as following,
$E_2 = 10.3GPa, E_2 = E_3, E_1 = 181GPa, G_{12} = 7.17GPa, G_{12} = G_{13}, G_{23} = 6.21GPa, \nu_{12} = \nu_{13} = 0.28,$
$\nu_{23} = 0.33, \alpha_0 = 10^{-6}, \alpha_1 = 0.02\alpha_0, \alpha_2 = 22.5\alpha_0$
Nitinol SMA Wires are utilized in the study, which has temperature dependent modulus as shown in Figure 2, adopted from the Tawfik [16]. $E_0 = 0.1GPa, G = 24.86GPa, \alpha_1 = 10.26x10^{-6}, \nu_{12} = 0.33$

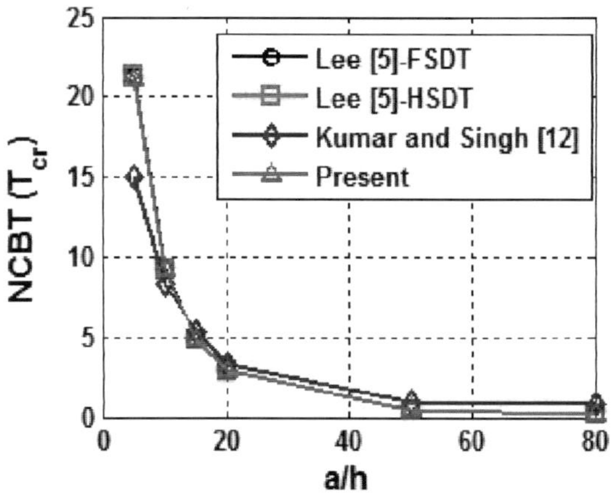

Fig. 3. Validation of NCBT with plate aspect ratio

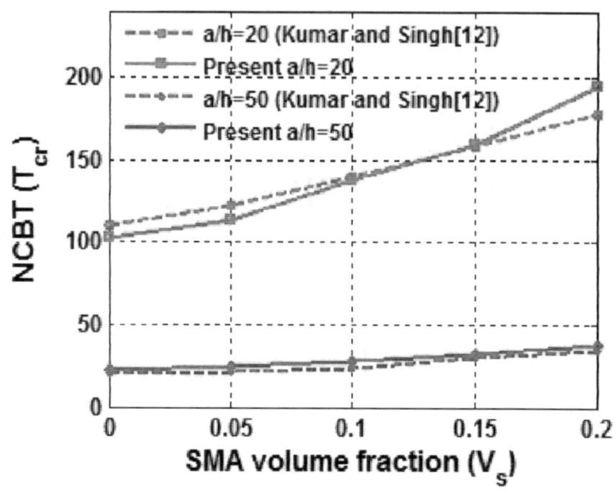

Fig. 4. Thermal post buckling variation with SMA volume fraction

Figure 5 shows the effect of different prestrains induced in SMA with SMA volume fraction variation for square smart sandwich composite plate (45/SMA/45) of a/h=50 in clamped BC. Figure shows the combined effect of SMA volume fraction and induced pre-strains over performance of sandwich composite structure. With 1% pre-strain, 23% increment in NCBT is noticed when comparing between without SMA and 5% SMA volume fraction, and also 57% variation observed for 5% to 15% volume fraction variation of SMA. Similarly 27% and 38% increment is observed between 5% SMA and without SMA results. Figure 6 shows the nature of variation of NCBT with the number of layers of SMA in laminated composites with 5% volume fraction under clamped BC and amplitude ratio of 0.1. As plate thickness varies from thick to thin, the resistance against buckling also get minimized. Figure clearly shows the effect of the number of layer on the performance of sandwich composite plate against critical buckling parameter, which explains, as thickness ratio varies from 5 to 100, NCBT increases by 6% to 46%.

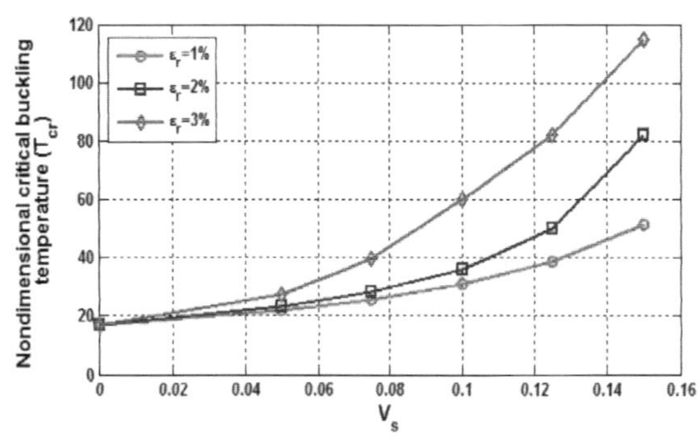

Figure 5. Effect of percentage strain variation with SMA volume fraction over NCBT

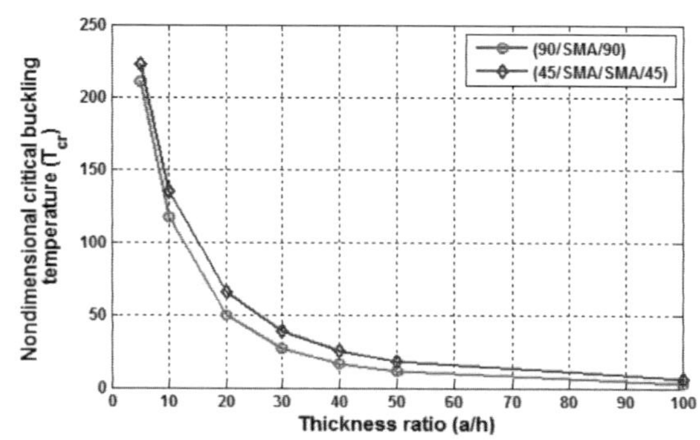

Figure 6. Effect of layer variation over NCBT

Figure 7 indicated the variation in NCBT for first four buckling modes for 1%, 2% and 3% strain rate with temperature variation. For this study sandwich composite square clamped plate (90/SMA/90) with thickness ratio of 100 and SMA volume fraction of 5% has been considered. Firstly, it is observed that, NCBT increased by 35%, 57% and 32% for Mode 1 to 2, Mode 2 to 3 and Mode 3 to 4 respectively. Secondly, whenever comparing ε_r=1% to ε_r=2% and ε_r=2% to ε_r=3% for 50^{0}C NCBT varies by 6.5% and

0.5% respectively whereas total increment of 7% noted down for ε_r=1% to ε_r=3%. Moreover, when comparing ε_r=1% to ε_r=2% and ε_r=2% to ε_r=3% for 60^0C, NCBT varies by 6.2% and 4.4% respectively whereas total increment of 10.4% noticed for ε_r=1% to ε_r=3%. Hence it is confirmed that, with temperature and strain rate increment, resistance against buckling, that is NCBT improves.

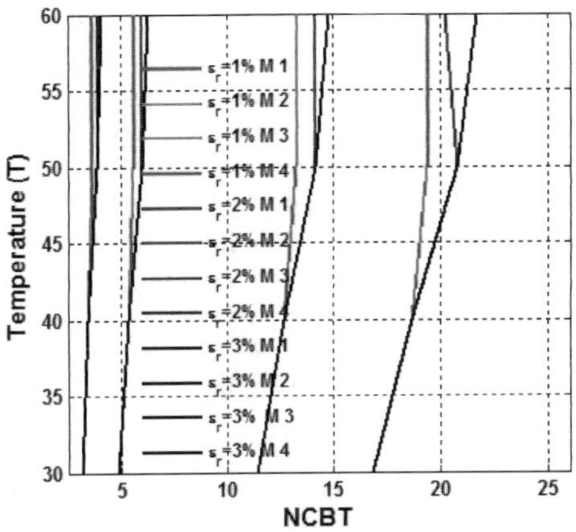

Figure 7. Variation in NCBT with strain rate for different Modes. (M stands for Mode)

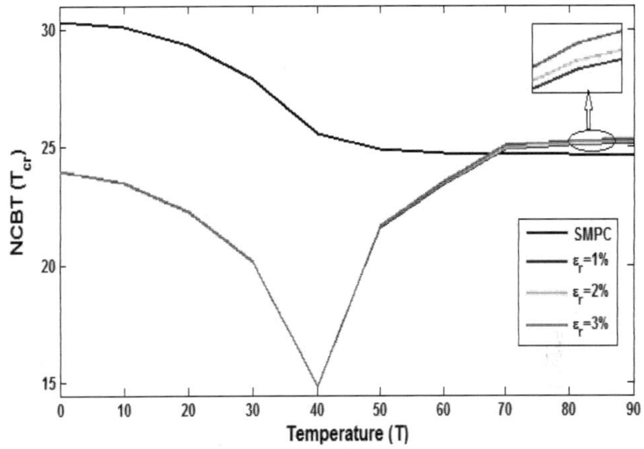

Figure 8. Variation in nondimensional critical buckling temperature under the influence of temperature and strain rate over SMPC and SMA

Figure 8 highlights the abrupt decline which is due to the volatile material behavior of SMP matrix and SMA fiber reinforcements in the similar temperature span. It can be observed that instantaneous drop is regulated by modulus of SMP matrix, which regulates the profile of buckling load. Tawfik [16] indicates that, in a region 0^0C to 30^0C, temperature dependent elastic modulus of SMA fiber doesn't show much variation as well as recovery stresses remains zero during same interval, but after this range dynamic change in modulus is takes place. Also at the same time, dynamic variation in elastic modulus of SMPC is observed as shown in Figure 1. Figure 8 indicates conclusive comparison of positive effect of combined shape memory properties of SMA and SMP on buckling resistance of material at elevated temperature. This is primarily due to the recovery stresses of SMA generated at elevated temperatures which enhances the stability of material as compared to normal SMPC or laminated composite structure.

4. Conclusions

Under induced uniform thermal loading, non-linear thermal post-buckling analysis of laminated composite plate, smart sandwich composite plate and for first time the combined effect of SMA and SMP is analyzed using higher order shear deformation theory and von Karman kinematics.

Study also shows the combined effect of SMA volume fraction and induced pre-strains over performance of sandwich composite structure. With increasing SMA volume fraction, increment is noted in NCBT. Study also confirmed that, with temperature and strain rate increment, resistance against buckling, that is NCBT improves.

References:

[1] Lagoudas D.C. 2008 *Shape memory alloys: modeling and engineering applications*. Springer Science & Business Media

[2] Asadi H, Akbarzadeh A H, and Wang Q 2015 Nonlinear thermo-inertial instability of functionally graded shape memory alloy sandwich plates Composite Structures 120 pp 496–508
[3] Mansouri M H, Shariyat M 2014 Thermal buckling predictions of three types of high-order theories for the heterogeneous orthotropic plates using the new version of DQM Composite Structures 113 pp 40–55
[4] Li Z M 2014 Thermal postbuckling behavior of 3D braided beams with initial geometric imperfection under different type temperature distributions Composite Structures 108 pp 924–936
[5] Lee J J, Choi S 1999 Thermal buckling and postbuckling analysis of a laminated composite beam with embedded SMA actuators Composite Structures 47 pp 695–703. https://doi.org/10.1016/S0263-8223(00)00038-6
[6] Lal A, Markad K 2020 Static buckling analysis of shape memory alloy reinforced composite laminated plate IOP Conf. Series: Materials Science and Engineering 814 p 012009
[7] Asadi H, Bodaghi M, Shakeri M, Aghdam M M 2013 On the free vibration of thermally pre/post-buckled shear deformable SMA hybrid composite beams Aerospace Science and Technology 31 pp 73–86
[8] Asadi H, Kiani Y, Shakeri M and Eslami M R 2014 Exact solution for nonlinear thermal stability of hybrid laminated composite Timoshenko beams reinforced with SMA fibers Composite Structures 108 pp 811–822. https://doi.org/10.1016/j.compstruct.2013.09.010
[9] Mahieux C A and Reifsnider K L 2001 Property modeling across transition temperatures in polymers: a robust stiffness–temperature model Polymer 42 pp 3281–3291.
[10] Gu J, Leng J, Sun H, Zeng H and Cai Z 2019 Thermomechanical constitutive modeling of fiber reinforced shape memory polymer composites based on thermodynamics with internal state variables Mechanics of Materials 130 pp 9–19. https://doi.org/10.1016/j.mechmat.2019.01.004
[11] Qi H J, Nguyen T D, Castro F, Yakacki C M and Shandas R 2008 Finite deformation thermomechanical behavior of thermally induced shape memory polymers Journal of the Mechanics and Physics of Solids 56 pp 1730–1751. https://doi.org/10.1016/j.jmps.2007.12.002
[12] Kumar C N and Singh B N 2009 Thermal Buckling and Post-Buckling of Laminated Composite Plates with SMA Fibers using Layerwise Theory International Journal for Computational Methods in Engineering Science and Mechanics 10(6) pp 423-429.
[13] Shen G L, Hu G and Liu B 2006 Mechanics of composite materials. Science and Technology, Beijing, China.
[14] Lal A, Singh B N and Kale S 2011 Stochastic post buckling analysis of laminated composite cylindrical shell panel subjected to hygro-thermo-mechanical loading Composite Structures 93 pp 1187–1200.
[15] Lee J 1997 Thermally induced buckling of laminated Composites by a layerwise theory Computers & Strucrures 65(6) pp 917-922.
[16] Tawfik M 1999 Suppression of post-buckling deflection and panel-flutter using shape memory alloy. MSc thesis, Aerospace department, old dominion university, Norfolk V A.

IOP Conf. Series: Materials Science and Engineering 1017 (2021) 012021 doi:10.1088/1757-899X/1017/1/012021

An Experimental Study on Machining of AISI H - 13 Steel Using Dimple-Textured and Non-Textured Tungsten Carbide Cutting Tools

G. Vignesh[1], Debabrata Barik[1], S. Aravind[1], P. Ragupathi[1], M. Arun[1]

[1]Department of Mechanical Engineering, Karpagam Academy of Higher Education, Coimbatore, India.

Email: gvcit2008@gmail.com

Abstract. Fabrication of micro-dimples at the rack surface closer to the cutting region of the tungsten carbide tool will induce the tribological modifications at the tribo-contact surfaces. The motive of this work is to find the influence of micro-dimples fabricated on the cutting tools during turning operations. Linear arrays of dimple-textures with various geometrical aspects (dimple diameter and depth) and area density ratio were prepared using laser marking technique. The influence of surface-texture pattern is experimentally analysed through dry machining of AISI H – 13 steel using a dimple-textured tungsten carbide cutting tool. The responses of cutting temperature and cutting force generated during the turning operation were analysed. It shows improvement in machinability in terms of wear rate reduction in dimple-textured tools. Optimizing the dimension of micro-dimples (diameter of 90μm, depth of 60μm) results in a 10% reduction in tool wear rate.

1. Introduction

In the machining area, dimple textured surfaces play a significant role in the tribological phenomenon. The essential thing about dimple texture in the machining area is the variation in frictional parameters that improve machinability. Dimple texture forming over the cutting inserts has several benefits attained by varying frictional factors at the tribo-contact surfaces. Alagan et al. [1] applied a laser ablation technique to produce dimple texture on the uncoated tungsten carbide round inserts (Grade H13A) for machining of superalloy 718. They suggested that micro-dimples on the sintered carbide round inserts increases the wear resistance without affecting the mechanical properties [2]. Darshan et al. [3] fabricated the dimple-shaped texture of an uncoated carbide insert using a multi-diode fiber laser technique. They found that micro-texturing on cutting inserts delivers the good lubricating phenomenon at tribo-contact surfaces [4]. Guarani, K.K et al. [5] compared the dimple-textured and coated dimple-textured inserts with the conventional device and they concluded that coated cutting tools performed better in the aspect of reduction in cutting force and friction coefficient. Lei, S et al. [6] analyzed the insert wear and chip morphology for the micro-textured inserts, and they found that a 20% reduction in friction force and coiled chips are produced. Li et al. [7] applied a fibre laser marking technique to make the micro-hole surface modification of PCBN cutting inserts. They stated that the surface quality of the workpiece was improved [8].

Niketh and Samuel [9] fabricated micro-dimples on the tungsten carbide drill bit using a laser ablation technique. Solid-powdered MoS_2 filled in the dimple-textures to make a coated drill tool. They observed that tool wear and cutting torques are reduced in the coated device compared with the uncoated tool [10]. Rajbongshi and Sarma [11] compared different samples of dimple-texture (dot and groove) on the flank surface in the machining of AISI D2 steel. They concluded that heat produced at the tool-work contact surface of the point-textured drill tool was exhibits better performance than the groove textured drill tool [12]. Sawant et al. [13] experimented on the influence of spot and dimple textured HSS insert in the turning of titanium alloy.

Using spot-textured insert, thrust and cutting forces, cutting temperature and tool wear were reduced [14-15]. Vignesh et al. [16-17] experienced the influence of dimple textured inserts in the turning of aluminum alloy. They found that chip adhesion was suppressed when the dimple-textures were fabricated straight to the cutting region. Xing et al. [18] fabricated three samples of dimple-textures (circular, linear, and rectangular) of the carbide inserts using a laser marking method for turning of aluminum alloys. They concluded that, among three kinds, circular textured inserts showed machinability improvement in cutting forces, frictional co-efficient decrement, and low surface region of the machined region [19].

It is found that the dimpled-texture on the rake surfaces of the inserts improves the cutting quality, and hence machinability is improved. The objective of this work is to analyse the dimple-textured coated and uncoated cutting tools over non-textured cutting tools. Furthermore, the effect of the dimple-texture dimension and the tool wear rate were also analysed.

2. Materials and Methods

2.1. Preparation of cutting inserts with dimple-textured surface

Tungsten carbide (WC) insert is selected as the cutting insert for dimple-texturing and machinability study. The tungsten carbide tool is used for turning rigid materials. The dimple-texture are formed on the rack face of the cutting tool using a fibre laser ablation technique. Figure 1. shows a tungsten carbide tool with a topography of the micro-dimples pattern. The dimple textures having a circular shape with various geometrical patterns were fabricated in the tungsten carbide cutting tool

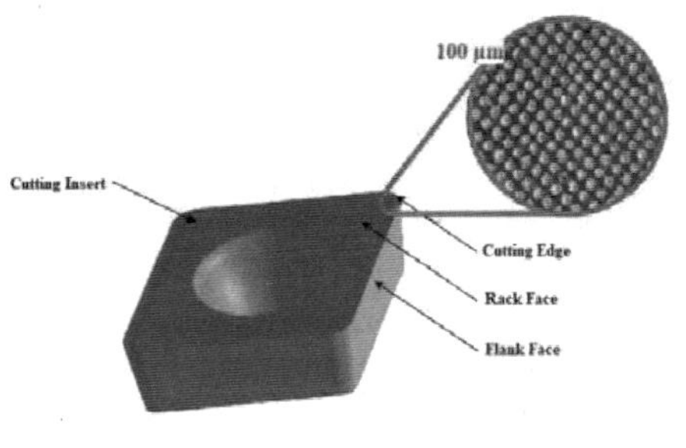

Figure 1. Tool with micro-dimples pattern

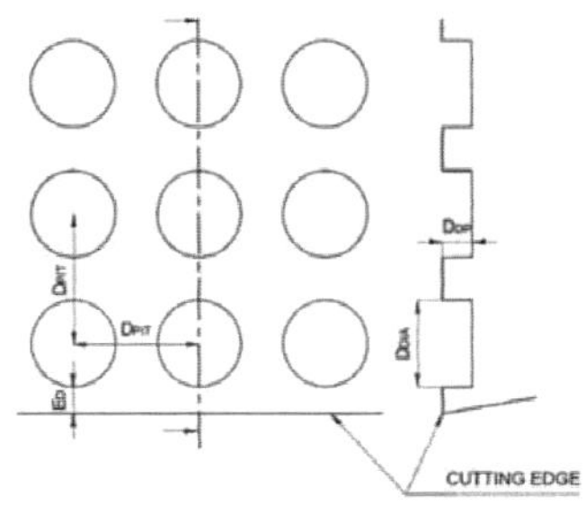

Figure 2. Geometrical dimensions of micro-dimples

To analyse the effect of micro-dimple on machinability, 12 types of inserts were prepared. Figure 2. illustrates the geometrical dimensions of micro-dimples, and Table 1 lists out the dimensional details of the micro-dimples, matched with the tool designated as Textured – A to Textured – L. In this work, circular shape micro-dimples are fabricated for all samples, and the dimple distance from cutting region was fixed as 150µm. The equation calculated the micro-dimple area density A_{dr} of all dimple-textured samples

$$A_{dr} = \left(\frac{A_T}{A_T + A_{UT}}\right) \times 100\% \qquad (1)$$

A_T refers to the sample area on the dimple-textured surface, and A_{UT} refers to the sample area of the non-textured surface of each insert sample. The coating of MoS_2 (solid lubricant) on a micro-dimple plays an important role in refining the frictional performance of inserts. Nano particles of MoS_2 packed in the micro-cavity to prepare a coated insert and analyses the cutting performance in contrast with the non-textured tools.

Table 1. Dimensional details of the micro-dimples

Cutting Tool Name	Dimple Size (D_{DIA}) (μm)	Dimple Height (D_{DP}) (μm)	Dimple Distance (D_{PIT}) (μm)	Area Density (D_R) (%)
Non-textured	-	-	-	0
Textured – A	80	60	135	45
Textured – B	80	60	150	40
Textured – C	80	70	135	45
Textured – D	80	70	150	40
Textured – E	90	60	135	35
Textured – F	90	60	150	30
Textured – G	90	70	135	35
Textured – H	90	70	150	30
Textured – I	100	60	135	25
Textured – J	100	60	150	20
Textured – K	100	70	135	25
Textured – L	100	70	150	20

2.2. Experimental Setup

The work material of AISI H - 13 steel was selected for this work because of high wear-resistant and enormous applications like propeller shaft, connecting rod, gear shaft, camshaft, etc., Table 2 lists out the property of AISI H – 13 steel. Both the tool and work material has been purchased from Coimbatore Metal Mart, Coimbatore 641018 with the testing certificate. The machining experiments were performed on the AISI H – 13 steel (Ø50mm × 130mm) using a CNC lathe. Table 3 provides the details on machining parameters, and Figure 3. details the arrangement of the machining setup. A machining force dynamometer was utilized to measure the cutting forces. The insert wear rate of each cutting tool sample after tuning was inspected through an optical microscope.

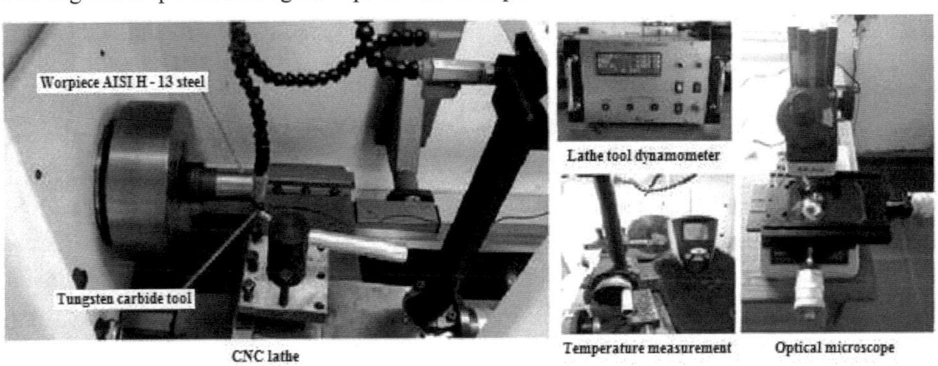

Figure 3. Machining set-up

Table 2. Properties of AISI H – 13 steel

Chemical Composition (%)		Mechanical Properties	
Cr	4.25 – 5.15	Impact Strength (J)	60 – 74
Mo	1.25 – 1.80	Yield Strength (MPa)	1000 – 1380
V	0.90 – 1.30	Elongation (%)	11
Si	0.90 – 1.30	Hardness (BHN)	290 – 370
C	0.22 – 0.63	Tensile Strength (MPa)	1200 – 1590
Cu	0.25		
Ni	0.3		
Mn	0.20 – 0.50		
S	0.03		
P	0.03		

Table 3. Machining parameters

Workpiece	AISI H – 13 steel Ø50mm × 130mm
Cutting Insert	Tungsten carbide cutting insert Nose radius – 0.8mm
Cutting RPM (N)	1500 rpm
Cutting depth (d)	3 mm
Cutting feed (f)	0.16 m/min
Machining condition	Dry condition

3. Results and Discussion

3.1. Micro-dimples on the cutting forces

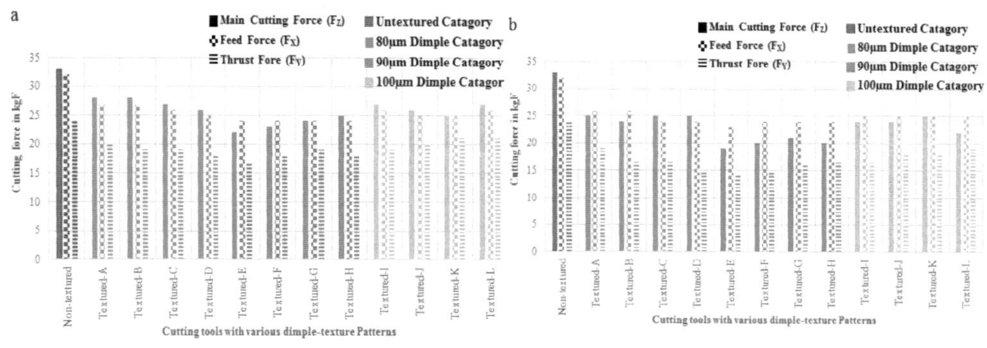

Figure 4. Cutting force for various dimple-textured cutting tools (a) uncoated (b) coated

Figure 4. shows the influence of various coated and uncoated dimple-textured inserts on the cutting force while turning of work material under dry turning conditions.

Effect on main cutting force (Fz)

Figure 4a shows the main cutting forces are reduced with dimple-textured tools, and dimple-textured tools were exhibited 33 % of reduction in the main cutting force as compared with the regular insert. Figure 4b illustrates the main cutting forces are minimized with MoS_2 coated tools, and uncoated tools exhibited a 42 % drop in main cutting force as compared with dimple-textured inserts and 16 % of reduction in main cutting force as compared with a regular insert. A peak main cutting force of 33kgf was noted with a regular insert that was minimized to 22kgf with uncoated Texture-E and 19kgf with coated Texture-E.

Effect on feed force (Fx)

Figure 4a shows the feed forces are reduced with dimple-textured tools, and dimple-textured tools were exhibited 25 % of reduction in the feed force as compared with the regular insert. Figure 4b illustrates the feed forces are minimized with MoS_2 coated tools, and uncoated tools exhibited a 28 % drop in feed force as compared with dimple-textured inserts and 4 % of reduction in feed force as compared with a regular insert. A peak feed force of 33kgf was noted with a regular insert that was minimized to 24kgf with uncoated Texture-E and 23kgf with coated Texture-E.

Effect on thrust force (Fy)

Figure 4a shows the thrust forces are reduced with dimple-textured tools, and dimple-textured tools were exhibited 29 % of reduction in the thrust force as compared with the regular insert. Figure 4b illustrates the thrust forces are minimized with MoS_2 coated tools, and uncoated tools exhibited a 14 % drop in thrust force as compared with dimple-textured inserts and 17 % of reduction in thrust force as compared with a regular insert. A peak thrust force of 24kgf was noted with a regular insert that was minimized to 24kgf with uncoated Texture-E and 14kgf with coated Texture-F.

Micro-textured cutting tools produce minimal cutting force when compared to untextured tools due to less surface contact at the cutting tool-chip interface. MoS_2 coated textured inserts outperforms when compared to dimple-textured tools due to the capillarity property of MoS_2 on the micro-dimples of dimple-textured tools with minimal cutting insert-chip contact area. From previous work [6-7] identified that with the help of dimple-textured tool 20% of reduction in overall cutting force but MoS_2 coated textured tools exhibits about 35% of reduction in overall cutting force.

3.2. *Micro-dimples on the heat generation*

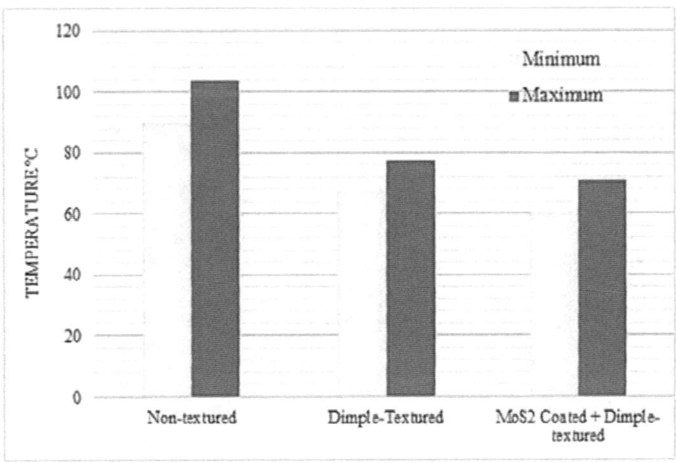

Figure 5. Heat generation at the cutting tool-work contacts

Figure 5. shows the heat generation during tuning operation at the tool-work contact surface for non-textured, MoS_2 coated dimple-textured, and uncoated dimple-textured cutting tools. At the spindle rpm of 1500, the low - high values of heat generated at the tool-work contacts of non-textured, MoS_2 coated dimple-textured and uncoated dimple-textured cutting tools was observed as 90.2°C - 103.5°C, 60.2°C - 70.5°C, and 67.5°C - 77.1°C respectively. The heat debauchery in dimple-textured cutting tools was much higher than non-textured tools. The close concave surface area of the micro-dimples exhibits much higher heat dissipation.

3.3. Tool wear mechanism

Figure 6. illustrates the images of non-textured, uncoated dimple-textured, and coated dimple-textured inserts after conducting dry machining experiments. Generally, cutting tool wear was observed in the rack face due to the development of the continuous chip, generation of heat at the cutting tool-work interface, and abrasive wear in cutting tool-chip contact area. Due to minimized tool-chip contacts in the dimple-texture, cutting tools exhibits minimal cutting tool wear compared with regular inserts. Furthermore, the wear rate further reduced in MoS_2 coated textured inserts due to the effect of thin layer formation and the micro-cavity lubrication effect of the solid lubricants.

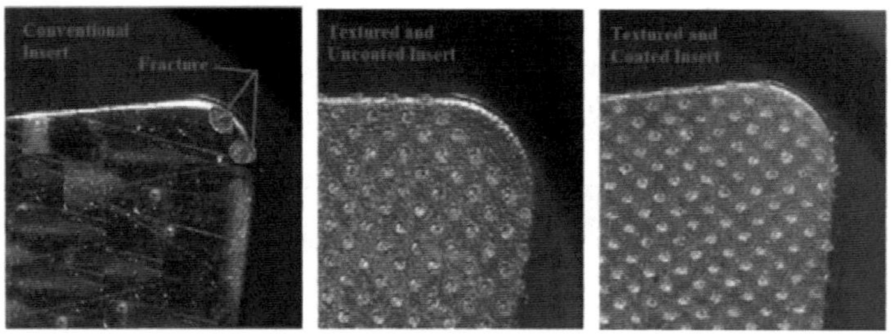

Figure 6. optical microscope images of various cutting tools

4. Conclusion

Using laser ablation technique, dimples are fabricated on the cutting tools. Machining is performed by non-textured, MoS_2 coated textured, uncoated textured cutting tools on AISI H – 13 steel in dry cutting condition. The machinability factor of the MoS_2 coated dimple-textured cutting tool is comparatively analysed with regular and dimple-textured inserts. The major findings observed through the experimental study was listed below

1. The micro-dimples influence the insert wear reduction rate as compared with regular inserts. The incremental order of insert wear as follows: MoS_2 coated dimple-textured tool < uncoated dimple-textured tool < regular non-textured tool.
2. Textured-E cutting tool (micro-dimple size of 90μm, the height of 60μm, a distance of 135μm, and density area of 35%) exhibits better performance.
3. Cutting tool wear rate further reduced in MoS_2 coated textured cutting insert due to the effect of thin layer formation and the micro-cavity lubrication effect of the solid lubricants.
4. Average machining heat generation in the coated textured cutting tool was reduced because the concave surface of the micro-dimples exhibits the much higher heat dissipation

References

[1] Alagan N T, Zeman P, Hoier P, Beno T and Klement U 2019 Investigation of micro-textured cutting tools used for face turning of alloy 718 with high-pressure cooling *Journal of Manufacturing Processes* **37** 606-616.

[2] Arulkirubakaran D, Senthilkumar V, Chilamwar V L and Senthil P 2019 Performance of surface textured tools during machining of Al-Cu/TiB_2 composite *Measurement* **137** 636-646.

[3] Darshan C, Jain S, Dogra M, Gupta M K and Mia M 2019 Machinability improvement in Inconel-718 by enhanced tribological and thermal environment using textured tool *Journal of Thermal Analysis and Calorimetry* **1** 1-13.
[4] Fatima A and Mativenga P T 2016 Performance of flank face structured cutting tools in machining of AISI/SAE 4140 over a range of cutting speeds *Proceedings of the Institution of Mechanical Engineers Part B: Journal of Engineering Manufacture* **230** 3-18.
[5] Gajrani K K, Reddy R P K and Sankar M R 2016 Experimental comparative study of conventional, micro-textured and coated micro-textured tools during machining of hardened AISI 1040 alloy steel *International Journal of Machining and Machinability of Materials* **18** 522-539.
[6] Lei S, Devarajan S and Chang Z 2009 A comparative study on the machining performance of textured cutting tools with lubrication *International Journal of Mechatronics and Manufacturing Systems* **2** 401-413.
[7] Li Q, Pan C, Jiao Y and Hu K 2019 Investigation on Cutting Performance of Micro-Textured Cutting Tools *Micromachines* **10** 352-364.
[8] Mishra S K, Ghosh S and Aravindan S 2018 3D finite element investigations on textured tools with different geometrical shapes for dry machining of titanium alloys *International Journal of Mechanical Sciences* **141** 424-449.
[9] Niketh S and Samuel G L 2017 Surface texturing for tribology enhancement and its application on drill tool for the sustainable machining of titanium alloy *Journal of cleaner production* **167** 253-270.
[10] Niketh S and Samuel G L 2018 Drilling performance of micro textured tools under dry, wet and MQL condition *Journal of Manufacturing Processes* **32** 254-268.
[11] Rajbongshi S K and Sarma DK 2019 Performance parameters studies in machining of AISI D2 steel with dot-textured, groove-textured & non-textured cutting tool at the flank face *International Journal of Refractory Metals and Hard Materials* **83** 104970.
[12] Ranjan P and Hiremath S S 2019 Role of textured tool in improving machining performance: a review *Journal of Manufacturing Processes* **43** 47-73.
[13] Sawant M S, Jain N K and Palani I A 2018 Influence of dimple and spot-texturing of HSS cutting tool on machining of Ti-6Al-4V *Journal of Materials Processing Technology* **261** 1-11.
[14] Sugihara T and Enomoto T 2017 Performance of cutting tools with dimple textured surfaces: a comparative study of different texture patterns *Precision Engineering* **49** 52-60.
[15] Sugihara T, Singh P and Enomoto T 2017 Development of novel cutting tools with dimple textured surfaces for dry machining of aluminum alloys *Procedia Manufacturing* **14** 111-117.
[16] Vignesh G, Prakash M, Selvam M D and Ragupathi P 2018 Frictional Performance of Dimpled Textured Surfaces on a Frictional Pair: An Experimental Study *i-Manager's Journal on Mechanical Engineering* **8** 18-24.
[17] Vignesh G, Barik D, Ragupathi P and Aravind S 2020 Experimental analysis on turning of AISI 4340 steel using non-textured, dimple textured and MoS_2 coated dimple textured carbide cutting inserts at the rack surface *Materials Today: Proceedings*.
[18] Xing Y, Deng J, Wang X, Ehmann K and Cao J 2016 Experimental assessment of laser textured cutting tools in dry cutting of aluminum alloys *Journal of Manufacturing Science and Engineering* **138** 1-10.
[19] Xie J, Luo M J, Wu K K Yang L F and Li D H 2013 Experimental study on cutting temperature and cutting force in dry turning of titanium alloy using a non-coated micro-grooved tool *International Journal of Machine Tools and Manufacture* **73** 25-36.

IOP Conf. Series: Materials Science and Engineering 1017 (2021) 012022 doi:10.1088/1757-899X/1017/1/012022

Dissimilar Metal Welds used in AUSC Power Plant, Fabrication and Structural Integrity Issues

Sanjay Singh[1], Alok Bihari Singh[1], Manoj Kumar[2], M L Meena[1] and G S Dangayach [1]

[1]Department of Mechanical Engineering, Malaviya National Institute of Technology, Jaipur, Rajasthan, 302017, India

[2]Department of Mechanical-Mechatronics Engineering, The LNM Institute of Information technology, Jaipur, Rajasthan, 302031, India

E-mail addresses: 2017rme9027@mnit.ac.in

ABSTRACT. In the current era, the Advanced Ultra Super-Critical (AUSC) plants have been used to substantially increased efficiency with environmental protection. AUSC components boiler header, re-heater and super-heater tube etc. are fabricated by using advanced materials to join by the welding process. Any failure of the power plant can lead to grave consequences, not only of substantial financial losses but additionally immensely treasured and irreparable human life loss. In this review, the dissimilar metal welds (DMWs) have many issues such as metallurgical issue, cracking, carbon migration, welding electrode selection, creep properties, suitable welding process and structure integrity issues which enhanced the failure of weldments. The results of our review showed that DMWs of AUSC power plant are joined by activated flux Tungsten Inert Gas (A-TIG) and Pulse A-TIG process to increase the productivity and cut the significant fraction of cost. However, A-TIG process with Post Weld Heat Treatment (PWHT) improved the mechanical properties and high-temperature creep resistance properties of DMWs. This state of the art is encouraged the researchers about the enormous future scope.

1. INTRODUCTION

Among all traditional energy sources, the coal remains the most important source of electricity in India. Recently, the AUSC power plants are developed all over the world with high efficiencies to improve human living standard. Hence, the most successful approaches to increase the efficiency of the power plant is developed the advanced materials. The material development for AUSC was initiated in 1998s by COMTES700 USA project started by China in 2011. Further, India started the AD700 project in 2012 with the development of new materials as shown in Fig 1 [1].

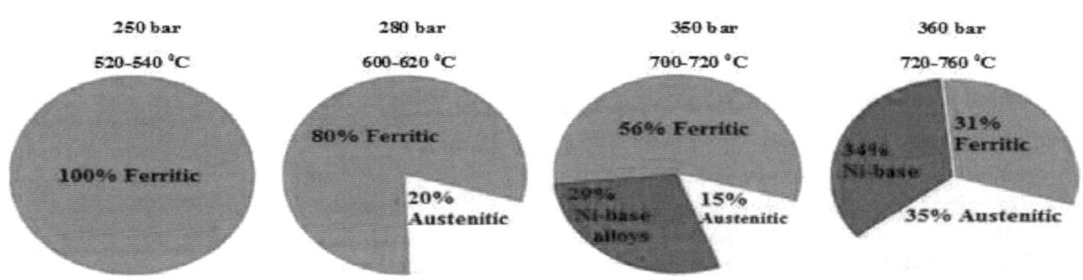

Figure 1. Materials used in SC, USC and AUSC boiler [1].

The selection of materials for AUSC components is based on high temperature creep properties, sufficient weldability, and industrial availability. The ferritic / martensitic steels such as ASME Gr. 91,

92,122 etc. are used in the thick section while super austenitic stainless steel as TP316 H, 304 H, TP321 H, TP347 H etc. and the Ni-base Alloys as 800 H, IN 617, 617 B, 740 H, 718 etc. are used in thin section of boiler but undoubtedly nickel alloy is more costly compared to the ferritic / martensitic steel [2]. Recently, the AUSC power plants in India are used various materials as shown in Table 1. The DMWs are mostly adopted in AUSC boiler headers, super-heaters and re-heater pipe or tube joined to make the whole system by welding together the components for transport the fluid. [3].

Table 1 AUSC advance materials with chemical compositions (wt. %) proposed by ASME [1, 3].

Elements	**TP23**	**TP91**	**Super304H$_{Cu}$**	**Super347H**	**IN 617**	**IN 740H**	**IN718**
Carbon (C)	0.04-0.10	0.07-0.14	0.07-0.13	0.0681	0.05-0.15	0.048	0.026
Manganese (Mn)	0.10-0.60	0.30-0.60	1.00 max	2.03	1.0 max	0.240	0.085
Phosphorous (P)	0.03 max	0.02 max	0.040 max	-	0.015 max	<0.001	0.009
Sulphur (S)	0.01 max	0.01 max	0.01 max	-	0.01 max	0.001	
Silicon (Si)	0.50 max	20-0.50	0.30 max	0.62	1.0 max	0.005	0.102
Nickel (Ni)	–	0.40 max	7.50-10.50	9.39	44.5 min	Bal.	Rem.
Chromium (Cr)	1.90-2.60	8.50-9.50	17.00-19.00	20.06	20.0-24.0	25.36	17.49
Molybdenum (Mo)	0.05-0.30	0.85-1.05	–	0.27	8.0-10.0	0.005	2.96
Cobalt (Co)	-	-	-	-	10.0-15.0	20.04	0.102
Copper (Cu)	-	-	2.50-3.50	-	0.5 max		0.025
Niobium (Nb)	0.02-0.08	0.06-0.10	0.30-0.60	0.44		1.56	5.03
Titanium (Ti)	-	0.01 max	-	-	0.6 max	1.36	0.905
Tungsten (W)	1.45-1.75	0.02 max	-	-	-	-	
Vanadium (V)	0.20-0.30	0.18-0.25	-	-	-	-	
Aluminium (Al)	0.03 max	0.07 max	0.003-0.030	-	0.8-1.5	1.39	0.572
Boron (B)	0.006 max	-	0.001-0.010	-	0.006 max	-	
Iron	Balance	Balance	Balance	Balance	3.0 max	0.23	

2. Problems associated AUSC dissimilar metal welds

2.1 Metallurgical problem and Imperfection during welding

The performance of DMWs in service basically depends upon the quality of weldments. The weld defects such as cracks, cavities, solid inclusion, incomplete fusion, unacceptable contour, arc strike etc. are causes failure of DMWs in service at high temperature [4]. Many researchers were reported that failure occurred in the welded joint due to metallurgical deterioration. The nickel base alloys showed laves phase (ϒ) in grain boundaries (GB) which reduced the fatigue strength and increased the yield strength due to ϒ combined with Cr and precipitates undesirable $M_{23}C_6$ phase[5]. Similarly, Austenitic steels is showed NbC, TiC and ϒ phase which improved the creep strength of welded joints but ϒ phase increased the solidification cracking sensitivity. The weldment solidification modes of steels are calculated by Cr_{eq}/Ni_{eq} ratio. However, Verma et al. [6] were studied constitution diagrams such as Schaeffler, WRC 1992 and Pseudo binary (Fe-Cr-Ni) diagram as shown in Fig. 2. For similar and dissimilar metals joint, these diagrams can be used to predict the microstructure of the DMWs.

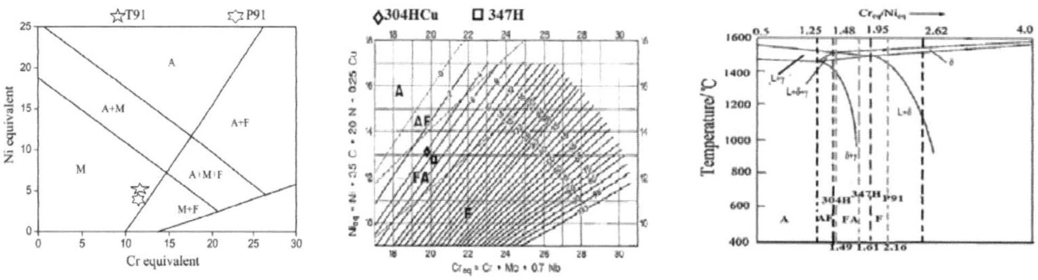

Figure 2. Constitution diagrams for dissimilar metal joining to predict the weld microstructure by Shaeffler, WRC-92 and Pseudo binary (Fe-Cr-Ni) diagram [6].

Further, Fu et al. [7] were reported that when Cr_{eq} / Ni_{eq} ratio is less than 1.2, thus the microstructure showed fully austenitic and there is no δ-ferrite in an equilibrium condition. The main constituents Cr, Mo, V, W, Co, N, Nb, B, Ni, C etc. are added to improve the mechanical properties in ferritic and martensitic steels. The W, Co increased the austenite phase and reduced the δ-ferrite phase in the steel [8]. The contents C, N are showed more solubility in austenitic steel than ferritic steel. The formation of laves such as Fe_2 (W, Mo) increased the hardness and short term creep properties in the steel weldment due to increase the lower critical temperature 810-825^{0}C or reduced the martensitic start (M_S) temperature at 390-400^{0}C. However, the V, Nb constitutes is formed MX phase which increased the corrosion and reduced the fracture toughness of the steel joints. The creep and fatigue properties of the steels are reduced when δ-ferrite increased in the weldment. Further the formation of δ-ferrite in WM and HAZ is calculated by using the Kaltenhauser ferrite factor (KFF) and Schneider Ferrite factor (SFF) Eq. (1) and (2) [9-10].

$$KFF = Cr + 6Si + 4Mo + 8Ti + 2Al + 4Nb - 2Mn - 4Ni - 40(C+N) \quad (1)$$

$$SFF= Cr+2Si+1.5Mo+5V+1.75Nb+0.75W-Ni-0.5Mn-30C-25N-0.3Cu \quad (2)$$

Also, Ramkumr et al. [11-12] were studied the microstructure of DMWs between the metals Nickel alloys (IN 718), martensitic steel (AISI 416) and Austenitic steel (316 SS). They were reported that the Nb and Mo secondary phase found in the heat affected zone (HAZ). In sequence, the laves phase (Ni_3Nb) with inclusion (FeS, MnS) was observed in the fusion zone (FZ) and the partially melted zone (PMZ). However, Dev et al. [13] were reported that the microstructure of weldments also showed δ-ferrite. Similarly, Ramkumar et al. [14] was investigated the DMWs of IN718 / SS 316L and found the Cr and Ni content in HAZ with the unmixed zone (UMZ). Additionally, Pavan et al. [15] were studied the microstructure of DMWs between IN 617 / 304 H. They were reported that Cr, Mo, C, Si contents found in the FZ and PMZ, which increased the $M_{23}C_6$ precipitation in the weldment. Along with, Falat et al. [16] were investigated the microstructure of DMWs T91 / TP316 H. It was reported that the weld metal (WM) consisted of Nb, Ti, and carbide phase (MC). Similarly, Mittal et al. [17] were found martensitic phase in WM and HAZ of T91 / 347 H DMWs. Briefly, the mechanical properties and service life of DMWs are influenced by phases, which developed the microstructure heterogeneity.

2.2 Cracking in AUSC welds

2.2.1 Hot cracking

Hot cracking have been found recently in weld metal (WM) and heat affected zone (HAZ) of nickel base alloys and austenitic steels. It is appeared during weld solidification or metal liquid inherent in between two solid grain boundaries. Kou et al. [18] were reported that nucleation and grain growth causes the hot cracking due to thermal stress and strain. Naffakh et al. [19] were studied the cracking susceptibility in DMWs of AISI 310 / IN 657 and found the Ti-rich phase, and segregation causes the solidification hot cracking.. However, Sayyer et al. [21] were examined the DMWs between nickel base alloy (IN825) and austenitic steel (AISI 321). They were reported that rich Ti-phase and low δ-ferrite promoted the solidification hot cracking. In sequence, Xin Ye et al. [33] were reported that sensitivity of hot cracking increased in IN 718 due to crater developed in the weld center by longitudinal and transverse micro-segregation by laves phase.

2.2.2 Hydrogen Induced Cracking (HIC)

HIC also known as cold cracking because, low-temperature (200 to 100^{0}C) martensite phase and hydrogen are combined and developed cracking in the HAZ. The various sources of hydrogen are moisture, oil, grease and shielding gas. Besides, the flux assisted SMAW, FCAW, SAW and GTAW or GMAW process with H_2 shielding gas introduced the hydrogen in WM. The hydrogen atom and molecules diffused and dissolved into the WM during welding process. The atoms of hydrogen easily penetrate the weld voids and caused cracking, blistering and hydrogen embrittlement. The hydrogen is transferred from high temperature (T_f) of the WM to low temperature (T_b) of the HAZ. Therefore, HAZ austenite converted into martensite because the ferrite is low solubility of hydrogen during welding. The combined effect of hydrogen and martensite caused a crack and developed HIC in HAZ as shown in Fig. 3 [22].

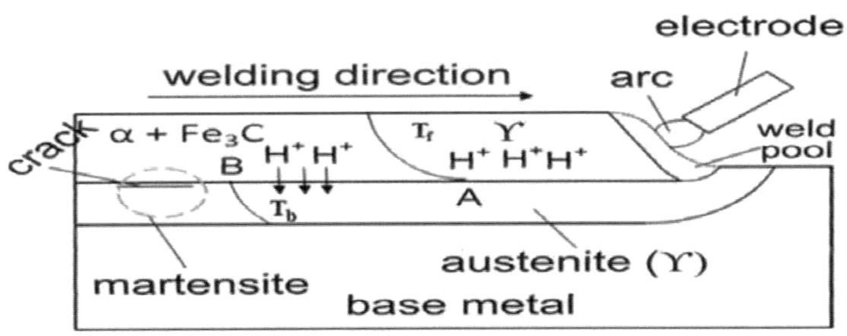

Figure 3. Schematic representation of diffusion of hydrogen from WM to HAZ [21].

The HIC susceptibility of steel is predicted by carbon equivalence (CE) in Eq. (3). The HIC susceptibility of the steel increased when CE value is high.

$$CE = \%C + \frac{\%Mn}{6} + \frac{\%(Si+Cu+Ni)}{15} + \frac{\%(C+Mo+\ )}{5} \quad (3)$$

Pandey et al. [23] were reported that hydrogen embrittlement increased when diffused hydrogen content more than 6.21ml/100 in WM. In sequence, the tensile and flexural strength of WM decreased when deposited WM content is a higher volume of hydrogen 12.54 ml / 100. However, Saini et al. [24] were found martensite in HAZ of P92 steel, then martensite phase increased the HIC susceptibility. Further, the HIC on AISI 4137M exhibited some evidence of hydrogen embrittlement during slow strain rate tests. Zhou et al. [25] were investigated the corrosion behavior in H_2S medium of DMWs of X60 / IN 625 and found high corrosion in FZ due to hydrogen embrittlement.

2.2.3 Stress corrosion cracking (SCC)

The SCC is considered combining interaction of the material, tensile stress-strain and corrosive environment as shown in Fig. 4. The SCC occurred trans-granular or branching type in HAZ into extent in BM due to C or Cr depletion.

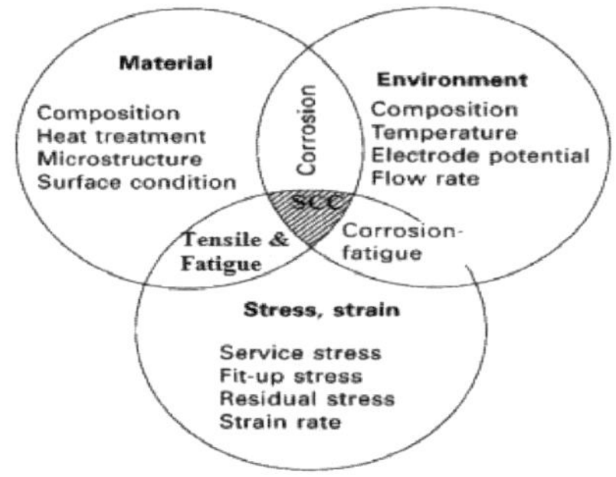

Figure 4. Conjoint interactions material, stress-strain and environment in SCC [54]

The constituent phases Fe-Ni, Fe-Cr, M_3O_4 (M=Ni, Cr, Fe) are produced intergranular SCC in AUSC materials at high temperature. The surface corrosion with stress-strain initiated the micro-cracks due to dislocation or defects then micro-voids coalesced and produced the cracking on surface or grain boundaries [26]. Dong et al. [27] were investigated the SCC behavior in DMWs of SA508 Cl. 3 / Alloy 52M with high pressurized water and found the reason of SCC due to Cr depletion and strength mismatch in fusion zone (FZ). Zhu et al. [28] were examined SCC of DMWs 316 L / 52M in borated and lithiated water solution. The cracking is occurred in FZ with mixed mode nature as well as trans-granular and intergranular. Hence, Vidyarthy et al. [29] were found the pitting corrosion in weld interface of DMWs between IN 718 / 316 L and P91 / 316 L which developed the stress corrosion cracking.

2.3 Segregation

The segregation is found in DMWs due to insufficient mixing and dilution of filler metal with BM during welding. Hence, Soyal et al. [30] were reported that the segregation occurred due to differences in liquids temperature of weld metal and base metal. Moosavy et al. [20] were found the segregation of Ti rich phase in inter-dendritic FZ region and developed the PMZ with laves phases in DMWs of IN 718 / IN 500. Similarly, Ramkumar et al. [31] were found the Mo, Cr micro-segregation in the FZ of DMWs IN 718 / IN 625. Further, Vashistha et al. [32] were found the C, Cr segregation in DMWs of austenitic stainless steel 304 SS / 201 SS.

2.4 Carbon Migration

The carbon migration occurred in the weld interface during PWHT or service at high temperature. Kulkarni et al. [34] were investigated the DMWs of ferritic steel P91 / P22 microstructure with carbon diffusion as welded and PWHT condition. Hence, the carbon migration mechanism reported that carbon migrated from high to low-level Cr zone and developed carbon enriched (dark) and carbon denuded (light) region. The migrated carbon combined with Cr and Mo particles which developed Fe_3C, MX, and $M_{23}C_6$ phases in both sides of carbon enriched interface zone as shown Fig. 5. Further, Jahanzeb et al. [35] were reported that microhardness heterogeneity developed due to carbon migration in DMW of 316L / SS400 steel.

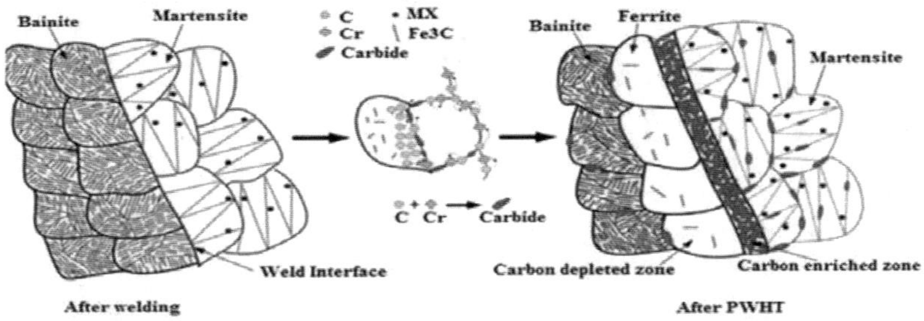

Figure 5. Schematic evolution of carbide precipitates during carbon migration [34].

2.5 Selection of filler wire

Recent, the AUSC materials are mostly joined by using GTAW, pulse TIG, SMAW or combination of two (hybrid) process. The selection of suitable filler metal for DMWs is very critical due to the difference in composition, thermal properties and mechanical behavior of BM. Some researchers are used the filler wire in DMWs as shown in Table 2.

Table 2: Filler metal composition (in wt. %) with welding process for DMWs.

Process	Filler	C	Si	Mn	Ni	Cr	Co	Al	Ti	Fe	Other	Ref.
TIG	IN A	0.10	1.00	3.0	Rem	15	-	-	-	12	A	[20]
GTAW/ SMAW	ERNiCr-3 (IN82)	0.035	0.30	3.0	72.6	20	-	-	-	3.0		[17]
GTAW/ SMAW	ERNiCr-1	0.050	1.0 max	3.0	70	16.5	-			6.50	B	[35]
GTAW/ SMAW	ENiCrFe-3 (IN182)	0.015	0.10	2.8	72.6	19.6	-	-	0.37	Bal.	Nb-2.68	[36]
SMAW	ENiCrFe-1	0.080	0.75	3.5	> 62	17				< 11	C	[55]
GTAW/GMAW	ERNiCrCoMo-1	0.06	0.10	0.1	Bal.	22	11	1.3	0.3	0.5		[36]
TIG	ERNiCrMo-3 (IN 625)	0.1	0.50	0.5	64	22-23	-	0.4	0.4	Bal.	D	[21]
TIG	ERNiCrFe-1 (IN65)	0.05	0.5	1	Bal.	23	-	0.2	1.1	23	E	[21]
Pulse TIG	ERNiCrMo-10	0.012	0.05	0.55	Bal.	22.4	0.22	-	-	2.5	F	[56]
TIG/P-TIG	ER2553	0.015	0.35	1.25	6.32	25.85	-	-	-	Bal.	G	[13]

P-TIG	ERNiCu-7	0.026	0.130	3.610	62.85	-	-	0.1	2.55	0.785	H	
P-TIG	ENiCrMo4	0.009	0.055	0.52	57.88	15.66	-	-	-	Bal.	I	

***A**: Mo - 1.5, Nb - 2.5, Cu - 0.5, **B**: Mo - 0.30, Nb - 2.6, **C**: Cu < 0.5, Nb + Ta - 1.5 - 4.0, **D**: Cu - 0.5, Nb - 3.6, Mo - 9.2, P - 0.015, **E**: Cu - 2, Mo - 3.5, **F**: S – 0.001, Co – 0.22; V – 0.01, W – 3.34, Si – 0.05, Cu – 0.2, **G**: Cu - 1.56, P - 0.02, S - 0.005, **H**: Cu - 29.94, S - 0.004, P - 0.005, **I**: Cu - 0.026, V - 0.041, W - 3.98, S - 0.004, P-0.007*

Dupont et al. [49] were reported that IN 622 produced better solidification cracking resistance than IN 625 due to dilution of Mo, Nb, σ phases, which increased the solidification temperature range. Further, Sayyar et al. [50] were reported that IN625 and ER347 filler developed the Nb-Mo-Ti rich phase with better hot cracking resistance but Mo phase developed MGBs. Thus, Wu et al. [50] were reported that the 310 SS showed high sensitivity of solidification cracking (SC) but IN 82 and IN 617 are developed hot cracking in WM due to low melting point secondary phase. Similarly, Dev et al. [11-13] were reported that ERNiCrMo-4, ER2553 and ERNiCu-7 witnessed the Mo, Nb phase with segregation and MGBs. Also, Pandey et al. [21-23] were found the certain variations in material chemistry of buttering due to Ni-Fe alloy buffer layer and enhanced the solubility of Nb and Ti content due to increased Fe content in DMWs of SA508Gr3Cl1 / SS316 L. Moreover, Mateshakker et al. [52] were reported that AWS ER 347 and 316 L consumables enhanced the Cr and Mo phase with δ-ferrite, which decreased the mechanical properties and corrosion resistance. From open literature, it is clear that selection of filler for DMWs is not standardized yet.

3. Creep characteristics for AUCS power plant

Recently, AUSC plant boiler and turbine section are designed with high creep rupture strength with allowable stress and time 100000 - 500000 h at high temperature. The creep rupture data for long-term creep for ferritic, martensitic, austenitic steels and nickel base alloys with operating temperature versus stress at 100000 h are shown in Fig. 6 [37].

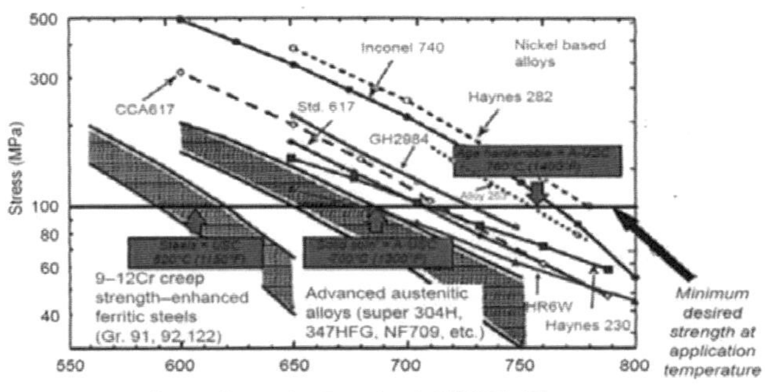

Figure 6. Application ranges of creep rapture for various materials [55].

Creep is a time-dependent progressive plastic deformation under an applied load at the high temperature about greater than $0.4T_m$. Hence, the rate of deformation of DMWs influenced by factors such as time of exposure, material, thermal and physical properties, stress-strain, fracture mechanism and microstructure. Furthermore, Chen et al. [38] were reported that fracture occurred in DMWs of T92 / TP347 in T92 HAZ zone during creep rupture test at stress 49 MPa, temperature 873K and time 10^5 h due to ferrite and carbide precipitation. Further, Falat et al. [53] studied the creep fracture behavior of

DMW in T91 / TP316H at temperature range 600-650^{0}C with stress 60-140MPa and fracture occurred at T91 HAZ side due to micro-void coalescence at grain boundaries. However, Shin et al. [39] investigated fracture during creep in DMWs 740 H / P92 for parameter range 650 °C / 80 MPa and 700 °C/ 35 MPa. It was concluded type IV cracking occurred at P92 side in CGHAZ due to poor grain boundaries strengthening by carbides. Along with, Matsunaga et al. [40] were studied the creep behavior of P91 / AISI 316 at stress 40 - 80 MPa and temperature 590 - 625 ^{0}C and reported that cavities nucleated at the top surface of the weld joint in the form of micro-cracks. The DMWs are mostly damaged in the outer surface of joint due to high creep strain appeared at the interface surface.

4. Structural integrity assessment issues

The structural integrity is considered scientific margin between safety and disaster for improved the life of existing structure by rehabilitation. Recently, the structural integrity assessment procedure has been based on the reliability of the structure stand in high temperature. However, the UK power generation industries developed the R6 procedure for defect assessment. The BS 7910 procedure is used for detecting the discontinuities in components used in high temperature and improved by the addition of SINTAP for assessing the structural integrity[42]. Moreover, Rathod et al. [43] were used FITNET, R6, SINTAP and BS 25 for analysis the yield strength ratio (YSR), plastic instability strength (PIS) and SENB toughness test in the DMWs of SA 508 / 304 LN. The structure integrity assessment are available for single joint. Recently, the DMWs design consideration very critical and essential for remnant life prediction of the plant with new testing and fabrication procedure.

5. The effect of Welding process and heat treatment on mechanical properties

The AUSC components are fabricated by using fusion welding process such as gas tungsten arc welding (GTAW), gas metal arc welding (GMAW), shielded metal arc welding (SMAW), Submerged arc welding (SAW). Recently, the researchers are diversified toward advance welding process as activated flux or autogenous A-TIG and pulse TIG process with selecting suitable parameters. Verma et al. [44] were studied the effect of the SMAW process parameters on DMWs of DSS 2205 / 316 L. The hardness and tensile value decreased as heat input increased. Further, Wang and Mittal et al. [54-17] were reported that GTAW process provided better mechanical properties than SMAW process in DSS 2205 / 16 MnR. and, T 91 / 347 H dissimilar joints. Along with, Rathod et al. [36] were investigated the effect of Ne - Fe buffer layer by using GTAW, GMAW, SMAW process with heat input 1.11, 2.46 and 3.47 KJ/mm respectively. It was reported that fracture toughness, weld strength and ductility of the GMAW is higher than SMAW process. Further, Naffakh et al. [23] were reported that WM with heat input 5.93 KJ/mm by GTAW process was showed better the hardness 240 HV, tensile strength 450 MPa and impact strength 121 J than hybrid process. Further, Vashistha et al. [31] were studied the effect of heat input (HI) of GTAW 0.34, 0.38, 0.48 KJ/mm and SMAW 0.72, 0.81, 0.97 KJ/mm process on the DMWs of AISI 304 / AISI 201steel. It was concluded that the GTAW process is produced better mechanical properties. Moreover, Pavan et al. [15] were studied the effect of the TIG process on DMWs of SUS 304 H / IN 617. It was found the hardness and impact strength better in WM than base metal (BM). It was concluded that TIG process suitable for AUSC boiler. Likewise, Ramkumar et al. [14-47] were studied the effect of continuous current CCGTAW and pulse current PCGTAW process on the mechanical properties of DMWs as IN 718 / AISI 316 L and SAF 2207 / IN 825. The PCGTAW was showed better tensile strength and impact toughness than CCGTAW. Likewise, Ramkumar et al. [11-12] were reported that joint efficiency of activated Flux A-TIG found higher than autogenous TIG such as 85% and 77% in DMWs of IN 718 / AISI 416. Also, Karthick et al. [44] were studied the effect of PWHT 760 ^{0}C / 2h on mechanical properties of P91 / 316LN. The notch tensile strength and impact toughness of WM and HAZ in P91 side was found higher 580-590MPa and 116-90J respectively. Further, Falat et al. [19] investigated the effect of PWHT 750 ^{0}C / 1h and found low creep strength with types IV cracking in HAZ. Similarly, Zang et al. [45] were reported that. WM of P92 / IN 617 showed better impact strength, tensile strength and hardness values after PWHT. Pandey et al. [10] were analyzed the effect of flux A-TIG process with PWHT on DMWs of P91 / P92. It was found that the tensile and impact strength increased after PWHT. Similarly they were reported that hardness of both welding processes after double PWHT 760 ^{0}C / 2h increased. Further, A-TIG process provided better mechanical properties than M-TIG [48] Even so, from the above section, it is clear that DMWs compatible with single and hybrid welding process with PWHT for evolution of microstructure and mechanical properties point of view, but in the case of creep properties with A-TIG and Pulse TIG, a lot of work is needed.

6. Conclusions

The DMWs of AUSC power plant is joined by hybrid or multi-pass welding process, these processes are developed some critical issues in DMWs, and decreased the productivity. However, A-TIG provided the superior results and cut the significant fraction of cost. The high-temperature properties of DMWs can be improved by suitable PWHT with A-TIG and pulse TIG process. Further, a structural integrity assessment procedure for DMWs of AUSC Power plant will be developed to adopt by the scientific community. This state of the art is encouraged the researchers about the enormous future scope.

References

[1] Gianfrancesco Di A 2016 Materials for USC and AUSC Power Plants, Woodhead Publishing Series in Energy Number 104.

[2] Abe F 2015 Research and Development of Heat-Resistant Materials for Advanced USC Power Plants with Steam Temperatures of 700 °C and Above, Engineering 1(2) pp. 211–24.

[3] Hosseini S, Shamanian, Kermanpur A 2016 Microstructural and weldability analysis of Inconel 617/AISI 310 stainless steel dissimilar welds, Int J Press Vessel Pipe 144 pp.18–24.

[4] Zerbst U, Ainsworth R A, Beier H T, Pisarski H, Zhang Z L, Nikbin K 2014 Review on fracture and crack propagation in weldments - A fracture mechanics perspective, Engg. Fract Mech 132 pp.200–76.

[5] Garakani A and Mehdizadeh M 2011 Effect of long-term service exposure on microstructure and mechanical properties of Alloy 617, Mat. Des. 32(5) pp. 695–700.

[6] Verma J and Taiwade R V 2017 Effect of welding processes and conditions on the microstructure, mechanical properties and corrosion resistance of duplex stainless steel weldments—A review, J Manu Process 25 pp.134–52.

[7] Fu J W, Yang Y S, Guo J J, Tong W H 2009 Formation of two-phase coupled microstructure in AISI 304 stainless steel during directional solidification, J Mat Res 24 pp.85–90.

[8] Arivazhagan B, Sundaresan S, Kamaraj M 2009 A study on influence of shielding gas composition on toughness of flux-cored arc weld of modified 9Cr-1Mo (P91) steel, J. Mat. Proc. Tech. 209 pp. 5245–53.

[9] Pandey C, Saini N, Mahapatra M M 2016 Hydrogen induced cold cracking of creep resistant ferritic P91 steel for different diffusible hydrogen levels in deposited metal, Int. J. Hydrogen Energy 41 pp. 17695–712.

[10] Pandey C, Mahapatra M M, Kumar P, Saini N, Thakre J G, Vidyarthy R S 2018 A brief study on δ-ferrite evolution in dissimilar P91 and P92 steel weld joint and their effect on mechanical properties, Arch Civ. Mech. Eng. 18 pp. 713–22.

[11] Ramkumar K D, Patel S D, Sri Praveen S, Choudhury D J, Prabaharan P, Arivazhagan N 2014 Influence of filler metals and welding techniques on the structure-property relationships of Inconel 718 and AISI 316L dissimilar weldments, Mat. Des. 62 pp. 175–88.

[12] Ramkumar K D, Dev S, Saxena V, Choudhary A, Arivazhagan N 2015 Effect of flux addition on the microstructure and tensile strength of dissimilar weldments involving Inconel 718 and AISI 416, Mat. Des. 87 pp. 663–74.

[13] Dev S, Ramkumar K D, Arivazhagan N, Rajendran R 2018 Investigations on the microstructure and mechanical properties of dissimilar welds of inconel 718 and sulphur rich martensitic stainless steel AISI 416, J. Mat. Proc. 32 pp. 685–98.

[14 Ramkumar T, Selvakumar M, Narayanasamy P, Begam A A, Mathavan P, Raj A A 2017 Studies on the structural property mechanical relationships and corrosion behaviour of Inconel 718 and SS 316L dissimilar joints by TIG welding without using activated flux, J. Man. Proc. 30 pp. 290–96.

[15] Pavan AHV, Vikrant KSN, Ravibharath R, Singh K 2015 Development and evaluation of SUS 304H - IN 617 welds for advanced ultra-supercritical boiler applications, Mat. Sci. Engg. A 642 pp. 32–41.

[16] Falat L, Svoboda M, Vyrostkova A, Petryshynets I, Sopko M 2012 Microstructure and creep characteristics of dissimilar T91/TP316H martensitic/austenitic welded joint with Ni-based weld metal, Mat. Char. 72 pp. 15–23.

[17] Mittal R and Sidhu B S 2015 Microstructures and mechanical properties of dissimilar T91/347H steel weldments, J. Mat. Pro. Tech. 220 pp. 76–86.

[18] Kou S 2015 A criterion for cracking during solidification, Acta Mat. 88 pp. 366–74.

[19] Naffakh H, Shamanian M, Ashrafizadeh F 2009 Dissimilar welding of AISI 310 austenitic stainless steel to nickel-based alloy Inconel 657, J. Mat. Proc. Tech. 209 pp. 3628–39.

[20] Naffakh M H, Aboutalebi M R, Seyedein S H 2013 Microstructural mechanical and weldability assessments of the dissimilar weld between γ' And γ''strengthened nickel base superalloys, Mat. Char. 82 pp. 41–49.

[21] Sayyar N, Shamanian M, Niroumand B 2018 Arc weldability of Incoloy 825 to AISI 321 stainless steel welds, J. Mat. Pro. Tech. 262 pp. 562–70.

[22] Pandey C., Mahapatra M.M., Kumar P., Saini N. 2018, Some studies on P91 steel and their weldments, J. Alloys Comp. 743, pp. 332–64.

[23] Pandey C, Mahapatra M M, Saini N, Srivastava A 2017 Microstructure and mechanical property relationship for different heat treatment and hydrogen level in multipass welded P91 steel joint, J. Mat. Proc. 28 pp. 220–34.

[24] Saini N, Pandey C, Mahapatra M M 2017 Effect of diffusible hydrogen content on embrittlement of P92 steel, Int. J. Hydrogen Energy 42 pp. 17328–38.

[25] Zhou C, Huang Q, Guo Q, Zheng J, Chen X, Zhu J 2016 Sulphide stress cracking behaviour of the dissimilar metal welded joint of X60 pipeline steel and Inconel 625 alloy, Co. Sci. 110 pp. 242–52.

[26] Toppo A, Pujar M G, Sreevidya N, Philip J, Xie X, Shen J 2017 Stress corrosion cracking in the heat affected zone of a stainless steel 308L-316L weld joint in primary water, Co. Sci. 388 pp. 243–48.

[27] Dong L, Peng Q, Han E H, Wang L 2018 Correlation of microstructure and stress corrosion cracking initiation behaviour of the fusion boundary region in a SA508 Cl. 3-Alloy 52M dissimilar weld joint in primary pressurized water reactor environment, Co Sci. 132 pp. 9–20.

[28] Zhu R, Wang J, Zhang Z, Han E H 2017 Stress corrosion cracking of fusion boundary for 316L/52M dissimilar metal weld joints in borated and lithiated high temperature water, Co Sci. 120 pp. 219–30.

[29] Vidyarthy R S, Kulkarni A, Dwivedi D K 2017, Study of microstructure and mechanical property relationships of A-TIG welded P91–316L dissimilar steel joint, Mat. Sci. Eng. A 695 pp. 249–57.

[30] Soysal T, Kou S, Tat D, Pasang T 2016 Macrosegregation in dissimilar-metal fusion welding, Acta Mat. 110 pp. 149–60.

[31] Ramkumar K D, Abraham W S, Viyash V, Arivazhagan N, Rabel A M 2017 Investigations on the microstructure tensile strength and high temperature corrosion behaviour of Inconel 625 and Inconel 718 dissimilar joints, J. Mat. Proc. 25 pp. 306–22.

[32] Vashishtha H, Taiwade R V, Sharma S, Patil A P 2017 Effect of welding processes on microstructural and mechanical properties of dissimilar weldments between conventional austenitic and high nitrogen austenitic stainless steels, J. Mat. Proc. 25 pp. 49–59.

[33] Ye X, Zhang P, Zhao J, Ma P 2018 Effect of macro- and micro-segregation on hot cracking of Inconel 718 superalloy argon-arc multilayer cladding, J Mat Pro Tec, 258 pp. 251–8.

[34] Kulkarni A, Dwivedi D K, Vasudevan M 2018 Study of mechanism, microstructure and mechanical properties of activated flux TIG welded P91 Steel-P22 steel dissimilar metal joint, Mat. Sci. Eng. A 731, pp. 309–23.

[35] Jahanzeb N, Shin J H, Singh J, Heo Y U, Choi S H 2017 Effect of microstructure on the hardness heterogeneity of dissimilar joints between 316L stainless steel and SS400 steel, Mat. Sci. Eng. A 700 pp. 338–50.

[36] Rathod D W, Singh R K R, Pandey S, Aravindan S, Singh P K 2017 Influence of graded compositions and carbon diffusivities in buttering on structural integrity of dissimilar metal welds, Mat Sci Eng A 702 pp. 289–300.

[37] Rathod D W, Singh R K R , Pandey S, Aravindan S 2016 Effect of buffer layered buttering on microstructure and mechanical properties of dissimilar metal joints for nuclear plant application Mat. Sci. Eng. 666 pp. 100–13.

[38] Rogalev N, Golodnitskiy A, Tumanovskiy A, Rogalev A 2014 A survey of state-of-the-art development of coal-fired steam turbine power plant based on advanced ultrasupercritical steam technology, Con. Eng. Sci. 34 pp. 1807–25.

[39] Chen G, Song Y, Wang J, Liu J, Yu X, Hua J 2012 High-temperature short-term tensile test and creep rupture strength prediction of the T92 / TP347H dissimilar steel weld joints, Eng. Fail. Anal. 26 pp. 220–9.

[40] Shin K Y, Lee J W, Han J M, Lee K W, Kong B O, Hong H U 2018 Transition of creep damage in dissimilar welds between Inconel 740H Ni-based superalloy and P92 ferritic/martensitic steel, Mat. Char. 139 pp. 144–52.

[41] Matsunaga T, Hongo H, Tabuchi M 2017 Interfacial failure in dissimilar weld joint of high boron 9% chromium steel and nickel-based alloy under high-temperature creep condition, Mat. Sci. Eng. A 695 pp. 302–8.

[42] Sharples J K, Ainsworth R, Budden P J, Dowfing R 2003 Defect Assessment Procedures R6 Revision 4 and Bs7910, Mat. Stru pp. 280–96.

[43]] Rathod D W, Pandey S, Singh P K, Kumar S 2017 Microstructure-dependent fracture toughness (JIC) variations in dissimilar pipe welds for pressure vessel system of nuclear plants, J. Nu. Mat. 493 pp. 412–25.

[44] Verma J and Taiwade R V 2016 Dissimilar welding behavior of 22% Cr series stainless steel with 316L and its corrosion resistance in modified aggressive environment, J. Mat. Proc. 4 pp. 1–10.

[45] Karthick K, Malarvizhi S, Balasubramanian V, Krishnan S A, Sasikala G, Albert S K 2017 Tensile and Impact Toughness Properties of Various Regions of Dissimilar Joints of Nuclear Grade Steels, Nu. Eng. Tech. 50 pp. 116–25.

[46] Zhang J, Huang B, Wu Q, Li C, Huang Q 2015 Huang, Effect of post-weld heat treatment on the mechanical properties of CLAM/316L dissimilar joint, Fus. Eng. Des. 100 pp. 334–9.

[47] Ramkumar K D, Naren S V, Karthik V R, Tiwari A, Arivazhagan N 2016 Development of pulsed current gas tungsten arc welding technique for dissimilar joints of marine grade alloys, J. Mat. Proc. 21 pp. 201–13.

[48] Sakthivel T, Vasudevan M, Laha K, Parameswaran P, Chandravathi K S, Mathew M D 2011 Comparison of creep rupture behaviour of type 316L(N) austenitic stainless steel joints welded by TIG and activated TIG welding processes, Mat. Sci. Eng. A 528 pp. 6971–80.

[49] DuPont J N, Banovic S W, Marder R 2003 Microstructural Evolution and Weldability of Dissimilar Welds between a Super Austenitic Stainless Steel and Nickel-Based Alloys, J. Weld 82 pp. 125-135.

[50] Sayyar N, Shamanian M, Niroumand B 2018 Arc weldability of Incoloy 825 to AISI 321 stainless steel welds, J. Mat. Proc. Tech. 262 pp. 562–70.

[51] Wu W and Tsai C H 1999, Hot cracking susceptibility of fillers 52 and 82 in alloy 690 welding, Met Mat Trans A 30 pp. 417–26.

[52] Moteshakker A, and Danaee I 2016 Microstructure and Corrosion Resistance of Dissimilar Weld-Joints between Duplex Stainless Steel 2205 and Austenitic Stainless Steel 316L, J. Mat. Sci. Tech. 32 pp. 282–90.

[53] Wang S, Ma Q, Li Y 2011 Characterization of microstructure, mechanical properties and corrosion resistance of dissimilar welded joint between 2205 duplex stainless steel and 16MnR, Mat. Des. 32 pp. 831–37.

[54] Shoji T, Lu Z, Peng Q 2011 Factors affecting stress corrosion cracking (SCC) and fundamental mechanistic understanding of stainless steels, Stress Co. Crack Theory Pra. pp. 245–72.

[55] Zhang Q, Zhang J, Zhao P, Huang Y, Yang Y 2015 Microstructure of 10% Cr martensitic heat-resistant steel welded joints and type IV cracking behavior during creep rupture at 650°C, Mat. Sci. Eng. A, 638, pp. 30–7.

[56] Pandey C, Mahapatra M M, Kumar P, Saini N 2018 Dissimilar joining of CFEF steels using autogenous tungsten-inert gas welding and gas tungsten arc welding and their effect on δ-ferrite evolution and mechanical properties, J. Mat. Proc. 31 pp. 247–59.

Combustion analysis of porous radiant burner for commercial cooking applications

Loveneesh Sharma[1], Satbir Singh Sehgal[1], Ravinder Tonk[1], Saurabh Chhabra[1]

[1]Chandigarh University, Gharuan Punjab, India

Email: loveneeshsharma7@gmail.com

Abstract. High efficiency and low efficiency were observed by using porous burner technology for commercial cooking. The whole research consists of survey of these three types of burners. Like burners diameter having convergent 100mm, straight 100mm and80mm divergent shape burner. LPG is the fuel i.e. liquid petroleum gas is used in the all experiment. The two zones combustion and preheating zones comprise by a porous burner. Preheating zone having steel balls of width 6.5mm where combustion zone has Sic matric of 10ppi was used. 150mm is the constant altitude of three burners comprise of combustion zone and having equivalence ratio from 0.10-0.60. The range at which burner was operated 79% is the maximum efficiency analyzed for porous burner having dimensions like 100mm diameter at power equal to 0.84 kW and equivalence ratio (Φ) equal to 0.49.

1. Introduction

Country having a second biggest engrossment of LPG (Liquid Petroleum Gas) is subsidized as the government of INDIA pay out large sum of money that is Rs. 630. Rs. 630 is the subsidized value of LPG (liquid petroleum gas) cylinder of 14.21kg whereas Rs 800 for non-subsidized cylinder. Due to this (GOI) which we also called as government of India which facing large financial crisis as they are also taking the responsibility of the rest measure of the cash of the oil organizations. So, there is great need for finding alternative fuel so that there will be raise in the domestic burner thermal efficiency and also used as cooking purposes. so that we can reduce the economic burden from (GOI). Therefore, in present work (PBT) which is also called as permeable burner innovation are tried for LPG (liquid petroleum gas) terminated burner for cooking purposes.

Combustion process required porous medium in porous burner technology. So that's why it is also known as PRB that is porous radiant burner and (PMC) porous medium combustion. Preheating zone and the combustion zone are the two zones of (PMC) porous medium combustion the first zone i.e. combustion zone contains Sic ceramic matrix as well as in preheating zone there is low porosity to ignore resulting flash back and combustion flash back [17]. We can see a two-layer porous burner in diagram i.e. Figure1. In pre-heating zone, air fuel mixture first preheating whereas combustion take place in the combustion zone. Convection conduction and radiation method of warmth move heat is transmitted in all the direction. As comparisons with the domestic burners available in the market PMC having two important characteristics that is less emission and good thermal efficiency. And to raise the value of thermal efficiency and two deplete the emission of burner many of researches are working and for hose hold applications combustion review paper was represented by scientist known as Isabel Malico et al. [17] Porous burner is best suitable for commercial applications like cooking .For household applications the scientist Isabel Malico et al. [17] introduced an audit paper on a permeable media ignition.

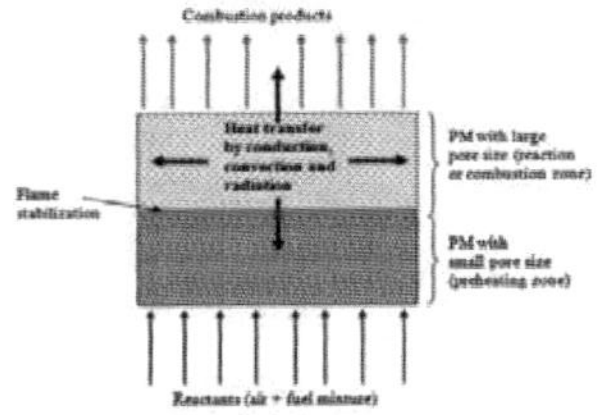

Figure 1. Schematic Diagram of A Two-Layer Porous Burner [2].

A review paper that was presented by M. Amujeebu et al. [7] on different materials that utilize for porous burner. Research and development for PMC was reflected in their paper. P Muthukumar et al. [7] also produce a burner that is known as commercial cooking porous burner for LPG. 80%, 85%, 90% are different porosities at which burners were tested for different equivalence ratio and wattages ratios and wattages. In experimental set up, 80-90% porosity of Sic foam was used in the combustion zone as well as in preheating zone 40% porosity used. Efficiency was 10% which is good then the commercial burner available in INDIAN market and that achieved efficiency was 75%. Purna et al. [13] formed porous burner i.e.self-aspirating porous radiant. 3.0 m/s, 3.6 m/s and 0.4 m/s is the range of different velocities is basically get from the experiment conducted by Purna etal and two things were measured for the burner i.e. temperature distribution analysis and thermal efficiency. 64.6 % for velocity of 3.5m/s is the thermal efficiency 55.46% for gas speed 3.0 m/s and 36.42 % as gas speed 0.4 m/s.

A study was performed by Layriss Iral et al. [19] that was basically for the porous radiant burner which was actually on the induced air at high altitude and serving commercial applications. At three different powers that is (i.e. 370 kW/ 2m , 480 kW/ 2 m and 670 kW/ 2m) burner was tested, with that we can likewise determine the warm productivity and outflows of the burner. 50% is the maximum thermal efficiency for the power 370kW/ 2 m was analyzed.

To check which one is gathering and furthermore estimated the warm productivity of the burner developed different various porous media burner was developed by Byeonghunet al. [12]. Metal fiber (MF), stainless steel fin (SF) and ceramic (CM) are three types of porous media which were basically utilized. Two different parameters were analyzed that is equivalence ratio and thermal efficiency of burner at different wattages and on the basis of that we can analyzed that MF burner have highest efficiency at 86.7% and $\Phi = 0.95$ at 25Kw whereas SF burner has efficiency 81.6% that is lowest and $\Phi = 0.80$ at 15Kw.

A self-aspirating porous burner that was built up by Wasan Yoksenakul et al. [4] is small scale enterprises and for medium one. Different practical work was analyzed at various power ranges as well as highest tower range from 21- 44 kW and variation of height was 75-125mm and for fuel LPG was used. We can analyze that there was decrease in thermal efficiency with increase in height with exception of the burner at 57.64% is the highest thermal efficiency.

For kerosene porous burner was developed and that was developed by Juga et al. [2]An experiment of porous insertion in kerosene pressure stove which is conventional one was performed by Sharma et al. [10]. The efficiency increased by 55% to 62% of stove by using porous insert. Porous radiant reticulated burner (PRRB) are designed for domestic application by Jagujai and Sanitjai [2], and by compared to other burner the (PRRB) Porous radiant reticulated burner is much efficient in term of efficiency, emission characteristics and combustion stability.

Cooking application for LPG (liquid petroleum gas) another burner is developed by V. K. Patagi and P. Muthu et al. [14], and for this experiment was performed by changing the porosities, equivalence ratio and power. When we compare it with commercial burner its efficiency was 71% and 72% which was 6% and 7% better respectively then the commercial burner.

Application in which porous burner technology (PBT) can we used to raise the efficiency and depletion of the emissions optimized shape need to be observed basically all done in the literature survey. Preheating and combustion zones are the two zones of PMC preheating zones have metal balls of diameter 6.5mm which have low porosity and the another zone known as combustion zone which have Sic ceramic matrix another point of the concern is using low porosity in the preheating zone so that combustion would not take place and to avoid resulting flashback. Thermal efficiency and temperature distribution were investigated [20].

2. Experimental approach and procedures

There are two zones in the burner one is Combustion zone and another is Preheating zone. Combustion zone contain the Sic ceramics foam hold dimensions 10ppi and stiffness 25mm on the other hand preheating zone contains steel balls having width 6.5mm each.
Two parameters thermal adaptability and also temperature distribution were determined for the two burners i.e. commercial and porous burner and with the help of water boiling test efficiency was calculated this is also shown in the above figure 1 Rotameter allowed LPG from cylinder of 14.21kg which is operated and then supplied to a burner. Rota-meter also aid to measure the fuel flow rate and flow testing two things are used like Aluminum pans along with the lid (0.5-1kg) of amount of water is required to filled the pan and at specific temperature i.e. (20-25°C). To check the repeatability quantity of water starting from 0.5-1 kg and varying a weight of water we can measured effect on efficiency with the help of weighing balance weight of pan with water can be calculated. The starting temperature and initial weight of the water and cylinder is calculated respectively.

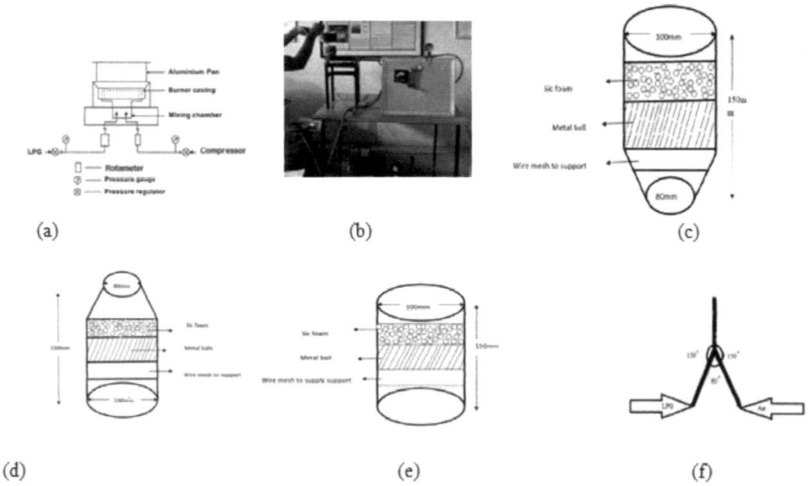

Figure 2. (A) Schematic Diagram Experimental Setup (B) Experimental Setup (C) convergent Porous Burner (D) Divergent Porous Burner (E) straight porous Burner (F) Schematic Y-Shaped Mixing Tube.

When the flame becomes stable then burner is used on which the pan was kept after that the temperature of water rises till it reaches to 90°C, burner was turned off at this stage, after this calculation was made on the LPG cylinder weight again. It is crucial to note the time that the burner take increase temperature which is of water from T1 that is starting temperature to 90°C. For more accuracy of the experiment review are taken at any rate multiple times and also to conclude the final result their unusual will be appropriated. According to a standard of IS 4246:2002[14]. Formula for calculating LPG (liquid petroleum gas) fired burner given by equation (1).

$$\frac{100(W + P)\,(T_{2} - T_{1})}{m_{f}.CV}$$

where W is the mass of water, mf is the mass of the LPG expended and CV is the calorific estimation of LPG and P is mass of the pan along with lid made up of aluminum.

(air / fuel) actual
(air/fuel) stoichiometric

Actual one part of the ratio that is air fuel ratio tells about fuel consume for the actual combustion quantity of the air utilize to increase the desired temperature of water. With the help of rotameter air flow is calculated as well as we can get rate of actual air consumed for ignition as well as weighing balance help to measure fuel with the help of which we can get the rate of actual quantity fuel utilize for ignition or increasing water temperature. For this Stoichiometric fuel air ratio appropriated as 15.6 (used by V.K. Patang) [12-18]

3. Result and Discussion

Different equivalence ratio at which burner was operated was (0.10-0.60) whereas power wattages like (0.4-1KW). The result of temperature distribution and domestic burner adaptability and porous burner is studied in this section.
Varied to calculate the distribution of temperature to parameters is taken into consideration are as follow;

i). Equivalence ratio
ii). Air flow ratio

3.1. Temperature Distribution

Conventional burner and facial of PRB temperature have been measured with the help of K- type of thermocouple. (DAS) Data acquisition system (make Aberrancies TC - 900) was used to acquire the output of the thermocouple mostly 1 hour is the time at which burner were operated. Different equivalence ratio we get when the at different equivalence ratio i.e. (0.11-0.16) we get when all burners temperature were calculated. Different maximum temperature (1297-1505) attained by energy burner.

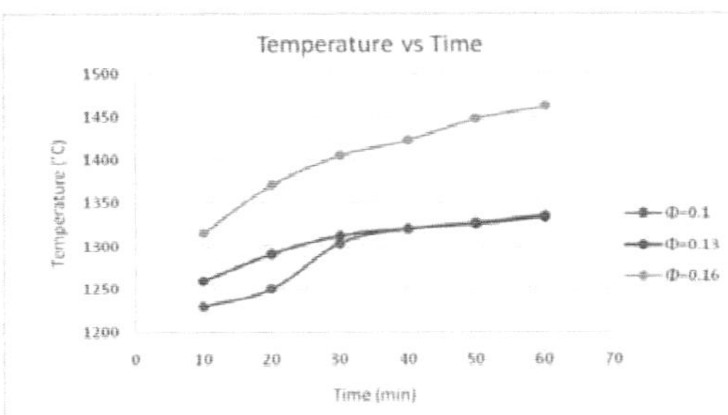

Figure 3. Variation of Temperature W.R.T. Time (Minutes) of PorousBurner 100 Mm at Different Equivalence Ratio of convergent.

At constant equivalence ratio the porous burner and increase in temperature as the time increase and max temperature attained by after 60 minutes that is 1462°C maximum temperature 13327°C when the temperature raise by 150°C then 5% increase in the mass fuel consumption at the equivalence of 0.11 and minimum temperature i.e. 1230°C toward the beginning of burning.

The difference in the porous burner surface temperature of 100mm diameter verses to air fuel ratio is shown by the figure.4. It was analyzed that as there is decrease in air to fuel ratio and it decline from 3.47 to 2.14 (38.2%). Then maximum raise of temperature was analyzed from 1227 to 1424°C. The temperature raise was less i.e. (1481 to 1490 °C) whereas the air fuel rate overload beyond 1.32 to 1.13 and this is because of saturation of heat in a porous medium.

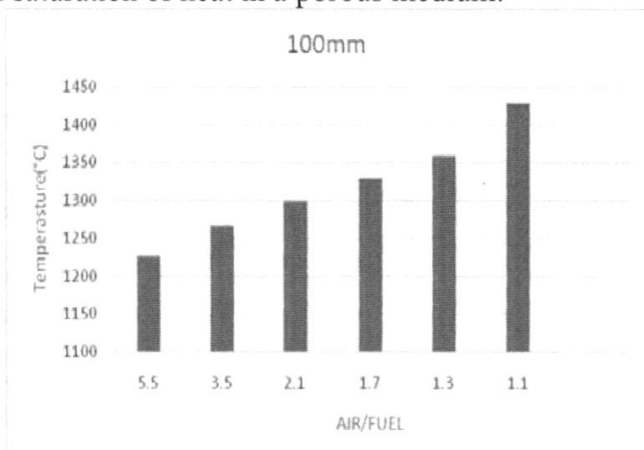

Figure 4. Variation of Temperature W.R.T. Air / Fuel Consumption ofPorous Burner 100Mm OF Convergent

In permeable disparate burner 90mm top diameter the temperature raised by some amount after 0f minutes as we increased the time. It was analyzed that 1502°C is the maximum temperature at equivalence ratio (Φ = 0.63). The mass fuel of expenditure increasing by 1.5 % and maximum temperature observed at equivalence ratio of 0.43 was 1442°C and that of 1227°C was least temperature at beginning of ignition.

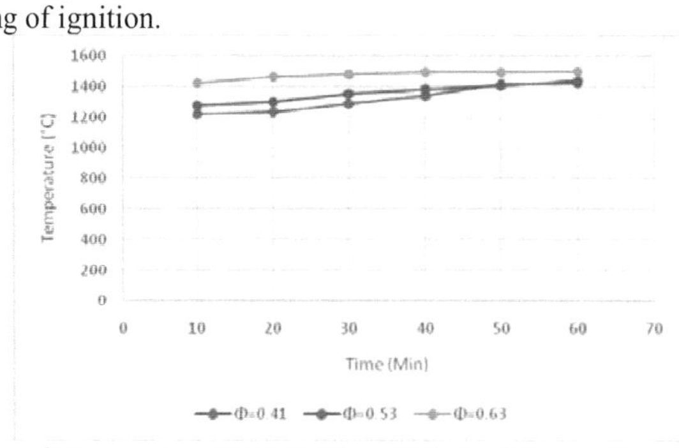

Figure 5. Variation of Temperature W.R.T. Time (Min.) of Porous BurnerDivergent 80Mm at Different Equivalence Ratio.

A porous divergent burner with dimension 90mm upper diameter with respect to air to fuel ratio and the its variation of surface temperature is shown by figure 6with the decline of air to fuel ratio raise in fuel utilization we analyzed that there is raise in the temperature (1190 to 1380°C) is the maximum raising range of the temperature was seen when there is decline in the air / fuel rate from 2.2 to 1.9 (13.6 %).

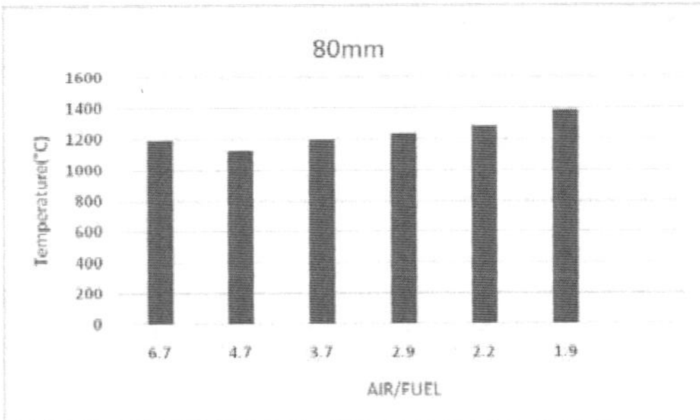

Figure 6. Variation of Temperature W.R.T. Air/Fuel Consumption of PorousBurner Divergent 80Mm.

Straight burner and facial of PRB temperature have been measured with the help of K- type of thermocouple. (DAS) Data acquisition system (make Aberrancies TC - 900). Was used to acquire the output of the thermocouple mostly 1 hour is the time at which burner were operated. Different equivalence ratio we get when the at different equivalence ratio i.e. (0.11-0.16) we get when all burners temperature were calculated. Different maximum temperature (1450-1485°C) attained by energy burner.

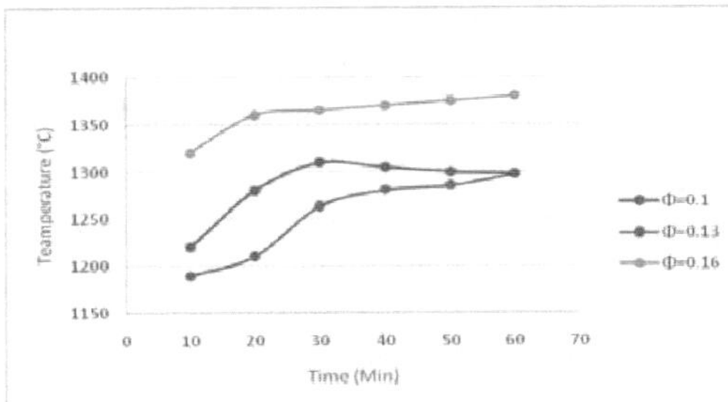

Figure 7. Variation of Temperature W.R.T. Time (Min.) of Porous Burner100Mm straight at Different Equivalence Ratio.

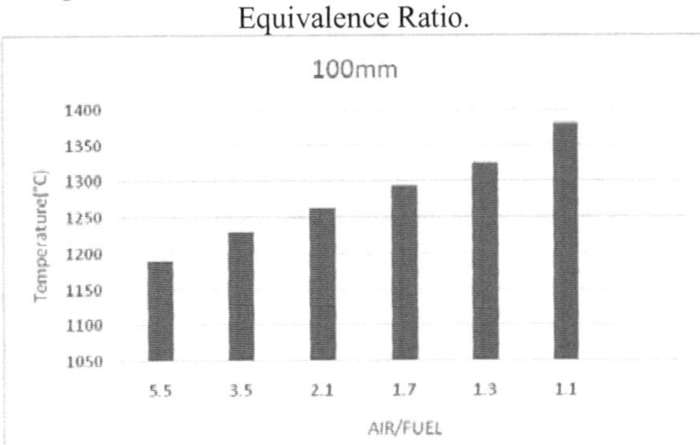

Figure 8. Variation of Temperature W.R.T. Air / Fuel Consumption ofPorous straight Burner 100 Mm.

Different temperature of the commercial burner v/s LPG fuel consumption is shown in Figure8. Can be analyzed in fig 9 and with the raise in the fuel utilization1230°C an is the most extreme temperature achieved by the commercial burner and 0.028 gm/sec consumption is required for this process requires. Another point we also analyzed that the higher estimation of greatest temperature that is in the range of (6.89 – 20.94 %) of porous burner and all this analyzed this for same fuel consumption.

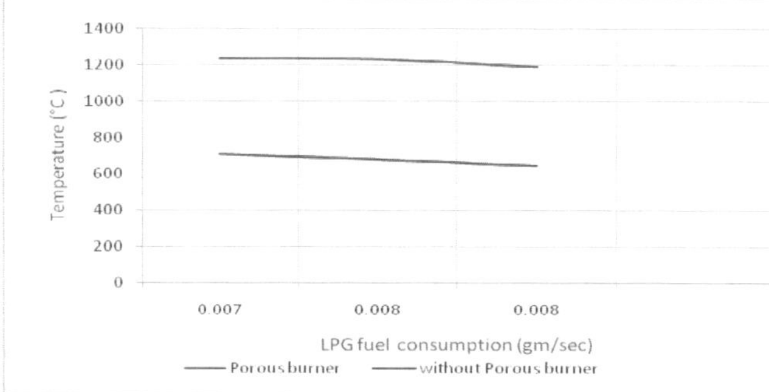

Figure 9. Variation of Temperature W.R.T. LPG Fuel Consumption (Gm/Sec) for porous burner and without porous burner.

3.2. Thermal efficiency

Thermal efficiency of porous burner as well as commercial burner is shown by fig 10. If we take comparison between porous and other burner, we can analyze that the porous burner of diameter 100mm has maximum efficiency. When there is a constant power of 0.6kW. We can analyzed that porous burner of 80mm diameter commercial burner as well as divergent burner all have thermal efficiency less than the 80mm porous burner they have thermal efficiency of 15.67 % , 22.13 % and 30.32 % respectively and the porous burner of width 150mm having maximum efficiency of 78.58 % at equivalence ratio of 0.48 and at power 0.84Kw.

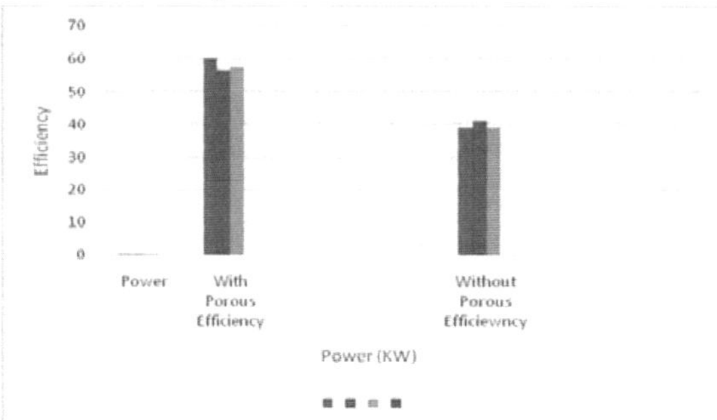

Figure 10. Variation of Efficiency of Porous and without porous Burner Vs. Power(Kw).

4. Conclusion

The porous burner and divergent shape of burner taken for an experimental analysis and their thermal efficiency were calculated and differentiate with commercial burner.

Following are the conclusions:

- 65 % is the higher capability of commercial burner.
- Porous burner has the thermal efficiency more than commercial burner that is 7 8.58 % and it is 13.5 % more than the domestic burner.
- 62.7 % is the maximum efficiency of porous burner 80mm that is 8 % more than the commercial burner.

- Porous burner which is divergent shape has maximum efficiency of 51.78%.
- 1462°C which is maximum temperature at 0.16 which is equivalence ratio (Ø) is of porous burner with diameter 100mm.
- 1380°C which is maximum temperature at 0.47 which is at equivalence ratio (Ø) of 80mm diameter porous burner.

After testing porous burner experimentally for the commercial applications and we can conclude that the thermal efficiency of porous burner with 100mm diameter was 78.58 % was highest among the entire burner. At the same time, fuel consumption. In future the computational analysis of work will be done.

References

[1] Nageshwar S et al. 2017 Performance analysis of porous radiant burner for cooking application International journal of Engineering &Technology **6** 65-70
[2] Jugjai S and Sanitjai S 1996 Parametric studies of thermal efficiency in a proposed porous radiant recirculated burner (PRRB) a design concept for the future burner, RERIC International Energy Journal, **18** 97-111.
[3] M. Amujeebu ,Durst F, Trimis D, and Dimaczek G 1996 Burner having material of varying porosity,United States Patent, No: 5522724
[4] Wasan Yoksenakul Indian standard, IS 4246, Domestic gas stoves for use with lique-fied petroleum gases specification (fifth revision) 2002
[5] Jugjai S and Pongsai C 2007 Liquid fuels-fired porous burner, Combust Sci. Technol. **179** 1823-1840.
[6] Sharma M et al.2009 Improvement of thermal efficiency of a conventional pressure stove using porous radiant in-sert, Int. J. Energy Res. **10** 247-254.
[7] Abdul Mujeebu M et al. 2009 Applications of porous media combustion technology – A review Applied Energy **86** 1365–1375
[8] Pantangi 2010 Development and performance analysis of porous radiant burners for cooking applications Ph.D Thesis. Department of Mechanical Engineering, IIT Guwahati.
[9] Sachdeva P 2011 Performance analysis of porous radiant burners used in LPG cooking stove In-ternational Journal of Energy and Environment **2** 367-374
[10] Yoksenakul W and jugjai S 2011 Thermal efficiency of self aspirating porous medium burner for small and medium scale enterprises (SMEs) The Second TSME International conference on Mechanical Engineering 19-21
[11] Yoksenakul W and jugjai S Thermal efficiency of self aspirating porous medium burner for small and medium scale enterprises (SMEs) The Second TSME International conference on Mechanical Engineering 19-21
[12] Byeonghun Y et al. 2013 Combustion characteristics and thermal efficiency for premixed porous media type of burner, Energy **53** 343-350.
[13] Purna C et al. 2013 Performance improvement of self-aspirating porous radiant burner by controlling process parameters Int. Journal of Engineering Research and applications **3** 978-984
[14] Muthukumar P and Shyamkumar P.I. 2013 Development of novel porous radiant burners for LPG cooking application Fuel **112** 562-566.
[15] Keramiotis C, Founti MA 2013 An experimental investigation of stability and operation of a biogas fuelled porous burner Fuel **103** 278-284.
[16] Boggavarapu P et al. 2014 Thermal Efficiency of LPG and PNG-fired burners: Experimental and numerical studies Fuel **116** 709-715.
[17] Malico I and Abdul M 2015 Potential of Porous Media Combustion Technology for Household Applications International Journal of Advanced Thermo fluid Research **1**
[18] Qu Z et al. 2015 Premixed Combustion in a Porous Burner with Different Fuels. Combustion Science and Technology **187** 489-504.

[19] Layrisser I and Amell A 2015 Performance study of an induced air porous radiant burner for household applications at high altitude Applied Thermal Engineering **83** 31-39.
[20] Isabel M and Abdul Mujeeb M 2015 Potential of porous media combustion technology for household applications International journal of advanced thermo fluid research **1** 50-69.

Optimization of FGM Composition for Better Environment Material

Mrinal Gautam[1] and Manish Chaturvedi[1]

[1]University Department, Mechanical Engineering, Rajasthan Technical University, Akelgarh, Kota, Rajasthan, 324010, India

E-mail: mrinal.phd18@rtu.ac.in

Abstract. In functionally graded heat-resisting material, the volume fraction of constituents varies functionally and continuously, and these are examined for high temperature engineering applications. By the spatial distribution of the volume fraction the thermal behaviour of FGMs is strongly influenced. Thus, the determination of volume fraction distribution becomes a critical part in the FGM design for a given loading condition and specification. In present work, averaging estimation approach viz, modified rule of mixture is considered. The present work is concerned with the Optimization of FGM Composition for better environment material. The investigation is carried out for uniform heating condition through the plate thickness. Classical laminate theory is used to determine residual stress. The code is developed on MATLAB for constrained minimization and the MATLAB function 'fmincon' is employed. The comparative study of three different thermal loading is presented. The effectiveness of residual stress minimization is explored for various boundary conditions.

1. Introduction

A novel set of materials named Functionally Graded Materials (FGMs) have newly been proposed that different material properties vary through the thickness and radial direction in a continuous behavior and, therefore free from interface weakness. Residual stresses, thermal stresses and stress concentrations are reduced by gradation of material properties. A functionally graded structure is described as, which to obtain a require function, two or more materials volume fractions are vary continuously as a function of location beside definite dimension (classically the thickness and radius) of the structure. Designers can get exceptional performance as well as modified response by the FGMs in thermal environment. In FGM the metal constituents provide stronger mechanical strength and reduce the risk of catastrophic fracture, even as in the ceramic constituents are resisting high-temperature environments for better thermal resistance characteristics. An Example, when Space Shuttle is entered into the Earth's environment, ceramic tiles applies while thermal protection as of heat generated. An FGM composed of metal on the inside surface and ceramic on the outside surface Structure. Classify the FGMs into; (a) stepwise-graded materials, as well as (b) continuously graded. In case of stepwise-graded materials, give rise to a multi-layered form by an interface there between the adjacent discrete layers. In case of continuously graded form, the transform in microstructure, composition, and thus in the properties take places with position. [1, 2]

An extensive literature review is presented detailing the state of the art in functionally graded material. Birman and Byrd [3] very recently recognized a comprehensive FGM research covering comprehensively the period 1997-2007. Na and Kim [4] proposed by considering stress reduction and developing behavior of thermo-mechanical buckling to optimize the volume fraction of FGM composite panels. Sadollah and Bahreininejad [5] employed multiobjective optimum design, for functionally graded dental implant by simulated annealing and genetic algorithms. Kou.et.al [6] performed by using a particle swarm optimization and procedural model of functionally graded materials. Noh.et.al [7] has presented to optimize volume fraction distribution of Reliability-based design in functionally graded composites in optimal design. Ashjari and Khoshravan [8] performed Volume Fraction optimization in FGM plates, with two evolutionary methods: (a) particle swarm optimization method (b) Real-coded genetic method, and. Taheri.et.al [9] proposed by an isogeometrical process for thermo-elastic optimization of material distribution of functionally graded structures, explaining the optimization complication by mathematical programming. Roque and Martins [10] developed the differential evolution to explain difficulties of natural frequencies to maximization, during optimization of functionally graded beams. Shi and Shemado [11] to optimize the interface shape, of functionally graded sandwhich structures for designing. Mohammadiha and Ghariblu [12] presented two process; (a) geometrical average and (b) multidesign objective methods, for functionally graded foam to optimize crush behavior by multi tubes filled.Shabana.et.al [13] presented the technique particle swarm optimization, in minimization the stresses in functionally graded cylinders under a pressure loading. Lieu and Lee [14] explained iso-geometric analysis, and adaptive hybrid evolutionary firefly method for modeling as well as optimization of functionally graded plates subjected to thermo-mechanical loading. Nguyen and Lee [15] analyzed lateral and, flexural-torisional buckling problems to optimize of thin- walled functionally graded beams. He and Sun [16] presented functionally graded material beam to optimize multi objective structural acoustic. Correia.et.al [17] developed a method of multi objective optimization for ceramic-metal functionally graded plates by higher order form.Cheng.et.al [18] employed the design of additive manufactured components, by stress constraints for topology optimization of functionally graded lattice structure. The objective of this work is to present the development of Optimization of FGM Composition for better environment material.

2. Formulation of FGM model

To study of FGMs, a model must be created that illustrates the function of composition throughout the material. In Figure 1, the volume fraction, V_c, describes the volume of ceramic at any point z throughout the thickness h according to a parameter n which controls the shape of the function (as seen in Figure 1) . V_c is specified by

$$V_c(z) = \left(\frac{z}{h} + \frac{1}{2} \right)^n \qquad (1)$$

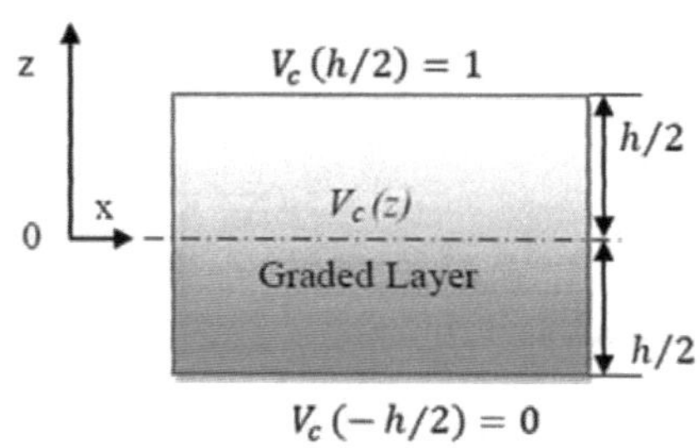

Figure1: Ceramic volume fractions across the FGM Layer

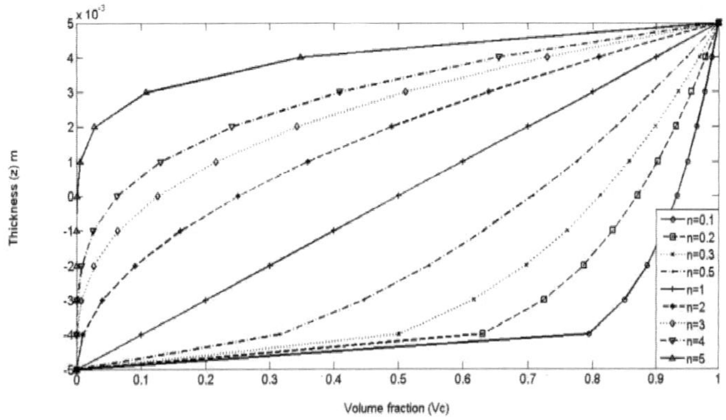

Figure 2: Effect of power law index (n) on volume fraction V_c

3. Temperature profile and heat conduction modelling

In present work, thermal loading are considered i.e. uniform heating or cooling of FGM. The estimation of the thermal residual stresses through the thickness is discussed. Temperature is assumed to be uniform through the plate. The thermal residual stresses at any position through the thickness z, σ are calculated using classical laminate theory [19].

$$\{\sigma\} = [\overline{Q}]\{\epsilon\} = [\overline{Q}]\{\epsilon^0\} + z\{k\} - \{\alpha\}[\Delta T] \quad (2)$$

4. Problem definitions

Figure 3 shows a simply supported plate-like FGM at the uniform initial temperature, $T_0 = 300$ K, where boundary conditions and geometry dimensions are specified. For temperature loading, uniformly decrease or increase their temperature. The volume fraction in the graded layer varies through the thickness, in the *x*-direction as invariant. In this problem, a dominant thermal stress component is became by axial stress σ_{xx} . With initial volume fraction $(V_m)_i = 0.5$, we divide the graded layer into 8 uniform homogeneous sub-layers.

Table 1: Thermo elastic material properties of Ni and Al_2O_3,

Material Property	Material	
	Ni	Al_2O_3
Young's modulus, *E* (*GPa*)	199.5	393
Poisson's ratio, υ	0.3	0.25
Thermal expansion coefficient, α (10^{-6}K)	15.4	7.4
Thermal conductivity, k (*W*/*m*·*K*)	90.7	30.1

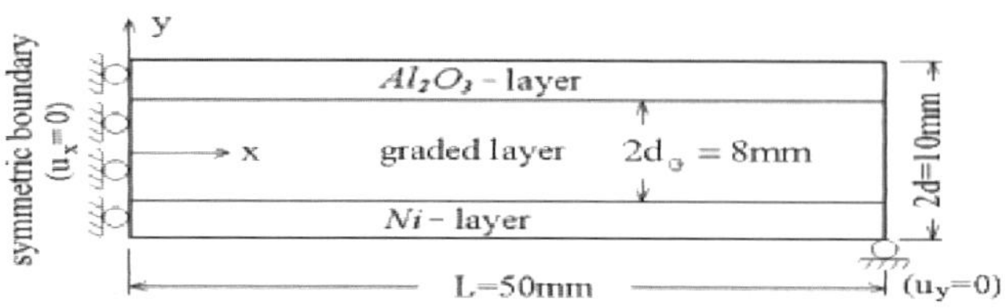

Figure 3: A simply supported plate-like FGM subject to uniform temperature change [20]

5. Residual Stress Minimization

5.1 Uniform Heating (or Cooling)

In this work, the design region is divided into 10 uniform homogeneous sub layers through initial volume fraction $V_m(i) = 0.5$. The total thickness of plate is 10 mm. The upper layer is pure ceramic and the lower layer is pure metal. The 8 middle layers are the FGM. Volume fraction in the FGM plate is being optimized towards the minimization of residual stresses (σ). Volume fraction optimization for minimizing thermal stress is assured as

Find: V_m (i), i=1,2,...,n (3)
Minimize: $f = \max(\sigma_i)$ (4)
Subjected to: $\max(\sigma_i) - \sigma_{per} \leq 0$ (5)
$0 \leq V_m(i) \leq 1,$

5.1.1 Validation of model

Considering modified rule of mixture to the estimation of the properties, the values of optimized metal volume fractions for the sub layers are presented in Table 2.

Table 2: Optimized Volume Fraction in all eight layers for Uniform heating (ΔT = 200°C), (using Modified rule of mixture)

Layers	Volume Fraction of Metal, V_m
V_1	0.8964
V_2	0.7974
V_3	0.7224
V_4	0.5889
V_5	0.4811
V_6	0.3692
V_7	0.2522
V_8	0.1298

The Table 2 shows the continuous volume fraction distributions, linearly interpolated through the centre points of sub layer volume fraction and the end points of Ni and Al_2O_3 – layer volume fractions. The optimized volume fraction, even as it varies almost linearly, presents steep gradient near Al_2O_3 – graded layer interface. It can also noted that the initial guess approaches 1 at Ni – graded layer interface and 0 at Al_2O_3 – graded layer interface. These results closely match with that of Cho [20].

5.1.2 Modified Rule of Mixture

The modified rule of mixture is considered for estimating the Young's modulus and other properties in different sub-layers.

Table 3: Optimized volume fraction in all eight layers for uniform heating (using Modified Rule of Mixture)

Layers	Volume Fraction of Metal, V_m for		
	$\Delta T = 200°C$	$\Delta T = 300°C$	$\Delta T = 400°C$
V_1	0.8964	0.8195	0.8957
V_2	0.7974	0.7232	0.8247
V_3	0.7224	0.7582	0.6949
V_4	0.5889	0.5247	0.5902
V_5	0.4811	0.4379	0.4820
V_6	0.3692	0.3976	0.3698
V_7	0.2522	0.2032	0.2529
V_8	0.1298	0.1007	0.1301

The results are presented in Table 3 and Figure 4. As seen from the Figure, the composition profile deviates from the linear. The profile also depends on the temperature difference to some extent.

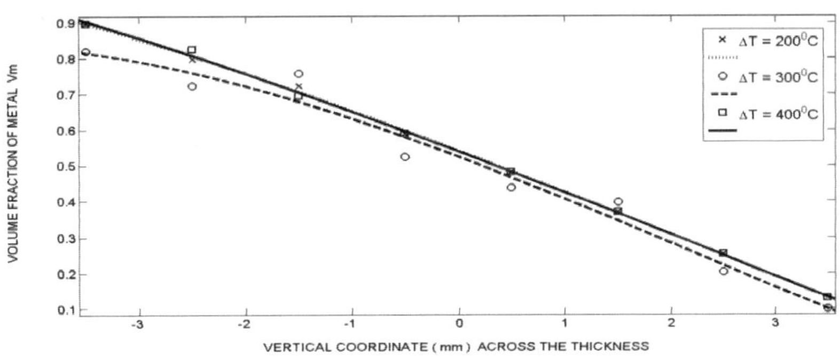

Figure 4: Distribution of Optimized volume fraction across the thickness for uniform heating (using Modified Rule of Mixture)

The residual stress distribution obtained for estimation approach considered in the present work. For modified rule of mixture, the residual stress shows fluctuations across the thickness but maximum value is not very high.

6. CONCLUSION

In present work, the optimization of volume fraction is carried out of thermal loading i.e. uniform heating. The residual stresses are considered as objective functions and estimation approach viz modified rule of mixture is explored. For modified rule of mixture, temperature change has significant effect on volume fraction distribution particularly that obtained with curvature minimization. From overall comparison, it can be said that modified rule of mixture estimation with residual minimization gives the best performance. In the above study a material considered is homogeneous having sufficient degree of thermal conductivity. However the analysis is presented in research article is not verified for heterogeneous material or for non thermal non conductive materials.

Refrences

[1] Kayikci R, Sava S (2014), "Fabrication and properties of functionally graded Al/AlB2 composites". *J Compos Mater,* **49(16)**:2029–2037. (https://doi.org/10.1177/0021998314541490)

[2] Mahamood RM, Akinlabi ET, Shukla M, Pityana S (2012), "Functionally graded material: an Overview". *In: Proceedings of the world congress on engineering*, **VIII**, London (http://hdl.handle.net/10204/6548)

[3] Birman V, Byrd LW. (2007), "Modeling and analysis of functionally graded materials and structures", *ApplMech Rev;* **60**:195–216 (DOI: 10.1115/1.2777164)

[4] Na, K.S., Kim, J.H (2009),"Volume fraction optimization of functionally graded composite panels for stress reduction and critical temperature", *Finite Elements in Analysis and Design*; **45 (11)**: 845-851. (https://doi.org/10.1016/j.finel.2009.06.023)

[5] Sadollah A, Bahreininejad A. (2011), "Optimum gradient material for a functionally graded dental implant using metaheuristic algorithms". *J Mech Behav Biomed Mater*; **4(7):**1384–95. (DOI- 10.1016/j.jmbbm.2011.05.009)

[6] Kou, X.Y., Parks, G.T., Tan, S.T. (2012), "Optimal design of functionally graded materials using a procedural model and a particle swarm optimization", *Computer-Aided Design*; **44 (4)**: 300-310. (https://doi.org/10.1016/j.cad.2011.10.007)

[7] Noh, Y.J., Kang, Y.J., Youn, S.J., Cho, J.R., Lim, O.K.. (2013), "Reliability-based design optimization of volume fraction distribution in functionally graded composites" *Computational Material Science*; **69**: 435-442. (https://doi.org/10.1016/j.commatsci.2012.12.003)

[8] Ashjari, M., Khoshravan, M.R. (2014), "Mass optimization of functionally graded plate for mechanical loading in the presence of deflection and stress constraints", *Composite Structures*; **110**: 118-132. (DOI: 10.1016/j.compstruct.2013.11.025)

[9] Taheri, A.H., Hassani, B., Moghaddam, N.Z.(2014), "Thermo-elastic optimization of material distribution of functionally graded structures by an isogeometrical approach", *International J. f solids and Structures*; **51**: 416-429. (https://doi.org/10.1016/j.ijsolstr.2013.10.014)

[10] Roque, C.M.C., Martins, P.A.L.S. (2015), "Differential evolution for optimization of

functionally graded beams", *Composite Structures*; **133**:1191-1197. (10.1016/j.compstruct.2015.08.041)

[11] Shi JX, Shimoda M. (2015), "Interface shape optimization of designing functionally graded sandwich structures". *Compos Struct*; **125**:88–95. (https://doi.org/10.1016/j.compstruct.2015.01.045)

[12] Mohammadiha O, Ghariblu H. (2016), "Crush behavior optimization of multi-tubes filled by functionally graded foam". *Thin-Walled Struct*; **98**:627–39. (https://doi.org/10.1016/j.tws.2015.10.025)

[13] Shabana, Y.M., Elsawaf, A., Khalaf, H., Khalil, Y. (2017), "Stresses minimization in functionally graded cylinders using particle swarm optimization technique", *Int. J. of Pressure Vessels and Piping,* **154**, 1-10. (https://doi.org/10.1016/j.ijpvp.2017.05.013)

[14] Lieu, Q.X., Lee, J. (2017), "Modeling and optimization of functionally graded plates under thermo-mechanical load using isogeometric analysis and adaptive hybrid evolutionary firefly algorithm", *Composite Structures*; **179**:89-106. (https://doi.org/10.1016/j.compstruct.2017.07.016)

[15] Nguyen TT, Lee J. (2017), "Optimal design of thin-walled functionally graded beams for buckling problems". *Compos Struct*; **179**: 459–67. (https://doi.org/10.1016/j.compstruct.2017.07.024)

[16] He MX, Sun JQ. (2018) "Multi-objective structural-acoustic optimization of beams made of functionally graded materials". *Compos Struct;* **185**:221–8 (https://doi.org/10.1016/j.compstruct.2017.11.004)

[17] Franco Correia VM, Aguilar Madeira JF, Araújo AL, Mota Soares CM. (2018), "Multiobjective optimization of ceramic-metal functionally graded plates using a higher order model".*Composite Structures*; **183**: 146-160. (https://doi.org/10.1016/j.compstruct.2017.02.013)

[18] Cheng L, Bai J, To AC. (2019), "Functionally graded lattice structure topology optimization for the design of additive manufactured components with stress constraints". *Comput Methods Appl Mech Eng*; **344**:334–59. (https://doi.org/10.1016/j.cma.2018.10.010)

[19] Shaw, L.L. (1998). "Thermal Residual Stresses in Plates and Coatings Composed of Multi Layered and Functionally Graded Materials", *Composites Part B*, **29**: 199–210. (https://doi.org/10.1016/S1359-8368 (97)00029-2)

[20] Cho JR, Ha DY (2002). "Volume fraction optimization for minimizing thermal stress in Ni-Al2O3functionally graded materials". *Mater Sci Eng A*; **334**:147–55. (https://doi.org/10.1016/S0921-5093 (01)01791-9)

IOP Conf. Series: Materials Science and Engineering 1017 (2021) 012025 doi:10.1088/1757-899X/1017/1/012025

Review: Technological Development of Heat Transfer Enhancement in Two Phase Flow by using Passive Method.

Gajanan Z. Jadhav[1], Dr. K K Gupta[1]

[1]SVKM'S, NMIMS, MPSTME, Shirpur, Maharashtra, India 425405.

Email: gajanan.jadhav@nmims.edu

Abstract. The Heating, Ventilation, Air conditioning & Refrigeration application always shows an opportunity to enhance the heat transfer rate. With these motives, researchers explored the use of passive methods in single phase flow which exhibits promising results of heat transfer enhancement. On the other hand, researchers investigated the effect of the passive method in Two phase flow, which also shows promising results in heat transfer enhancement. Two phase flow includes Boiling, Evaporation & condensation phenomenon. This paper focused on the review of the experimental studies conducted on the use of heat transfer enhancement in the condensation process mostly. Out of this passive method, Micro fin tube shows better heat transfer enhancement and less pressure drop compared to other methods. It has been observed that the use of porous media also showing promising effects in heat transfer enhancement and it's fetching the attention of researchers. Therefore, this review work also reviewed the numerical studies on the use of porous media in two phase flow. It is needed to explore the different structure and optimum design of porous media, which have high heat transfer enhancement with low pressure drop penalty. Also, the use of low Global Warming Potential refrigerant effects needs to be explored, which could be a better alternative for existing refrigerants.

1. **Introduction** Two phase flow is a combination of Liquid-Gas, Liquid-Liquid, Liquid-Solid & Gas-Solid flows. Here we focused on Gas-Liquid Flow, again Gas-Liquid flow combination involves mainly Boiling, Condensation & Evaporation phenomenon. This review mainly features the studies and investigations performed on condensation process. Condensation refers to the process where saturated vapour comes in contact with the wall which is maintained at below saturation temperature. Vapour releases its latent heat and gets converted into the liquid. Since decade, researchers exploring the use of passive techniques in condensation, Evaporation and Boiling. Use of this techniques reveals satisfactory results in heat transfer enhancement, which encourages to researchers, to explore the effect of various parameters on HTC (Heat Transfer Coefficient) and pressure drop. With this aim, this review work focused on use of passive methods in two phase flow to improve the heat transfer rate. In the passive method researchers tried geometrical changes of tube and its parameters to improve heat transfer. Some of the active and passive techniques are shown in fig. 1. But the use of active method was not explored as much as the passive method explored therefore, most of researchers concentrates on the use of passive techniques rather than use of active techniques. This reviews considered the use of passive techniques. Researchers performs the experimental and numerical investigation to find the effect of these passive techniques. Fig. 2 Shows the main passive methods with the parameters that causes heat transfer increased in two phase flow.

On the other hand, most of researchers used numerical studies for single phase flow, but only few of researchers focused on use of these numerical studies in two phase flow [1]. With this objective, we reviewed, numerical studies on the use of metal foam in two phase flow in last section of paper.

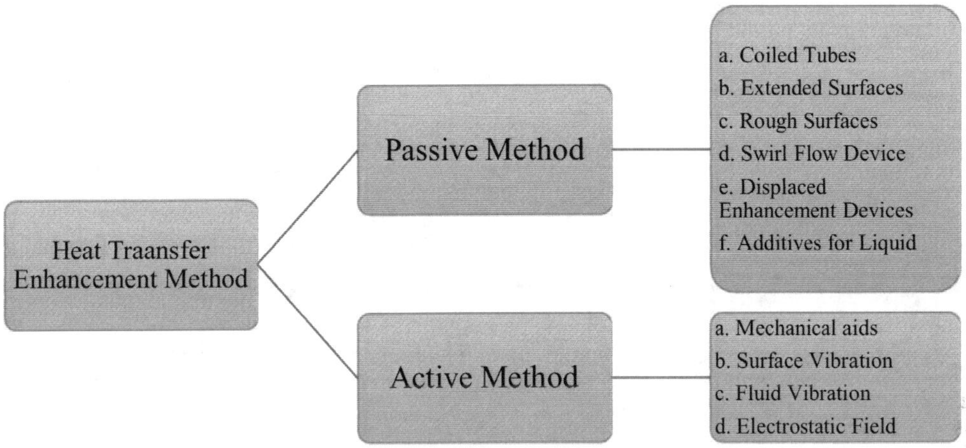

Figure 1. Heat Transfer Enhancement Method. [35].

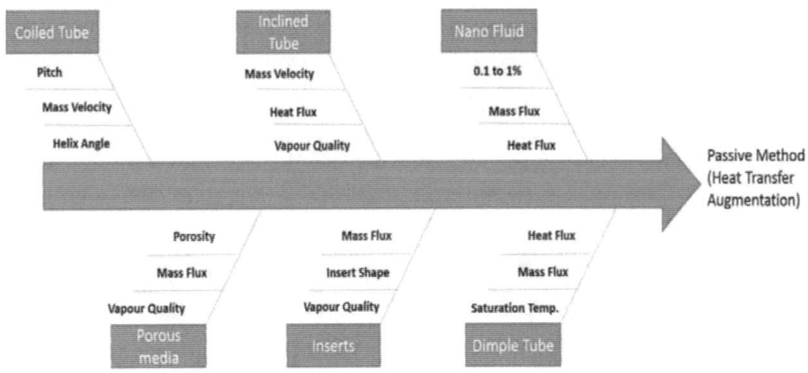

Figure 2. Passive Method with Parameters studied.

2. Effect of Inclination

An experimental investigation was performed Kundu et al. [2] on refrigerants in a smooth copper tube by varying five different angles between 0 and 90^0. Heat transfer coefficient for the evaporation process were compared with the variables mass velocity, heat flux, and vapour quality. Lips et al. [3] also performs experimentation on R134a condensation to find the inclination effect "from vertical downward to vertical upward" [3].

Experimental results found, the maximum heat transfer coefficient for downward flow with an optimum inclination angle in an inclination dependent zone. Khoeini et al. [4] Another study, which also varies inclination from -90 to +90°, investigates, the effect of mass flux and vapour qualities with corrugated tube on HTC and pressure drop. Results shows that, highest heat transfer coefficient observed at +30°, which were more than 1.27 to 1.41 times of tube with -90° with low vapour quality.

3. Effect of Dimple Tube

Aroonrat & Wongwises [5] This study experimentally investigates the effect of heat flux, saturation temperature and mass flux on the heat transfer coefficient & pressure drop. Comparison of smooth & enhanced tube were obtained, it had been observed that HTC and pressure drop increased over smooth tube. Subsequently, [6] in this experimental investigation, variation of helical pitch as well as dimple pitch with the effect of heat flux, mass flux & saturation temperature on the heat transfer coefficient and pressure drop were reported. As the helical pitch and dimple pitch decreases, it exhibits increase in HTC & pressure drop. At 5.08mm of helical pitch & 3.24 mm of dimple pitch maximum 88% & 630% higher HTC & pressure drop observed compared to smooth tube.

Solanki & Kumar [7] In this study, researchers compared the performance of straight smooth tube, helical coiled smooth tube, & helical coiled dimple tube. Effect of mass flux, vapour quality & saturation temperature on the HTC and pressure drop. Researchers also reported that smooth helical coiled tube shows high HTC than straight smooth tube. Aroonrat & Wongwises [8] Authors investigates the effect of dimple depth(0.5,0.75,1.0mm) on the HTC and pressure drop. It had been observed that, for all dimple depth, HTC & pressure drop increases significantly compared to smooth tube, highest HTC & pressure drop of 83% & 892% reported compared to smooth tube. Author also proposed correlation for Nusselt No. and friction factor.

4. Effect of Use of Micro fin

Sapali & Patil [9] Researchers conducted experimental investigation on HFC 134a & R-404A with micro fin tube, in this experimentation, effect of mass flux and condensing temp. drop were examined on pressure drop. Result shows that, the Pressure drop increased with mass flux, & condensing temperature for both smooth & micro fin tube. Sapali & Patil [10] also conducted experimental investigation work on micro fin tube with refrigerant HFC 134a, R404A. In this study researchers varies condensing temperature from 35°C to 60°C and found its effect on HTC. Results reveals that HTC improves with improving mass flux but it decreases by increasing condensing temperature. The average HTC with micro fin tube was 1.5-2.5 times greater than smooth tube for HFC 134a, whereas 1.3-2 times higher with micro fin tube of R404A compared to smooth tube.

Yang et al. [11] Examined the HTC & pressure drop through vertical upward flow in smooth and micro fin tube for the refrigerant R410A. Heat transfer enhancement factor investigated, which were reported in the range of 1.3-2.05 at saturation temp. of 40°C & 1.4-2.2 at 48°C. Researchers also developed revised correlation for vertical upward flow. Li et al. [12] In this experimental study, Authors examined four micro fin tube and compared its performance with smooth tube. Authors investigates the effect of mass velocity, Reynolds No. & condensation temp. on HTC & pressure drop. Micro fin tube exhibits better performance over smooth tube was the conclusive remark of study.

Solanki & Kumar [13] also examined the effect of mass flux, condensing temperature, Vapour quality on HTC & pressure drop with helical coiled micro fin tube. Its performance was compared with straight smooth tube.

It is reported that micro fin tube had higher HTC of 160-255% than smooth tube and 69-155% higher pressure drop than smooth tube. Ji et al. [14] In this study, condensation HTC outside plain tube & enhanced tube were investigated with the Refrigerants R134a, R1234ze (E) & R290. It had been reported that, performance of R134a was better than other two refrigerants over smooth and enhanced tube made from Titanium material. Later on, [15] HTC of R134a, R1234ze(E) & R1233zd (E) were experimentally investigated on two different enhanced tube. Here, fin on outside as well as inside of tube were used for experimentation. In this experimental work, Condensation HTC of R134a found higher than R1234zd (E).

5. Effect of Use of Porous Media

Shi et al. [16] In this experimental investigation, effect of Vapour mass flowrate, cooling water temp., inlet vapor pressure with different foam size enhanced tube on HTC and pressure drop were reported. Tube with 10PPI metal foam, exhibits high performance compared to micro fin tube. Author also obtained correlation for HTC & pressure drop for foam tube. Subsequently [17], investigates the effect of metal foam surface wettability on HTC & pressure drop with annularly partial filled hydrophobic metal foam inside tube. It is reported that, 30.158KW/m^{2}K of HTC were obtained with mass flux of 65.8Kg/m^{2}S and cooling water temperature of 55°C.

Nasr [18] numerically investigates the effect of porosity & thickness of porous layer on heat transfer and mass transfer for liquid film condensation. It is reported that decrease in porosity & porous layer thickness improve heat & mass transfer performances at liquid –gas interface. Wang et al. [19] In this study, researchers examined the selective laser melting fabricated enhanced tube, and compared its performance with plain tube. Effect of fin height, refrigerant flow direction and mass flux on HTC & pressure drop were investigated. HTC of foam tube were higher than that of plain tube. Preston et al. [20] In this experimental work, investigation of enhanced condensation heat transfers by wick condensation film through porous metal were examined. This study reveals that, the thermal resistance of wick condensation film is lower than the filmwise condensation which leads to higher magnitude of heat transfer enhancement.

6. Effect of Use Inserts

Moghaddam & Shafaee [21] In this experimental investigation, effect of wire coil with its diameter and three different pitch with mass flux, Vapour quality on HTC were examined. Highest HTC of 107% over plain tube were reported for coil index of 0.0185. Subsequently, Moghaddam et al. [22] researchers investigate the effect of twisted tape inserts with twist ratio of 4,10,15 with mass flux & vapor quality on HTC & pressure drop. It found that, performance factor varies from 0.39 to 1.05, which depend upon operating conditions & inserts type.

Hejazi et al. [23] In this experimental study, researchers examined the condensation HTC & pressure drop for the tube with insert tape of 6,9,12 &15 twist ratio. It is reported that HTC & pressure drop increased by 40% & 240% over the plain tube. Sajadi et al. [24] Also, examined the twisted tape with twist ratio of 6,9,12 on R1234y, which is alternative to R134a with low GWP. Effect of mass flux, vapor quality of HTC & pressure drop was observed. An increment of 42% & 235% of HTC & pressure drop respectively over plain tube had been observed.

Table 1 Experimental studies on two phase flow by using passive method.

Author	**Enhanced Tube**	**Working Fluid**	**Test Section**	**Mass flux/Heat Flux/Vapor Quality**	$T_{sat.}$
[5]	Dimple Tube	R134a	Tube in Tube ϕ=8.1mm, l=1.5m	300,400,500 Kg/m^{2}s	40,45,50 ^{O}C
[6]	Helical Dimple Tube	R134a	Tube in Tube ϕ=8.1mm, l=1.5m	300-500 Kg/m^{2}s	40-50 ^{O}C
[7]	Helical coiled Dimple tube	R134a	Tube in Shell type	75,115,156,191 Kg/m^{2}s	35 & 45 ^{O}C
[8]	Dimple tube with diff. depth	R134a	Tube in Tube ϕ=8.1mm, l=1.5m	300-500 Kg/m^{2}s	40-50 ^{O}C
[9]	Microfin tube	R134a	Tube in Shell ϕ=8.96mm	90-800 Kg/m^{2}s	35-60 ^{O}C
[10]	Microfin Tube	HFC-134a, R404A	ϕ=8.96mm	90-800 Kg/m^{2}s	35-60 ^{O}C
[11]	Microfin	R410A	Tube ϕ=8.76mm	80-345 Kg/m^{2}s	40,45,48 ^{O}C
[12]	Microfin four tube	R134a	Tube in Tube	400-1100 Kg/m^{2}s	35 – 45 ^{O}C
[13]	Microfin	R134a	ϕ=8.58mm	75,115,156,191 Kg/m^{2}s 1.5-17.25 Kw/m^2 0.1-0.8	35-45 ^{O}C
[15]	Microfin	R134a, R1234ze (E), R1233zd (E)	Tube ϕ=17.14 & 17.12mm	20-90 Kw/m^2	36 ^{O}C
[16]	Foam Tube (10,15,20PPI)	Water	Tube in Tube type ϕ=22mm	20-100 Kg/hr 1-3 m^3/h	
[17]	Treated Foam Tube (10,15,20PPI)	Water	Tube in Tube type ϕ=22mm	20-100 Kg/hr 1-3 m^3/h	
[19]	Foam Tube	R134a	Tube ϕ=8.7mm	50-150 Kg/m^{2}S 0.9,0.3	
[25]	Tube (5PPI)	R1234yf & R1234ze (E)		50-200 Kg/m^{2}S 50,75,100 Kw/m^2 0.2 to 0.95	30 ^{O}C
[26]	Foam Thickness 1.6mm & 2.5mm, 40,80,130PPI	R134a		10-60 Kw/m^2	35 & 45 ^{O}C
[27]	30,60,90 PPI	DI Water		30-200 Kg/m^{2}S	100 ^{O}C

IOP Conf. Series: Materials Science and Engineering 1017 (2021) 012025 doi:10.1088/1757-899X/1017/1/012025

7. Numerical Studies

In previous section, we reviewed, mostly experimental studies on the use of passive method in two phase flow, while in this section, specifically, numerical studies on the use of porous media in two phase flow are reviewed. Modelling of Two phase flow in porous media is complicated phenomenon. Here the table 2 represents the numerical studies performed with its computation domain.

Table 2 Numerical studies on two phase flow in porous media.

Author	Computation Domain	Remark
[28]	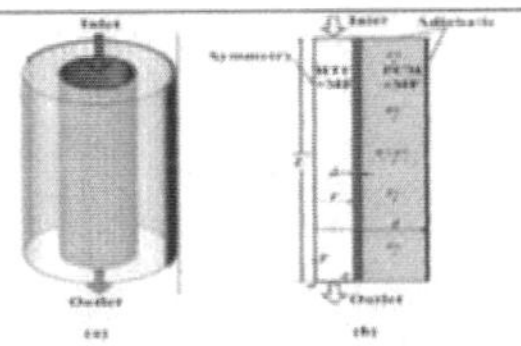	A copper foam of 0.94 porosity were used in thermal storage system. Effect of metal foam in heat transfer fluid and phase change material investigated. Use of metal foam shows enhancement in heat transfer.
[29]	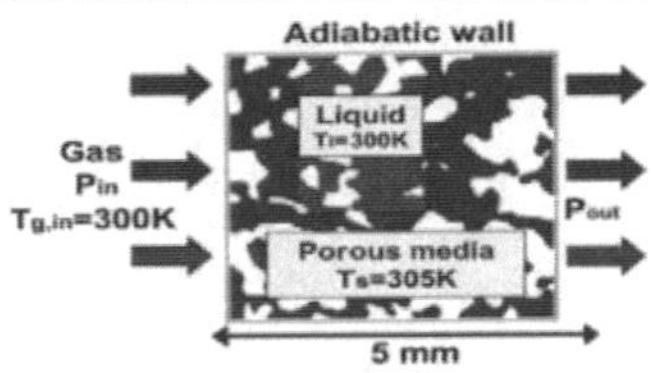	The 3D numerical model of porous media developed by using Micro-CT. A Lattice-Boltzmann method at pore scale were used to obtained velocity and temperature field with the help of contact angle.
[30]	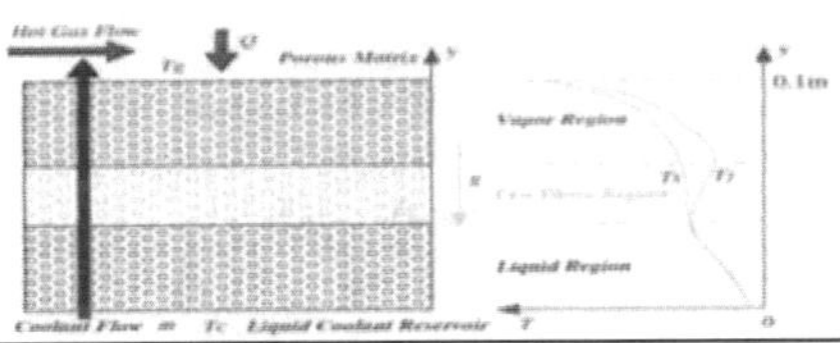	A Modified Separated Flow Model developed. Effect of capilarity, foam porosity and other paramters has great effect on heat &fluid flow.
[31]	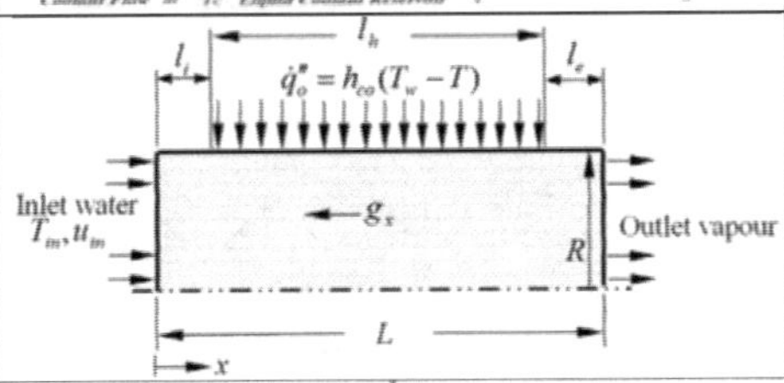	An efficient smoothing algorithm developed for effective diffusivity. Finite volume method adopted for discretization.
[32]	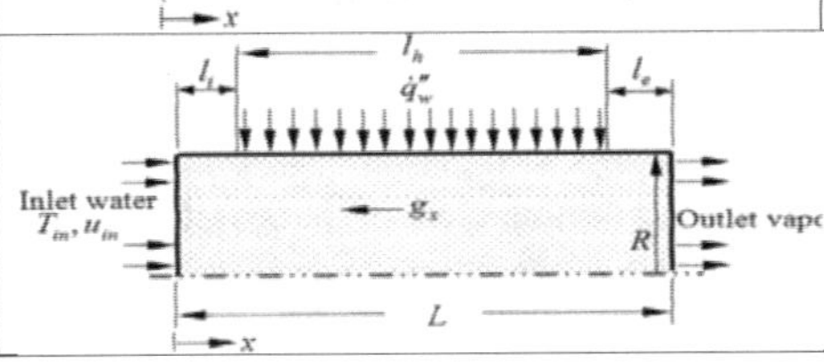	One dimensional problem was developed for phase change heat transfer in porous media with local thermal equilibrium and Local non thermal equilibrium conditions.

IOP Conf. Series: Materials Science and Engineering 1017 (2021) 012025 doi:10.1088/1757-899X/1017/1/012025

[33]	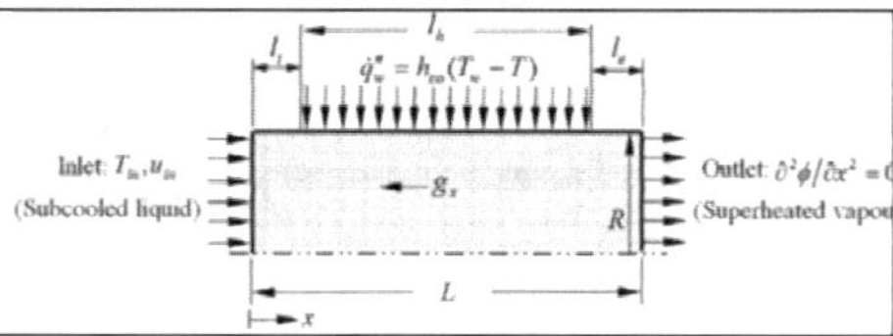	Modified enthalpy formation method were adopted
[34]	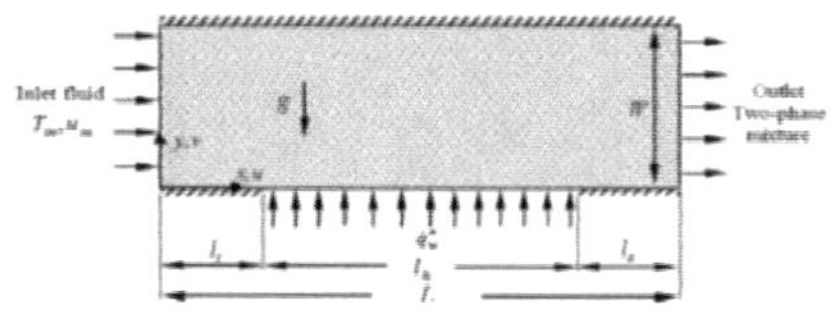	A modified enthalpy formation method was used to analyse the variable porosity, thermal dispersion and local thermal non equilibrium.

8. Conclusion

1.Use of passive techniques in two phase flow are increasing at higher rate, it exhibits beneficial results in term of heat transfer enhancement with a penalty of pressure drop.
2.Out of the various passive method of heat transfer enhancement, comparatively micro fin tube was investigated extensively. It's clearly indicates other techniques need to be explored.
3.Experimental studies and numerical studies on the use of porous media in two phase flow reveals good heat transfer enhancement, a more study is required to find optimum parameters of metal foam and effect of use tailored porous media.
4.Refrigerants R1234yf, R1234ze (E) with low GWP could be a good alternative for R134a.

9. References

[1] Baloyo, J. M. (2017). Open-cell porous metals for thermal management applications: fluid flow and heat transfer. Materials Science and Technology, 33(3), 265-276.

[2] Kundu, A., Kumar, R., & Gupta, A. (2014). Evaporative heat transfer of R134a and R407C inside a smooth tube with different inclinations. International Journal of Heat and Mass Transfer, 76, 23-533.

[3] Lips, S., & Meyer, J. P. (2012). Experimental study of convective condensation in an inclined smooth tube. Part I: Inclination effect on flow pattern and heat transfer coefficient. International Journal of Heat and Mass Transfer, 55(1-3), 395-404.

[4] Khoeini, D., Akhavan-Behabadi, M. A., & Saboonchi, A. (2012). Experimental study of condensation heat transfer of R-134a flow in corrugated tubes with different inclinations. International Communications in Heat and Mass Transfer, 39(1), 138-143.

[5] Aroonrat, K., & Wongwises, S. (2017). Experimental study on two-phase condensation heat transfer and pressure drop of R-134a flowing in a dimpled tube. International Journal of Heat and Mass Transfer, 106, 437-448.

[6] Aroonrat, K., & Wongwises, S. (2018). Condensation heat transfer and pressure drop characteristics of R-134a flowing through dimpled tubes with different helical and dimpled pitches. International Journal of Heat and Mass Transfer, 121, 620-631.

[7] Solanki, A. K., & Kumar, R. (2018). Condensation of R-134a inside dimpled helically coiled tube-in-shell type heat exchanger. Applied Thermal Engineering, 129, 535-548.

[8] Aroonrat, K., & Wongwises, S. (2019). Experimental investigation of condensation heat transfer and pressure drop of R-134a flowing inside dimpled tubes with different dimpled depths. International Journal of Heat and Mass Transfer, 128, 783-793.

[9] Sapali, S. N., & Patil, P. A. (2009). Two-phase condensation heat transfer coefficients and pressure drops of R404a for different condensing temperatures in a smooth and micro-fin tube. Int J Eng Sci Technol, 1(2), 43-58.

[10] Sapali, S. N., & Patil, P. A. (2010). Heat transfer during condensation of HFC-134a and R-404A inside of a horizontal smooth and micro-fin tube. Experimental Thermal and Fluid Science, 34(8), 1133-1141.
[11] Yang, Y., Jia, L., & Peng, Q. (2017). Investigation on condensation heat transfer and pressure drop of R410A during upward flow in vertical smooth and micro-fin tube. International Journal of Heat and Mass Transfer, 108, 2293-2302.
[12] Li, Q., Tao, L., Li, L., Hu, Y., & Wu, S. (2017). Experimental Investigation of the Condensation Heat Transfer Coefficient of R134a inside Horizontal Smooth and Micro-Fin Tubes. Energies, 10(9), 1280.
[13] Solanki, A. K., & Kumar, R. (2018). Condensation of R-134a inside micro-fin helical coiled tube-in-shell type heat exchanger. Experimental Thermal and Fluid Science, 93, 344-355.
[14] Ji, W. T., Chong, G. H., Zhao, C. Y., Zhang, H., & Tao, W. Q. (2018). Condensation heat transfer of R134a, R1234ze (E) and R290 on horizontal plain and enhanced titanium tubes. International Journal of Refrigeration, 93, 259-268.
[15] Ji, W. T., Lu, X. D., Yu, Q. N., Zhao, C. Y., Zhang, H., & Tao, W. Q. (2020). Film-wise condensation of R-134a, R-1234ze (E) and R-1233zd (E) outside the finned tubes with different fin thickness. International Journal of Heat and Mass Transfer, 146, 118829.
[16] Shi, J., Zheng, G., & Chen, Z. (2018). Experimental investigation on flow condensation in horizontal tubes filled with annular metal foam. International Journal of Heat and Mass Transfer, 116, 920-930.
[17] Shi, J., Zheng, G., Chen, Z., & Dang, C. (2019). Experimental study of flow condensation heat transfer in tubes partially filled with hydrophobic annular metal foam. International Journal of Heat and Mass Transfer, 136, 1265-1272.
[18] Nasr, A. (2018). Heat and mass transfer for liquid film condensation along a vertical channel covered with a thin porous layer. International Journal of Thermal Sciences, 124, 288-299.
[19] Wang, X. W., Ho, J. Y., Leong, K. C., & Wong, T. N. (2018). Condensation heat transfer and pressure drop characteristics of R-134a in horizontal smooth tubes and enhanced tubes fabricated by selective laser melting. International Journal of Heat and Mass Transfer, 126, 949-962.
[20] Preston, D. J., Wilke, K. L., Lu, Z., Cruz, S. S., Zhao, Y., Becerra, L. L., & Wang, E. N. (2018). Gravitationally driven wicking for enhanced condensation heat transfer. Langmuir, 34(15), 4658-4664.
[21] Moghaddam, H. A., Sarmadian, A., & Shafaee, M. (2019). An experimental study on condensation heat transfer characteristics of R-600a in tubes with coiled wire inserts. Applied Thermal Engineering, 159, 113889.
[22] Moghaddam, H. A., Sarmadian, A., Asnaashari, A., Joushani, H. A. N., Islam, M. S., Saha, S. C., & Shafaee, M. (2020). Condensation heat transfer and pressure drop characteristics of Isobutane in horizontal channels with twisted tape inserts. International Journal of Refrigeration, 118, 31-40.
[23] Hejazi, V., Akhavan-Behabadi, M. A., & Afshari, A. (2010). Experimental investigation of twisted tape inserts performance on condensation heat transfer enhancement and pressure drop. International Communications in Heat and Mass Transfer, 37(9), 1376-1387.
[24] Sajadi, B., Soleimani, M., Akhavan-Behabadi, M. A., & Hadadi, E. (2020). The effect of twisted tape inserts on heat transfer and pressure drop of R1234yf condensation flow: An experimental study. International Journal of Heat and Mass Transfer, 146, 118890.
[25] Diani, A., Mancin, S., Doretti, L., & Rossetto, L. (2015). Low-GWP refrigerants flow boiling heat transfer in a 5 PPI copper foam. International Journal of Multiphase Flow, 76, 111-121.
[26] Ji, W. T., Li, Z. Y., Qu, Z. G., Guo, J. F., Zhang, D. C., He, Y. L., & Tao, W. Q. (2015). Film condensing heat transfer of R134a on single horizontal tube coated with open cell copper foam. Applied Thermal Engineering, 76, 335-343.
[27] Ji, X., & Xu, J. (2012). Experimental study on the two-phase pressure drop in copper foams. Heat and Mass Transfer, 48(1), 153-164.

[28] Yu, J., Yang, Y., Yang, X., Kong, Q., Yanhua, L., & Yan, J. (2018). Effect of porous media on the heat transfer enhancement for a thermal energy storage unit. Energy Procedia, 152, 984-989.
[29] Liu, Z., & Wu, H. (2016). Numerical modeling of liquid–gas two-phase flow and heat transfer in reconstructed porous media at pore scale. International Journal of Hydrogen Energy, 41(28), 12285-12292.
[30] Xin, C., Rao, Z., You, X., Song, Z., & Han, D. (2014). Numerical investigation of vapor–liquid heat and mass transfer in porous media. Energy conversion and management, 78, 1-7.
[31] Alomar, O. R., Mendes, M. A., Trimis, D., & Ray, S. (2014). Numerical simulation of complete liquid–vapour phase change process inside porous media using smoothing of diffusion coefficient. International journal of thermal sciences, 86, 408-420.
[32] Alomar, O. R., Mendes, M. A., Trimis, D., & Ray, S. (2015). Simulation of complete liquid–vapour phase change process inside porous evaporator using local thermal non-equilibrium model. International Journal of Thermal Sciences, 94, 228-241.
[33] Ray, S., & Alomar, O. R. (2016). Simulation of liquid–vapour phase change process inside porous media using modified enthalpy formulation. International Journal of Thermal Sciences, 105, 123-136.
[34] Alomar, O. R. (2019). Analysis of variable porosity, thermal dispersion, and local thermal non-equilibrium on two-phase flow inside porous media. Applied Thermal Engineering, 154, 263-283.
[35] Bejan, A., & Kraus, A. D. (Eds.). (2003). Heat transfer handbook (Vol. 1). John Wiley & Sons.

Study & Analysis of Tire Wear Characteristics of TATA LPO 1618.

Gaurav Gohane[1], Anmol Indurkar[1], Palash Chimurkar[1], Sandip Patle[1] Suraj Kumar Nagina[1], Swapnil Tirpude[1] and Ujwal Bawane[1]

[1] Mechanical Engineering Department, JD College of Engineering and Management, Nagpur, India.

Email: gmgjdcoe@gmail.com

Abstract. The tires are a key sub-system of vehicles that have a big responsibility in comfort, fuel consumption and traffic safety. As we know that current tires are just passive rubber elements which do not contribute actively to improve the driving experience or the vehicle safety. During the driving of vehicle, we get few and unsatisfied information from tires. Therefore, this is the main reason to develop an intelligent tire, which could provide useful information to other systems and become an active safety system. In the Tata (LPO 1618) Vehicle an experimental tire strain-based system is used to measure tire tread deformation by means of various aids. Tests under different working conditions such as vertical load or slip angle considering a certain camber angle have been carried out using an indoor tire test (Slightly). By this project and research work, we are trying to introduce an automated tire monitoring and inflation system, which can ensure adequate tire pressure to better accommodate handling requirements. In this paper the numerical method for introducing the prediction tire where account has been introduced to reduce break down time.

1. Introduction

For our Live Project, we have chosen An Automobile Sector, for the same purpose we had reported certain Bus Maintenance and Repairing Centre at which maintenance of Shiv-Shahi (Public Transport Vehicle under MSRTC) Bus is carried out. In the Maintenance Centre there are buses of Ashok Leyland & Tata. But there are majorities of Tata Buses of Tata LPO 1618, which is commercial vehicle used for public transport by MSRTC (Maharashtra State Road Transport Corporation). Tata LPO 1618 comes fitted with 5883 cc Cummins 6BT 5.9-180-30 – BS4 Water cooled, Turbocharged Diesel Engine. A max power of 131 Kw at 2500 rpm and max torque of 700 Nm at 1400 rpm gives enough strength to the bus to carry load full to its capacity. Like other Tata trucks in this range, thus bus also has a 400 Liters of fuel tank capacity. The tires for this vehicle are available in two options - 10.00 x 20 -16 PR Diagonal Ply - 10R20-16PR and 10.00 x 20 -16 PR Diagonal Ply - 11R22.5-16PR. This vehicle has a GVW of 16200 kg and a max speed of 110 Km/h. It is quite a dependable bus for long distance commutation with those numbers.

While Consulting and Observing the vehicle and its performance experienced by the drivers and people working in the center, we came to know various problems arising in the TATA LPO 1618 Vehicle such as Chassis Issue, Front Axle Issue, Tire Wear Issue, Abnormal Body Length, Impact of Retarder Weight, Misalignment of Kingpin due to sudden high Impact of Shock offered by wheel, Increment offered in temperature of Tire, Tire tread due to surface heating of road due to friction. Due to over sizing of front body, tires wear increases.

While consulting with the employees working in the center, we came to know that their huge amount of investment is spent on the tire and its maintenance. So, to reduce their Investment and tire wear we had focused light on this major arising issue in the maintenance center.

Tire wear is a complex and important part of each and every vehicle that is affected by tire materials, structures, vehicle dynamics, and road conditions. In this paper we basically focused on the tire wear of vehicle, because tire is very important part for driving any vehicle and everyone wants that tire of their vehicle should execute less damage and in result it will become more durable. So, basically, we had observed five major and main types of wearing of tire and they are as follow- Camber wear (one-sided wear), Feathered Directional Wear, Cupped Wear (Cupping Wear), Centre wear, Shoulder wear etc. The paper discusses various theoretical models for tire wear in various commercial and non-commercial vehicles and factors affecting the tire wear. A brief review of tire wear characteristics is shown in this paper. Theoretical models for analysis of tire wears are helpful in understanding numerical results with physical mechanisms which are responsible for the generation of tire wear and their behaviour at different values of tire parameters

2. Literature review

2.1. Measurement and analysis using a tire sensing approach

When we drive the vehicle, various forces and moments are applied on the tire which leads to rolling deformation of tire by application of braking, cornering and accelerating. Due to this rolling resistance is decreased which tends to increase in fuel consumption and carbon emission. Tire tread plays the main role for improving the rolling resistance of tire. Therefore, we have to focus on measuring this. To measure this, there is provision for optical laser sensors which measure the deformation such as inner profile, loaded radius, Sidewall and tread deformations of heavy vehicle. It is a 2D laser profilometer with an ultra-high-speed sampling rate. It consists of Two triangular laser sensors for measuring the rolling deformation of tire: Laser sensor 1 (Keyence LJ-V7300) and Laser sensor 2 (Keyence LK-H150). In this, sensor module is embedded into the sensing house and mounted on a steel rim by means of tread attachment. To measure the angular position of 2D laser profilometer, an optical encoder (2048 counts/revolution) with in a slip ring nut (Michigan Scientific SR 20AW/T1024) is used. It is used for new and remolded tire too. Laser1 provide the direct measurement of inner profile of tire rolling deformation due to which unreasonably many high and low values is occurred. To eliminate this, filter is used for showing the clear peaks, when tire is being actually deformed. The actual arrangement is somewhat like that the sensor equipped tire is used in the truck and data acquisition system I.e. hardware is placed on the fuel tanker of the vehicle and the software I.e. the display system is on the driving board. The measurement signals which we are getting from the Laser Sensor 2 include both periodic errors and random noise. The main reason for the periodic errors is mainly the eccentric alignment between the rotating axes of the tire and slip ring. To reduce such type of error, a pattern recognition procedure is applied to minimize this type of error. Now for the random noise, the main reason behind this error is road roughness (texture less than 1 mm) which can be smoothed by a Butterworth low-pass fitter. Fully integrated miniature module:

- Size: 4.9x2.5x1.56 mm.
- Emitter: 940 nm invisible laser (Class1).
- SPAD (single photon avalanche diode) receiving array with integrated lens.
- Low-power microcontroller running advanced digital firmware.

IOP Conf. Series: Materials Science and Engineering **1017**(2021) 012026 doi:10.1088/1757-899X/1017/1/012026

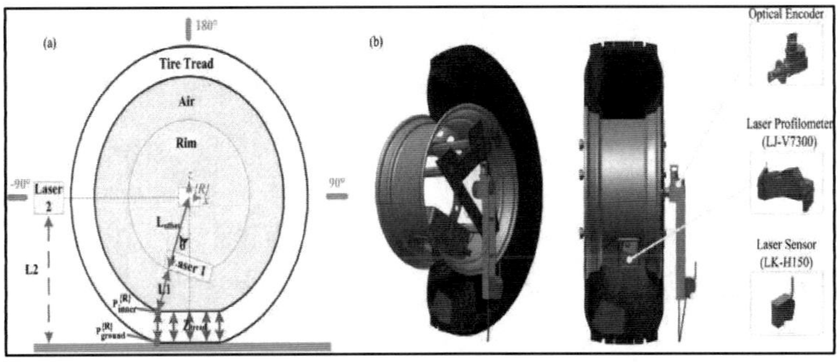

Figure 1. (a) Methodology for the tire sensor system; (b) prototype for the tire sensor system

There is also presence of Pin-to-pin compatible with the VL53L0X Flight Sense™ ranging sensor Fast and accurate long distance ranging up to 400cm distance measurement up to 50Hz ranging frequency Typical full field-of-view (FoV): 27° Programmable region-of-interest (ROI) size on the receiving array, allowing the sensor FoV to be reduced Programmable ROI position on the receiving array, providing multimode operation control from the host.

2.2. Numerical optimization of wear performance – Utilizing a metal model based friction law

In the present contribution, Aleksandra Serafinska, Nermeen Hassoun and Michael Kaliske had developed a finite element wear analysis methodology, which aspires to fulfill several requirements on a realistic modeling of this phenomenon. Pronounced and uneven wear lead to significant durability and service life reduction of tire and its surface structures. Thus, the development of methods for the enhancement of wear performance is of prime importance. Since the reduction of wear is a significant goal also for the pneumatic tire design, several tire optimization approaches can be found in the literature.

To reduce the future tire wear and uneven characteristic tire wear we can use the predicted technique of search, analyze and run technique in which we have to pre-analyses the path on which we have to drive the vehicle. By doing this, we are able to predict the tire wear based on the experiment results that we will carry out in further experiment or we can choose alternate way of path to drive the vehicle.

In this brief and as per our review, a finite element wear analysis is developed in the research paper, which aspires to fulfill several requirements on a realistic modeling of this phenomenon. First, the ability to perform an iterative adaption of the geometrical configuration due to material removal, based on a wear rate calculation within a constitutive wear equation is given. Second, the friction-wear interaction is considered within the FE (Finite Element) wear analysis, since wear and friction phenomena exhibit a strong correlation. Third, the influence of local frictional phenomena on wear is taken into account in friction-wear interaction since local stress concentrations are identified as the origin of wear.

The initial pioneering work in this field is done by "Reye", who discovered that the depth of the worn material is proportional to the tangential stress and therefore, to the contact pressure. Furthermore, he concluded, that the volume of the worn material V is linearly dependent on the work W done by the tangential force. The above theory will be going to consider in further analysis and experiment that will be going to carry out.

2.3. Abrasive wear amount estimates for 3D pattern tire utilizing frictional dynamic rolling analysis.

In this paper the numerical method for introducing the prediction tire where account has been introduced in this paper to reduce break down time. The frictional energy dissipation through the tire footprint containing with the abrasive ground was computed by fictional dynamic simulation off the 3D pattern tire model. As a result this paper introduced the fictional energy rate was correlated with the thread wear read by a power function wear rate model that was derived on an experimental basis as illustrated through numerical experiments threadbare amount can be estimated and also this numerical method can be useful to determine the rubber compound the tread pattern and the material and structure composition in the design of the tire model which satisfy the preset rear performance.

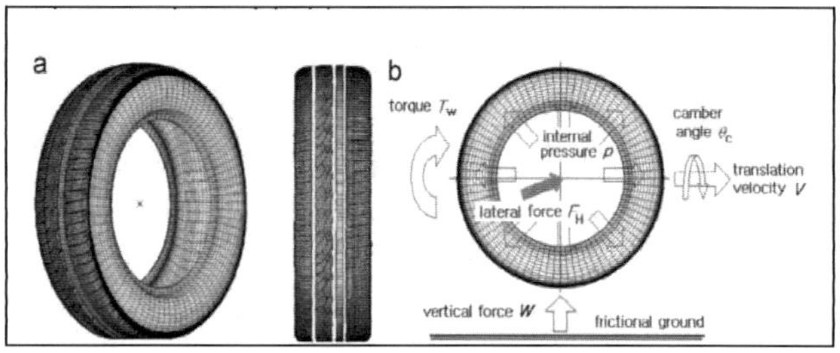

Figure 2. A simulation tire model: (a) FEM mesh and (b) specification of loading and driving condition.

2.4. Tire wears particle hotspots: A review of influencing factor.

From this research paper we understood that, Author found the major influencing parameter on the amount of tire wear. First is the rolling resistance offered by road it is mainly depends upon the surface of road with varying density, it is found that low density roads cause higher amount of wear rate and it also occurs the highly rolling resistance which leads to tire wear and the other is due to the uneven and occasional seasonal change. It is found that wear rates in summer increases by 1 to 4 % per degree Celsius which reduces the lifespan of tire by 9% for commercial vehicles other parameter also causes the tire wear of the commercial vehicle like tire pressure, tire geometry, tire age and the balancing of wheels-Imbalance causes wear.

b a

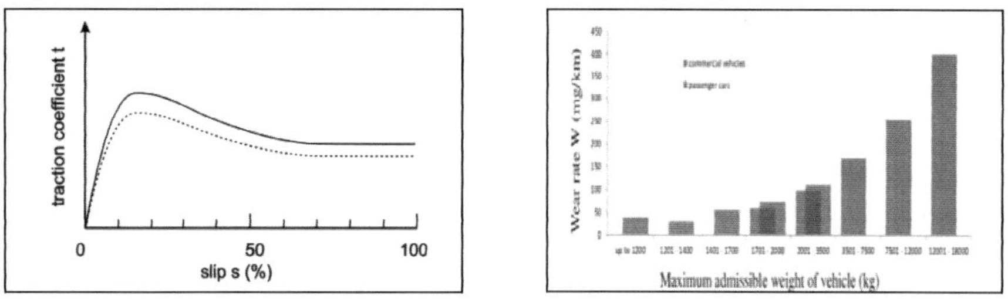

Figure 3. Graphical representation of (a) Traction coefficient (b) mean values for wear rates of vehicles as a function of the maximum admissible weight.

IOP Conf. Series: Materials Science and Engineering **1017**(2021) 012026 doi:10.1088/1757-899X/1017/1/012026

2.5. Study and analysis of passenger bus tie rod.

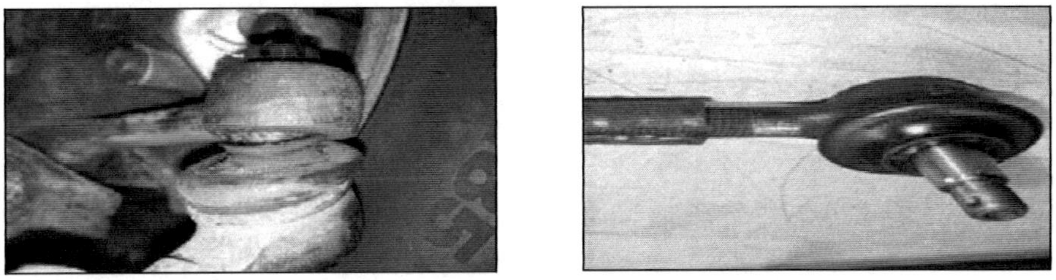

Figure 4. (a) Worn out tie rod; (b) Tie rod of TATA LPO 1618

Prashant Vithalkar, R. R. Gawande – the authors of the research paper of above research area, the main objective of this paper on study of buckling load on tie rod of steering system because of the external factors like road conditions, vibrations, sudden jerks are having great impact on the tie rod. The tie rod generally buckles under the action of the compressive forces when it becomes worn out the steering became difficult and hard to steer, which ultimately leads to the tire wear. From the observation, they have taken readings and measurements of the tie rod of the passenger bus and by that reading they have prepared the 3D CAD model of the tie rod using the software Pro-E.

2.6. ECU controlled intelligent lift axle dropping and lifting system for heavy trucks

In India, heavy transportation using trucks are used for transferring the goods to their final destination. It is an expensive mode of transportation. While transporting through heavy vehicles, over loading of these vehicles are one of the main reasons of pavement damage and tire wear due to the excessive pressure on the road at the tire road interface. Restoring the wear damage caused by overweight trucks requires huge capital investments and also reduces the capacity of traffic flow due to construction. In this paper I studied about an automatic lift axle dropping and lifting system to help improve the tire life and to adhere to the legal per axle weight and weight distribution limits. The main aim of an automatic lift axle dropping system is to drop the appropriate axles automatically based on the loading condition, instead of being manually controlled and lifted by the driver. We know about the conventional axle lifting system. In manual controls and usages, solenoid valves are controlled by switches mounted in the instrument panel in the truck cabin. In order to lift/drop the axle, the driver pushes the corresponding switch and the axle is actuated directly, regardless of the vehicle condition. But it has some disadvantages, because it is sometimes ignored by the driver and leads to damage the tire particles and pavement too. The purpose of an automatic lift axle lifting and dropping system is to automate the dropping action. It aims to lower damages to the truck, as well as damages to vehicle parts and road pavement due to loading conditions.

The main requirement for the Auto-Drop system is the measurement of axle loads. The system must drop the required axles automatically based on the loading condition. If the second axle is lifted and if the rear axle group load exceeds 18.0 tons, the system will drop the nearest axle, which is second axle, for sharing overload and maintain axle loads under permissible limits and does not allow driver to lift axle again. In this paper, The Auto-Drop system algorithm developed for the 8×2S test truck runs in a PI INNOVO M250 ECU module using MATLAB/Simulink as the software development platform a. ECU parameter calibration and embedded program variable monitoring is provided with ATI VISION based on CCP (CAN Calibration Protocol). Automatic tag axle dropping system has been developed for mechanical suspended vehicles for the first time to reduce tire pressure by preventing to get damaged and worn out. Measurement of integrated axle weight via electronic sensors has been developed in mechanical suspended vehicles with tag axle lifting feature for the first time. The axle load approximate error per axle was decreased to 0.4 tons from 2 tons as compared to conventional methods. By ensuring that axle loads remain within permissible limits, service maintenance costs are reduced. We found that here the system allowed the transmission of vehicle weight data to Electronic Brake System (EBS) with higher accuracy. Tag axle Lift/Drop Function is connected to Parking Brake in order to activate parking brake actuators located at tag axle. First time implemented in Turkey 2020 by using the EU Directive 1230/2012 EEC Mass and Dimensions Policy.

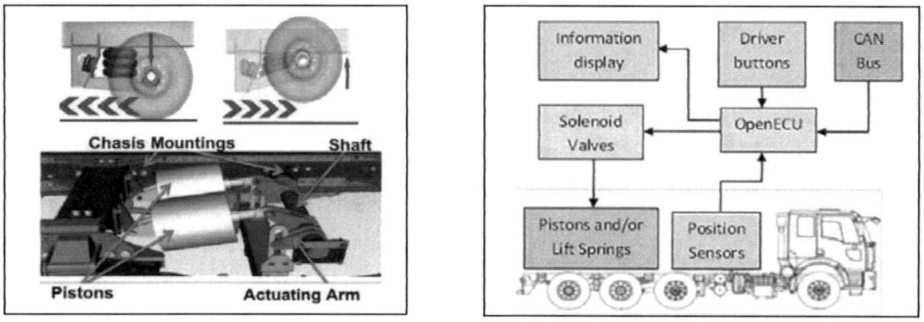

Figure 5. (a) Block diagram & (b) Mechanism of truck Axle lifting mechanisms.

3. Working Methodology

For the project on which we are working, we had chosen bus of TATA company having TATA LPO 1618, Tata LPO 1618 comes fitted with 5883 cc Cummins 6BT 5.9-180-30 – BS4 Water cooled, Turbocharged Diesel Engine. Here we know that a maximum power of 131 KW at 2500 rpm and maximum torque of 700 Nm at 1400 rpm gives enough strength to the bus to carry load full to its capacity. Like other Tata trucks in this range, thus bus also has a 400 Liters of fuel tank capacity. The tires for this vehicle are available in two options - 10.00 x 20 -16 PR Diagonal Ply - 10R20-16PR and 10.00 x 20 -16 PR Diagonal Ply - 11R22.5-16PR. This vehicle has a GVW of 16200 kg and a max speed of 110 Km/h. With these numbers, it becomes a dependable bus for long distance commutation. But it consists of major problem that is tire wear due to various parameters like friction, overheating of surface of tire due to heating of road too etc. For the proper analysis of the commercial vehicle (TATA LPO 1618), we had done regular visit to the maintenance center. And while doing regular visits we had found almost five to six major types of tire wear in the tire of the commercial vehicle.

Tire wear is a complex phenomenon that is affected by tire materials, structures, vehicle dynamics, and road conditions. In this paper we basically focused on the tire wear of vehicle, because tire is very important part for driving any vehicle and everyone wants that tire of their vehicle should execute less damage and in result it will become more durable. So, basically, we had observed main types of wearing of tire and they are as follow-

Camber wear (one-sided wear), Feathered Directional Wear, Cupped wear, Center wear, Shoulder wear etc.

For reviewing and analysis of alignment issue which irrelevantly arises in the vehicle, we had visited MHKS alignment center, and we got to know some probable solutions on reduction of tire wear causing due to the misalignment in the vehicle arising due to the road conditions, sudden breaks, over friction between the road surface and the surface of the tire etc. And also, we have to do re-alignment of the vehicle in the span of every six months whether the vehicle is in running or in stationary condition in addition with this, we also have to do the re-alignment of the king-pin, so that it will help to execute less wearing of tire while driving. The surface of the road on which the vehicle is going to be run. While doing alignment we have to check proper measures of alignment parameters like "Toe in" & "Toe out".

IOP Conf. Series: Materials Science and Engineering **1017** (2021) 012026 doi:10.1088/1757-899X/1017/1/012026

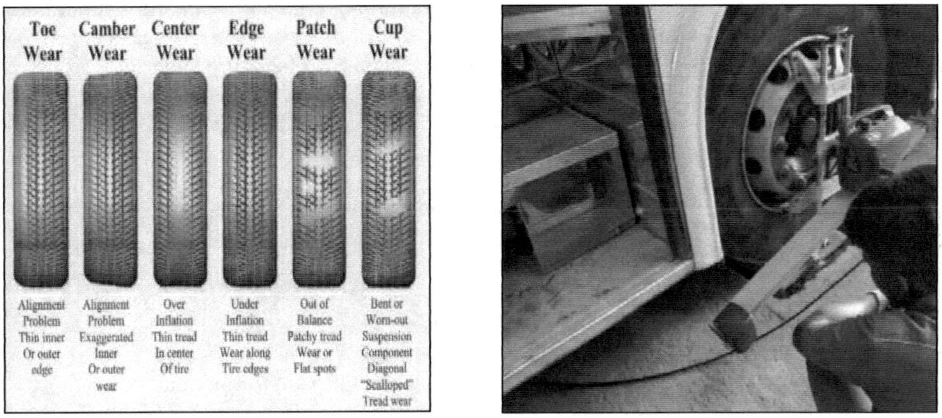

Figure 6. (a) Tire Wear Observed in TATA LPO 1618; (b) Alignment procedure of MHKS.

Table 1. Tire wears readings taken by using depth Gauge at Nagpur bus depot.

Readings of tyre wear at ShivShahi Bus Depot.						
Bus No.	Front RHS	Front LHS	Rear RHS out	Rear RHS in	Rear LHS out	Rear LHS in
806	4mm	4mm	Remould	Remould	2mm	Reject
1335	5mm	6mm	3mm	Remould	2mm	2mm
905	4mm	3mm	4mm	4mm	6mm	6mm
3346	6mm	7mm	7mm	7mm	7mm	7mm
978	2mm	3mm	4mm	4mm	3mm	Reject
1068	6mm	6mm	6mm	Remould	7mm	7mm
1066	7mm	7mm	4mm	6mm	5mm	5mm
1018	6mm	7mm	6mm	5mm	4mm	4mm
903	7mm	7mm	6mm	6mm	5mm	5mm
All buses are in the series of **MH 29 BE**						

4. Problem Description

4.1. Uneven Tire Wear

Abnormal wear, which is more critical for its safety implications, includes two types: uneven wear and irregular wear. Uneven wear describes non-uniform distribution of wear across tread pattern and irregular wear mainly characterizes circumferential wear vibration. Uneven tire wear is usually caused by improper alignment, over-inflation, under-inflation or a worn-out suspension. Tires are an extremely important component of a Commercial vehicle (Shivshahi Bus). Excess wear on the inner or outer edge of the tire, known as "toe wear" or, in more extreme cases, as "camber wear", suggests something may be wrong with the wheel alignment. Toe or camber wear at one end of the strip indicates wheel alignment problem, center wear in the center indicates over-inflation, edge wear at both ends indicates under-inflation, and patch or cup wear around the edge on one side and a sharp side wear.

Setting proper algorithm in FUZZY LOGIC and doing proper maintenance in suitable and preferred time interval to avoid accidental and random tire wear. Using proper monitoring aid or providing tire wear reading meter display on the suitable space area available on the dashboard in front of the driver seat, so that he can decide the proper speed and way of driving on the road surface, to avoid tire wear.

4.2. Tire wear offered due to misalignment

Due to improper alignment of the alignment parameters like TOE IN & TOE OUT, CAMBER AND CASTER, tire wear offered. But in major cases, we have seen the tire wear is more offered due to TOE IN & TOE OUT. This misalignment happens due to the sudden impact of shock offered due to the sudden application of the brakes while running the vehicle on the road surface. In result, the material of the tire also starts removing by its own due to the friction occurs while driving, and gradually the degradation of the quality of the tire initiates.

The industry does not focus on the tire and tire wear until and unless it's gotten ruptured or worn out, that why it causes critical condition and more investment on the tire maintenance of the vehicle. If we find out any instrument or any proper measuring device that will directly provide the proper measures and reading to the driver during the running of the vehicle on the road. Then, they can easily save the tire to be worn out & send the tire of the vehicle to the industry for remolding and its proper maintenance. In remolding, by adding new tread on the tire surface which will increase the life of the tire and saves the money which is spent on new tire .40 % of money we can save from remolding instead of purchasing new tire.

4.3. Measurement and analysis of the tire wear

To predict the probable and accurate solution of the tire wear, we will be requiring proper readings and data of the tire wear. There is a huge and measure requirement of a proper aid for measuring the accurate data of tire wear in the maintenance center and same in the vehicle itself. So, it will be easier for us to do the analysis of tire wear using various engineering analysis software like ANSYS Workbench, Solid works SW16 or CATIA etc.

For the probable solution of the above-mentioned problem, we have opted measurement technique using sensing approach in which we have to choose 2-D laser profilometer, which can give us ultra-high-speed sampling worn out rate of tire wear. In this device, we can use two types of laser sensor which laser 1 provide the direct measurement of inner profile of tire rolling deformation and laser 2 provides side wall deformation of tire. We can use optical encoder which can display the wearing rate of tire to the driver during the running of the vehicle on the road surface by which we can save the tire to be worn out and send this for remolding.

4.4. Tire Wear due to worn out suspension

This type of wear is majorly occurred in those vehicles whose suspension got worn during driving on irregular and uneven road surface. At the high, when the vehicle experiences sudden impact of speed breaker or even small stops, the padding of the suspension system gets ruptured and in result it gets damaged. And ultimately, the suspension deformed straight and the tensions in the suspension loosen out which results the bending of I-beam. And hence increases the tires wear and the tire wear offered due to this reason is also known as "Scalloped Tire Wear". To reduce such tire wear, we have to regularly check the width and dimension of the suspension using proper measuring instrument/aid like 'Depth Gauge'. Using sensors instead of manual measuring method can increase the precision of getting reading of the wearing of suspension directly on the display unit which is mounted on the surface of the dashboard in front of the tire wear.

5. Conclusion

In this paper, we had discussed the tire wear offered in the commercial vehicle that we have chosen (TATA LPO 1618 - SHIVSHASHI), which runs under MSRTC. Abnormal wear, which is more critical for its safety implications, includes two types: uneven wear and irregular wear. Uneven wear describes non-uniform distribution of wear across tread pattern and irregular wear mainly characterizes circumferential wear vibration. While doing survey we also came to know that the tire on the driver side that is the front right side of the commercial vehicle had 1mm of wear after running operation of completing every 12000 Km and correspondingly the tire on the co-driver side that is the front left side of the commercial vehicle had 1mm of wear after running operation of completing every 11000 Km. We can reduce the tire wear using various techniques mentioned in the above section like tire sensing approach, Alignment approach and also using Fuzzy Logic. We can use advanced tire wear measuring aid using advanced sensor which will not only gives tire wear reading but also provide relevant basic solution so that the driver of the vehicle can apply the provided solution in the vehicle, this can be used as a future scope development.

6. Appendices

Appendix A Authors

Prof. Gaurav M. Gohane
Assistant Professor, Mechanical Engineering JD College of Engineering and Management, Nagpur.

Anmol Indurkar, Palash Chimurkar, Sandip patle, Suraj kumar Nagina, Swapnil Tirpude, Ujwal Bawane.
(B.tech Under Graduate Students in Mechanical Engineering from J D college of engineering and management, Nagpur)

Appendix B Literature Review

Here all the reviews are written in this section, by which we got an idea about moving forward on this project by doing analysis and comparison between leaf and air suspension of our vehicle which directly related with our tire wear.

References

[1] Grosch, K.A. (1992) *Abrasion of Rubber and Its Relation to Tire Wear. Rubber Chemistry and Technology.*

[2] Pauwelussen, J.P. (2004) *The Local Contact Between Tire and Road Under Steady State Combined Slip Conditions.* Vehicle System Dynamics.

[3] Lupker, H., Cheli, F. and Braghin, F. (2004) *Numerical Prediction of Car Tire Wear, Tire Science and Technology.*

[4] Sawant, P.J. and Joshi, S.G. (2005) *Theoretical Analysis of Unstable Vibration of Tyre Mass-Suspension System with Cab of a Typical Road Vehicle,* Journal of the Institution of Engineers (India).

[5] Zhou, Z.H., Zuo, S.G. and Feng, Q. (2008) *Research on Polygonal Wear of Automotive Tire Based on Self Excitation Theory,* System Simulation Technology.

[6] ABAQUS (2010*) Analysis User's Manual,* **Version 6.10**. Dassault Systems Similia, Inc.

[7] Cho, J.R., Choi, J.H., Yoo, W.S., Kim, G.J. and Woo, J.S. (2006*) Estimation of Dry Road Braking Distance Considering Frictional Energy of Pattered Tire*. Finite Element Analysis and Design.

[8] Cho, J.R., Kim, K.W. and Yoo, W.S. (2004*) Mesh Generation Considering Detailed Tread Blocks for Reliable 3-D Tire Analysis,* Advances in Engineering Software, **35**, **105-113.**

[9] Stalnaker, D.O. and Turner, J.L. (2002) *Vehicle and Course Characterization Process for Indoor Tire Wear Simulation*. Tire Science and Technology, **30, 100-121**.

[10] Garrick J. Forkenbrock, Garrott W. Riley, Heitz Mark, Bryan C. O'Harra, *An Experimental Examination of J-Turn and Fishhook Maneuvers That May Induce On-Road, Untripped, Light Vehicle Rollover*, **SAE 2003-01-1008, 2003**.

[11] Hinch, *WHTSAYs Rollover Rulemaking Program – Results of Testing and analysis*, **SAE 92058.**

[12] K.D. Lampert, *A study of vehicle properties that influence rollover and their effect on electronic stability controllers*

[13] L. Zhang, H. Liu, Y. Xu, S. Wu, *Study on Modeling Method of Anti-roll Bar Using One Dimensional Beam Element,* 2009.

[14] E. Sert, S.D. Kizilca, S. Dileroglu, *Ride Comfort Sensitivity Analysis of Portal Axle Using Adams/Ca*r, ISITES, 2014.

[15] E. Sert, S. Dilerog˘lu, B. Kaya, *the rollover control design for commercial truck using different controller applications,* Int. J. Autom. Eng. Technol. (2015).

[16] S.R. Ribeiro, M.E. Silveira, *Application of Finite Element Method in the Study of Variables that Influence the Stiffness of the Anti-Roll Bar and the Body Roll*, SAE 2013-36-0643, 2013.

Acknowledgments
Firstly, we would like to thank our teachers, HOD "Dr. Bhushan Mahajan" and guide professor "Prof. Gaurav Gohane" who gave us their valuable suggestions and ideas, when we were in need of them. They encouraged us to work on this project. We would also thank to all of them who helped us to complete this project.

IOP Conf. Series: Materials Science and Engineering 1017 (2021) 012027 doi:10.1088/1757-899X/1017/1/012027

Comparative Experimental Investigation of Erosion behaviour of AA2024/B_4C and AA6061/B_4C Metal Matrix Composites with Particulate coating

S Vijay[1,2], Ch Srinivasa Rao[2], J Laxmi Lalitha[1]

[1] Bapatla Engineering College, Bapatla-522101, India
[2] AU college of Engineering, Vishakapatnam-560037, India

Email: vizzu515@gmail.com

Abstract. Different amounts of B4C reinforcement with 5% and 10% are coated with Nickel using Electroless coating technique and are used for reinforcement in Aluminium alloys to produce aluminium metal-matrix composites (AMC). The composites with/ without nickel coating thus produced were tested for erosion rate at various pressure conditions and impinging angle. The addition of certain amount of nickel coating provides not only the wettability effect of B4C, it also increases the strength of parent materials (pure Aluminium, AA2024 and AA6061). SEM micrographs confirmed the formation of a Nickel layer on Boron Carbide. Sand erosion tests are designed using L27 orthogonal array with factors as four having three levels. External factors such as, type of matrix and the weight-percentage of particulate are selected. Internal factors viz., the pressure of the air/sand mixture and the angle of impingement. Three levels for every factor used in the experimentation and out of which, Impingement angle shows the predominant significant effect on the erosion rate. It is found that the impingement angle of level 1 (i.e. 30o) has the max influence of 39.75 % contribution on erosion rate. The Regression model is developed for estimating the erosion rate, in terms of the levels of all factor considered. From these models, it is observed that the 5% B4C reinforcement with nickel coating gives minimum erosion rate for aluminium alloys.

1. Introduction

Present work aims at investigating the erosion wear performance of aluminium metal matrix composite (AMC) with different matrix and particulate materials. Nickel coated B4C particles having typical grain size of 40μm (325 mesh) is used as a particulate with diverse weight percentage. Boron carbide (B4C) is found to be a capable ceramic material due to its high strength, low density (2.52 g/cm3), exceptionally high hardness and better chemical stability. Extreme hardness of B4C makes it a better alternative to silicon carbide (SiC) and aluminium oxide (Al2O3) for reinforcement with aluminium alloys. A aluminium-matrix composite added the high strength and hardness of reinforcing materials with the ductility and toughness of these light metals.

2. Literature review

The investigations of various researchers in this area reveal how the properties of MMC's alter as per variation in composition and type of manufacturing process. Qing-Jun Zhou [1] et al studied the consequence of charging and out-gassing on Ni-P amorphous coating and observed improvement in wear durability for hydrogen as charging environment. C.S.Ramesh [2] et al., investigated the abrasive wear behavior of Ni-P layered Si3N4 reinforced Al-6061 composites. They observed grain refinement with increased content of reinforcement. Ayfer Kilicarslan [3] et al., worked on Ni-P coating onto B4C particles and observed a homogeneous and continuous layer onto B4C particles. Alireza Abdollahi [4] et al., studied dry sliding behavior of Al2024-5%B4C nano-composites manufactured using hot extrusion and finds enhancement of mechanical and wear properties. K. Umanatha [5] et al., investigated on metal-to-metal wear performance of AA6061 hybrid composites cast by stir- casting technique and observed improved wear properties with the reinforcement of SiC and Al2O3. Adarsh. H [6], et al., studied the effect of carbon fibre rod reinforcement on slurry erosive behaviour of Al6061 composites and finds the progress in the slurry wear-resistance properties contrast to the basic matrix alloy. A less work has been investigated on sand erosion behaviour on hybrid MMCs with nickel coated B4C as the reinforcement of different matrix materials.

3. Methodology

Hybrid aluminium composites with boron carbide (B4C) in various amounts of reinforcement were cast using the stir-casting process [7]. The boron-carbide particles with the average particle size of 40 μm were preheated up to 300 °C. Degassing tablets are added for the period of the melting to remove gaseous molecules. Magnesium (w=2 %) is added to the liquefied metal for improving the wettability [8] between the reinforcements and aluminium. The mixture of the liquefied alloy is mechanically stirred, by means of a steel stirrer for 10 min, to get hold of a homogenous mixture. A constant speed of 400 RPM and bucketing temperature of 710°C is maintained. Three dissimilar matrix materials for composites, pure aluminium (A), AA2024 (B) and AA6061 (C) are selected. Hard ceramic, Boron carbide (B4C) powder of average size of 40 μm, is used as reinforced particulate. For improving the wet angle and bonding strength among matrix and reinforcement, B4C particles are coated with Nickel, having a thickness of nano-layer. Electro-less plating technique is used for coating B_4C with Nickel. Deposition of Nickel layer, on the B4C particles is observed with SEM images and confirmed the optimum deposition of Nickel using EDAX images. Specimens are prepared in accordance with ASTM standards and tested with solid particle erosion test rig [9]. The weight % of B4C is taken as 5% without nickel coating, 5% B4C with nickel coating and 10%B4C with nickel coating. Specimens are tested for hardness and found enhancement of about 10% hardness in all types of AA/MMC. The current work is focused on comparative investigation of sand erosion behaviour of different Aluminium Alloys Metal Matrix Composites. Commercially available three different aluminium alloys, pure aluminium, AA2024 and AA6061 (table 1) are used as matrix material for the preparation of hybrid aluminium alloy metal matrix composites (hybrid AA/MMC).

Table 1. AA 2024 and AA 6061 chemical composition

AA2024 Alloy	Si	Fe	Cu	Mn	Mg	Cr	Zn	Ti	Al
	0.5	0.5	3.8-4.9	0.3-0.9	1.2-1.8	0.1	0.25	0.15	Balance
AA 6061 Alloy	0.4-0.8	0.7	0.15-0.4	0.15	0.8-1.2	0.04-0.35	0.25	0.15	

Boron Carbide (B4C) with typical particle size of 40 μm is used as particulate for the preparation of AA/MMC. To increase the wettability of the ceramic particles with molten aluminium alloy B4C is coated with the Nickel using electroless process. The weight-percentage of the Nickel metal ions on the Boron Carbide surface is optimized by changing the bath conditions by means of Design of Experiments. Optimum bath conditions for the electroless process are provided in table 2.

Table 2. Different baths and their chemical composition for electrolysis process

Bath	Reagents	Weight %
Sensitization	SnCl2+ HCl	10gm/l + 40 ml/l
Activation	PdCl2+ HCl	0.2 gm/l + 2.5 ml/l
Plating	$NiSO_4.6H_2O$ + $NaH_2PO_2.H_2O$ + $Na_3C_6H_5O_7$. $5H_2O$+ NaOH	60gm/l + 20gm/l + 122gm/l + p^H-8

Figure 1. and Figure 2. are the SEM images obtained using VEGA3TESCAN for B4C particles before and after the electroless process with intensification of 264X and 1.51kX respectively. The weight percentage of Nickel deposited on the B4C particles is observed at 58%.

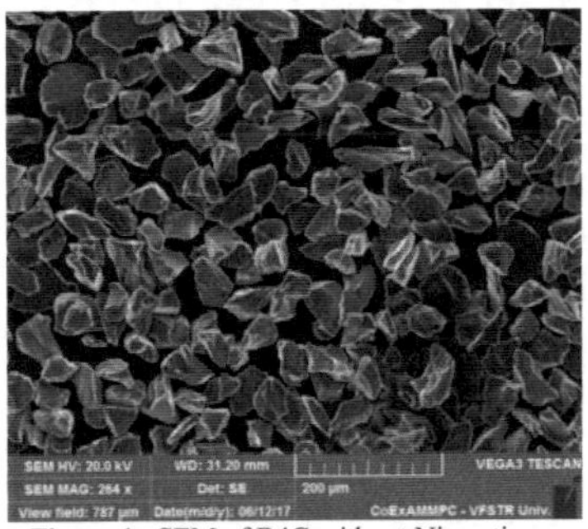

Figure 1. SEM of B4C without Ni coatings

Figure 2. SEM of B4C with Ni coating

3.1. AA/B4C casting preparation. Aluminum alloy metal matrix composites with nickel coated B4C particle is prepared using Stir Casting (Liquid state process). The stir casting furnace used which is equipped with automatic powder feeding with preheating. The operating conditions of the furnace and mould are shown in the table 3. Hybrid AA/MMC castings include three special matrix materials namely pure Aluminum, AA2024 and AA6061. The particulates have three different compositions of 5% B4C without Ni coating, 5% B4C with Ni coating and 10% B4C with Ni coating.

The present work focuses on the rate of sand erosion in Hybrid AA/MMC. The specimens are prepared from the castings shown in Figure 3.

Figure.3. Castings of AMCs

Table 3. Casting conditions of Stir casting furnace

Parameter	Material/Operating conditions
Crucibles	Graphite
Mould	Cast Iron (size 150 × 75 × 15 mm)
Furnace Temp.	750 ± 50°C
Preheat of B4C	200°C
Pressure	Atmospheric pressure
Stirring Speed	400 RPM
Stirring Time	15 min.

Prismatic portion of the casting is employed for preparing the specimens for sand abrasive test as per the ASTM standard. The size of specimens for hardness test is ϕ25X25 mm and the dimensions for sand abrasion test are 25 X 25 X 20 mm. The specimens for erosion test are shown in Figure 4.

3.2. Sand erosion test. Sand erosion wear tests of hybrid AA/MMCs are carried out on an erosion test rig, as shown in Figure 5, as per the ASTM G76 standard. Testing machine consists of different units. Drying unit for making the sand particles, moisture-less and free flow Compression unit for increasing the pressure of the air, Mixing chamber for mixing the high pressure air and sand particles and the Feeder unit to control the flow of erodent.

Figure 4. Specimen for erosion test

Figure 5. Sand erosion apparatus

The high impact of air mixture is made with convergent brass nozzle, having 3 mm internal diameter. The angle of impact onto the specimen is changed by tilting the specimen holder. The erosion test rig conditions for experimentation are shown in the Table 5. The specimens are rinsed with Acetone and dried for 5 min after each experiment, for taking the readings of erosion rate. Electronic balance with accuracy ±0.01 mg is used for weighing the specimens after testing. For investigating the effect of different parameters on the erosion rate, Taguchi orthogonal array [10] (L27) is used. After screening, two qualitative and two quantitative factors were selected. External variables selected are type of the matrix (A) and the weight % of particulate (B). Internal variables are pressure of the air/sand mixture (C) and angle of impingement (D). Three levels of every factor used for the experimental run and are listed in response table. Controllable factors that are considered for investigation of erosion rate are listed in the table 4. Three different levels were selected for each factor depending on the range

Table 4. Factors and their levels for L_{27} orthogonal array

Factors	Levels		
	1	2	3
Matrix (A)	Pure Al	AA2024	AA6061
Particulate (B)	5% B_4C with Ni	10%B_4C with Ni	5% B_4C without Ni
Air Pressure (C)	1 bar	2 bars	3 bars
Impingement angle (D)	30°	60°	90°

4. Results and Discussion

To run the experiments for different matrix and different composition, on the sand erosion test rig L27orthogonal array design is selected. The weight of the sample before (Wi) and after (Wf) the each experiment is noted, to the accuracy of 0.01 mg. The erosion rate of the material is calculated using the equation [4] (1)

$$\textbf{Erosion rate} = \frac{(W_i - W_f)}{W_i} \quad (1)$$

The response table of erosion test is given in the table 5. The obtained erosion rates are analyzed for identifying the most affecting factors is done with MINITAB-18 software. The objective function is to minimize the erosion rate; smaller-the-best signal-to-noise (S/N) ratio [6] is selected. Since the erosion rate values are of 5th decimal, for observing the sum of square of variables, the data entered is multiplied by 10^5.

Table 5. Response table for Taguchi orthogonal array of erosion test

Run order	Matrix (A)	Composition (B)	Air Pressure (C)	Impingement Angle (D)	Erosion Rate1 X 10^{-5}	S/N Ratio
1	1	1	1	1	1.42	-3.0458
2	1	1	2	2	2.91	-9.2779
3	1	1	3	3	3.48	-10.8316
4	1	2	1	2	4.30	-12.6694
5	1	2	2	3	4.59	-13.2363
6	1	2	3	1	2.46	-7.8187
7	1	3	1	3	5.12	-14.1854
8	1	3	2	1	2.94	-9.3669
9	1	3	3	2	4.62	-13.2928
10	2	1	1	1	2.24	-7.0050
11	2	1	2	2	3.72	-11.4109
12	2	1	3	3	4.11	-12.2768
13	2	2	1	2	5.11	-14.1684
14	2	2	2	3	5.50	-14.8073
15	2	2	3	1	3.28	-10.3175
16	2	3	1	3	5.97	-15.5195
17	2	3	2	1	3.75	-11.4806
18	2	3	3	2	5.21	-14.3368
19	3	1	1	1	0.30	10.4576
20	3	1	2	2	1.78	-5.0084
21	3	1	3	3	2.17	-6.7292
22	3	2	1	2	3.17	-10.0212
23	3	2	2	3	3.54	-10.9801
24	3	2	3	1	1.34	-2.5421
25	3	3	1	3	4.03	-12.1061
26	3	3	2	1	1.81	-5.1536
27	3	3	3	2	2.30	-7.2346

Table 6. Response table for signal-to-noise ratios (smaller is better)

Level	Matrix (A)	Particulate (B)	Air Pressure(C)	Impingement Angle (D)
1	-10.414	-6.125	-8.696	-5.141
2	-12.369	-10.729	-10.080	-10.824
3	-5.480	-11.408	-9.487	-12.297
Delta	6.889	5.283	1.384	7.156
Rank	2	3	4	1

Table 7. Analysis of Variance for erosion1, using Adjusted SS

Source	DF	Seq SS	Contribution	Adj SS	Adj MS	F-Value	P-Value
Matrix (A)	2	19.2617	36.06%	19.2617	9.6308	213.10	0.000
Particulate (B)	2	11.7075	21.91%	11.7075	5.8537	129.53	0.000
Air Pressure (C)	2	0.4058	0.76%	0.4058	0.2029	4.49	0.026
Impingement Angle(D)	2	21.2344	39.75%	21.2344	10.6172	234.93	0.000
Error	18	0.8135	1.52%	0.8135	0.0452		

Model Summary

S	R-sq	R-sq(adj)	PRESS	R-sq(pred)
0.212588	98.48%	97.80%	1.83035	96.57%

From the table 7, it is observed that the fisher test value is more for factor D (234.93), the larger the F-value indicates the larger the internal variance of response between the levels of the same group [6]. Factor A (213.10) is also having the same effect on their erosion rate. Factor C with the minimum F-value shown less variation of response within the levels of C. From the Fig.6, main effect plots for the S/N and response table shown in table 6, confirmed that Factor D (impingement angle) is the most influencing factor on the erosion rate with the 39.75% contribution with the rank 1. And Factor A (matrix) is the also a significant factor with the % contribution of 36.06% with the rank 2. Particle condition (C) has 21.91% contribution with the rank 3. Considering the above analysis, the regression equation is generated which includes the contribution of each factor with their levels on the minimization of the erosion rate.

$$\text{Erosion rate } (10e-5) = 3.3767 + 0.1611 * A1 + 0.944 * A2 - 1.1056 * A3 - 0.9178 * B1 + \\ +0.3222 * B2 + 0.5956 * B3 + 0.1411 * C1 + 0.0167 * C2 - 0.1578 * C3 - 1.2056 * D1 + \\ 0.3033 * D2 + 0.9022 * D3 \quad (2)$$

Equation (2) gives the regression equation in terms of coded valued of factors. Factors A (matrix) and factor B (composition) are the qualitative factors and Factors C and D are quantitative. The factors with negative coefficient values give the minimum erosion rate. From the equation, it is observed that A3 (matrix AA/6061), B1 (5% B4C with Ni) and D1 (30o impingement angle) have large negative values. That means these levels minimize the erosion rate.

5. Conclusion

Sand erosion behaviour of Aluminium metal matrix (AMC) composite with B4C as reinforcement with two different conditions was investigated. Three diverse matrix materials (Al, AA/2024 and AA/6061) are selected. MMCs were made using different composition of B4C with Nickel coating and with Nickel coating. Under tribological behaviour sand erosion of MMCs were investigated. For identifying the most influencing factors on the erosion rate ANOVA has conducted. And the conclusions are as follows

(i) Influencing factors in order, on the erosion rate of hybrid AA/MMC are impingement angle (D), matrix type (A), composition (B) and Air pressure (D) respectively.

(ii) Impingement angle of 30o show least erosion value compared to other levels and factors.

(iii) The Regression equation is generated for estimating the erosion rate. The equation also studied the influence of different factors at different levels.

Acknowledgements
Author is grateful to Dr. G. Raghavendra, Asst. Professor at NIT- Warangal, for giving the permission to make use of sand erosion test rig facility.

References

[1] Qing-Jun Zhou, Jian-Qiang Zhoe, fei Nie, Wu-Yang Chu, Jin-Xu, Li-Jie Qiao 2001 Wear **266** 810
[2] C.S. Ramesh, R. Keshavamurthy, S. Pramod, Praveennath G.Koppad 2009 Mater. Sci. Eng. A **502** 99.
[3] Ayfer Kilicarslan, Fatih Toptan, Isil Kerti 2012, Mater. Lett **76** 11.
[4] Alireza Abdollahi, Ali Alizadeh, Hamid Reza Baharvandi 2014, Mater. Des. **55** 471.
[5] K. Umanath, S.T.Selvamani, K.Palanikumar and D. Niranjanavarma 2014, Procedia Eng **97** 703.
[6]Adarsha.H, R.Keshavamurthy, Ramesh S, Naveen P Noronha 2016, Procedia Technology **25** 916.
[7] David RajaSelvam. J., Robinson Smart D.S. Dinaharan I. 2013, Energy Procedia **34** 637
[8] B Ravi, B. Balu Naik, J. Udaya Prakash 2015, Mater.Today: Proceedings **2** 2984.
[9] Manoj Panchal, G. Raghavendra M. Om Prakash, S.Ojha 2018, Silicon **10** 627.
[10] V C Uvaraj, N Natarajan, I. Rajendran, K. Sivakumar 2013, J. Tribol **135** 021101.

Effect of Concentration of NaOH Treatment on Mechanical Properties of Epoxy/Sisal Fiber Composites

D Yadav[1], G R Selokar[1], A Agrawal[2], V Mishra[3], I A Khan[4]

[1]Department of Mechanical Engineering, Sri Satya Sai University of Technology and Medical Science, Sehore-46601, India
[2]Department of Mechanical Engineering, Sagar Institute of Research and Technology-Excellence, Bhopal-462041, India
[3]Department of Mechanical Engineering, Indore Institute of Science and Technology, Indore-453331, India
[4]Department of Mechanical Engineering, Sagar Institute of Research and Technology, Bhopal-462041, India

Email: alokag03@gmail.com

Abstract. In recent years, natural fibers incorporation in polymeric resin has received huge attention among the research community. The reasons of demand are multiple that includes its light weight, environmental friendly nature; non-toxic, low cost, easy availability, low processing cost and most importantly they possess characteristics which are comparable to conventional material. With this approach, present work comprises of fabrication of new category of natural fiber reinforced composites with polymer as base matrix. Sisal fiber is selected as reinforcing phase with epoxy matrix. Four different combination of composites are prepared with sisal fiber loading varies from 2.5 wt. % to 10 wt. % using well-known hand lay-up method. Sisal fibers were treated with NaOH at varied concentration to observe the effect of surface modification and its concentration on the developed material. Three different concentration of NaOH is used i.e. 2 mole, 4 mole and 6 mole for preparing three sets of composites. One set is prepared with raw sisal fiber to make total of four sets of composites. Mechanical properties under investigation are tensile behaviour, compressive behaviour and flexural behaviour. The experimental results obtained are compared for optimizing the concentration of NaOH. From the analysis it is seen that composite with surface modified sisal fiber yield better results and further fiber treated with 2 mole NaOH concentration is superior among their counterpart. The maximum tensile strength, compressive strength and flexural strength obtained are 31.5 MPa, 72.5 MPa and 37.8 MPa respectively. All these values are obtained for fiber treated with 2 mole NaOH aqueous solution.

1. Introduction

In past few years, there has been a strong awareness for products from renewable sources. Environmental friendly products are in demand mainly because of availability of new derivatives for recycling and also due to social influencing. In this regards, composite materials are being established to improve and adapt outdated products and present fresh products in a sustainable and answerable way [1]. Polymer composite materials derived from some sort of natural resources are successfully replacing the non-biodegradable composite material and even various conventional materials. The reasons behind this are the multiple benefits provided by them which include its light weight, low cost, ease and cheap manufacturing, high flexibility, high specific strength, biodegradable and environmental friendly in nature [2]. Apart from this, natural fibers do not generate any harmful gases during it processing. But the development of high performance composite material derived from natural resources is an ambitious goal which is presently of interest among the scientific community all across the world. The main concern with natural fibers is their hydrophilic nature as it restricts them to be the first choice as reinforcement with polymeric resin. Because of this hydrophilic nature, the composite fabricated had high water absorption rate which has several drawbacks like poor matrix-fiber interfacial adhesion, improper distribution of fiber within polymeric matrix and low stress transfer efficiency [3]. A significant work is in progress to overcome this problem by modifying the surface of the fiber with chemical treatment process. It has been observed that there is tremendous enhancement in various properties of the natural fiber composite material developed by this method [4, 5].

Surface modification of natural fibers results improved adhesion between matrix and fiber and simultaneous reduce the hydrophilic nature of the fiber. The various chemical treatment methods implemented are alkaline treatment [6], silane treatment [7], acetylene treatment [8], benzene treatment [9], peroxide treatment [10], potassium permanganate treatment [11], steric acid treatment [12]. Among them the most widely used treatment methods are alkaline treatment as it is simple, economic and effective. It has been reported that the speed of fiber fragmentation and disaggregation also got increased with this method. During the process of alkanization, hydroxyl group moves away from the structure of fiber and is replaced by –O-Na group. With the removal of hydroxyl group from the surface of the fiber, fiber resistance against moisture increases. Alkaline treatment also cleans the surface of the fiber and makes them uniform which results in enhanced stress transfer capability of the fiber. Proper alkalization reduces the diameter of fiber as well and with reduced diameter, for given weight fraction, surface area of fiber increases which provide better adhesion with continuous phase [13]. Saravanakumaar et al. [14] reinforce alkali treated kenaf fiber in polypropylene matrix and compared the mechanical properties of composite with and without fiber treated composite. They reported to found significant improvement in mechanical properties when treated fiber was used. In more recent work, Asumani et al. [15] treated papaya fibers with NaOH aqueous solution at room temperature and varied the time of treatment. In their study they concluded that treatment time also govern the mechanical property of the developed material.

Among the various natural fibers, sisal fiber is considered to be the best option because of multiple benefits. A sisal fiber composite has various commercial uses mainly due its above average specific values, low density, environmental friendliness and cost-effectiveness. As compared to different NFCs, polymer composites prepared with sisal fiber as reinforcement delivers greater impact strength with reasonable tensile and flexural strength. This makes sisal fiber a potential candidate to be used as reinforcement in polymeric resin and used in applications where improved specific properties are required. But directly usage of raw sisal fiber as reinforcement will not fulfill the need and hence proper surface modification is required prior to use it as reinforcement. In this regards, the work concentrates on modifying the surface of the sisal fiber with aqueous solution of NaOH for fabrication of polymer composites with epoxy as base matrix. The main aim of the work is to present the influence of fiber modification on response to mechanical loading for the fabricated samples. Also the work consists of optimizing the concentration of NaOH to be used with sisal fiber for providing better results as under and over treatment both are undesirable. The different mechanical properties under consideration are tensile strength and extension, flexural strength and compressive strength.

2. Material and Methods

2.1. Raw Material Used

Thermoset epoxy resin LY 556 in combination with triethylene tetramine HY951 is selected as continuous phase in present work. Epoxy is readily available at low cost and processing is also quite easy which make this polymer suitable for present investigation. Further it possesses good mechanical properties as compared to other polymers and has excellent thermal properties as well. The sisal fiber used as reinforcement is obtained from the local market as it is used in rural areas for making rope, mat etc. Among the various natural fibers, sisal fiber is the most promising to be used as reinforcement in polymer composites as it is easily and cheaply available. Also it possesses reasonably good mechanical properties. In present investigation, short sisal fiber of approx. 4 mm length with 230 mm diameter is used for composite fabrication. A NaOH flake is supplied by Rankem Corporation Limited; New Delhi, India is used for modification of fiber surface.

2.2. Composite Fabrication

Epoxy based composites are prepared by simple hand lay-up method. Sisal fibers are treated with NaOH solution before processing. The concentration level of NaOH considered in the study is 2 mole, 4 mole and 6 mole concentration. Epoxy composites with both treated (three sets) and non-treated (one set) sisal fibers are prepared at four different fiber loading i.e. 2.5 wt. %, 5 wt. %, 7.5 wt. % and 10 wt. %. In total sixteen sets of composites are fabricated in present work. The list of fabricated composites is shown in table 1.

Table 1. Combination of epoxy/sisal fiber samples prepared.

S. No.	Composition			
	SET A	SET B	SET C	SET D
1	Epoxy + 2.5 wt. % raw sisal fiber	Epoxy + 2.5 wt. % sisal fiber (Surface modified with 2 mole NaOH aqueous solution)	Epoxy + 2.5 wt. % sisal fiber (Surface modified with 4 mole NaOH aqueous solution)	Epoxy + 2.5 wt. % sisal fiber (Surface modified with 6 mole NaOH aqueous solution)
2	Epoxy + 5 wt. % raw sisal fiber	Epoxy + 5 wt. % sisal fiber (Surface modified with 2 mole NaOH aqueous solution)	Epoxy + 5 wt. % sisal fiber (Surface modified with 4 mole NaOH aqueous solution)	Epoxy + 5 wt. % sisal fiber (Surface modified with 6 mole NaOH aqueous solution)
3	Epoxy + 7.5 wt. % raw sisal fiber	Epoxy + 7.5 wt. % sisal fiber (Surface modified with 2 mole NaOH aqueous solution)	Epoxy + 7.5 wt. % sisal fiber (Surface modified with 4 mole NaOH aqueous solution)	Epoxy + 7.5 wt. % sisal fiber (Surface modified with 6 mole NaOH aqueous solution)
4	Epoxy + 10 wt. % raw sisal fiber	Epoxy + 10 wt. % sisal fiber (Surface modified with 2 mole NaOH aqueous solution)	Epoxy + 10 wt. % sisal fiber (Surface modified with 4 mole NaOH aqueous solution)	Epoxy + 10 wt. % sisal fiber (Surface modified with 6 mole NaOH aqueous solution)

2.3. Characterization

Mechanical characterization of the samples includes response of material on tensile loading, flexural loading, compressive loading and indentation. All the experimentation under consideration are conducted in computerized Instron 3382 Universal testing machine. Tensile tests were performed in accordance with

ASTM D638 procedure. For conducting the compressive test, the fixture of UTM needs to be change. The grip for holding the specimen is changed to 50 mm diameter hardened-steel compression platens. The samples are of cylindrical shape with length to diameter ratio of 2. The compression test is performed as per ASTM D695 standard. The flexural test was carried out as per ASTM D2344-84 standard in the same universal testing machine. The size and shape of the samples used for testing are also as per the ASTM standard.

3. Results and Discussion

3.1. Tensile Strength

Ultimate tensile strength of all sets of material developed in present work is presented in figure 1. Figure represents the change in tensile strength of the prepared composites at varied fiber content and concentration of NaOH used for fiber treatment. It is clear that inclusion of fibers in small fraction i.e. 2.5 wt. % reduces the tensile strength drastically irrespective of untreated or treated fibers. Later, when loading of fiber increases, increase in value of tensile strength is noticed. The trend is similar for all composites under investigation. Ultimate tensile strength of neat epoxy is 27 MPa as per the measurement. With the addition of raw sisal fiber, the value reduces to 16.4 MPa for a combination of epoxy with 2.5 wt. % raw sisal fiber.

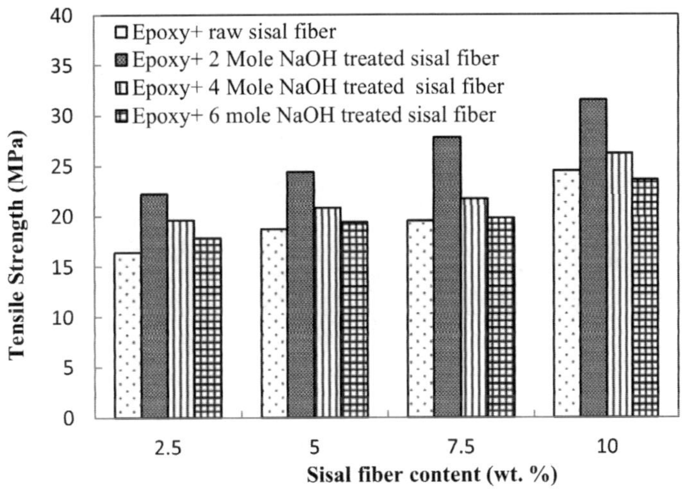

Figure 1. Ultimate tensile strength of samples under study.

With sisal fiber added further, tensile strength increases with fiber content and maximum tensile strength of 24.5 MPa is obtained with 10 wt. % of sisal fiber. Tensile strength of the composites increases with fiber content due to reinforcing effect of sisal fibers. But still the maximum value obtained for raw sisal fiber with maximum fiber content is less than that of the tensile strength of neat epoxy. This can be explained due to lower interfacial adhesion and compatibility between the two phases. Further, surface modification of sisal fiber using aqueous solution of NaOH results in improvement in tensile strength of the fabricated composites. Alkaline treatment of sisal fiber removes wax and impurities present over the surface of the fiber. Alkalization also helps in removing some hemicellulose from the fiber surface. This results in generation of rough surface on fiber followed by fiber fibrillation. With this, effective surface area of fiber increases drastically to adhere with epoxy matrix. Hence, adhesion between fiber and matrix improves via mechanical interlocking and thus for same fiber content, alkaline treated sisal fiber/epoxy shows improved tensile strength as compared to raw sisal fiber/epoxy.

Tensile strength also varies with concentration of NaOH used. Fibers surface modified with 2 mole NaOH aqueous solution delivers better results irrespective of content of fiber as compared to their counterparts. Maximum tensile strength of 31.5 MPa is obtained when 2 mole concentration of NaOH treated 10 wt. % sisal fiber is added in epoxy. The improvement is around 16 % over neat epoxy but 28.5 %, 20.2 % and 33.7 % over set A, set B and set C composites respectively. It is observed that when the concentration of NaOH for treatment increases, tensile strength decline. It can be explained as when concentration of NaOH is low i.e. 2 wt. %, than the surface material present over sisal fiber dissolves completely. Due to this rough surface is presented for epoxy resin for better mechanical interlocking, thus increases the tensile strength. When the concentration of surface increases, sisal fibers were severely corroded. This damage the surface of the fiber and severe fiber fibrillation occurs which reduces the mutual integrity of the fiber. Further, excessive fiber treatment partially removed the cellulose from fiber surface. Hence, limited concentration treatment of sisal fiber is suggested for better output.

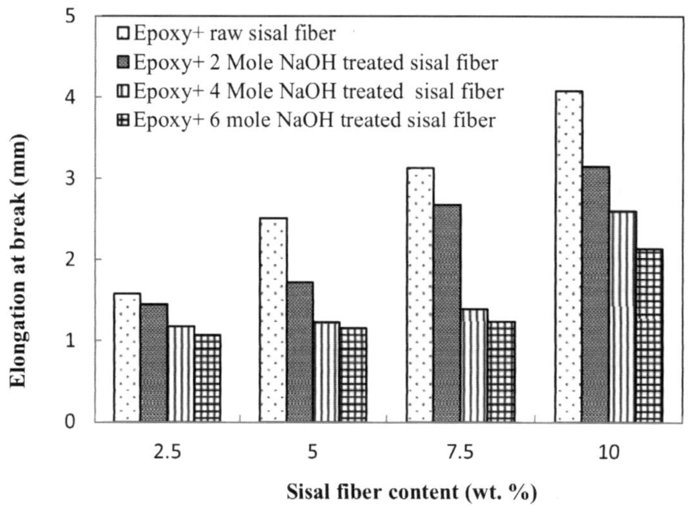

Figure 2. Elongation at break of samples under study.

Figure 2 shows the dependency of extension at break variation with fiber content and concentration of fiber treatment. Epoxy is highly brittle in nature with extension of only 0.88 mm before it breaks. This limited the application of epoxy as highly brittle nature is not desirable in many applications especially where repeated loading took place. It can be seen that incorporation of raw sisal fiber changes its behavior from being highly brittle to slightly ductile where extension at break increases as the fiber loading increases. Elongation at break increases to 4.08 mm with 10 wt. % of raw sisal fiber. This is an increment of 363.6 %. The similar trend is obtained when sisal fiber are surface modified. Though, raw sisal fiber imparts more ductility as compared to treated fiber. It is mainly due to the improvement of single sisal fiber after treatment with NaOH. Also, elongation at break reduces as the concentration of treatment increases. This is due to sudden decreases in strength and severe fibrillation with excessive fiber treatment.

3.2. Compressive Strength

The bonding between the matrix and the fiber mainly govern the compressive strength of the composite material. The dependency of compressive strength with varying content of sisal fiber is plotted in Figure 3. It is clear from graph that, compressive strength of the material decreases with increase in fiber content. The decreasing trend in the value of compressive strength with increase in fiber content is for all sets of composites irrespective of treated or untreated fiber. Compressive strength of neat epoxy is measured to be 82.5 MPa. When raw sisal fiber incorporated in it, compressive strength reduces to 73.7 MPa with 2.5 wt. % of fiber. This decrement continuous and compressive strength reduces to 56.8 MPa when 10 wt. % of sisal fibers are added. This is a decrement of 31.15 %. Similarly, when 10 wt. % of treated fibers are incorporated in epoxy resin, compressive strength reduces to 72.5 MPa, 74.9 MPa and 67.4 MPa for 2 mole, 4 mole and 6 mole NaOH concentration respectively. The decrement is limited to 12.1 %, 9.2 % and 18.3 % for 2 mole, 4 mole and 6 mole NaOH concentration treated fibers respectively against the decrement of 31.15 % with raw sisal fibers. The phenomenon of decrease of compressive strength can be explained as addition of sisal in epoxy increases the brittleness of the material which has been seen in figure 3. This reduces the deformation of material significantly under similar loading condition. With sudden change of behavior, relative deformation reduces which allow early rupture of material and result in reduction in the value of compressive strength. Again, treated fibers show comparatively better results as compared to raw fibers. This is mainly because of the enhanced compatibility between the fiber and matrix post treatment of fibers.

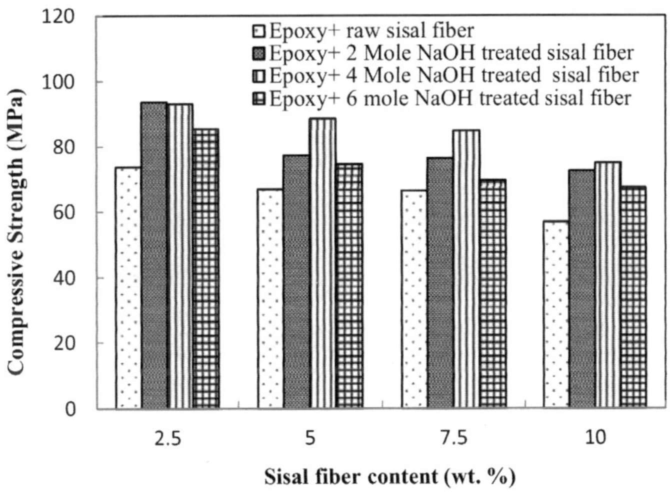

Figure 3. Compressive strength of samples under study.

3.3. Flexural Strength

The flexural strength of samples with and without surface modification of fibers is shown in figure 4. The trend obtained with treated fibers is different from those with untreated fibers when flexural strength is under consideration. Flexural strength of neat epoxy is 27.3 MPa. When raw sisal fibers are added in epoxy resin, flexural strength of composites increases when small quantity of fibers are added. When 2.5 wt. % of fibers are added, flexural strength increases to 28.1 MPa. With further addition of raw sisal fiber, flexural strength reduces with raw sisal fiber content. When 10 wt. % of raw sisal fibers are added, flexural strength reduces to 26.9 MPa. Though the decrement is very less but decreasing trend is obtained with raw sisal fibers. First, the flexural strength increases with fiber content due to proper wetting of fibers and when loading of raw fibers increases, low interfacial adhesion and improper fiber wetting decreases the flexural strength of the material. Against that, when treated sisal fibers are incorporated in epoxy resin, flexural strength of the composites increases and the trends continue as the fiber content increases. Similar trend is obtained with all treated fibers irrespective of the concentration used. The enhancement in flexural strength with treated fiber is due to strong fiber-matrix interfacial bonding. With surface modification, hydrophilic nature of the fiber reduces as seen from the result obtained from water absorption test. With reduction in hydrophilic nature, load bearing capacity increases.

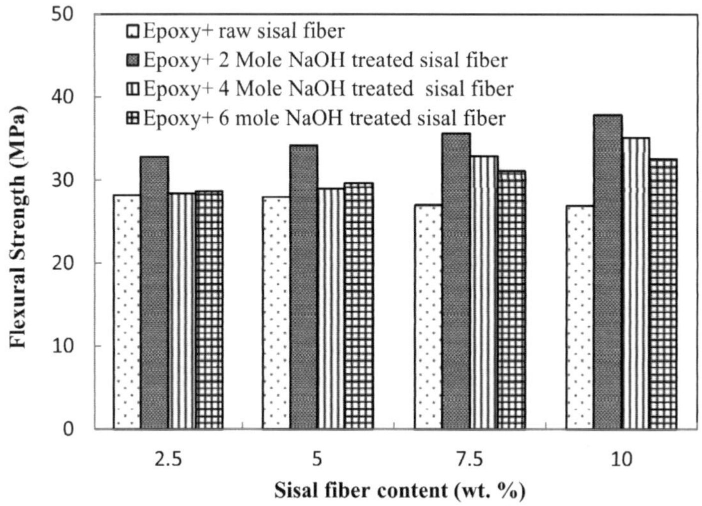

Figure 4. Flexural strength of samples under study.

With this, material withstand high bending load as compared to its counterpart. Further, sisal fiber treated with mole NaOH concentration yield better result as compared to other similar sets. It is again due to damage of fiber when over treatment occurs. Maximum flexural strength of 37.8 MPa is achieved with 2 mole NaOH treated fiber at 10 wt. % of sisal fibers. The appreciable increment of 38.5 % is observed with this combination. Again, when 4 mole NaOH treated fiber is used, 35.1 MPa of flexural strength is obtained and when 6 mole NaOH treated fiber is used, 32.5 MPa of flexural strength is noticed. This is an increment of 28.5 % and 19.2 % respectively.

4. Conclusion

The surface modification effect of sisal fiber with aqueous solution of NaOH on various mechanical properties under consideration is studied and following conclusion is drawn from the investigation.

1. With the addition of sisal fiber, tensile strength of composites reduces for a combination of epoxy with 2.5 wt. % sisal fiber. With further addition of sisal fiber, tensile strength increases.
2. Tensile strength also varies with concentration of NaOH used for alkaline treatment of fibers. Fibers treated with 2 wt. % NaOH aqueous solution delivers better results.
3. Incorporation of raw sisal fiber changes epoxy behavior from being highly brittle to slightly ductile where extension at break increases with increase in fiber content and hence extension at break decreases with increase in fiber content. Further, elongation at break reduces as the concentration of treatment increases.

4. Compressive strength of the material decreases with increase in fiber content. The decreasing trend in the value of compressive strength with increase in fiber content is for all sets of composites irrespective of treated or untreated fiber.
5. When raw sisal fibers are added in epoxy resin, flexural strength of composites increases when small quantity of fibers are added. With further addition of raw sisal fiber, flexural strength reduces with raw sisal fiber content. Against that, when treated sisal fibers are incorporated, flexural strength increases and the trends continue as the fiber content increases.
6. From the above investigation it is seen that modification of surface of sisal fiber using aqueous solution of NaOH improves the various mechanical properties of composite and best result is obtained when fiber surface are modified with 2 mole NaOH aqueous solution.

5. Acknowledgements
The authors acknowledge the help of the Department of Science and Technology (DST), Government of India to grant a project titled "Design and development of polymer composite for microelectronic applications" in this line through ECR/2017/000540 for continuing research activity in this direction.

References
[1] Mitra B C 2014 *Def. Sci. J.* **64** 244-252.
[2] Gupta M K and Srivastava R K 2015 *Mater. Today Proc.* **2** 2909-2917.
[3] Mochane M J et al 2019 *Express Polym. Lett.* **13** 159-198.
[4] Saba N et al 2016 *Constr. Build. Mater.* **124** 133-138.
[5] Kiruthika A V 2017 *J. Build. Eng.* **9** 91-99.
[6] Atiqah A, et al 2018 *J. Nat. Fibers* **15** 251-261.
[7] Tserki V et al 2005 *Compos. Part A Appl. Sci. Manuf.* **36** 1110-1118.
[8] Nair K M et al 1996 *J. Appl. Polym. Sci.* **60** 1483-1497.
[9] Sreekala et al 2000 *Appl. Compos. Mater.* **7** 295-329.
[10] Zaman et al 2010 *Fibers and Polym.***11** 455-463.
[11] Dolez et al 2017 *Procedia Eng.* **200** 81-88.
[12] Annace et al 2001 *J. Appl. Polym. Sci.* **79** 1084-1091.
[13] Mwaikambo L Y and Ansell M P 2002 *J. Appl. Polym. Sci.* **84** 2222-2234.
[14] Saravana kumar A et al 2018 Int. J. *Polym. Anal. Charact.* **23** 529-536.
[15] Asumani et al 2012 *Compos. Part A Appl. Sci. Manuf.* **43** 1431-1440.

IOP Conf. Series: Materials Science and Engineering 1017 (2021) 012029 doi:10.1088/1757-899X/1017/1/012029

Usance of industrial 4.0 technique to overcome the pandemic situation of COVID-19

Vaibhav Bhardwaj[1], Pankaj Agarwal[1], Soumya Ranjan Nayak[2], Mangal Singh Sisodiya[1] and Vijander Singh[3]

[1]Department of Mechanical and Automation Engineering, Amity University Rajasthan, Jaipur – 302006, Rajasthan India
[2]Amity School of Engineering and Technology, Amity University Uttar Pradesh, Noida- India
[3]Department of Computer Science and Engineering, Manipal University Jaipur-303007 India

Email id: vijan2005@gmail.com

ABSTRACT. COVID 19 (Coronavirus) pandemic has created surge demand for essential good and other items like medical equipment and industries which were shut and need to revive back themselves again with new strategies and plan to bring the situation back towards normalcy the study also aimed to propose new technique including advantage of Industry 4.0 ethics to reduce the impact and tackle the situation of COVID-19 and post COVID-19.A detailed review of the literature is done on the technologies along with their Industry 4.0 applications during this COVID-19 pandemic, The review holds some basic and advance technologies application to enhance industrial safety and boosting the work culture to normal and efficient. Using appropriate search words on the databases of IEEE papers, Google Scholar and Research Gate. We found several useful applications which can be inculcated along with modern human - machine interfacing which can help for proper control and management of COVID-19 pandemic and these techniques are been discussed in this paper. Available technologies blend along with Industry 4.0 technique and procedure could also help the society to work and understand smartly and get aware of this methodology which can lead them to stay safe and secure along with maintaining working efficiency. Industry 4.0 can fulfil the requirements of society/community during this crisis's situation. The new technique which firstly make assure that the induvial and belonging health and safety is first priority and their demand are need to be fulfil on time so that no one suffers. These Industry 4.0 technologies could provide a lot of innovative ideas and solution at from base local areas to global platforms. The supply chain partners and government bodies should work along with new technologies wisely for improving the services during COVID-19 and of any similar situations happening anywhere anytime.

1. Introduction

COVID 19 (Coronavirus) pandemic has affected almost all nations across globe and has made a significant effect on everything either human life or industrial point of view in which the available facilities and systems has come to halt. After sudden nationwide lockdowns for many months in almost every country the normalcy cycle of every induvial has also came into impact. An attempt to stop spreading of this dangerous virus the governing bodies of nation has order for complete halt of everything at sudden in which the major supply chain also came to stop by which supply and demand system stop .After months of observation this led to the conclusion that everyone has to live with this only for further more time till vaccine is not discovered but the situation of country depends upon its economy and economy is not a constant factor. It all depends upon the movement of trade but since due to this lockdown trade was being paused there was downfall in economy .Now the challenges arose is that to revive back the situation of nation along with human safety and health So this all lead to be understand that everyone has to adapt this new change and need to work within this situation maintain your moral health and other also. So, there is a requirement of various advance technologies to tackle various problems during this pandemic. Industry 4.0 is also known as the fourth industrial revolution, which have advance manufacturing and information technologies, to fulfil the requirement of different sector of the human being in lesser time along with fast productivity rate. These can be wireless connectivity in the manufacturing and service sector to enhance the concept of automation which is vital in today's scenario this totally means less human hinderance or contact but greater the productive rate.

Industry 4.0 have smart machines which are supported by high speed Internet ,wireless connectivity ,quicker response time .These sensors are connected to a system that can visualize and monitor the entire procedure from statistical demand to supply of actual product in a single database .The smart database application now a day can be a game change for social community which can make appropriate analysis to take step to ensure alert for future demands. Industry 4.0 uses smart manufacturing processes for the manufacturing. Involvement of techniques with old industries will provides a smart supply chain system during this crisis by which the customer can receive the required essential good at time and ensuring their safety at priority [5]

This paper aimed to confirm the utility and applications of Industry 4.0 technologies for the management of different sector during this COVID-19 pandemic, in this comprehensive review.

2. MAJOR CHALLENGES REPORTED DURING COVID-19 OUTBREAK:

1. There was a drastic increase in demand for items (e.g... masks, sanitizer, essential commodities) that was not fulfilled by limited supply as this epidemic hit the most countries together past 4-5 months before this pandemic this material like mask and sanitizer was just part of keeping induvial hygienic but now it has become mandatory commodity. So sudden demand of this wasn't fulfil which lead to only hiking of prices of commodity. The advance infrastructure then has become only assets when no smart or proactive produce are taken.
2. As this virus is spreading mainly through human contact and contaminated things, trust issues become the biggest challenge face by retail industries. "Touch-less deliveries" became new norms ,every induvial has now fear inside him So everyone is trying to as much apart from each other .The concept of Bio bubble came into effect which maintain the induvial isolated environment which minimize the chance of getting affected . Everyone in todays world are avoiding to get contact with each other and also from machine which are in public for public use like ATM machines
3. Supply and demand can only be compensated by transportation system of any organization but this can only possible when working environment is possible every organization tried to accelerate there shipment process to fulfil demand but were unable to do so due to lack of other pause at different stages of processing or storage space un availability if manufactured in surplus and travel restriction by different governing authority.

3. SIGNIFICANT TECHNOLOGIES WAY WHICH MAY HELP IN COVID 19 OUTBREAK:

As we know how crucial role the technologies play in our day-to-day lives but the perspective of technology in assisting the mitigation of infection and in controlling the situation like the COVID-19 pandemic is something we fail to notice [2]. Therefore, we conceptualized the aspects of technology utilization as various strategies to provide a helping hand in an epidemic state of affairs. For overcoming the epidemic circumstances, the contribution of various technologies can be identified as directly influencing and indirectly influencing [1]. This research work is looking to identify the challenges faced by different industries and retailers during an epidemic outbreak and explicitly for the COVID-19 pandemic.

Major sectors where this combination and interaction of consumer and industries can work through the new technologies-based system which is incubated in shell of industry 4.0

A. Airport/Air travel- Airport and air travel where among those industries which saw drastic downfall due to this pandemic air travel where restricted so that spread of virus doesn't take place but now since unlock procedure are being implemented to boost economy of nation smart moves need to be taken by the organization which can help them to revive ,here technologies can play a good and make travel for consumer smooth One of the common technology/IT solutions that the airport can implement and conduct during this period is the use of biometric solutions (facial/iris recognition) which also include the paperless process. By using this technique of solutions, it will largely reduce contact at all passenger touchpoints such as check-in counter (including baggage drop), immigration clearance. Accelerate implementation of biometric capabilities that significantly reduce human to human interaction points from curb to gate This would make hassle free travel and safe for commutators [6].

B. Remote area/Community - Advance digital technologies provide telemedicine and mobile app-based system which can provide service for proper preventive and control of this virus and can fulfill the requirement of community from anywhere in country to any location. Digital technologies are helpful for distance education, remote and online learning during the emerging of COVID-19 pandemic. During the lockdown, these technologies are helpful for teaching and learning process in remote areas.so that in any situation from best to worst the learning process of induvial should not stop Collection of data can be done easily of everyone with the help of sensors incorporated in mobile phones, robots etc. The data collected is then sent for analytics and decision making to the central cloud server eg Aaroyga setu application which is connected to everyone in present day this application can help you in identify the situation of this pandemic this is because of big data storage and quick response to real time action. Big data can be highly useful for analyzing and forecasting the reach and impact of the coronavirus on people and subsequently equip the scientists, doctors, epidemiologists and policymakers with the latest information which can be very helpful to make better decisions in order to fight against the virus.

C. Logistics/ Storage/Transportation-In these places an autonomous robot can be deployed for patrolling the areas to confirm that there is no delay in shipping and supply items so that there could not any disturbance in any part of supply chain The autonomous may also be deployed in the big warehouses in order to help the staff to perform their duties without any disruption and will increase the efficiency and trust of consumer of contact less delivery and items. On taking hand with technologies Failure of supplies and uncertainty in material supply would decrease with appropriate procedure like using of IOT and big data method at suppliers end and always ensuring real time tracking of supply and demand which will be Impacting ability to ship and receive products on time before the shortages arrives.

4. INVOLVE BENEFITS WITH TECHNOLOGIES:

When involving new techniques of industrial 4.0 during this pandemic situation

1. Providing a better experience without imposing the risks to healthcare and other workers since in present scenario the requirement of induvial cannot be bring to halt for growth and development the industries has to fulfil the demand of community therefore it is essential to build trust towards there worker because in present time due to this pandemic the fear of this virus is so much that no individual wants to take any risk of his healthcare

2. Increasing asset utilization and efficiency. Since the industries cannot work with there full manpower to maintain the parameter of social distancing but therefore it is mandatory to maintain the production level to fulfil the demand so utilizing the industries assets like machine with up to full and optimal strength and try to minimize shutdown of plant so that production don't halt. In case of public area like airport smart robots can be efficient which can provide information on different parameter as per induvial convenience

3. Improved quality of services the presented case study displays the rework percentage was much higher earlier as compare to present time, that is, 15%. but from specialize machines and cost-effective techniques the firm are now able to sustain the rework cost and the man-hours required in the reworking has decreased subsequently. The industries reduce its cost of production by a high percentage of its assets, which will decrease the chances of human error and thus making it cheap for the organization to operate. The sole expenses for the machines are its acquirement and its maintenance which is far much less than what would be the cost of paying to human labor.

4. These digital technologies help people to perform daily life work during the lockdown and restriction as the movement of public was restricted because of spreading of this virus. The concept of work from home came into light, an induvial can now connect himself from his home with any risk and starts to work freely. The digital technologies like video conferencing and online group meets help the community to stay connect without any hinderance. This not only help working official but the school and colleges student to take back their classes from home through these platforms as that there is no gap and time throwaway.

5. By means of technologies accessibility to remote areas has been establish and earlier there were no statical data there to define their living standard, there were no information about what they need ,what are their requirement during any situation throughout the year but now scenario has change with efforts of establishing new technique and advance service with advance technologies their problems and requirement can solved with fast decision making.

6. Overcoming the information communicating challenges ,nowadays due to availability of internet access to everyone now information can be easily broadcasted to everyone very rapidly in earlier times

it was difficult to broadcast the information as limited resources where there but now in today's advance world it has now become easier to stay connected to world for latest update and get aware.

5. MAJOR HIGHLIGHTS FROM ABOVE TECHNIQUES

1. Technology based approach to support society during this pandemic any other future pandemic
2. Key Technologies include Human-machine interaction, hassle free supply chain management system
3. Major role of gadgets for surveillance and provide security during action example drone
4. Influence today's society towards future technologies for betterment.

6. CONCLUSION:

The present study aims to identify the operational challenges and other basic problems faced during the COVID-19 outbreak. This section of paper briefs about research scope in any pandemic situation in present or future. Industry 4.0 aims to create intelligent factories where technologies are transformed and upgraded using IoT, CPSs, big data analytics, Industry 4.0 can be helpful to other sectors like service, healthcare, agriculture, and food. This discusses the use of Industry 4.0 technologies for COVID-19 challenges. Along with each challenge, possible research scopes are also explained in brief. Based on the research scope, a roadmap for implementation of Industry-4.0 was proposed.

7. REFERENCES

[1] Javaid, Mohd, et al. "Industry 4.0 technologies and their applications in fighting COVID-19 pandemic." *Diabetes & Metabolic Syndrome: Clinical Research & Reviews* (2020 Apr 24)

[2] Elavarasan, R. M., & Pugazhendhi, R. (2020). Restructured society and environment: A review on potential technological strategies to control the COVID-19 pandemic. *Science of The Total Environment*, 138858.

[3] Kanitkar, T. (2020). The COVID-19 lockdown in India: Impacts on the economy and the power sector. *Global Transitions*, *2*, 150-156.

[4] Kumar, A., Luthra, S., Mangla, S. K., & Kazançoğlu, Y. (2020). COVID-19 impact on sustainable production and operations management. *Sustainable Operations and Computers*, *1*, 1-7.

[5] Kumar, M. S., Raut, R. D., Narwane, V. S., & Narkhede, B. E. (2020). Applications of industry 4.0 to overcome the COVID-19 operational challenges. *Diabetes & Metabolic Syndrome: Clinical Research & Reviews*, *14*(5), 1283-1289.

[6] Serrano, F., & Kazda, A. (2020). The future of airport post COVID-19. *Journal of Air Transport Management*, *89*, 101900.

[7] Shafi, M., Liu, J., & Ren, W. (2020). Impact of COVID-19 pandemic on micro, small, and medium-sized Enterprises operating in Pakistan. *Research in Globalization*, *2*, 100018.

[8] Gupta, A. K., Singh, V., Mathur, P. & Travieso-Gonzalez, C. M., (2020) Prediction of COVID-19 pandemic measuring criteria using support vector machine, prophet and linear regression models in Indian scenario. Journal of Interdisciplinary Mathematics DOI : 10.1080/09720502.2020.1833458.

[9] Singh, V., Poonia, R. C., Kumar, S., Dass, P., Agarwal, P., Bhatnagar, V. & Raja, L. (2020). Prediction of COVID-19 corona virus pandemic based on time series data using support vector machine, Journal of Discrete Mathematical Sciences & Cryptography DOI : 10.1080/09720529.2020.1784535

[10] Kumari, R., Kumar, S., Poonia, R. C., Singh, V., Raja, L., Bhatnagar, V. & Agarwal, P. (2020). Analysis and Predictions of Spread, Recovery, and Death Caused by COVID-19 in India. Bit Data Mining and Analytics

[11] Bhatnagar, V., Poonia, R. C., Nagar, P., Kumar, S., Singh, V., Raja, L., & Dass, P. (2020). Descriptive analysis of COVID-19 patients in the context of India. *Journal of Interdisciplinary Mathematics*, 1-16.

IOP Conf. Series: Materials Science and Engineering 1017 (2021) 012030 doi:10.1088/1757-899X/1017/1/012030

Investigation of rheological and performance characteristics of oxidized polyethylene polymer blended with SBS modified bitumen

Ajay Kumar Mandrawalia[1*], Arun Gaur[1], Abhishek Mittal[2]

[1]Civil Engineering Department, MNIT, Jaipur,302017, Rajasthan, India

[2]Flexible Pavement Division, CRRI, New Delhi 110025, India

*E-mail: 2016rce9507@mnit.ac.in

Abstract. This research article investigates the influence of oxidized polyethylene (PE) polymer on rheological and performance properties of SBS modified bitumen at an optimum dose of 0.5 % by weight. To evaluate the impact of PE polymer on SBS modified bitumen different performance testing such as indirect tensile strength and resilient modulus was conducted on PE polymer blended bituminous concrete mixes and rheology of modified bitumen was determined using dynamic shear rheometer. It was concluded that the intrusion of PE polymer in SBS modified bitumen improved elastic recovery and viscosity. The findings of resilient modulus test indicated that the addition of PE polymer enhanced the response of modified bituminous mixes under repeated loading (M_r value) by 11%, 20% and 16% at 25°C, 35°C and 45°C respectively as compared to SBS bituminous mix.

1. Introduction

India has mixed traffic in terms of both volume and axle loads. About 98 % of the total Indian roads network is comprised of bituminous roads. After laying, most of the roads develop distress conditions like ravelling, rutting, cracking, bleeding, shoving and potholing. Different studies indicated that useful lifespan of road structures have reduced to four to six years due to increased traffic density, overloading of vehicle beyond permissible limits, higher tyre pressure, varying pavement temperatures and an insufficient degree of maintenance [1],[2]. The accelerated deterioration of bituminous pavement affects the availability of funds for new construction. It is a global view that the higher lifespan and performance of flexible pavements must be ensured in order to reduce the cost of maintenance. Therefore, modification of conventional bitumen with natural and synthetic polymer binder gives a way to meet the new required traffic demands.
Increased shear modulus, better resistance to thermal fracture at low temperatures and reduced plastic flow at high temperatures can be achieved by proper modification of conventional bitumen using different polymers [3]. The results of various studies revealed that the performance of conventional 60/70 bitumen can be improved by SBS polymer modification. The various bituminous mixes

prepared with SBS modified bitumen has shown improved resistance to rutting and moisture sensitivity as compared to conventional bitumen. Pavement construction technique requires the binder to remain stable previously and during construction of road. Therefore, for improved stability and workability SBS modified bitumen may be used in conjugation with other polymer additive such as PE polymer for better elastic and tensile properties to bituminous mixes.

2. Materials

2.1. Bitumen

For this study, SBS modified bitumen (STYRELF PMB-40) was used as control binder which was provided by TOTAL Group situated at Rohut-Jodhpur (Rajasthan). PE polymer was provided by Honeywell Group. Worldwide their additive is known as TITAN (7686). The optimum dose of PE polymer determined was 0.5% by weight of control binder and blended with SBS modified bitumen using melt blending technique. For melt blending, SBS modified bitumen was heated and maintained at a temperature of 160°C and the optimum dose of PE polymer was added steadily. The contents were continuously stirred at 1000 rotation per minute for about 60 minutes at 160°C temperature to produce homogenous mixture. The blended bitumen was cooled and stored for further physical testing, results of physical properties of PE polymer blended bitumen and SBS modified bitumen are presented in Table1.

Table 1. Physical properties of control SBS binder and PE modified SBS binder

Sl. No	Physical properties	Control SBS binder	PE modified SBS binder	Specification
1.	Viscosity at 150°C Poise	8.65	8.85	ASTM D2983 [4]
2.	Specific gravity at 27°C	0.98	0.99	IS 1202 [5]
3.	Softening point ^{O}C	68	70	IS 1205 [5]
4.	Elastic recovery %	70	71	IRC SP 53 [6]

2.2. Aggregate

Locally available crushed granite coarse aggregate and stone dust as fine aggregate was used to design bituminous concrete (BC) mix as per MoRTH specifications [7]. Maximum nominal aggregate size of 20mm, 10mm and 6mm were used. The physical properties of coarse and fine aggregate as per IS: 2386-1963 (Part 1-6) [8] are represented in Table 2.

Table 2. Physical properties of aggregate

Sl. No	Physical properties	Observed Value
1.	Specific gravity	
	20mm	2.64
	10mm	2.65
	6mm	2.58
	Stone dust	2.80
	Lime	2.80
2.	Water absorption	0.12

	20mm % 10mm %	0.20
3.	Flakiness and elongation index %	28
4.	Aggregate impact value %	18
5.	Stripping [9] value %	100

3. Mix composition

To evaluate the influence of PE polymer on bituminous mixes, BC mix for 50-65 mm thickness was designed as per MoRTH specifications [7]. Marshall method of bituminous mix design was adopted to determine optimum binder content (OBC) for the required BC mix. Between (160-170)^{O}C mixing temperatures, different Marshall samples were prepared with varying SBS modified binder content at 0.5% incremental increase. Proportioning of different aggregates for BC mix design are shown in figure 1. and observed Marshall parameters are recorded in Table 3.

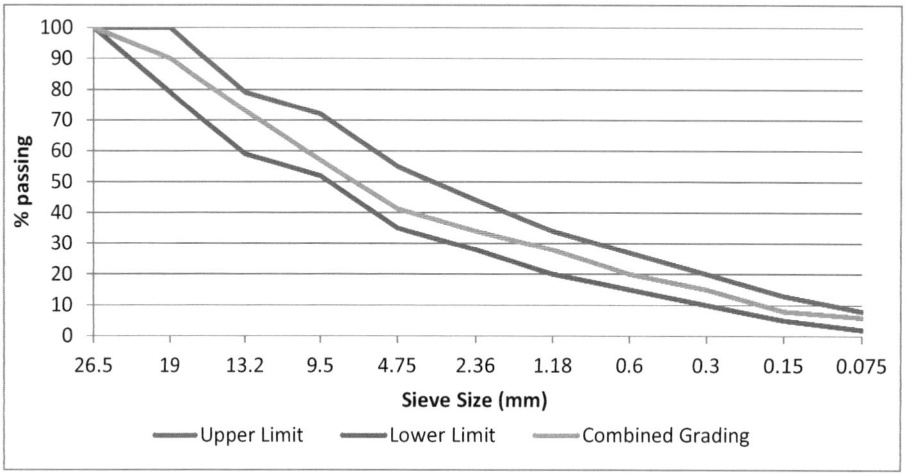

Figure 1. Combined aggregate gradation chart

Table 3. Marshall parameters

% Binder	Bulk Density, (Gb) (gm/cc)	Stability, Kg(f)	Flow (mm)	Air voids, (Vv)%	Voids in mineral aggregates (VMA)%	Voids filled with bitumen (VFB)%
5.0	2.34	2093.48	2.91	5.67	17.28	66.98
5.5	2.35	2187.29	3.17	4.64	17.17	73.13
6.0	2.36	2018.02	4.08	3.64	17.40	79.11
6.6	2.35	1565.26	5.14	3.32	18.10	81.68

OBC determined for required BC mix was 5.8% by weight of aggregate and different properties of mix at 5.8% OBC are as follows:

- Bulk density(gm/cc) = 2.356
- Stability(kN) at 60^0C = 20.45
- Flow (mm) =3.7
- %Voids = 4.04
- %VFB = 77

4. Methodology

An experimental investigation was carried out to evaluate the influence of PE polymer, which was broadly divided into binder and bituminous mix characterization. Dynamic shear rheometer (DSR) test was performed for binder characterization whereas, resilient modulus and indirect tensile strength test was conducted on Marshall samples for bituminous mix characterization.

5. Results and discussion

5.1. Dynamic shear rheometer

As per AASHTO T 315 [10] specification two parallel plate oscillating dynamic shear test at (60-110)^{O}C temperature range was performed to measure the failing temperature of PE polymer blended bitumen and SBS bitumen, which is represented in figure 2 &3 at 10 rad/sec oscillations with 25 mm plates and 1 mm gap. Therefore, the various DSR parameters for PE polymer blended bitumen and SBS bitumen are presented in Table 4. Improvement in DSR parameters due to at low content of PE polymer (0.5%) shows a continuous dispersion of PE particles in binder medium, which were also reported by [11-13] in their studies, with the incorporation different additives in conventional binder.

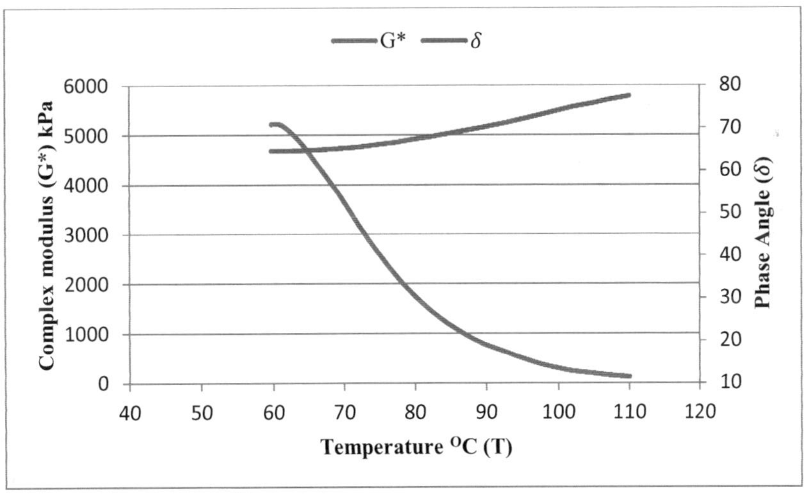

Figure 2. DSR graph (G* vs T and δ vs T) for SBS modified bitumen

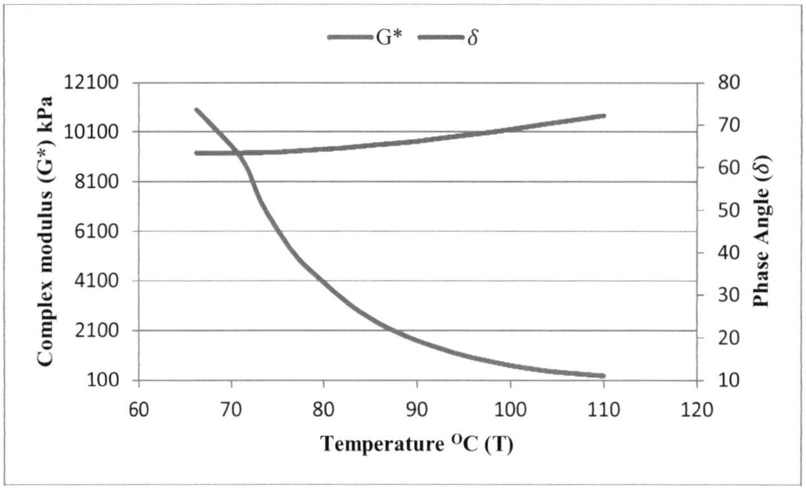

Figure 3. DSR graph (G* vs T and δ vs T) for PE polymer blended SBS bitumen

Table 4. DSR parameter at $G^*/\sin\delta$=1.0 kPa

Binder	Failing Temperature (°C)	Phase Angle (δ)	$G^*/\sin\delta$ (kPa)
Control binder	101	74.5	1.0
PE polymer blended binder	109	72.0	0.9

5.2. Resilient modulus

To evaluate performance and response of bituminous concrete under repeated traffic and temperature condition, resilient modulus (M_r) test was performed as per the guidelines of ASTM D 4123 [14]. Improved resilient modulus (M_r) values of PE polymer modified as compared to SBS bituminous samples at 25°C, 35°C and 45°C temperatures are recorded in Table 5. Similar enhancement of M_r values was reported by [15-17]

Table 5. Resilient Modulus (Mr) values of samples at 25°C, 35°C and 45°C

Mix	Temperature		
	Avg. M_r Value At 25°C MPa	Avg. M_r Value At 35°C MPa	Avg. M_r Value At 45°C MPa
SBS modified BC mix	4804	1700	802
PE blended BC mix	5317	2212	929

5.3. Indirect tensile strength and tensile strength ratio

To measure moisture sensitivity and adhesiveness of bituminous mix in the presence of water in the compacted BC mix, indirect tensile test (ITS) test was conducted as per the detailed guidelines of AASHTO T 283 [18]. Observed results for ITS and tensile strength ratio (TSR) values for PE polymer and SBS modified compacted bituminous samples are recorded in Table 6. From Table 6.

improvement in ITS values is observed, which is due to cross linking ability [19] of PE polymer with SBS binder.

Table 6. TSR values of compacted Marshall samples for PE blended binder and SBS modified binder

Mix	Avg. Dry ITS (Kg/cm^2) T1	Avg. Wet ITS (Kg/cm^2) T2	Tensile Strength Ratio T1/T2	Design Requirement
SBS modified BC mix	8.77	7.50	85.50 %	Minimum 80% (as per MoRTH Table 500-17) [7]
PE blended BC mix	11.30	10.44	92.30 %	

6. Conclusion

The addition of PE polymer (Honeywell Titan 7686) to the SBS modified bitumen has affected the binder properties such as increased softening point, elastic recovery and viscosity profile of the control SBS bitumen. The following conclusion from experimental investigations are:

- PE polymer can be used in conjugation with the SBS modified bitumen to enhances the elastic properties of binder as well as of mix. PE polymer blended BC mixes shows enhanced engineering properties of mix such as stability, workability and resistance to deformation.
- Bituminous concrete mix prepared with PE polymer blended with SBS binder have performed better under the repeated loading. Increased resilient modulus (M_r) values were observed at 25°C, 35°C and 45°C temperatures by 11%, 20% and 16% respectively, as compared to the mix prepared with SBS modified bitumen which will further reduce the required thickness of bituminous layers in flexible pavement.
- Other engineering properties of SBS binder such as viscosity, elastic recovery and critical failing temperature were also found to be improved with the addition of PE polymer.
- Increased TSR values of compacted Marshall samples prepared with PE polymer blended binder was recorded. Increased ITS results of compacted bituminous mix further reduces cracking potential, therefore better mix and pavement life.
- In the present study, the value of G* (complex modulus), G*/sin δ (shear modulus) and G*sin δ (loss modulus) for PE polymer blended SBS binders observed to be higher than control SBS modified binder.

7. References

[1] Airey, G.D. and Brown S.F., "Rheological Performance of Aged Polymer Modified Bitumen's", *Journal of the Association of Asphalt Paving Technologists*, (1998) Vol. 67, pp 66-94.

[2] Airey, G.D., "Rheological Properties of Styrene Butadiene Styrene Polymer Modified Road Bitumen's", *Fuel*, Vol 82, (2003), pp 1709-1719.

[3] Gonzalez, O., Munoz, M.E., Santamaria, A., Garcia, M.G, Navarro, F.J. and Partal, P. "Rheology and Stability of Bitumen/EVA Blends" *European Polymer Journal*, Vol. 40, (2004), pp 2365-2372.

[4] ASTM D2983-19e1, Standard Test Method for Low-Temperature Viscosity of Automatic Transmission Fluids, Hydraulic Fluids, and Lubricants using a Rotational Viscometer, ASTM International, West Conshohocken, PA, 2019.

[5] IS: 1205-1978 Softening Point Test and IS: 1202-1978 Specific Gravity Test.

[6] IRC: SP-53-2010. "Guidelines on Use of Polymer and Rubber Modified Bitumen in Road construction", Indian road Congress, New Delhi.

[7] Ministry of Road Transport and Highways (MoRTH 2013), "Specifications for Road and Bridge Works (Fifth Revision)", Indian Roads Congress, New Delhi, Section 508, Bituminous concrete, Table 500-19.

[8] IS: 2386 (Part 4)-1963 Impact Test, IS: 2386 (Part 1)-1963 Shape Test, IS: 2386 (Part 3)-1963 Specific Gravity & Water Absorption Test.

[9] IRC: SP-53-2010. "Guidelines on Use of Polymer and Rubber Modified Bitumen in Road construction", Indian road Congress, New Delhi.

[10] AASHTO T 315-10, Edition, 2010 - Standard Method of Test for Determining the Rheological Properties of Asphalt Binder Using a Dynamic Shear Rheometer (DSR)

[11] Bulatović, V.O., Rek, V. and Marković, K.J., 2012. Polymer modified bitumen. *Materials research innovations*, 16(1), pp.1-6.

[12] Bhat, F.S. and Mir, M.S., 2020. Rheological investigation of asphalt binder modified with nanosilica. *International Journal of Pavement Research and Technology*, pp.1-12.

[13] Nguyen, V.H. and Le, V.P., 2019. Performance evaluation of sulfur as alternative binder additive for asphalt mixtures. *International Journal of Pavement Research and Technology*, 12, pp.380-387.

[14] ASTM D4123-82(1995), Standard Test Method for Indirect Tension Test for Resilient Modulus of Bituminous Mixtures, ASTM International, West Conshohocken, PA

[15] M. Enieb, A. Diab, Characteristics of asphalt binder and mixture containing nanosilica, *Int. J. Pavement Res. Technol.* 10 (2) (2017) 148- 157.https://doi.org/10.1016/j.ijprt.2016.11.009.

[16] Ganesh, K. and Prajwal, D.T., 2020. Studies on fatigue performance of modified dense bituminous macadam mix using nano silica as an additive. *International Journal of Pavement Research and Technology*, 13(1), pp.75-82.

[17] Bindu, C.S., Joseph, M.S., Sibinesh, P.S., George, S. and Sivan, S., 2020. Performance evaluation of warm mix asphalt using natural rubber modified bitumen and cashew nut shell liquid. *International Journal of Pavement Research and Technology*, 13(4), pp.442-453.

[18] AASHTO T 283, 2014 Edition, 2014 - Standard Method of Test for Resistance of Compacted Asphalt Mixtures to Moisture-Induced Damage.

[19] M. Saltan, S. Terzi, S. Karahancer, Examination of hot mix asphalt and binder performance modified with nano silica, *Constr. Build. Mater.* 156 (2017) 976-984.https://doi.org/10.1016/j.conbuildmat.2017.09.069.

A literature review on injection moulding process based on runner system and process variables

P D Kale[1], P D Darade[2] and A R Sahu[1]

[1]G H Raisoni College of Engineering and Management, Pune, India
[2]Sinhgad Institute of Technology and Science, Narhe, Pune, India

Email: prachidkale@gmail.com

Abstract. Injection moulding is the fabrication process which is the most widely used for polymeric. This paper describes Injection moulding process in detail along with its process parameters and their effects on the moulding component. Runner system is playing a significant role in the quality of the moulded part. Every injection moulding method works when a plastic substance flows from the sprue of the runner system, through a gate into the mould cavity. Runner system is of two types, Hot runner and cold runner system, both having some advantages and disadvantages of each. Research has been carried out which concludes about the effectiveness of hot runner over the cold runner. It has been observed that although hot runner gives better quality of the product, many of the industries use cold runner due to the high cost and complexity of the design. The paper deals with the illustration of mould defects related to the process parameters and the need for its Optimization. This paper focuses on the effectiveness of hot runner systems and scrap reduction of the process by replacing the cold runner system with the hot runner
Keywords: Injection moulding, Process parameters, Runner system, Mould defects.

1. Introduction

Injection moulding is the most traditional plastic parts manufacturing procedure. Using injection moulding, a large range of items are made, varying in scale, complexity and implementation. Hot runner is better than cold runner but is seldom utilised due to high cost and difficulty. Every substance needs a complicated set of parameters such as injection temperature, injection pressure, flow rate, mould temperature, ejection temperature, cooling rate and cycle time. Improper set of parameters leads to many flow lines, burn marks, warping, vacuum voids / air pockets, sink openings, weld lines, jetting. Few defects, including discoloration, plastic use and delamination storage. Quick shots and flash triggers faulty construction or fix. These criteria include process optimization and defect-removal collection. The impact of these criteria on moulding process and recommendations to generate defect-free components needs more study.

2. Literature review

Mr. P. Vinod, Mr. K. Vijaykumar [7] have designed multi-cavity injection mould with HRS and CRS. By comparing both designs, they researched the impact of runner systems, mould cooling and venting. Moulding analysis is carried out using ANSYS.

G. Rajendra Prasad, Dr S. Chakradhar Goud [8] Studied dynamic characteristics like hot runner nozzle strain using analysis in FEA.

N. Divya, Dr. S. N. Malleswara Rao, Dr. V. S. Parameswara Rao, [9] concluded in their paper as the hot runner system accommodates the molten plastic. The runner method determines mould component consistency and productivity. They performed structural and thermal analysis of the mould 's original and updated designs concluding that the modified design produces the best performance.

Rashi A. Yadav, S. V. Joshi, N. K. Kamble [5], Recent studies to design and determine injection moulding method parameters. In parameter environment for injection moulding, several test works were performed focused on various approaches. In the plastic injection moulding (PIM) industry, optimum process parameter settings are critically affected by performance, consistency, and cost of output.

A. Demirer, Y. Soydan [3], Unlike the conventional runner approach, hot runner machine effects on injection moulding method and injection product properties were studied. They used data from method parameters experiments. Injection pressure and temperature change a broad variety. For the hot runner process, injection pressure was marginally lower for higher weight samples. If the temperature of the process increases, shrinkage and warpage increased, reducing with increased injection pressure and happening at a low level where the sample weight was high.

Gurjeet Singh, Ajay Verma [10] Studied primary moulding conditions from design creation to product manufacturing. They studied different factors based on processing parameters. It is concluded that efficiency declines when channelizing efforts to improve quality. Parameters must be optimised to ensure good quality and efficiency. Authors analysed different responses to injection moulding process quality based on output parameters and methods.

Mehdi Moayyedian, Kazem Abhary [11], the injection moulding method implemented a new gate geometry. It was observed that current edge gates corners create turbulence of molten plastic leading to internal and external defects. New geometry was introduced to reduce injected parts' internal and external flaws. The study's goal was to make the final piece easier geometry, which eliminates the last part 's apparent blemish after de-gating.

V.Chandra Sekhar, N. Jaya Krishana [12] Suggested design for two circular flat plate 1 mm wide. Contribution of this research was to change the existing edge gate geometry by eliminating rectangular edge gate corners to minimise scrap occurrence of injected bits. Smooth plastic flow through cavities often prevents internal and exterior defects. The result reveals no shot-filling cavities. No weld lines, meld lines or sink marks were identified with new edge gate design. The experiment ends with an added portion of the initial edge gate step, less noticeable than the current edge gate.

Harold Godwin [13], Mold Making Technology — Competition and price pressure in the injection moulding industry affected many manufacturers of OEMs, moulders, and moulds. The required operating costs were due to increased resin, electricity, human capital, resulting in profit pressure. The decision to use a hot runner was noticeable for medium and large moulded component volumes.

Sal Benenati,[14], Hot runner systems are so popular injection moulds. Hot runner suppliers must adjust part designs to satisfy more strict output and material demands, sometimes adding complexity to products. Some techniques , for example, use moving motors to control valve gates, some need specific heaters, and some require unique mould tolerances to function properly. Standard convection, low shear velocity, and high viscosity-dependent effects were studied.

Lee and Lin et al. [15] Built a multi-cavity mould runner and gating system. Use Finite Element Theory (FEM) network. Optimal runner unit parameter used to minimise injection mould warp, FEM, Taguchi phase and adductive framework. Processes during mould filling, enhancing moulding condition. Model injection mould simulation at steady flow rate. Finite differences method offers strong consent for methodological solutions.

Different gate sizes and locations using flow simulation have been detected for defects reduction such as weld lines and air traps, air traps and warpage can be managed by varying process parameters [17]. Part flow-reducer studied using Autodesk Mould Flow tools. The mould flow analysis is used to predict the

piece's deformation and change the design accordingly. [18]. The plastic toy building block section is analysed; cycle time is successful for four cavities as filling and cooling time for four cavities does not improve in single cavity configuration. The outcome indicates that both parameters influence product consistency, Plastic Flow Simulation analyses the flow of molten plastic to optimise part and mould designs, reducing possible part defects [19]. Simulation research was undertaken for optimum gate location. Comparing two gate positions was achieved. Comparative analysis found the optimal gate location. Several product flaws decrease due to improper eating. [20]. Comparative analysis of the use of different gate types and runner system for the same job, resulting in a satisfactory reduction of moulding defects and increased pressure in the component, which is also within limits [21].
Vikas B J and Chandra Kumar R [22] Optimizing scrap processing period. Manual gui plays a key role in producing plastic components without compromising product consistency. This paper explains gate position and size effect by repetitive analyses. Plastic flow advisor programme prevents fill, scrap and automated degassing. Simulation analyses method parameters including fill time, shrinkage, weld lines, pressure decrease, and air traps in successive tests. Experimental verification for current improved gate location in the injection moulding method. Reduced shrinkage and ventilation, minimising defeats.
Simulation software has greatly impacted the injection moulding industry [23]. Cheaper simulation simulations are an alternative to traditional factory-floor experiments. Study on developing plastic injection moulding method has now expanded a tonne. Sadeghi's [24] Built a neural network model to predict consistency or soundness of injected plastic parts based on major process variables and differing material grade. The system shortened preparation period, improving process requirements or operational parameters. Lee and Kim [25] Used a revised, complicated warpage reduction method by optimising various surface thicknesses, further growing the warpage by reaching optimal processing conditions. Ozcelik and Erzurumlu [26] Comparison acrylic warpage moulding for ANOVA. Neural, genetic algorithm. Sahputra [27]. The results of the injection moulding system and simulation analysis of MPI software were contrasted with those of a limited manufacturing process MPI was found to measure the optimum pressure required to correctly fill the mould cavity and graphically track weld lines. MPI offered further simulation analysis.

3. Moulding machine

The various components of the in-line injection screw moulding system are shown in the following figure.

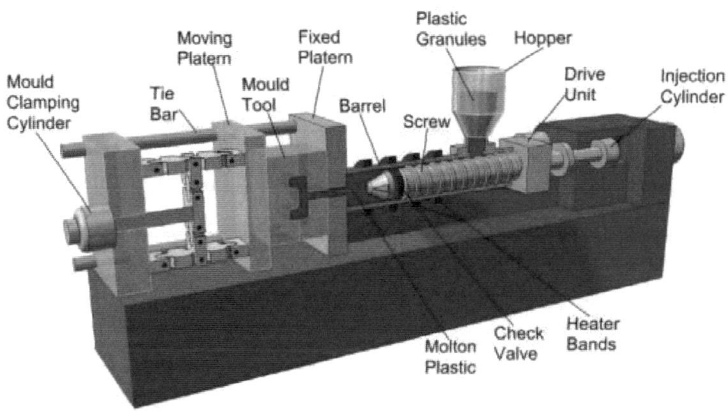

Figure: 1: Major parts of Injection moulding machine [1]

Clamping device, plasticizing device, and drive unit are a standard injection moulding machine 's key units; as seen in Fig.1. Mold-containing clamping system. For locking, clamping and moulding. Its key components are shifting and fixed frames, tie bars, opening, closing and clamping mechanisms.
Heats plastic and injects it into the mould. The drive unit involves plasticizing and clamping. Ultimate clamping force also classifies moulding devices of the system. The force that forces the two moulds together to avoid mould opening due to plastic melting in the mould. The clamping intensity of standard injection moulding machines varies from 200-100,000 kN.

4. The runner system [2]

Sprue, racer, gate and cavity are main mould sections. The sprue is the channel, cut from the nozzle to the runner in the stationary plate. The plastic melt enters the mould, running through a runner system, into the mould cavities through the gates. The runner's gate connects. The gate cross-section is usually small, so the runner can be quickly extracted from the piece and leaves no broad gate mark on the part. The runner device is part of the injection moulding machine since it plays an essential function in moulding.
The injection moulding method will use two different moulding systems — cold runner moulds and hot runner moulds. Each injection moulding system operates as sprue plastic content flows through the mould cavity.

4.1Cold Runner Molds[1]

This process holds runners at the same temperature as moulds. Cold runner system consisting of two or three surfaces. The two-platform system is simple. It needs an ejection device to remove both the runner and the component. Three-plate device requires the area to be ejected separately. The three-plate moulds make construction flexible. In a cold runner method, each moulding cycle moulds a new runner and the runner is ejected along with the moulded pieces. Runner plastic can also be reprocessed and re-developed. The aim of runner device design is to make plastic hit all gates simultaneously. An significant problem in multi-cavity moulds is the amount of cavities in a rectangular runner. Circular runner can use any number of cavities. In the cold runner method, the runner dimension must be larger than part. Ensure the fluid fills the piece properly and the mould is not underfilled.

4.1.1 Advantages:

- It is accessible to maintain and inexpensive
- Due to less heat sensitivity, it is suitable for more types of polymers.
- Colour changes can be made quickly.

4.1.2 Disadvantages:

- It is a Longer process compare to the hot runner system.
- Robotics or a person to remove runners is needed.
- If the runners cannot be melted down properly and recycled, then it Creates waste.

4.2 Hot Runner Molds [1]

Hot runner moulds are two surfaces. These plates are heated by multi-system. There are two types of hot runners, insulated internally. Externally heated moulds are suitable for heat-sensitive materials, whereas internal heated moulds are ideally flow-stable. Runners can be heated by including coils. We can use heating rods and heating pipes as well. Throughout the runner process, the material remains molten until it enters the mould cavity.

4.2.1 Advantages:

- Cycle time is Faster.
- Waste is less
- laminated because of no runners

- Adapts bigger components and greater manufacturing rates
- Quality of the components is better with the right consistency
- Less friction to force the liquid mixture into the mould cavity

4.2.2 Disadvantages:

- Higher cost to produce moulds and maintain equipment
- Colour changes cannot be made quickly.
- It is not suitable for some thermally sensitive material.

5. Process Variables [3]

Process variables are divided into five major types including speed, pressure, time , temperature, and stroke. These relationships cannot be readily separated. They boost hydraulic back pressure, adjust linear retraction speed. It allows improve screw recovery time, melt temperature, and homogeneity. Mould fill time varies, injection strength, mould temperature, ejection temperature and product measurements may be changed as melt temperature rises. It concludes as the pressure variable shift impacts three other variables and affects the consistency of mould components.
Changes in process parameters influence the reliability of the moulding process, which can lead to faulty parts. To rectify this disruption, process variables must be monitored. The incorrect hopper throat temperature has been found to cause short mouldings. It misleads the moulder to alter other variables including holding pressure, shot length, mould filling speed, etc.
The following are common process variables to track and manage throughout each step.

5.1 process variables related to velocity

1. Opening and closing velocities
2. Injection velocity
3. Screw velocity
4. Retraction of components

5.2 Pressure related process variables

1. Pressure at the Injection
2. pressure needed for holding
3. back pressure value for hydraulic

5.3 Time-related process variables

1. Time of Injection
2. Time for Holding pressure
3. Lead time
4. Time for Cooling
5. Time for total Cycle

5.4Temperature related process variables

1. Melting temperature
2. Mould Inside temperature
3. temperature of Barrel
4. Temperature of Cooling water

6.Summary

Injection moulding is the most common production process of plastic pieces. Using injection moulding, a broad variety of products are produced, varying greatly in size, sophistication and implementation. Hot runner is stronger than cold runner, but is seldom used due to its high cost and complexity nature. Each material requires a complicated collection of parameters such as injection temperature, injection speed, flow rate, mould temperature, ejection temperature, refrigeration rate and cycle time. Improper collection of flow patterns, burn marks, warping, vacuum voids / air pockets, sink gaps, weld lines, jetting. Few faults, including discoloration, plastic storage and delamination. Fast shots and flash cause bad design or repair. These parameters include process optimization and selection to remove defects. The effect of these requirements on the moulding method and guidelines to manufacture defect-free components need further research.

References

[1] N.Divya, Dr.Ch.V.S.Parameswara Rao,Dr. S.S.N.MalleswaraRao, (March 2017) " Multiintegral Analysis Of Injection Mould With Hot Runners For Gate", International Journal of Scientific and Research Publications, Volume 7, Issue 3, ISSN 2250-3153

[2] A. Demirer, Y. Soydan, et al. (2007) "An experimental investigation of the effects of hot runner system on injection moulding process in comparison with conventional runner system", Science Direct, Materials and Design 28 ,1467–1476

[3] PK Bharti, M. I. Khan, 2010, "Recent Methods For Optimization Of Plastic Injection Molding Process –A Retrospective And Literature Review", International Journal of Engineering Science and Technology, Vol. 2(9), 4540-4554

[4] Rashi A.Yadav, S.V.Joshi, N.K.Kamble, (December-2012), "Recent Methods for Optimization of Plastic Injection Molding Process - A Literature Review", International Journal of Scientific & Engineering Research Volume 3, Issue 12, ISSN 2229-5518

[5] Ivan W.M. Chana,*, Martyn Pinfold a, C.K. Kwongb, W.H. Szeto, (2014) "Automation and optimization of Family Mould Cavity and Runner Layout Design (FMCRLD) using genetic algorithms and mould layout design grammars", Elsevier, Computer-Aided Design 47 118–133

[6] Mr P. Vinod, Mr K. Vijaykumar, May -2017 Multi-cavity hot runner injection mould tool polymer", International Research Journal of Engineering and Technology (IRJET), e-ISSN: 2395 -0056, Volume: 04 Issue: 05,

[7] G. Rajendra Prasad, Dr S. Chakradhar Goud, Multiintegral May -2017, "Analysis Of Injection Mould With Hot Runners For Gate", International Journal Of Engineering Sciences & Research Technology, ISSN: 2277-9655,

[8] N. Divya, Dr Ch. V. S. Parameswara Rao, Dr S. S. N. Malleswara Rao, March 2017, "Multiintegral Analysis Of Injection Mould With Hot Runners For Gate", International Journal Of Engineering Sciences & Research Technology, Volume 7, Issue 3, ISSN 2250-3153

[9] Gurjeet Singh, Ajay Verma, (2015) , "A Brief Review on injection moulding manufacturing process", 5th International Conference of Materials Processing and Characterization (ICMPC 2016), ScienceDirect Materials Today: Proceedings00

[10] Mehdi Moayyedian, Kazem Abhary, Romeo Marian, February 2015, 2nd International Materials, Industrial, and Manufacturing Engineering Conference, MIMEC2015, 4-6, Bali Indonesia,

[11] V.Chandra Sekhar, N. Jaya Krishana, Hot Runner Mould Design and Plastic Flow Analysis for CAP, International Journal & Magazine of Engineering, Technology, Management and Research, ISSN No: 2348-4845

[12] Peter Unger, 2006 —Hot Runner Technology, Hanser Gardener Publication,.

[13] D.Frenkler, H. Zawistowski, — Hot Runners in Injection Moulds, Rapra publishers, Shawbury 1998, p.267.

[14] Spina R., (2004), Injection moulding of automotive components: comparison between hot runner systems for a case study.J. Mat. Process. Tech. 155-156 1497-1504.

[15] Tang SH, Kong YM, Sapuan SM, Samin R and Sulaiman S. (2006) ,Design and thermal analysis of plastic injection mould.J. Mat. Process. Tech. 171(2) 259-267.

[16] Vikas. B. J and Chandra Kumar. R, 2013 "Influence of Feeding System in Injection Moulding for Lower Washer of Bearing," International Journal of Research in Engineering and Technology, pp. 396-399.
[17] V. V. Andhalkar and Dr S. R. Dulange, 2017 "Injection Molding Methods Design, Optimization, Simulation of Plastic Flow Reducer Part by Mold Flow Analysis," International Research Journal of Engineering and Technology, vol. 4, no. 6, pp. 1742-1746.
[18] M. Salunke, R. Kate, V. Lomate and G. Sopal, 2015 "Injection Moulding Design Methods Optimization", Simulation of Plastic Toy Building Block by Mold Flow Analysis, International Journal of Mechanical Engineering and Technology, vol. 6, no. 6, pp. 33-42.
[19] S. R. Pattnaik, D. B. Karunakar and P. K. Jha, 2012 "Application of Computer Simulation for Finding Optimum Gate Location in Plastic Injection Moulding Process," International Journal of Advanced Engineering Research and Studies, vol. 1, no. 2, pp. 159-161,.
[20] K. R. Yathish Kumar and P. Nagaraja. R, 2014 "Significance of Mold Filling Analysis for Finding Optimal Gate Location in Injection Molding Process for Bobbin.," International Journal of Scientific and Research Publications, vol. 4, no. 4, pp. 1-8.
[21] Vikas B J, Chandra Kumar R, Influence Of Feeding System In Injection Moulding For
[22] Lower Washer Of A Bearing", International Journal of Research in Engineering and Technology eISSN: 2319-1163 | pISSN: 2321-7308
[23] Manzione, L.T. Applications of Computer-Aided Engineering in Injection Moulding. Hanser, New York,1987.
[24] Sadeghi, B.H.M., 2000, A BP-neural network predictor model for the plastic injection moulding process. Journal of Materials Processing Technology, , 103, 411-416.
[25] Lee, B.H., Kim, 1995,B.H. Optimization of part wall thickness to reduce warpage of injection-moulded parts based on the modified complicated method. Polymer-Plastic Technology and Engineering, 34, 793–811.
[26] Ozcelik, B., Erzurumlu, T., 2006, Comparison of the warpage optimization in the plastic injection moulding using ANOVA, neural network model and genetic algorithm. Journal of Materials Processing Technology, 171,437–445.
[27] Sahputra, I.H., 2-4 Dec. 2007, Comparison of two flow analysis software for injection moulding tool design. Proceedings of the International Conference on Industrial Engineering and Engineering Management , Singapore.
[28] Chen, Wu-Lin, Chin-Yin Huang, and Ching-Ya Huang, (2013), "Finding an efficient frontier of process parameters for plastic injection moulding." Journal of Industrial Engineering International 9, no. 1 1-11.
[29] Stanek, Michal, David Manas, Miroslav Manas, and Jakub Javorik, 2011,"Simulation of the injection moulding process." In Proc. 13th WSEAS International Conference on Automatic Control, Modelling & Simulation, pp. 231-234.
[30] Islam, Aminul, Hans Nørgaard Hansen, N. M. Esteves, and T. T. Rasmussen. (2013), "Effects of holding pressure and process temperatures on the mechanical properties of moulded metallic parts." ANTEC 2013
[31] A. RiazAhamed, A.K. ShaikDawood, R. Karthikeyan", December 2013, Designing and Optimizing the Parameters which affect the Molding Process using Design of Experiment". Volume 1, Issue 2, October- pp.116-122.
[32] Pareek, Rishi, and Jaiprakash Bhamniya, (2013), "Optimization of Injection Moulding Process using Taguchi and ANOVA." Journal of Scientific and Engineering Research 4, no. 1.

[33] Rahimi, Mohammad, Mohsen Esfahanian, and Mehran Moradi, (2014) "Effect of reprocessing on shrinkage and mechanical properties of ABS and investigating the proper blend of virgin and recycled ABS in injection moulding." Journal of Materials Processing Technology 214, no. 11 2359-2365.
[34] Müller, Egon, Rainer Schillig, Timo Stock, and Miriam Schmeiler. (2014) "Improvement of Injection Moulding Processes by Using Dual-Energy Signatures." Procedia CIRP 17 704-709.2
[35] Chen, Wu-Lin, Chin-Yin Huang, and Ching-Ya Huang. (2013) "Finding an efficient frontier of process parameters for plastic injection moulding." Journal of Industrial Engineering International 9, no. 1 1-11.
[36] Mehta, Nik Mizamzul, Shahrul Kamaruddin, and Abdul Rahim Othman, 2012, "Reducing the Shrinkage in Plastic Injection Moulded Gear via Grey-Based-Taguchi Optimization Method." In Proceedings of the World Congress on Engineering, vol. 3
[37] Fei, Ng Chin, Nik Mizamzul Mehat, and Shahrul Kamaruddin, 2013 "Practical applications of Taguchi method for optimization of processing parameters for plastic injection moulding: a retrospective review." ISRN Industrial Engineering 2013.
[38] Tang SH, Kong YM, Sapuan SM, Samin R and Sulaiman S. Design and thermal analysis of plastic injection mould.J. Mat. Process. Tech. 171(2) (2006) 259-267.
[39] Ozcelik, B., Erzurumlu, T. 2006, Comparison of the warpage optimization in the plastic injection moulding using ANOVA, neural network model and genetic algorithm. Journal of Materials Processing Technology, 171,437–445.
[40] Saman, A.M., Abdullah, A.H., 2009, MAM Computer simulation opportunity in plastic injection mould development for an automotive part. International Conference on computer technology and Development,13-15 Nov. Kota Kinabalu, Malaysia.

Comparative Structural Analysis of Various Composite Materials based Unmanned Aerial Vehicle's Propeller by using Advanced Methodologies

Ramesh M[1], Vijayanandh R[1], Raj Kumar G[1], Vijayakumar Mathaiyan[1], P Jagadeeshwaran[2], Senthil Kumar M[1]

[1]Department of Aeronautical Engineering, Kumaraguru College of Technology, Coimbatore, Tamil Nadu, India
[2]Department of Physics, Rajalakshmi Institute of Technology, Chennai, Tamil Nadu, India

E-mail: vijayanandh.raja@gmail.com

Abstract. In the recent technology over the Unmanned Aerial Vehicle (UAV), the implication of propeller contributes a principal part in the production of thrust by pushing the air with the rotating blades. Especially, propeller's incomparably works perfect in the advanced UAVs, because of its huge contribution to UAV's construction. Analyzing the structural behavior of propeller and its rotodynamic effects with the help of advanced engineering approaches can provide better performance and lifetime in advanced operating conditions. The composites of CFRP (Carbon Fiber Reinforced Polymer) and GFRP (Glass Fiber Reinforced Polymer) are widely used for the propellers, due to the advantage of load resisting capacity, lightweight, higher production. This work deals with the comparative studies of the propeller with the different composites such as CFRP and GFRP to inculcate the structural characteristics of deformation with respect to the applied load, maximum stress, and normal stress by coupled cum advanced simulations for different rotational velocities. The primary outcome of these comparative studies is providing the optimized performance as well as its ultimate lifetime of the rotor component for an advanced UAV. Finally, material optimizations are executed for UAV's propeller by using two advanced numerical methodologies that are FSI (Fluid-Structure Interaction) for the evaluation of structural parameters and coupling of CFD (Computational Fluid Dynamics) - MRF (Moving Reference Frame) for the representation of rotating nature of propeller. Through these approaches and grid convergence test, the best composite material is finalized, which is GFRP. Especially, Epoxy-E-Glass UD and Epoxy-S-Glass-UD are performed better than other composite materials, wherein the Epoxy-E-Glass-UD is reacted low deformed value of 3.7829μm and Epoxy-S-Glass-UD is induced low stress value f 55556Pa than other composites at average loading conditions.

1. UAV's Propeller and its Materials

Generally, the propulsive system of a multi-rotor UAV is focally depended upon its propeller. The endurance is comparatively low in the electric propulsive system based multi-rotor UAVs [1]. Hence, the useful research on propeller can able to increase the performance of a multi-rotor UAVs. Nowadays, the important ongoing research on propellers are, material optimizations for high-speed applications, noise reduction in UAV's propeller, lift enhancement techniques in the propeller and drag reduction techniques in UAVs, etc [2]. The fundamental aims of all these researches are to provide high lifetime and endurance in UAVs. In this work, analyze the structural performance of UAV's propeller in various dynamic environments using advanced and integrated numerical simulation techniques [3]. Currently, the glass and carbon fibers based propellers are drastically used in the UAVs because of its advanced properties. In general, the stiffness to weight ratio is quite high in the

Polymer matrix composite, which can able to provide better performance at critical phases such as high forward and vertical speed conditions, high gust load environments, and large endurance applications [4]. Also, rotodynamic nature-based propellers have severely occurred structural failure, and thereby the performance may chance to collapse, this complicated condition is perfectly tackled by Polymer matrix composite based propellers [5]. In polymer matrix composite, various varieties are available which are, uni-directional fiber, bi-directional fiber, prepreg production, wet conditional product, etc [6]. All these variations of glass and carbon fibers are implemented in propellers and analyzed by using ANSYS Workbench to find suitable material to withstand effectively at high loading conditions. Finally, the suitable material with its unique property is to be found for all the loading conditions [7].

2. Problem identification and Solution Techniques

The fundamental platform of this work is the propeller, which has 6 inches in diameter and 4.5 inches in pitch. The conceptual design of the used propeller is shown in Figure 1, in which CATIA is used as a modeling tool. The fundamental platform of this research is to select suitable composite materials for UAV's propeller under rotodynamic loading conditions. Because of the critical working nature of the propeller, the attainment process of the desired output and its reliability is quite complicated. To attain reliable output for this critical and complicated problem, the advanced numerical simulation approach is used here because it has a flexible facility at any stage and it is an easier adaptation with the advanced techniques [8]. Therefore, in this work, ANSYS Workbench based numerical analyses are conducted on the UAV's propeller for different composite materials. In this simulation, two different environments are used separately and then coupled with the help of one way coupling FSI to solve complicated rotodynamic and thereby its structural effects [9]. The fluid environmental conditions are analyzed on propeller with various RPMs by using the CFD tool, i.e., ANSYS Fluent. To represent the RPM of the propeller, the MRF (Moving Reference Frame) approach is implemented in the CFD tool. With these advanced implementations, the CFD analyses are executed on the UAV's propeller and thereby the fluid pressures are estimated. The estimated total pressures are transferred into the structural environmental analysis as an external load category on the UAV's propeller with the help of a one-way coupling FSI methodology. The one-way coupling FSI is one of the advanced numerical techniques, this is used for load sharing between two different environments to provide reliable output. In this work, the fluid load generation on the propeller to be captured perfectly because of the unpredictable nature of the fluid behavior and that can be transferred to structural analysis without any adjustments. Hence, the advanced numerical tool like one-way coupling FSI is mandatory to work as a coupling tool in-between dynamic fluid analysis and structural analysis [10]. The FEA tool, ANSYS Structural, is used for the estimation of structural parameters such as strain energy, various stresses, and total deformation. Finally, the FSI analyses are executed on the UAV's propeller for eight different materials, which are Epoxy Carbon Woven – 230 GPa – Wet, Epoxy Carbon Woven–230 GPa–Prepreg, Epoxy Carbon UD–230 GPa–Prepreg, Epoxy Carbon UD–230GPa–wet, Epoxy S–Glass UD, Epoxy E-Glass Wet, Epoxy E-Glass UD, and Aluminium alloy. In which, the Aluminium alloy is used as a reference element, thus the structural outputs are compared with Aluminium alloy, and then optimization is finalized.

3. Material Optimization based on FSI

3.1. FSI

FSI is a kind of integrating and advanced methodology, which is used to couple data between two different working environments. The fundamental objective of this work is to select the best composite material based on rotodynamic com structural effects [10 - 11]. Thus, the representation of the rotodynamic effect and thereby the structural effects on UAV's propellers are unavoidable ones, which provided the path to use one-way coupling FSI [12]. Based on the nature of load sharing, the FSI is

IOP Conf. Series: Materials Science and Engineering 1017 (2021) 012032 doi:10.1088/1757-899X/1017/1/012032

primarily classified into two types, which are one-way coupling FSI and two-way coupling FSI. In one-way coupling, the loads are shared through direct coupling whereas, in two-way coupling FSI, the loads and data are shared with the help of a system coupling system [13 - 14].

3.2. CFD

The first and foremost data of this work is aerodynamic pressure acting on the propeller due to its rotational, which is executed with help of the CFD tool, i.e., ANSYS Fluent. The average speed of the civilian UAVs is flying at the speed of 10 m/s so the same one is used as velocity inlet for this fluid analysis [15]. Thus, the aerodynamic fluid density does not vary and thereby the pressure plays a vital role in the fluid analysis so the pressure-based solver is implemented. Because of the critical work nature of the propeller, the turbulent flow is used at an acceptable rate [16]. A separate comparative analysis is completed with respect to RPM versus thrust generation on the UAV's propeller, in which 2300 RPM can have the capability to provide the thrust of 500 grams. Apart from these primary conditions, the fixed rectangular control volume is used, which has the dimensions of 20x5 inches in the 'X' direction, 10x5 inches in both 'Y' and 'Z' directions. The rotating control volume is also implemented to represents MRF regions with respect to Propeller, in which the dimensions are used as 6X5 inches in diameter. The fine tetrahedral 3-D elements are used in which the finely tuned conditions are used at no-slip regions [17]. Figure 2 is revealed the finite volume model of the propeller, which can able to predict the fluid properties in and around the propeller, and then further needful results are executed. The pressure and velocity variations on the propeller are revealed in Figures 4 and 5. The grid convergence study is organized in this CFD simulation to enhance the reliability of its outcomes. The comprehensive report is generated and revealed in Figure 3.

Figure 1. Conceptual design of a 6-inch propeller

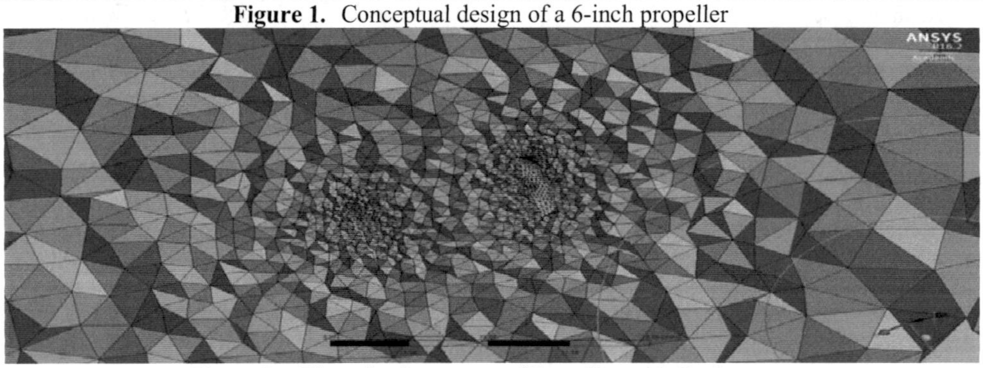

Figure 2. Discretized structure of Propeller with Enclosure

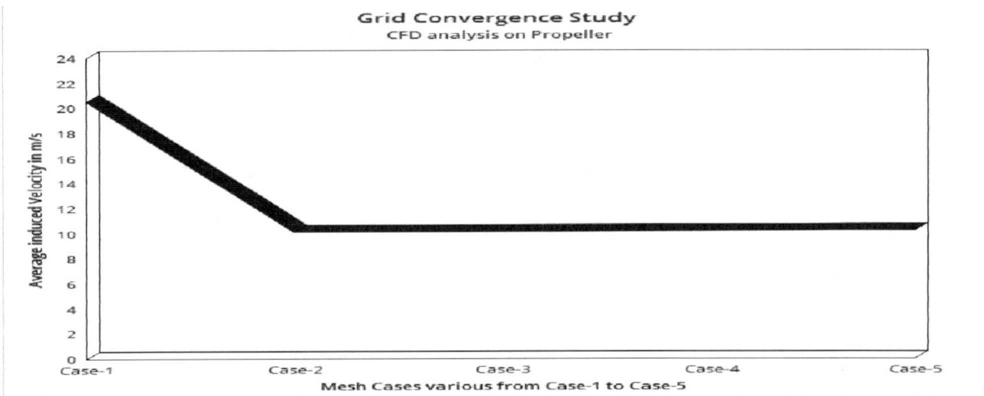

Figure 3. Grid Convergence Test

In this grid convergence study, five mesh cases are implemented, in which Case-2 is short-listed as an optimized case based on the induced velocity outcome. All the five cases are constructed based on fine mesh setup as primary facility and secondary facilities such as inflation, face mesh, refinement, etc.

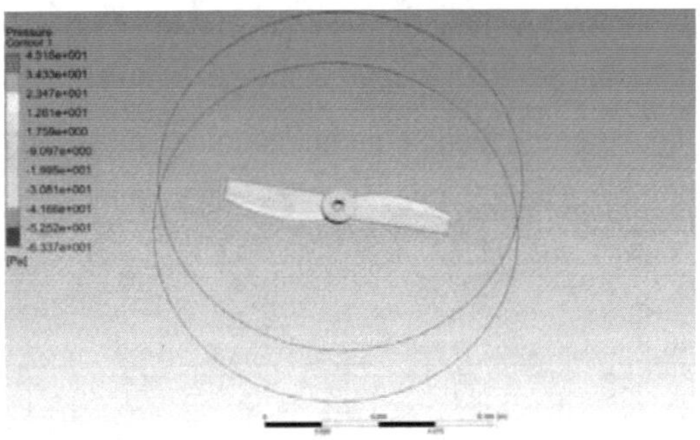

Figure 4. Pressure variations on the propeller with MRF region

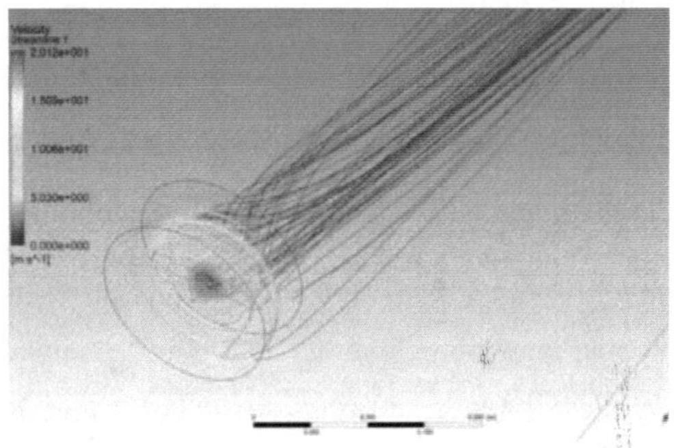

Figure 5. Velocity variations on the propeller with MRF region shown in black color

3.3. Computational Structural Results

The structural parameters such as deformation and stresses are estimated with the help of the FEA numerical tool, in which the roller supports are provided at the hub section of the propeller. The external fluid loads are transferred and applied at the FSI region on the propeller through a one-way coupling FSI facility [16].

3.3.1. Results – Aluminium Alloy – Reference Components

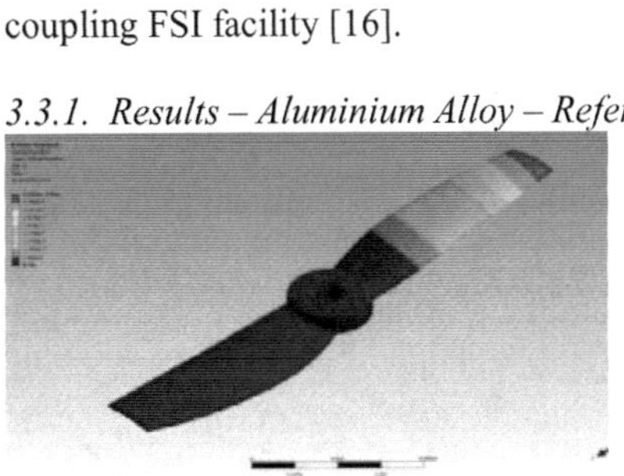

Figure 6. Deformed structure of Aluminium Alloy's Propeller

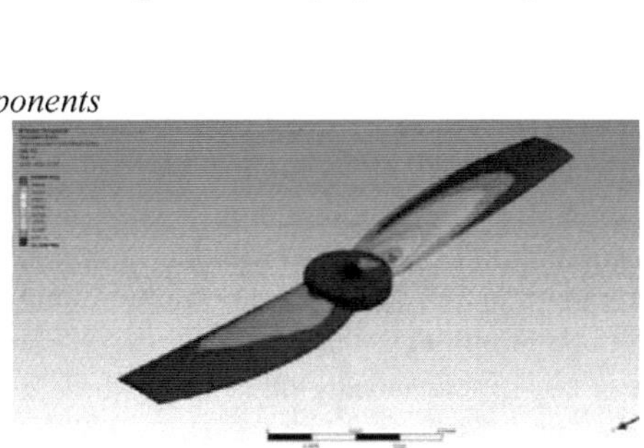

Figure 7. Equivalent Stress variations of Aluminium Alloy's Propeller

3.3.2. Numerical Results – Epoxy Carbon UD – 230 GPa – Prepreg

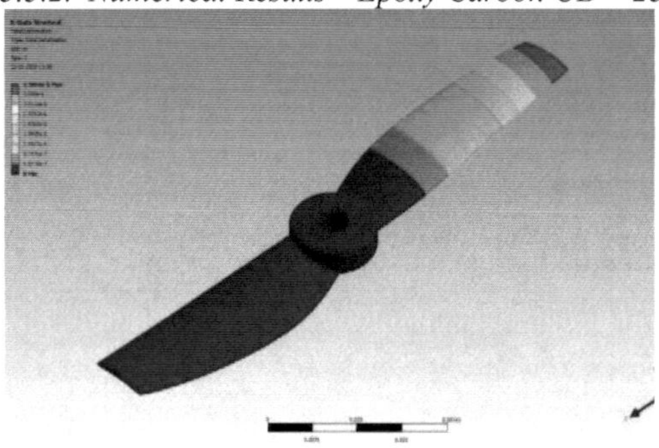

Figure 8. Variations of deformation of Epoxy Carbon UD – 230 GPa – Prepreg's Propeller

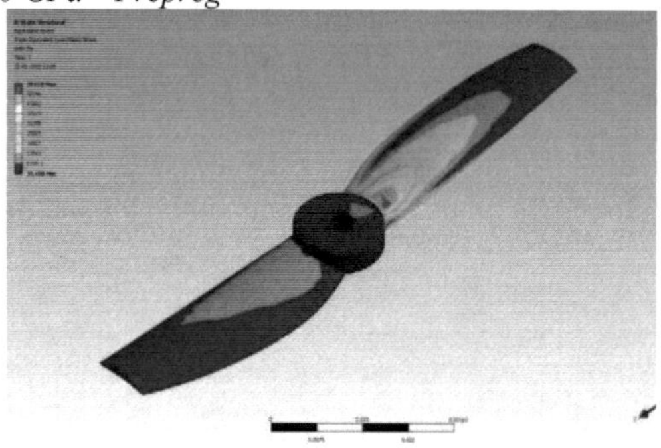

Figure 9. Variations of equivalent stress of Epoxy Carbon UD – 230 GPa – Prepreg

Primarily, four materials are short-listed and analyzed, in which one material is used for reference. The entire materials' details and the structural results are listed in Table 1. The deformations, normal stress, and equivalent stress variations of Aluminium Alloy [reference material] are shown in Figures 6, 7, and 10, in which the values are considered as maximum output data. Apart from this primary material, the structural results of Epoxy Carbon UD-230 GPa-Prepreg are also revealed in Figures 8, 9, and 11.

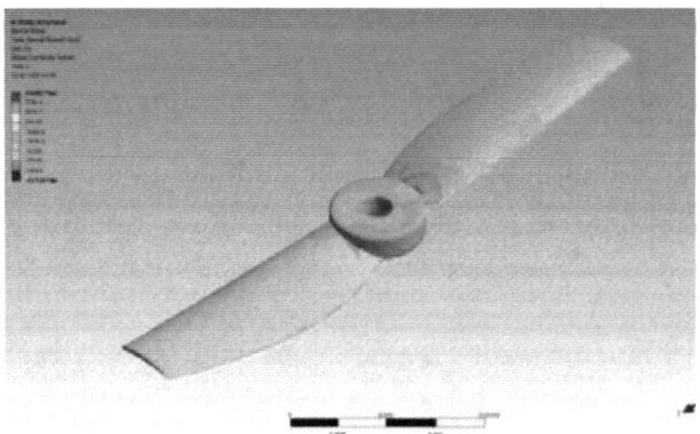

Figure 10. Normal Stress variations of Aluminium Alloy's Propeller

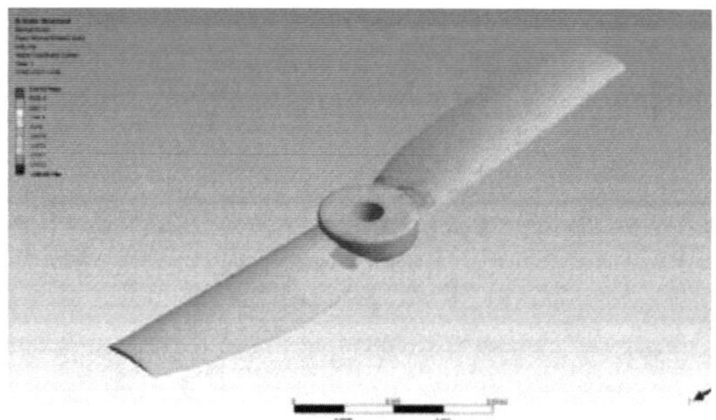

Figure 11. Variations of normal stress of Epoxy Carbon UD – 230 GPa – Prepreg

Table 1. Comparative structural outputs of primary materials

Sl. No	Material Name	Equivalent Stress (Pa)	Total Deformation (μm)	Normal Stress (Pa)
1	Aluminium Alloy	55589	0.53514	22739
2	Epoxy Carbon Woven – 230 GPa – Wet	55960	5.0932	25512
3	Epoxy Carbon UD – 230 GPa – Prepreg	56410	4.3864	28658
4	Epoxy S – Glass UD	55556	4.6575	26250

From Table 1, it is understood that Epoxy-S-Glass fiber based composite is reacted with low cumulative cum equivalent stress, thus the glass fiber is more suitable to work as a manufacturing material for UAV's propeller.

3.4. Computational Structural Results of various Glass fiber results

Owing to the top role in the primary tests, the structural analyses are extended to all the other good glass fibers such as Epoxy E-Glass UD and Epoxy E-Glass wet [11, 12].

3.4.1. Numerical results of Epoxy S-Glass UD

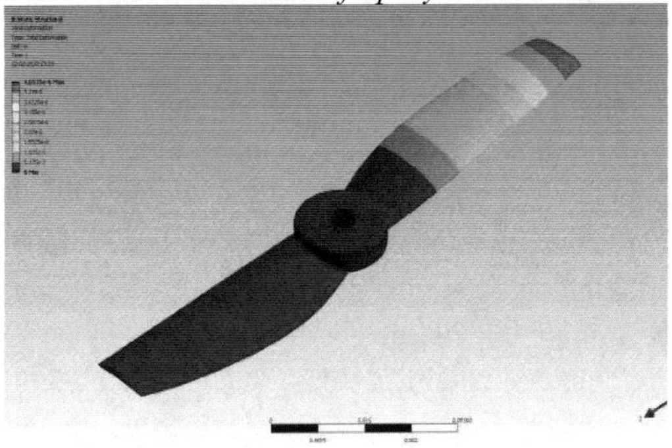

Figure 12. Deformed structure of Epoxy S-Glass UD's Propeller

Figure 13. Equivalent Stress variations of Epoxy S-Glass UD's Propeller

The entire structural outputs of glass fibers based composites are analyzed and listed in Table 2. For the sample case, the structural results of the Epoxy S-Glass-UD based propeller are shown in the Figures 12, 13, and 16, wherein Figure 12 is revealed the variations of deformation, Figure 13 is revealed the variations of equivalent stress, and Figure 13 is revealed the variations of normal stress.

Table 2. Comparative structural outputs of GFRP materials

Sl. No	Material Name	Equivalent Stress (Pa)	Total Deformation (μm)	Normal Stress (Pa)
1	Epoxy E-Glass UD	55725	3.7829	24279
2	Epoxy E-Glass Wet	55638	4.1951	23060
3	Epoxy S-Glass UD	55556	4.6575	26250

3.5. Computational Structural Results of various Carbon fiber-based composites results

Comparatively, the deformation of the propeller is also having capable to judge the best material. Thus the low deformed perspective also analyzed, in which carbon UD-Prepreg is reacted with low deformation compared to other materials therefore the FSI analyses are extended for all the Carbon fiber based composite materials. The entire structural results are analyzed and listed in Table 3. The bi-directional carbon fiber-based composite structural results are revealed in Figures 14, 15, and 17, in that the corresponding parameters are deformation, equivalent stress, and normal stress respectively.

3.5.1. Numerical results of Epoxy Carbon Woven – 230 GPa – Prepreg

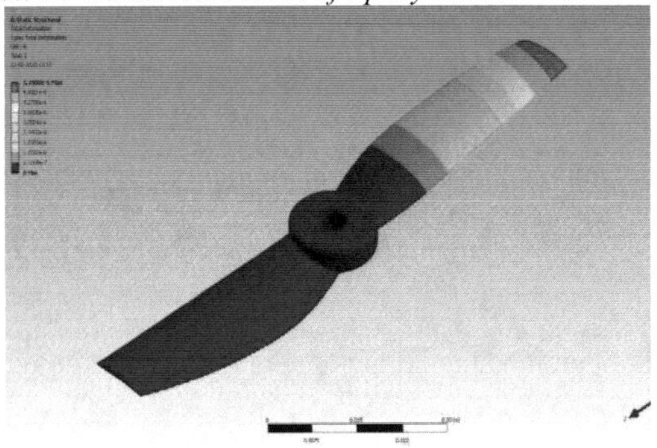

Figure 14. Variations of deformation of Epoxy Carbon Woven – 230 GPa – Prepreg's Propeller

Figure 15. Variations of equivalent stress of Epoxy-CFRP-Woven-230-GPa-Prepreg

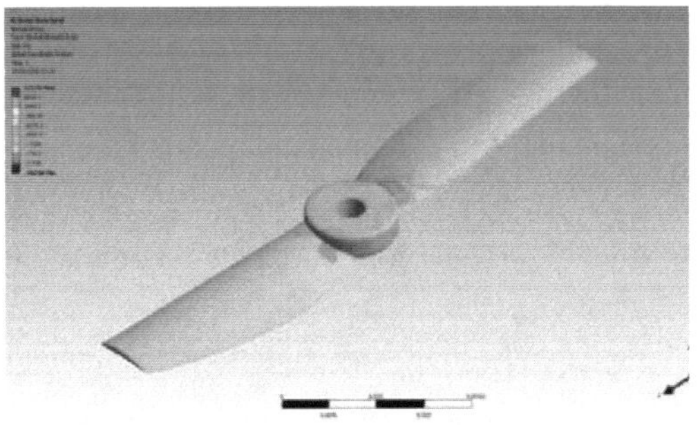

Figure 16. Normal Stress variations of Epoxy S-Glass UD's Propeller

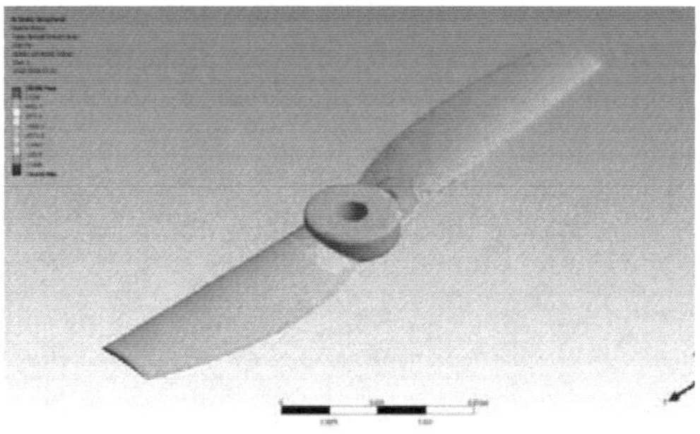

Figure 17. Variations of normal stress of Epoxy Carbon Woven – 230 GPa – Prepreg

Table 3. Comparative structural outputs of CFRP materials

Sl. No	Material Name	Equivalent Stress (Pa)	Total Deformation (μm)	Normal Stress (Pa)
1	Epoxy Carbon UD – 230 GPa – Wet	56119	4.8002	30515
2	Epoxy Carbon UD – 230 GPa – Prepreg	56410	4.3864	28658
3	Epoxy Carbon Woven – 230 GPa – Wet	55960	5.0932	25512
4	Epoxy Carbon Woven – 230 GPa – Prepreg	55590	5.4908	26458

4. Conclusions

The UAV's propeller is used as a platform for this entire comparative work, in which the diameter is calculated as 6 inches, the pitch is estimated as 4.5 inches and the complete model is designed through CATIA. The shortlisted propeller can able to provide an average thrust of 500 grams, which is the maximum thrust requirement by the real time UAVs. The pressure acting on the propeller surface, velocity variations on the propeller are predicted through the CFD tool, ANSYS Fluent for the given boundary conditions of 1 m/s as fluid input velocity and 2300 RPM is given as rotational velocity. The stresses induced in the composite propellers and its displacement variations are predicted by using FEA numerical tool, ANSYS Structural. Finally, the Glass fiber-based composite is selected as a more suitable material for UAVs, based on its low structural outputs and the results are compared with theoretical outputs for validation purposes. The displacement of the Epoxy E-Glass-UD is 3.8 μm and Epoxy-E-Glass UD is created a low stress of 55556 Pa at peak and average loading conditions, which means the stiffness to weight ratio is high in this material. Because of this high stiffness to weight ratio, the nature of the propeller's profile does not affect with respect to working time. Therefore the shortlisted material is fit to provide high structural performance at all the critical environments.

References

[1] Rajagurunathan. M, Raj Kumar. G, Vijayanandh. R, Vishnu. V, Rakesh Kumar. C & Mohamed Bak. K, The Design Optimization of the Circular Piezoelectric Bimorph Actuators Using FEA, International Journal of Mechanical and Production Engineering Research and Development, ISSN(E): 2249-8001, Vol. 8, Special Issue 7, 2018, 410-422

[2] Raj Kumar. G, Senthil Kumar. M, Vijayanandh. R, K. Raja Sekar, Mohamed Bak. K & Varun. S, The Mechanical Characterization Of Carbon Fiber Reinforced Epoxy with Carbon Nanotubes, International Journal of Mechanical and Production Engineering Research and Development, ISSN(E): 2249-8001 Vol. 9, Special Issue 1, 2019, 243-255.

[3] Vijayanandh R., M. Senthil Kumar, K. Naveenkumar, G. Raj Kumar, R. Naveen Kumar, Design Optimization of Advanced Multi-rotor Unmanned Aircraft System Using FSI, Springer Series Title - Lecture Notes in Mechanical Engineering, eBook ISBN - 978-981-13-2718-6, DOI 10.1007/978-981-13-2718-6, 18, pp. 299 – 310, 2019

[4] Vijayanandh R, Ramesh M, Raj Kumar G, Venkatesan K, Senthil Kumar M, Research of Noise in the Unmanned Aerial Vehicle's Propeller using CFD, International Journal of Engineering and Advanced Technology, 8, 6S, 2019, pp.145-150, DOI:10.35940/ijeat.F1031.0886S19.

[5] Senthil Kumar. M, Vijayanandh. R & Gopi. B, Numerical Investigation on Vibration Reduction in Helicopter Main Rotor Using Air Blown Blades, International Journal of Mechanical and Production Engineering Research and Development, Vol. 8, Spl Issue 7, pp.152-164, 2018.

[6] Vijayanandh. R, Ramesh. M, Senthil Kumar. S, Raj Kumar. G, Senthil Kumar. M & Naveen Kumar. R, The Conceptual Design of the Tilt-Copter Based On the Speech Control (A Theoretical Approach), International Journal of Mechanical and Production Engineering Research and Development, Vol. 8, Special Issue 7, pp. 423-439, 2018.

[7] K. Naveen Kumar, R. Vijayanandh, G. Raj Kumar, Hariharan B, S. Guru Prasad, Comparative Approaches for Fatigue Life Estimation of Aluminium Alloy for Aerospace Applications, Int. J. Vehicle Structures & Systems, 10(4), 282-286, doi: 10.4273/ijvss.10.4.11

[8] Arul Prakash. R, Sarath Kumar. R, Vijayanandh. R, Darsi Venkata Praveen Raja Sekar. K & Ananda Krishnan. C, Design Optimization Of Convergent - Divergent Nozzle Using Computational Fluid Dynamics Approach, International Journal of Mechanical and Production Engineering Research and Development, Vol. 9, Spl Issue 1, pp, 220-232, 2019.

[9] R. Vijayanandh, M. Ramesh, K.Venkatesan, G.Raj Kumar, M.Senthil Kumar, and R. Rajkumar, Comparative Acoustic Analysis of Modified Unmanned Aerial Vehicle's Propeller, Advances in IC Engines and Combustion Technology, Lecture Notes in Mechanical Engineering, 45, pp. 557-571, 2021, https://doi.org/10.1007/978-981-15-5996-9_45

[10] P. Jagadeeshwaran, Dr. V. Natarajan, Vijayanandh R, Senthil Kumar M, Raj Kumar G, Numerical Estimation Of Ultimate Specification Of Advanced Multi-Rotor Unmanned Aerial Vehicle, International Journal Of Scientific & Technology Research, ISSN 2277-8616, Volume 9, Issue 01, 2020, pp. 3681 – 3687.

[11] Naveen Kumar K, Vijayanandh R, Bruce Ralphin Rose J, Swathi V, Narmatha R ,Venkatesan. K, Research on Structural behavior of Composite Materials on different Cantilever Structures using FSI, International Journal of Engineering and Advanced Technology, ISSN: 2249 – 8958, 8, 6S3, pp: 1075 - 1086, 2019, DOI: 10.35940/ijeat.F1178.0986S319

[12] Vijayanandh R, Venkatesan K, Ramesh M, Raj Kumar G, Senthil Kumar M, Optimization of Orientation Of Carbon Fiber Reinforced Polymer Based On Structural Analysis, International Journal of Scientific & Technology Research, ISSN 2277-8616, Volume 8 - Issue 11, November 2019.

[13] G. Raj Kumar, Vijayanandh R, Experimental and Numerical Studies on Mechanical characterization of EPDM/S-SBR with Nanoclay Composites, IOP Conference Series: Materials Science and Engineering, 912, 052016, 2020, pp. 1-11, doi:10.1088/1757-899X/912/5/052016

[14] Vijayanandh R, Venkatesan K, Senthil Kumar M, Raj Kumar G, P. Jagadeeshwaran, Raj Kumar R, Comparative fatigue life estimations of Marine Propeller by using FSI, IOP - Journal of Physics: Conference Series. 1473 (2020) 012018, doi:10.1088/1742-6596/1473/1/012018

[15] K. Venkatesan, K. Ramanathan, R. Vijayanandh et al., Comparative structural analysis of advanced multi-layer composite materials, Materials Today: Proceedings, Volume 27, Part 3, 2020, Pages 2673-2687, https://doi.org/10.1016/j.matpr.2019.11.247

[16] Ramesh Murugesan, Vijayanandh Raja, Acoustic Investigation On Unmanned Aerial Vehicle's Rotor Using CFD-MRF Approach, Vol.-2, Proceedings of the ASME-2019, V002T08A005, 7 pages, ISBN: 978-0-7918-8353-2, https://doi.org/10.1115/GTINDIA2019-2430.

[17] Balaji Sonaimuthu, Prabhagaran Panchalingam, Vijayanandh Raja, Comparative Analysis Of Propulsive System in Multi-Rotor Unmanned Aerial Vehicle, Volume-2, Proceedings of the ASME 2019, V002T08A004; 8 pages, ISBN: 978-0-7918-8353-2, https://doi.org/10.1115/GTINDIA2019-2429.

IOP Conf. Series: Materials Science and Engineering 1017 (2021) 012033 doi:10.1088/1757-899X/1017/1/012033

Overview on Hardfacing Processes, Materials and Applications

R.R. Garbade[1], N.B. Dhokey[1]

[1]Department of Metallurgy and Materials Science, College of Engineering Pune 411005, India

Email: rrg20.meta@coep.ac.in

Abstract. Hardfacing alloys are mainly obtained through atomization of powders and by determining of alloying elements in pure form or ferro alloys. They are widely used materials as weld overlay to improve wear and corrosion resistance in applications such as agricultural, construction, mining, etc. Wear is an important factor that controls life of any component. Different types of wear such as corrosion, impact, abrasion, metallic, etc are responsible for worn out of components. Many researches have been done to either reduce the wear or improve the wear resistant of the existing materials. Among these, hardfacing technique has made many developments since last few years. As in nuclear reactors, hardfacing of the grid plate through Laser cladding and Plasma Transferred arc welding, to improve the life of the components and resistance from wear and corrosion. Hardfacing involves applying a consumable material with desired wear properties over a soft base metal to enhance resistance to different wear mechanism. Using Calphad technique for development of desired composition and application of Schaffler diagram has been attempted successfully. In this paper a review of hardfacing processes and materials, highlighting the use of Laser cladding and Plasma transfer arc for overlaying hardfacing alloys has been attempted.

1. Introduction

Hardfacing is a process or surfacing operation where harder or tougher material is applied to a base metal, to extend the service life of new components, or worn down surface of old components. Critical components of machineries used in mining, mineral processing, construction, agriculture and many industrial processes are subjected to severe wear in service.[2] Machine parts deteriorate and fracture early from their intended life not only because of poor operation but also because of wear and abrasion. Degradation of the component surface due to wear has lead to downtimes and increased production cost. Many types of wear exists, some of them are – Abrasion, Impact, Metallic, Heat, Corrosion etc. In agricultural tools, mining and earthmoving equipments the problem is same, where the tools have to work on the hard surface. Welding deposits can make a surface function and reclaim the components by extending their service life. [6] Welding is a key technology to fulfill these requirements and to apply hardfacing alloys. [7]

Hardfacing is a method to improve surface properties of tools and components in which an alloy is homogeneously deposited onto the surface of a base material by different welding processes, with the purpose of increasing hardness and wear resistance.[8] Selection of the material is done on the basis of hardness, mechanical properties , microstructure and wear resistance. Hardfacing can be applied to a new component during its manufacturing or to restore a worn-down surface.

2. Hardfacing processes

Hardfacing is generally applied through welding processes; it is also better from other processes as it can be suitable for almost all alloys. The other methods include Thermal spraying and Cladding. The selection of the welding process depends on size and shape of component, nature of work to be hard-faced, application where it is going to be used and base metal composition. Thermal spraying processes are preferred for applications where thin and hard coatings are required with minimum thermal distortion of the component and good process control. Cladding processes are used to bond bulk materials to the substrate to provide good tribological properties. The cladding process is used for precise coating and where thermal spraying and welding are not possible. The welding processes can be grouped as follows:

- Arc welding : shielded metal arc welding (SMAW), submerged arc welding (SAW);[15-16]
- Gas welding: oxyacetylene gas welding (OGW);
- Combination of arc and gas: tungsten inert gas welding (TIG), gas metal arc welding (GMAW); [9]
- Powder spraying: flame spraying, high velocity oxyfuel process (HVOF), electric arc spraying and plasma transferred arc welding (PTAW); and laser hardfacing/laser cladding (LC).[10,11,13]

Some of the welding methods have been successfully used to apply hardfacings on substrates such as: gas tungsten arc welding (GTAW), plasma arc welding (PAW), laser beam welding, tungsten inert gas welding (TIG), gas welding, manual metal arc welding (MMAW), shielded metal arc welding (SMAW), and flux cored arc welding (FCAW). The rate of dilution depends on materials and on the welder's skill.

Table 1: Comparison of welding processes used for hardfacing [2]

Welding Process	Auto/Manual	Precautions	Weld pool	Dilution	Deposit rate
Shielded metal arc	Manual	Electric arc baking	Slag	15-30%	1-3 kg/h
Submerged arc	Automatic	Flux baking	Slag	30-50%	5-20kg/h
Gas tungsten arc	Manual/ Automatic	Electric arc	Gas	5-15%	0.5-1.5kg/h
Gas shielded metal arc	Semi auto/Automatic	Electric arc	Gas	15-35%	3-10kg/h

Table 2: Comparison of Thermal spraying processes used for hardfacing

Spraying Process	Coating material form	Heat source	Flame temperature °C	Gas velocity mm/s	Porosity %	Coating adhesion MPa
Wire arc spray	Wire	Electric arc	5000-6000	<300	5-10	28-41
Wire flame	Wire	Oxy-fuel combustion	3000	<300	5-10	14-21
Plasma spray	**Powder**	Plasma flame	12000-16000	500-600	2-5	40-70
High Velocity Oxy-Fuel	**Powder**	Oxy-gas fuel combustion	3200	1200	1-2	>70
Laser cladding	**Powder**	Laser	Upto 30000	25-30	1-2	-

3. Selection of Substrate material

The most common substrate material used for producing hardfaced parts are Steels. Substrates are chosen according to the application required such as high temperature application or abrasive wear environment, etc. Some of them are listed as follows:

(i) High carbon steels, (ii) Medium carbon steels, (iii) Low carbon steels, (iv) Low alloy steels,
(v) High speed steels, (vi) Stainless steels, (vii) Manganese steels, (viii) Low nickel chrome steels,
(ix) Cast iron (grey and white)

4. Hardness evaluation of hardfacing alloys

Hardness can be described as the resistance to plastic deformation by indentation, scratching or other frictional means. Dilution increases with an increasing welding current which leads to decrease in hardness, however, lower welding current increases the hardness of the hardfacing.
The resultant hardness of some hardfacing alloy when deposited on substrate by various welding processes is shown in Table 3

Table 3: Comparison of Hardfacing alloy, process, and substrate by various researchers

Author	Hardfacing alloy	Process	Substrate	Hardness
Yuksel N. et. Al.	Fe-Cr-C	Open arc welding	AISI 1020 steel substrate	1450-1700 HV[25]
B. Maroli et al.	Fe-Cr-V-Ti-C-B	Plasma transfer arc welding	EN S235JR mild steel plate	60-61 HRC (600 HV) [5]
B.K. Sreedhar	Stellite6*, Colomonoy5*	Vibratory cavitation equipment	316L Austenitic stainless steel	393 HV[27]
Kown-yeong Lee et al.	Fe-Cr-C-Si	Gas tungsten arc welding	AISI 304 plate	600 HV[24]
J. Zeisig et al	Fe-Cr-Mo-V-C	Laser cladding Nd:YAG	High performance tool steel (X155CrMo12-1)	666 HV[26]
E. Badisch et al	Fe-Cr-C-B	Gas metal arc welding	M.S plate (DIN 1.0038)	175 HV[21]
DashuangLiy et al	Fe-Cr-Ti-C	Self shielded metal cored arc welding	Mild steel	54-64 HRC (420-560 HV) [22]
Wang Xin-hong et al	Fe-Ti-V-C	Shielded metal arc welding	1045 steel substrate	60-62 HRC (570-610 HV) [23]
Agustin Gualco et al	Fe based	Flux cored arc welding	AISI 1010 carbon steel plate	920 HV[28]

5. Hardfacing alloy production methods

Methods of producing hardfacing powders include:
(i) Powder metallurgy, followed by pulverization, grinding and sieving of the sintered alloy [21,22].
(ii) Atomization, which produces powders in the range of few microns to 1 mm. Methods of atomizing liquid metal are by water atomization, gas/air atomization and centrifugal atomization[19].
(iii) Granulation, followed by grinding and sieving of the granules to obtain the alloy powder. The granulation process results in particle sizes of millimeter range sometimes greater than 10 mm [20].

6. Classification of Hardfacing and their applications

Hardfacng improves the properties of the surface of a component without influencing its bulk properties. It is used where the surface is subjected to wear, oxidation, and corrosion. Hardfacing alloys are mainly catogoriesed as Iron-based, Cobalt-based and Nickel-based alloys.

The applications vary widely, ranging from rock crushing to those which minimize metal-to-metal wear such as control valves. Cobalt and nickel-based hardfacing alloys are used extensively at elevated temperatures and in corrosive environments. [31]

6.1 Iron based Hardfacing alloys

Large amount of Chromium and Molybdenum are present in high alloy irons which help in formation of carbides for resisting abrasion. The matrix may be austenitic, martensitic, pearlitic, ferrite or some combination of these phases. They show excellent low-stress abrasion resistance since they have a smaller volume fraction of brittle carbides. Martensitic irons which are cheaper show good high-stress abrasion resistance because of the high hardness of the martensitic matrix. Some of the Fe based hardfacing alloys available are mentioned below along with their composition and applications. [31]

- Work hardening Manganese alloys- (46-48 HRC)

Chemical composition; C- 0.1-1%, Mn- 6-16%, Cr- 14-19%, Ni-9%
Properties- High resistant to cracking, High rate of work hardening
Application- Retouching of casting imperfections.

- Low and Medium alloys- (55-60 HRC)

Chemical composition: C- 0.1-0.5%, Mn- 0.7-2%, Cr- 1-9.5%, Mo-0.2-0.7%, V- 0.3%
App - Transmission shaft, rolls, Cams, raceways, press and transport screws.

- Heat Treated steels for tooling- (50-60HRC)

Chemical composition: C- 0.2-1.1% Mn- 0.3-2% Cr- 2-10% Mo- 1.5-7% V- 0.6-1.1% Ti- 0.3
App- Mill roll, hopper seats in blast furnace, Steam valves, press tooling, moulds and ceramic tiles, Milling cutters, knives, forging tooling.

- Anti abrasion- (61-67 HRC)

Chemical composition: C- 0.5-5.5%, Mn- 0.2-2%, Cr- 10-27%, Mo-3-5%, Ni- 2% B- 2.2-4.5%
App- Equipments used in agriculture, Quarrying, mining, Screw conveyers, mineral conveying equipment, vertical crushers, extractor fans, grinding mills.

- Ferritic and Martensitic stainless steels - (42-54 HRC)

Chemical composition: C- 0.04-0.3%, Mn- 0.8-1.5%, Cr- 12-17%, Ni- 1.2-5% Mo-0.5-1.2%, V- 0.2-0.5% Co- 0.2%

App - Anti corrosion coating, valve seats, pump bodies and rotors, hydraulic rams, hot rolling mills, static brakes for railway.

6.2 Nickel-Based Hardfacing Alloys

Ni-based alloys are generally deposited to improve the wear resistance during service at high temperatures. Carbon containing alloys are popular as a replacement for cobalt-based alloys. Molybdenum and tungsten additions improve the hardness and high temperature strength. Boron containing Ni-based alloys is also available in which the presence of hard chromium borides provide exceptional abrasion resistance. They have poor impact resistance due to the large volume fraction of hard inter-metallic precipitates.

- Nickel based- (200-250 HB) (as welded)

Chemical composition: C- 0.05-0.06%, Mn- 0.2-1%, Cr- 13-16%, Fe- 2.2-5% Mo- 6-16%, W- 0.8-4.5% Co- 2.3-11.5% Ti- 3%

App- High speed forging tools, tube extrusion mandrels, pumps and valves for chemical and petrochemical industries

6.3 Cobalt-Based Hardfacing Alloys

Cobalt-based hardfacing alloys have been used widely for over 50 years because of their good wear, oxidation, and corrosion resistance combined with high hot hardness at temperatures sometimes approaching 980°C.The most commonly used Co-based hardfacing alloys are of the "Stellite" variety, with a nominal composition of Co-28Cr-4.0W-1.1C (wt%).

- Cobalt based- (40-47HRC)

Chemical composition: C- 0.15-2.3%, Mn- 1-1.5%, Cr- 2.-30%, Fe- 4% Mo- 5.50%, W- 4.5-14% Ni- 3-9.5%

App- Vertical mill rolls, forging dies, hot shearing blades, Petrochemical and industrial valves, Valve seats for marine engines, wood and paper industries, rubber kneaders.

General Discussion:

Various methods of Alloy development have been adopted to get the required composition, for example Calphad method, Thermocalc software, Schaffler diagram, etc. In the CALPHAD method, one collects and assesses all available experimental and theoretical information available on phase equilibria and thermo chemical properties in a system. The thermodynamic properties of each phase are then described through the Gibbs free energy, applying a mathematical model containing adjustable parameters. These parameters are evaluated by optimizing the fit of the model to all the assessed information, also involving coexisting phases. Following this it is possible to recalculate the phase diagram, as well as the thermodynamic properties of all the phases and the system as a whole. The CALPHAD method is used to obtain a consistent description of the phase diagram and the thermodynamic properties, to reliably predict the set of stable phases and their thermodynamic properties in regions without experimental information. It is also used for predicting metastable states during simulations of phase transformations. Thermocalc is widely used for a variety of calculations including calculating Stable and meta-stable heterogeneous phase equilibrium, amounts of phases and their compositions, phase diagrams (binary, ternary and multi-component), Thermodynamic properties of chemical reactions, etc.

A Schaeffler diagram is used to represent the effect of the proportion of two elements (and therefore the composition of the alloy) on the structure obtained.

The current trend is toward the use of semiautomatic and automatic welding processes using FCAW and GMAW. GMAW uses either a solid wire or metal-cored welding wire, requires a gas shield, whereas FCAW uses welding wires that are used open-arc or gasless, as well as with a gas shield. SMAW with flux-coated electrodes is still popular for field on-site hardfacing applications, because the equipment is inexpensive and portable.

Laser cladding and Plasma transfer arc welding though being costly processes are very popular in terms of low dilution rates and higher power density, which are useful for powder metallurgy route.

Conclusion:

Hardfacing is the most versatile and economical process these days for reducing the cost of replacement and improving the life of worn out components. Wide variety of welding processes can be done on steel materials by hardfacing. Different alloying elements can be introduced in to the base metal in the form of weld consumables like tube rod, wire electrode or powder to achieve the desired properties like hardness, wear resistance, abrasive resistance, corrosion resistance, crack resistance etc.

- Melting of alloy followed by atomization or granulation, grinding and sieving are the process routes to achieving homogeneous hardfacing consumable powder with fine carbides in the microstructure.
- High hardness of carbides of carbide forming elements has direct relationship with their wear resistance for example Chromium. The smaller the carbide particles in the microstructure , better is the wear resistance.
- Both traditional and sophisticated welding methods can be used for hardfacing; however, dilution must be kept as low as possible to obtain optimal properties.
- Laser cladding and PTAW gives the lowest dilution rates which are used for Powder metallurgy routes only.
- Hardness of hardfacings vary with the welding method, this may be due to variation in dilution, which increases with increasing current or heat input and vice versa.
- Hardfaced parts reduce downtime, resulting in reduced maintenance cost, increased plant availability and improved organizational productivity.
- Hardfacing can be carried out as many times as possible and component can be reused before the wear gets to the substrate.

References:

[1] Surface Engineering for Enhanced Performance against Wear- *Manish Roy, Defense Metallurgical Research Laboratory, DRM Laboratory, Hyderabad, India*

[2] Welding Metallurgy- *Sindo Kou, Professor and Chair Department of Materials Science and Engineering, University of Wisconsin*

[3] Doctoral thesis: Laser cladding: An experiment and therotical investigation- '*Hans Gedda*'

[4] A.S.C.M. d'Oliveira, R. Vilar, C.G. Feder, '*High temperature behaviour of plasma transferred arc and laser Co-based alloy coatings'*,Applied surface science 201 (2002) 154-160

[5] B. Maroli, Höganäs AB, Sweden, S. Dizdar, Höganäs AB, Sweden S. Bengtsson, Höganäs AB, Sweden,' *Iron based hardfacing alloys for abrasive and impact wear*'Int. Thermal spray conf., June 7-9 2017, Germany

[6] Agustin Gualco, Hernan Svoboda, Estela Surian, Luis Vedia, *"Effect of welding procedure on wear behaviour of a modified martensitic tool steel hardfacing deposit"* Materials and Design 31(9):4165-4173 October 2010

[7] M. Kirchgabner, E. Badisch, F. Franek, *"Behaviour of iron based hardfacing alloys under abrasion and impact"*, Wear 265 (2008) 772-779

[8] Xinhong Wang, Fang Han, Xuemei Liu, Shiyao Qu, Zengda Zou, *"Microstructure and wear properties of the Fe– Ti–V–Mo–C hardfacing alloy"* , Wear 265 (2008) 583–589.

[9] Moselli, P. C., de Oliveira, M. F. and Moreno, J. R. S. "*Wear resistance in hardfacing applied in substrate SAE 1020 using welding process gas tungsten arc welding alloy stellite6 in powder form*", Scientific Research and Essays , 8(36), pp.1730 – 1740, 2013.

[10] Zikin, A., Ilo, S., Kulu, P., Hussainova, I., Katsich, C. and Badisch, E. *"Plasma transferred arc hardfacing of recycled hard metal reinforced Nickel-matrix surface composites*", Materials Science , 18(1), pp. 12 – 17, 2012.

[11] Klimpel, A., Janicki, D., Lisiecki, A. and Rzeźnikiewicz, A. "*Laser repair hardfacing oftitanium alloy turbine*", Journal of Achievements in Materials and Manufacturing Engineering , 49(2), pp. 400 – 411, 2011.

[12] Rakić, S., Lazić, V., Aleksandrović, S., Mitrović, S., Nikolić, R., Milosavljević, D. and Čukić, R. "*Reparatory and manufacturing hard-facing of working parts made of stainless steels in confectionary industry*", Tribology in Industry , 34(3), pp. 138 – 151, 2012.

[13] Xianfen Li, Haojie Chu, yuanting Chen, Peng Hua, Guoping Wang, *'Microstructure and properties of the laser cladding ODS layer on CLAM steel'*, Surface and coating technology 357 (2019) 172-179

[14] Bęczkowski, R. and Gucwa, M. "*Defects appearing in the surfacing layers of abrasion resistant*", Archives of Foundry Engineering , 16(4), pp. 23 – 28, 2016.

[15] Coronado, J. J., Caicedo, H. F. and Gomez, A. L. "*The effect of welding processes on the abrasive wear resistance for hardfacing deposits*", Tribology International , 42, pp. 745 – 749, 2009.

[16] Morsy, M. A., Khafagy, S. M. and Zaghlool, B. "*Repair welding restoration of the screw conveyor for resin extruder*", Journal of American Science , 7(1), pp. 313 – 320. 2011.

[17] Shivanath, R. and Jones, P. "*Powder metal alloy process*", US Patent No. 5,476,632, 1995.

[18] Causton, R. J. Steel powders composition, US Patent No. 5,217,683, 1993.
[19] Dolman, K. F. "*Hardfacing ferroalloy materials*", US Patent No. 20080251507A1, 2008.
[20] Dunkley, J. J. and Norval, D. "*Atomisation of ferroalloys*", Proceedings: Tenth International Ferroalloys Congress , Cape Town, South Africa, February 1 - 4, 2004, pp. 541 – 547, 2004.
[21] E. Badisch , C.Katsich , H.Winkelmann , F.Franek , Manish Roy, '*Wear behaviour of hardfaced Fe-Cr-C alloy and austenitic steel under 2-body and 3-body conditions at elevated temperature*' Tribology International 43 (2010) 1234-1244
[22] Dashuang Liu, Renpei Liu∗, Yanhong Wei, Yan Ma, Kun Zhu, '*Microstructure and wear properties of Fe–15Cr–2.5Ti–2C–xB wt.% hardfacing alloys*' Applied Surface Science 271 (2013) 253-259
[23] Wang Xin-hong, Zow Zeng-da, Qu Shi-yao, '*Microstructure of Fe-Based Alloy Hardfacing Reinforced by Tic-VC Particles*', Journal of iron and steel research, International 2006, 13(4) 51-55
[24] Kownyeong Lee, Sung hoon Lee, Yangdo Kim, Hyun seon Hong, Young min Oh, Seonjin Kim, '*The effect of additive elements on the sliding wear behaviour of Fe-basehardfacing alloys*', Wear 255 (2003) 481-488
[25] Yuksel N., Sahin S., '*Wear behavior-hardness-microstructure relation of Fe-Cr-C and Fe-Cr-C-B based hardfacing alloys*' ,Materials and Design 58 (2014) 491-496
[26] J. Zeisig, N. Schadlich, L. Giebeler, J. Sander, J. Eckert, U. Kuhn, J. Hufenbach, '*Microstructure and abrasive wear behavior of a novel FeCrMoVC laser cladding alloy for high performance tool steels*', Wear 382-383 (2017) 107-112
[27] B.K. Sreedhar, S.K. Albert, A.B. Pandit, '*Improving cavitation erosion resistance of austenitic stainless steel in liquid sodium by hardfacing- comparison of Ni and Co based deposits*', Wear 342-343 (2015) 92-99
[28] Agustin Gualco, Cristian Marini, Hernan Svoboda, Estela Surian, '*Wear resistance of Fe-based Nanostructured Hardfacing*',International congress of 9science and technology of Metallurgy and Materials, SAM-CONAMET 2013
[29] B. Venkatesh, K.Sriker, VSV Prabhakar, '*Wear characteristics of hardfacing alloys: state of the art*', 2nd International Conference on Nanomaterials and Technologies (CNT 2014)
[30] A.S.C.M. d'Oliveira, R. Vilar, C.G. Feder, '*High temperature behaviour of plasma transferred arc and laser Co-based alloy coatings*',Applied surface science 201 (2002) 154-160
[31] Serdar atamert Darwin college, cambridge, A dissertation submitted for Doctor of Philosophy, "Stability, wear resistance, and microstructure of Iron, cobalt and nickel-based hardfacing alloys"

Process Improvement Using Six-Sigma (DMAIC Process) in Bearing Manufacturing Industry: A Case Study

Manish Bhargava[1], Sanjay Gaur[2*]

[1]Dept.of Mechanical Engineering, National Institute of Technology Agartala, Tripura, India
[2]MNIT Innovation and Incubation Centre, MNIT Campus, Jaipur, India

*Email - sanjaygaur29@yahoo.com

Abstract. Propagation, development of technologies and increasing customer demands switch the approach of existing work in the industries. To overcome barricades, the six-sigma DMAIC (define, measure, analyse, improvement, control) approach is trendy along with being advantageous. This approach decreases the variation and set up the way for up gradation in the manufacturing companies. This paper contains the Six-Sigma DMAIC approach that was used to decrease the process variation of inner and outer races of ball bearing for enhancing product quality. Define phase of DMAIC approach begins by difficulty detection through the voice of internal and external customers. The later stage constitute of measuring the data of bearing parts of existing process. This stage followed by the analyze as well as improvement stage, where the Six-Sigma quality improvement tools i.e. statistical process control (SPC), Control charts, MINITAB 18.0 software, fish bone diagram along with significant study of alive process were imposed to identify root causes and minimizing process variability. The improvement stage minimized the assignable causes for variability. The control phase called to maintain the improved process till further improvement.

This work expected that the Six-Sigma DMAIC methodology was effective for increasing the sigma level, decreasing the value of standard deviation and also decreasing the expected part per million (PPM) out of specification limits. Values of capability indices i.e. (Cp, Cpk and Cpm) were improved. A Six-Sigma DMAIC methodology is well-known and is capable of playing an efficient role in manufacturing industry by reducing variability in the bearing part process.

Introduction

In manufacturing, bearing sector is the most demanding in our country. Manufacturer faces critical challenge to minimize variability in dimensions of bearing components manufacturing. Over 95% of equipment use bearings and the failure might possibly go ahead to breakdown of that equipment. Bearing is considered as a critical part in industrial appliances. For that reason, the bearing quality plays a key role in the high production volume system.

If a firm wants to improve their product quality, process control play a key role [1]. Manufacturing industries need to analyze, watch and make up improvement of their existing processes to meet the requisites with the market competition. The industries use numerous techniques, approaches and tools for implementation of programs used for regular enhancement in terms of quality [2]. The industries faced the problem of variations in each stage of process of every product. The variability that occurs during the manufacturing process is investigated in any organization using quality tools like Six-

Sigma DMAIC, process capability analysis (PCA), control charts, process capability indices (PCIs), which assist the organization to examine and assess the potential of the process [3].

PCA provides a baseline intended for us to be aware of how the process is performing relatively to the pre specified limits. PCA is the initial concept, which gives information that how variability affects the process. Process capability studies make us aware about the best situations of process performance for achieving target [4]. PCA helps in statistical analysis of PCIs. If the value of these PCIs is low, it means higher rejection rate and higher values of PCIs means minimum rejection rate [5].

Literature review

Industries use various approaches, techniques and tools for regular enhancement of their overall quality. More to point these, every organization chooses an accurate mixture of these tools, and techniques [1]. If a firm works in the direction of implementation of six-sigma practically with enhancement in process capability and consumer happiness it is considered six sigma implementing firms [6]. Some well known successfully six sigma implemented industries like Motorola, General Electric, Honeywell, Sony, L&T, TCS etc. Are a motivation to others [7], [8]. Six-Sigma purpose is to attain excellence in every stage of production in an industry [9], [10]. Six-Sigma means DPMO or rate of success with 99.997%. [11]. Many organizations work according to level of 3σ, that means 93% success rate or 66,800 DPMO. According to present market demand organizations need to review their existing process. Six- sigma DMAIC approach helps them to reduce rejection rate and enhance their success rate up to 99.997% [12]. Six-sigma DMAIC is a benchmark to check the process or product quality, also having ability for improving efficiency and quality of product [13]. DMAIC is a systematic and fact based methodology and provides best results when the process is flexible. Literature review shows the six- sigma DMAIC methodology is the outstanding exercise for enhancing the quality of process in manufacturing industries [14], [15]. Six-sigma DMAIC methodology gives an effective results in printed circuit boards (PCB) and integrated circuits (IC) [16], [17]. Six sigma DMAIC approach applied in various processes for improvement in manufacturing industries i.e. food processing industry, IC engine parts manufacturing firms, TFT LCD screen manufacturing firms [18], [19], [20], [21]. Six-sigma DMAIC approach is extremely important for Telecom and other service sectors for process and quality improvement [22]. Ford automobile company saying 'Quality Job is first' and taken an initiative of six sigma in year 2001. Companies reported significant economic benefits and product or process quality improvement using six sigma DMAIC approach [23]. Hence, the recent work concentrate on the effective utilization of six-sigma DMAIC methodology expected for enhancement of capability of the bearing part manufacturing industry.

Company Profile: Firm A was the topmost manufacturer and supplier of ball bearing components in Rajasthan. Firm started in the year 1995, its first and second both manufacturing plant are located at Jaipur.

Methodology: In this work, using six-sigma DMAIC methodology focus is to get better quality level and process capability for bearing part manufacturing. If DMAIC approach is properly applied it gives targeted products at right time with minimum cost [24]. DMAIC steps in this work were explained:

Define phase

The aim of this phase is to improve the existing process. The main goal was acquired using the customer and firm employee voice method. This would be useful for the quality improvement of the firm. In addition, the goal will minimize the defect level or reduce process variability and improve productivity for a particular process. The present work focused to enhance the process of CNC machining where the forge bearing races were machined with specified size with tolerance ±0.05mm given by customers. CNC machining plays an important role in the process of bearing race manufacturing, because it is shaping the critical dimensions of bearing races. Both races are shown in fig. 1.

Figure 1: Inner race (IR) and outer race (OR) of ball bearing

The basic operations parameter taken in this work is bore diameter of IR and OR of ball bearing.

Measure Phase

In measurement phase of DMAIC measured existing process data of ball bearing bore diameter of IR and OR which are shown in Table 1. This collection of data has 200 observations. These 200 observations converted into 40 samples. Each sample contains 5 observations.

Table: 1 Observations of bore diameter of inner and outer races of existing process

S. No.	Bore diameter of inner race 34.40±0.05 mm					Bore diameter of outer race 67.00±0.05 mm				
	1	**2**	**3**	**4**	**5**	**1**	**2**	**3**	**4**	**5**
1	34.40	34.41	34.40	34.42	34.38	67.01	67.02	67.04	66.98	66.96
2	34.40	34.37	34.41	34.42	34.43	66.95	66.99	67.00	67.02	67.01
3	34.44	34.38	34.39	34.42	34.38	67.03	67.01	67.00	66.97	67.00
4	34.42	34.41	34.38	34.39	34.41	66.99	66.95	66.96	66.98	67.03
5	34.38	34.42	34.38	34.41	34.42	67.04	67.01	67.00	66.99	67.01
6	34.38	34.39	34.36	34.43	34.39	67.00	67.02	66.99	66.98	67.02
7	34.40	34.45	34.41	34.38	34.39	67.03	66.96	66.98	66.99	67.00
8	34.37	34.38	34.40	34.41	34.40	67.01	67.02	67.02	67.02	67.04
9	34.42	34.42	34.43	34.41	34.45	67.00	67.01	67.02	66.99	67.00
10	34.41	34.39	34.40	34.45	34.39	66.99	67.02	66.98	67.01	66.99
11	34.41	34.39	34.40	34.41	34.39	67.00	67.02	66.97	66.99	67.01
12	34.40	34.42	34.41	34.44	34.40	67.00	67.00	66.99	67.00	67.00
13	34.38	34.40	34.37	34.40	34.38	67.01	67.00	66.99	66.98	67.00
14	34.40	34.41	34.42	34.39	34.40	67.03	67.01	67.01	66.95	67.03
15	34.39	34.36	34.40	34.38	34.41	67.01	66.97	67.00	66.98	67.01
16	34.38	34.39	34.40	34.41	34.40	67.02	67.00	67.03	67.01	66.98

17	34.38	34.39	34.41	34.39	34.38	67.02	67.00	66.99	67.04	67.01
18	34.43	34.35	34.39	34.40	34.42	67.05	67.00	66.99	67.00	66.98
19	34.40	34.42	34.41	34.38	34.35	67.01	67.00	67.01	66.98	67.01
20	34.41	34.40	34.40	34.37	34.39	67.03	67.02	67.00	67.02	66.99
21	34.38	34.40	34.39	34.41	34.40	67.00	66.96	66.99	67.01	67.02
22	34.44	34.42	34.39	34.40	34.41	67.00	66.97	67.00	67.02	67.04
23	34.40	34.39	34.42	34.38	34.42	67.00	67.01	67.02	67.00	66.99
24	34.39	34.38	34.38	34.39	34.36	67.00	66.98	67.02	67.00	67.00
25	34.37	34.38	34.37	34.38	34.42	66.99	67.00	66.98	67.01	67.02
26	34.43	34.42	34.37	34.40	34.40	67.05	67.02	66.99	67.00	66.99
27	34.39	34.40	34.41	34.43	34.41	66.96	66.98	66.95	66.99	67.02
28	34.39	34.36	34.38	34.40	34.38	67.01	66.99	67.00	67.01	66.95
29	34.42	34.35	34.38	34.39	34.41	66.98	67.04	67.01	67.00	67.00
30	34.42	34.39	34.41	34.42	34.40	66.99	67.03	67.01	67.03	67.02
31	34.41	34.40	34.39	34.41	34.40	66.98	67.02	67.01	67.02	66.99
32	34.39	34.40	34.42	34.36	34.39	67.00	66.97	67.05	66.99	67.01
33	34.42	34.41	34.39	34.37	34.41	66.98	66.99	66.99	67.00	67.01
34	34.44	34.43	34.39	34.42	34.39	67.00	67.00	67.01	66.97	67.00
35	34.38	34.37	34.45	34.40	34.40	66.99	66.98	67.01	67.00	66.99
36	34.38	34.35	34.40	34.41	34.39	67.00	67.05	67.01	67.02	66.98
37	34.44	34.39	34.41	34.37	34.41	66.99	67.00	66.98	67.00	66.96
38	34.40	34.39	34.40	34.40	34.42	66.98	67.01	67.02	67.00	67.00
39	34.41	34.42	34.38	34.40	34.44	66.99	67.00	67.00	67.03	67.01
40	34.40	34.41	34.36	34.40	34.38	67.04	67.00	67.01	66.98	67.00

After observing above data, for diagnosing process to see whether the process is under statistical control or not and for assessing process capability, X-bar R charts were prepared for IR and OR using MINITAB 18.0 software shown in figure 2.

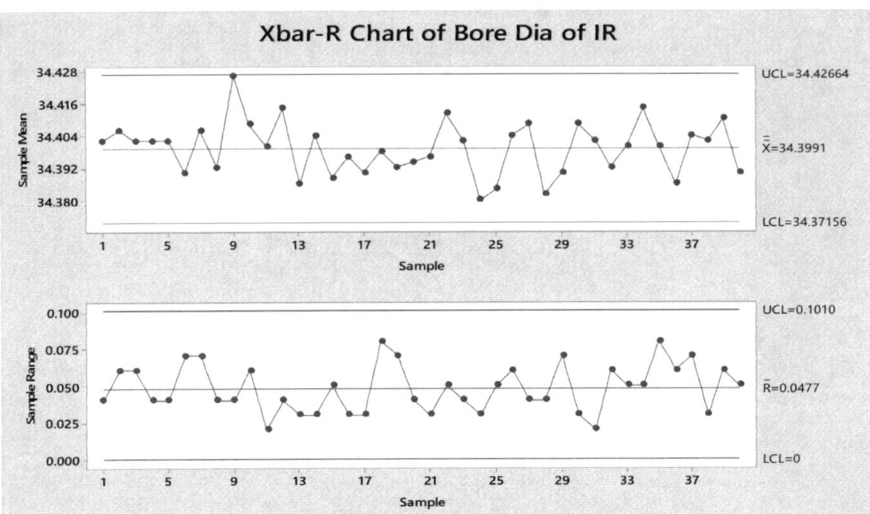

(a)

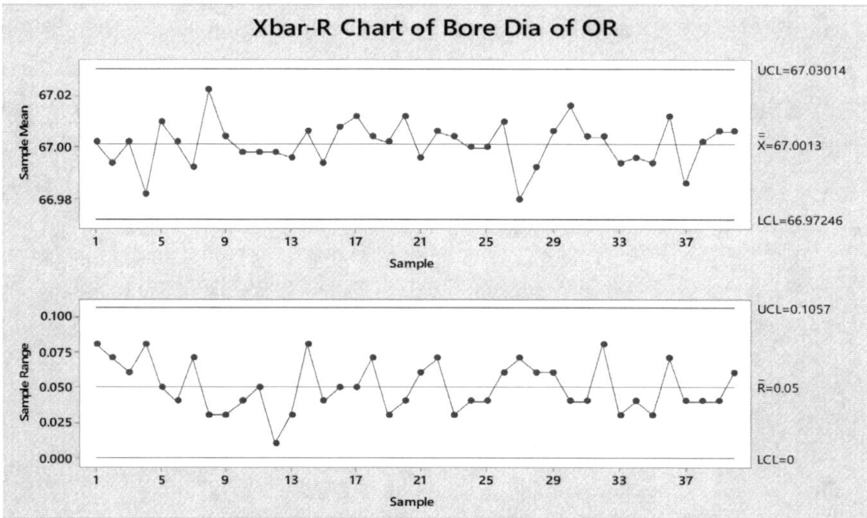

(b)

Figure 2: X-bar R charts of bore diameter of (a) inner and (b) outer race for existing process

From X-bar R chart analysis showed that the process is under statistic control, all sample points were inside the lower and upper control limits furthermore go for PCA using again MINITAB 18.0 software. The analysis of process is shown in figure 3.

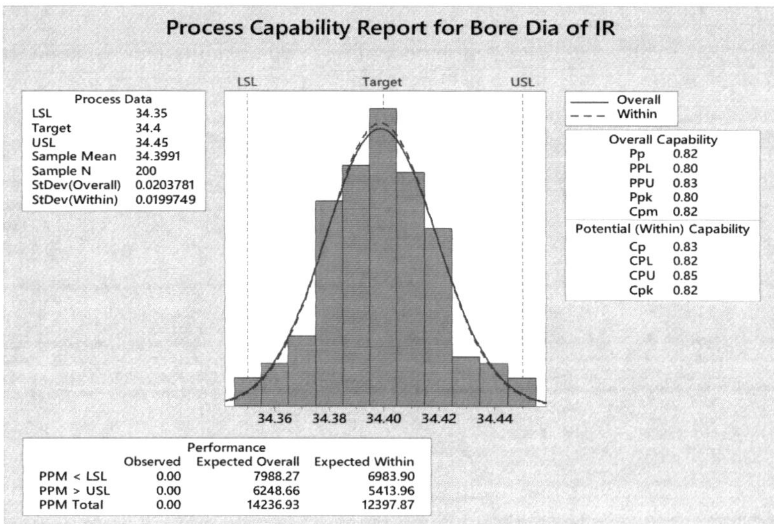

(a)

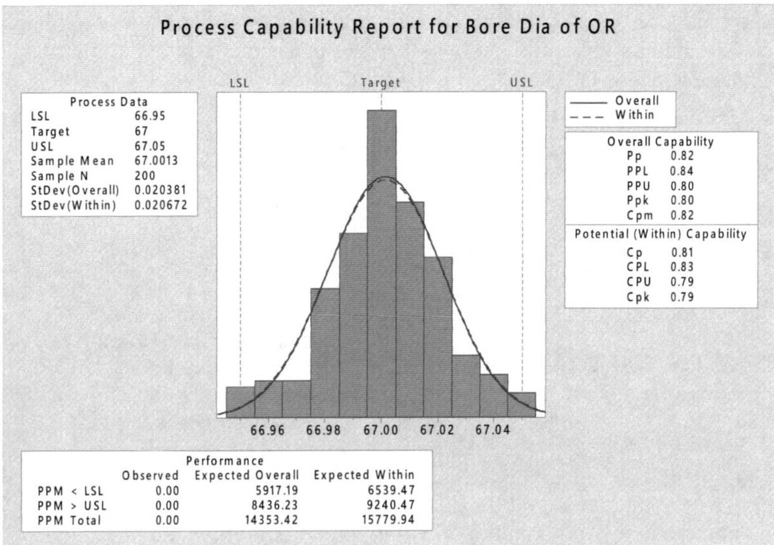

(b)

Figure 3: Process capability report of bore diameter of (a) inner race and (b) outer race for existing process

From the PCA, found the values of PCIs (Cp = 0.81 & 0.83, Cpk = 0.79 & 0.82 and Cpm = 0.82) and standard deviation (0.02067 & 0.01997). The values of capability indices are less than one (<1), that means the process is not capable.

Analysis Phase

In this phase of DMAIC identify responsible assignable causes for poor quality or variability in the existing process. These responsible causes are mentioned in cause and effect diagram figure 4.

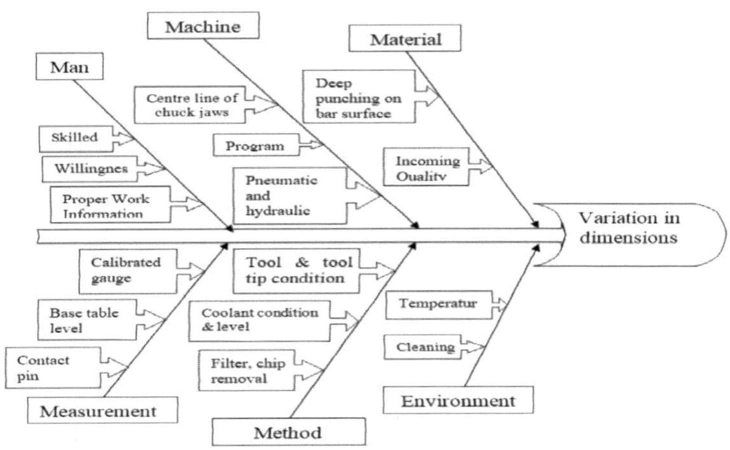

Figure 4: Cause and effect diagram

The role of possible causes identified in cause and effect diagram for variability were analyzed using survey opinion of 60 employees of the various bearing part manufacturing companies. Analysis of survey questionnaire is shown in figure 5.

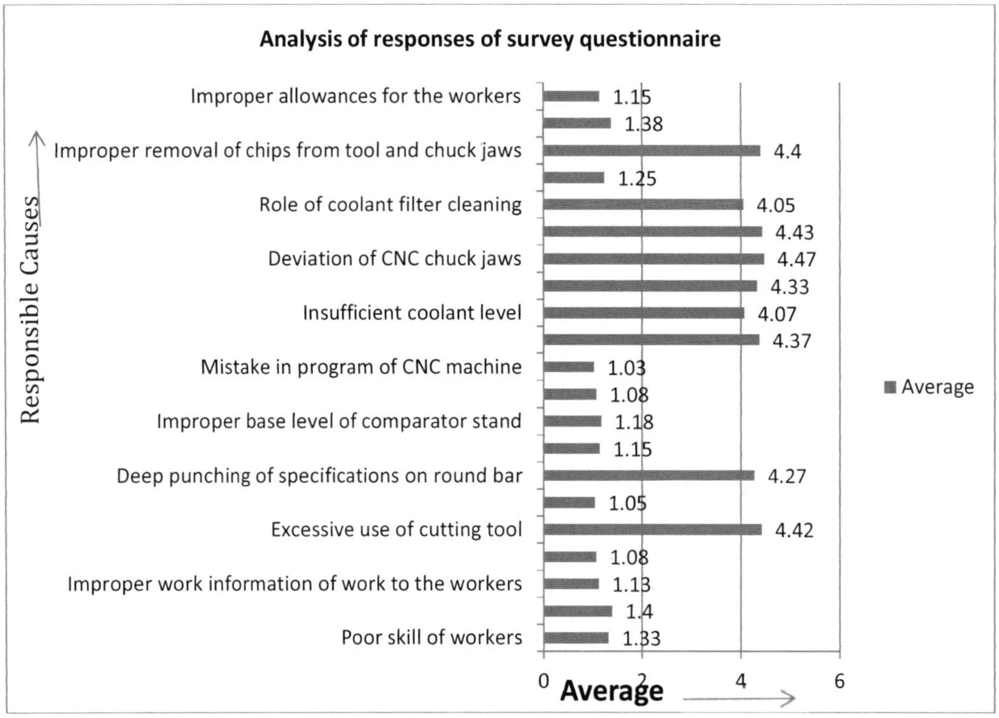

Figure 5: Analysis of responses of survey questionnaire

On the basis of analysis of survey questionnaire, following causes have been found playing major role in variations of dimensions of bearing components as shown in table 2.

Table 2: Major responsible causes

S. No.	Major responsible cause of variation
1	Excessive use of cutting tool
2	Deep punching of specifications on round bar
3	Improper tool and tool tip condition of CNC machine
4	Insufficient coolant level
5	Excessive use of coolant
6	Deviation of CNC chuck jaws
7	Role of coolant filter mesh size
8	Improper removal of chips from tool and chuck jaws

Improve Phase

In this phase of DMAIC, after finding the responsible causes some corrective actions were implemented.

- Proper removal of chips from cutting tool and chuck jaws after every machined product shown in figure 6 (a)
- The tool life should be closely monitored. Tool should be changed not after ongoing production batch, in between the ongoing production batch to minimize the possible variation.
- Due to tiny chip particles mixed in coolant shown in figure 6 (b), replacement of existing coolant filter with smaller mesh size (5 microns to 4 microns). The coolant and coolant filter changed after specified duration or machine run time. Also clean the coolant filter after every production batch and check the level of coolant in tank.
- Before start the production check all the setting of machine every day. Specifically check chuck jaws were centered or not to prevent ovality.
- Use sticker on raw material steel round bar for specifications shown in figure 6 (c).

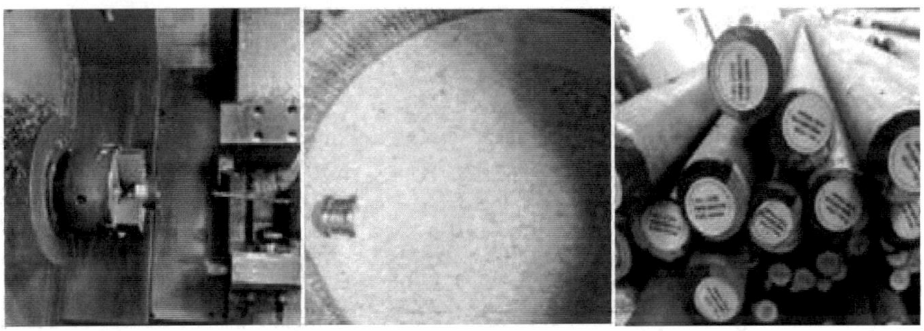

(a) (b) (c)

Figure 6: (a) Cutting chips, (b) chips mixed in coolant and (c) raw material with stickers

After taking above corrective actions may be the improved process is better than existing process in terms of productivity, quality and cost effective. Machining data were again collected and analyzed for bore diameter of inner race. Collected data for bore diameter of inner race and outer race after taking corrective actions has shown in table 3. This collection of data has 200 observations.

Table 3: Observations of bore diameter of inner and outer races of improved process

S. No.	Bore diameter of inner race 30.60±0.05 mm					Bore diameter of outer race 62.95±0.05 mm				
	1	2	3	4	5	1	2	3	4	5
1	30.61	30.60	30.62	30.59	30.62	62.96	62.97	62.95	62.96	62.94
2	30.58	30.61	30.60	30.60	30.58	62.93	62.95	62.96	62.93	62.95
3	30.59	30.59	30.60	30.58	30.59	62.94	62.95	62.94	62.94	62.93
4	30.58	30.60	30.60	30.60	30.62	62.97	62.95	62.95	62.93	62.95
5	30.59	30.59	30.58	30.59	30.59	62.94	62.93	62.94	62.94	62.95
6	30.60	30.61	30.62	30.60	30.61	62.96	62.97	62.94	62.95	62.95
7	30.60	30.60	30.60	30.59	30.61	62.96	62.95	62.95	62.95	62.94
8	30.62	30.60	30.61	30.60	30.57	62.92	62.96	62.95	62.97	62.95
9	30.60	30.62	30.59	30.57	30.59	62.94	62.96	62.97	62.95	62.92
10	30.60	30.60	30.58	30.59	30.60	62.95	62.93	62.95	62.95	62.94
11	30.61	30.60	30.59	30.61	30.62	62.97	62.94	62.95	62.96	62.96
12	30.62	30.63	30.59	30.60	30.60	62.95	62.94	62.98	62.97	62.95
13	30.60	30.61	30.61	30.60	30.60	62.95	62.96	62.96	62.95	62.95
14	30.58	30.60	30.60	30.61	30.59	62.94	62.95	62.95	62.93	62.96
15	30.59	30.61	30.61	30.60	30.59	62.94	62.96	62.96	62.94	62.95
16	30.59	30.60	30.58	30.59	30.61	62.96	62.93	62.95	62.92	62.94
17	30.58	30.58	30.62	30.62	30.60	62.95	62.97	62.93	62.93	62.97
18	30.61	30.61	30.60	30.58	30.61	62.96	62.95	62.96	62.96	62.93
19	30.60	30.61	30.60	30.60	30.59	62.94	62.95	62.96	62.95	62.95
20	30.57	30.62	30.60	30.57	30.57	62.92	62.95	62.97	62.92	62.92
21	30.62	30.59	30.59	30.62	30.61	62.96	62.94	62.94	62.97	62.97
22	30.61	30.57	30.61	30.60	30.60	62.95	62.96	62.92	62.96	62.95
23	30.59	30.62	30.62	30.60	30.60	62.95	62.97	62.97	62.94	62.95
24	30.61	30.61	30.61	30.59	30.58	62.94	62.96	62.97	62.96	62.94
25	30.61	30.60	30.61	30.59	30.60	62.95	62.96	62.95	62.96	62.94
26	30.59	30.59	30.61	30.61	30.60	62.95	62.96	62.93	62.94	62.96
27	30.59	30.58	30.62	30.60	30.57	62.92	62.97	62.93	62.94	62.95
28	30.60	30.60	30.60	30.58	30.60	62.95	62.95	62.95	62.95	62.93
29	30.59	30.62	30.60	30.61	30.63	62.98	62.95	62.97	62.94	62.96
30	30.60	30.60	30.60	30.58	30.63	62.98	62.95	62.95	62.95	62.93
31	30.58	30.59	30.59	30.61	30.61	62.96	62.94	62.94	62.93	62.96
32	30.59	30.59	30.61	30.61	30.59	62.94	62.96	62.94	62.94	62.96
33	30.60	30.59	30.58	30.61	30.59	62.94	62.93	62.94	62.95	62.96
34	30.60	30.58	30.59	30.60	30.62	62.97	62.94	62.93	62.95	62.95
35	30.62	30.59	30.59	30.60	30.61	62.96	62.94	62.94	62.97	62.95
36	30.61	30.61	30.61	30.60	30.60	62.95	62.96	62.96	62.96	62.95
37	30.61	30.59	30.63	30.60	30.59	62.94	62.98	62.94	62.96	62.95
38	30.58	30.58	30.59	30.60	30.59	62.94	62.94	62.93	62.93	62.95
39	30.59	30.61	30.60	30.59	30.59	62.94	62.95	62.96	62.94	62.93
40	30.61	30.60	30.58	30.59	30.60	62.95	62.93	62.95	62.96	62.94

After observing above data prepared X-bar R charts for IR and OR using MINITAB 18.0 software shown in figure 7.

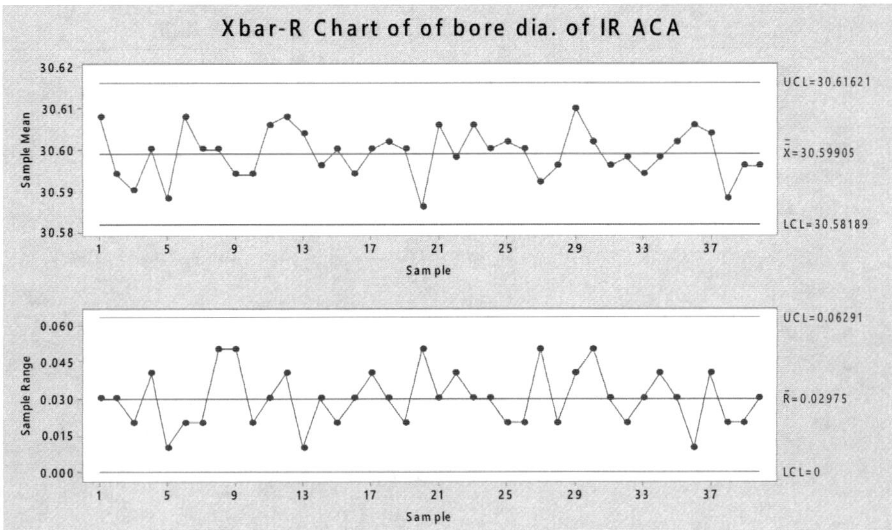

(a)

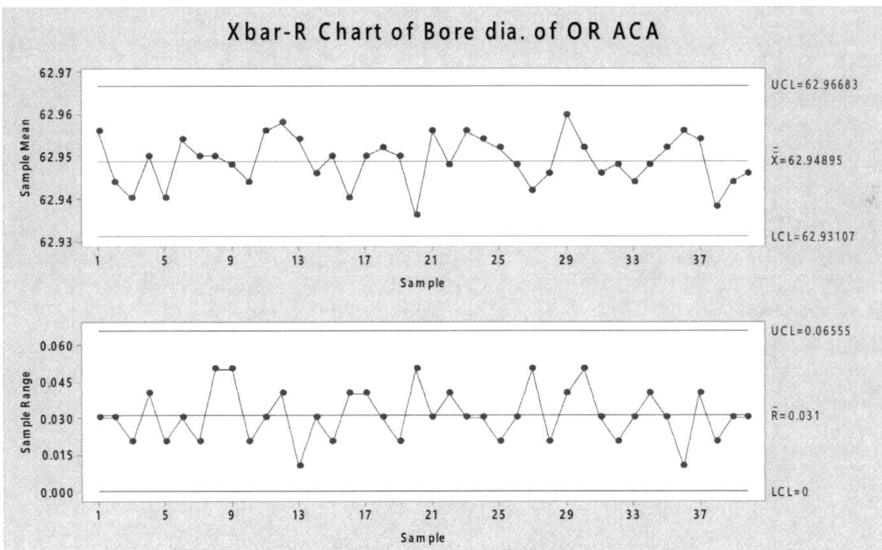

Figure 7: X-bar R charts of bore dia. of (a) inner race and (b) outer race for improved process

From X-bar R chars analyzed that the process is under statistic control, furthermore go for PCA using MINITAB 18.0 software.

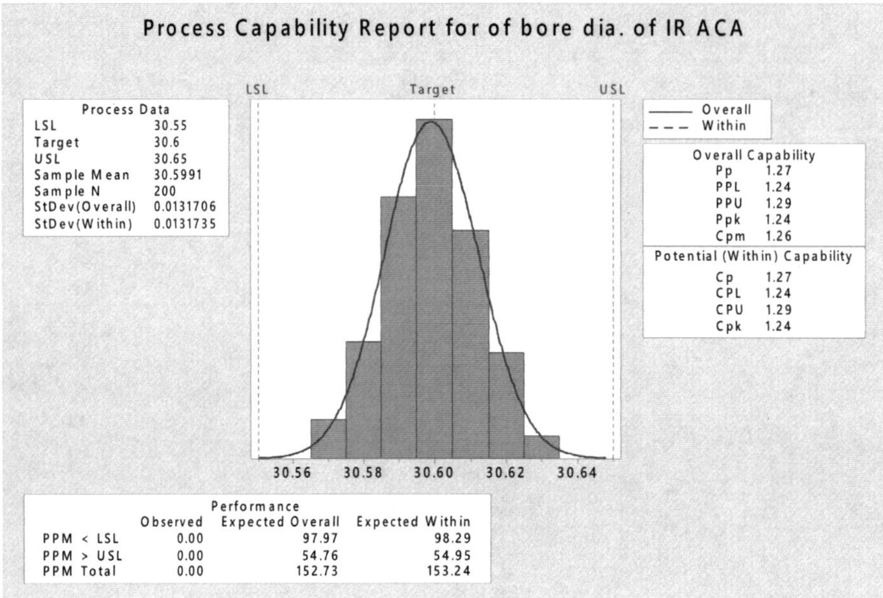

(a)

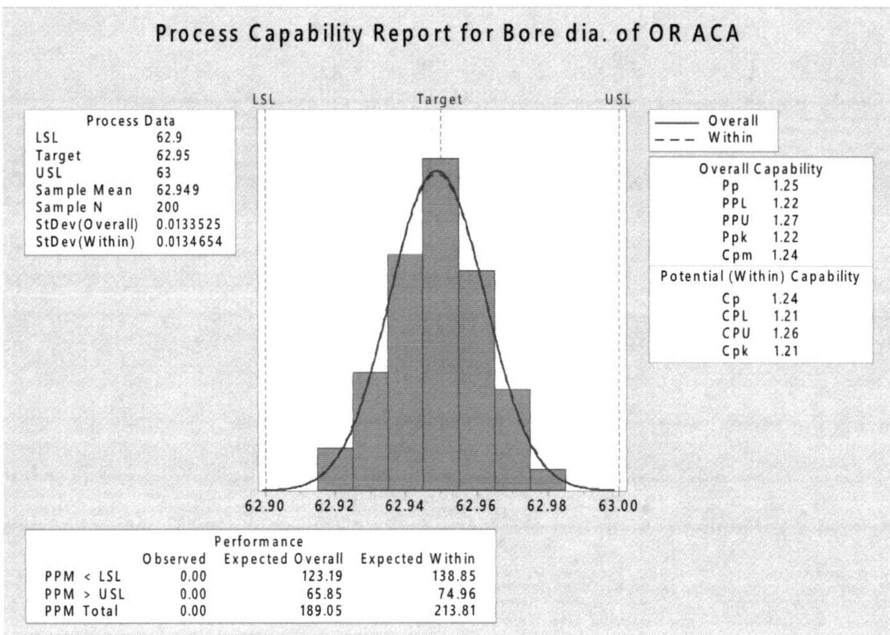

(b)

Figure 8: Process capability report of bore diameter of (a) inner race and (b) outer race for improved process

Found the values of process capability indices (Cp = 1.24 & 1.27, Cpk = 1.21 & 1.24 and Cpm = 1.24 & 1.26) and standard deviation (0.01346 & 0.01317) using MINITAB 18.0. The values of capability indices were improved up to greater than one (>1), that means the improved process is better than the existing process. Comparison of PCIs before and after improvement of the process is shown in table 4.

Table 4: Comparison of process capability indices before and after improvement

		Cp	Cpk	Cpm	Std. Dev.
For Inner Race	Before Improvement	0.83	0.82	0.82	0.01997
	After Improvement	1.27	1.24	1.26	0.01317
For Outer Race	Before Improvement	0.81	0.79	0.82	0.02067
	After Improvement	1.24	1.21	1.24	0.01346

Control Phase

Maintain and continue the improved process for sustaining the firm in competitive market. The improved process runs till further improvement. From the results process/ product variability is reduced and quality and customer satisfaction will be improved.

Results

Using Six-Sigma DMAIC methodology increased the approximate sigma level from 2.5 to 3.5, decrease the value of standard deviation from 0.02067 to 0.01317 and decreased the expected part per million (PPM) out of specification limits from 15780 to 154. Values of capability indices i.e. (Cp, Cpk and Cpm) were improved from 0.81 to 1.27, 0.79 to 1.24 and 0.82 to 1.26 respectively.

Conclusion

Six-Sigma DMAIC is an efficient methodology to find out the actual needs of a process for improvement. Six-Sigma DMAIC methodology also provides practical solution for analysis of data. Results with successful implementation of Six-Sigma in different applications are:

- Reduced costs of poor quality
- Improve process capability
- Process improvement

In this research, Six-Sigma DMAIC methodology was implementing for process improvement in bearing part manufacturing industry. In the first step, values of process capability indices C_p, C_{pk}, C_{pm} and standard deviation for the existing process were calculated. These values were found to be less than one, so to improve the values of process capability indices by reducing the responsible causes of the process variability with the help of cause and effect diagram. In improvement phase, the value of process capability indices were improved and found greater than one after taking corrective actions. We can conclude from this research that quality level or performance level of a process can be improved by implementing Six-Sigma DMAIC methodology. The aim of this research was to reduce the process variability of bearing part manufacturing, which contributes to betterment of the performance. This work provides information about the procedure for reducing the process variability in manufacturing of bearing components and other products also.

References

[1] G. V. S. S. Sharma, P Rao Srinivasa, B Babu Surendra, "Process capability improvement through DMAIC for aluminium alloy wheel machining," J Ind Eng Int, vol. 14, pp 213–226, 2018.
[2] M. Sokovic, D. Pavletic, K. K. Pipan,"Quality improvement methodologies–PDCA cycle, RADAR matrix, DMAIC and DFS," Journal of Achievements in Materials and Manufacturing Engineering, vol. 43(issue 1), pp. 476-483, 2010.
[3] Ronald Blank, 'Cross-Functional Productivity Improvemen," ISBN 9781466510739, CRC Press Taylor and Francis group, pp. 77-94, 2013.
[4] Y. Wooluru, D. R. Swamy, P. Nagesh,"The Process Capability Analysis-A Tool for Process Performance Measures and Metrics–A Case Study," International Journal for Quality Research, vol. 8 (issue 3), pp.399-416, 2014.
[5] I. Lasa Serrano, R. D. Castro, C. O. Laburu, "Extend of the use of Lean concepts proposed for a value stream mapping application," Production Planning & Control, vol. 20 (issue 1), 82-98, 2009.
[6] M. E. Kabir, S. M. I. Boby, M. Lutfi, "Productivity improvement by using Six-Sigma," Int J Eng Technol, vol. 3 (issue 12), pp. 1056–1084, 2013.
[7] M. Hekmatpanah, M. Sadroddin, S. Shahbaz, F. Mokhtari, F. Fadavinia, "Six Sigma process and its impact on the organizational productivity," World Acad Sci Eng Technol, vol. 43, pp. 2070–3740, 2008.
[8] B. Charles, Q. Patrick, "Streamlining enterprise records management with Lean Six-Sigma," The Information Management Journal, pp.58–62, 2005.
[9] V. Narula, S. Grover, "Six Sigma: literature Review and Implications for future research," Int J Ind Eng, vol. 26 (issue 1, pp. 13–26, 2015.
[10] J. Antony, "Design of Experiments and its Role within Six Sigma," Design of Experiments for Engineers and Scientists 2nd ed., pp.201–208, 2014.
[11] J. Antony, R. Banuelas, "Key ingredients for the effective implementation of Six Sigma program," Meas Bus Excell, vol. 6 (issue 4), pp. 20–27, 2002.
[12] H. Erbiyik, M. Saru, "Six Sigma Implementations in Supply Chain: An Application for an Automotive Subsidiary Industry in Bursa in Turkey," Procedia-Social and Behavioral Sciences, vol. 195, pp.2556-2565, 2015.
[13] V. Gupta, R. Jain, M. L. Meena, G. S. Dangayach, "Six-sigma application in tire-manufacturing company: a case study," J Ind Eng Int, vol. 14, pp. 511–520, 2018.
[14] P. Rathore P, S. Kota Chakrabarti, "Sustainability through remanufacturing in India: a case study on mobile handsets," J Clean Product, vol. 19 (issue 15), pp. 1709–1722, 2011.
[15] K. Govindan, K. M. Shankar, D. Kannan, "Application of fuzzy analytic network process for barrier evaluation in automotive parts remanufacturing towards cleaner production–a study in an Indian scenario," J Clean Product, vol.114, pp. 199–213, 2016.
[16] J. P. C. Tong, F. Tsung, B. P. C. Yen, "A DMAIC approach to printed circuit board quality improvement," Int J Adv Manuf Technol, vol. 23, pp. 523–553, 2004.
[17] C.T. Su, T. L. Chiang, K. Chiao, "optimizing the IC delamination quality via six-sigma approach," IEEE Trans Electron Packag Manuf, vol. 28 (issue 3), pp. 241–248, 2005.
[18] J. C. Chen, Y. Li, R. A. Cox, "Taguchi-based Six Sigma approach to optimize plasma cutting process: an industrial case stud,". Int J Adv Manuf Technol, vol. 41, pp. 760–769, 2009.
[19] G. V. S. S. Sharma, P. S. Rao, "Process capability improvement of an engine connecting rod machining process," J Ind Eng Int, vol. 9, pp. 37, 2013.
[20] G. V. S. S. Sharma, P. S. Rao, "A DMAIC approach for process capability improvement an engine crankshaft manufacturing process," J Ind Eng Int, vol. 10, pp. 65, 2014.
[21] D. A. Desai, P. Kotadiya, N. Makwana, S. Patel, "Curbing variations in packaging process through six sigma way in a large-scale food processing industry," J Ind Eng Int, vol. 11 (issue 1), pp. 119–129, 2015.

[22] M. Bhargava, A. Bhardwaj, A. P. S. Rathore, "Prediction model for telecom post-paid customer churn using Six-Sigma methodology," Int. J. Manufacturing Technology and Management, Vol. 31 (issue 5), pp. 387-401, 2017.
[23] J. Antony, "Design of Experiments and its Role within Six Sigma," 2nd ed., pp.201–208, Design of Experiments for Engineers and Scientists, 2014.
[24] C. G. Smith, "A Cell Phone Virus that Will Wreck Your Life," April, p.45, Industry Technology and Standards Technology Issues, Washington, 2005.

Increase productivity of vertical roller mill using seven QC tools

Praveen Pareek[1] and Dr. Vijayendra Singh Sankhla[1]

[1] Department of Mechanical Engineering, GITS, Udaipur, India.

Email: praveen.parieek@gmail.com

Abstract. Cement sector in india is enjoyign success and capacity enhancement due to the rapidly growing demand in various sectors but at the same time a threat is approaching to cement industry that its input cost is increasing day by day like power cost, fuel cost, raw material cost. Also cement is a highly energy consuming industry. Hence in order to survive and sustain in the market its profitability need to be enhanced by increasing productivity and reducing power consumption. High productivity and low power costs can be achieved by increasing output, lowering breakdowns and optimizing the grinding process which eats almost 60% of power cost. In recent years vertical roller mill (VRM) has proved to be a popular choice for finished cement grinding due to low power consumption but it is also very sensitive to vibrations and can deteriorate productivity if process optimization is varied slightly. Objective of this research paper is process optimization of Vertical roller mill, improvement in productivity and reduction in vibration breakdowns using seven QC tools. The research explains complete process of identifying possible causes and final root causes, finding possible solutions and implementing them in order to increase the output of vertical roller mill. This study can benefit the organizations using VRM and are not able to utilize its full productivity due to some bottlenecks or constraints.

1. Introduction

India is the world's second largest producer of cement and produces more than 8 per cent of the total installed capacity globally. Due to the rapidly growing demand in various sectors such as defense, housing, commercial construction and industrial construction, it is expected to touch 550-600 million tones per annum (MTPA) by the end of year 2025.[11]

With recent growth and success journey there is also a threat approaching to cement industry that its input cost is increasing day by day. Power cost, fuel like coal cost, raw material cost have doubled up in recent years whereas cement price has not hiked in that fashion. Also cement is a highly energy consuming industry and 40% of cement production costs are energy costs out of which more than 60% of the total electricity is used in grinding circuits and hence bears most of the manufacturing cost. Hence in order to survive and sustain in the market they need to increase their profitability which can only be achieved by increasing productivity and reducing power consumption. These can be achieved by increasing output, lowering breakdowns and optimizing the energy consuming grinding process. For the past three decades the vertical roller mill has emerged to be the preferred choice for grinding raw materials[12]. It has advantages like low power consumption, process simplifications and compactness but it is also very sensitive to vibrations and can deteriorate productivity if process optimization is varied slightly.

The problems of high vibration and low productivity of VRM can be addressed using seven QC tools by identifying possible and responsible causes and working on to those causes by modifications & Kaizens[9]. Productivity can be enhanced by use of innovative products like grinding aids.

The future of grinding is high productivity and lower specific energy consumption machines and machines having capabilities of adapting to complex processes and new advancements in technologies.[1]

2. Research Study

The study and research was conducted in a cement industry where Vertical roller mill was employed in cement grinding section. In this case problem was, Cement Mill was not performing as per its design parameters. It was a Vertical Roller Mill & is designed for rated capacity 215 TPH but its output was very low so specific power consumption was higher and contributing to loss in terms of production and in turn profitability.

2.1. Tools & Techniques Used

2.1.1. Data Collection

- To increase productivity of VRM - For primary pre-improvement phase data collection, SAP reports and monthly & daily MIS reports were used. Data of last 6 months was taken to analyse the exact nature. Data for production rate and specific power consumption were taken.
- To reduce vibration breakdowns of VRM - Last six months history of breakdowns related to vibration were collected with cause of vibration.

2.1.2 Defining & Solving the Problem

We have adopted following methodology & tools step by step:-

- Analysis of pre-improve phase data
- Brain storming sessions & defining the problem statement
- Captured all the probable causes.
- All probable causes put in the Ishikawa (Fish bone) diagram
- Critical issues were identified by Pareto Analysis chart.
- Brainstorming session to solve to critical issues and make action plan accordingly.
- Followed the PDCA cycle.

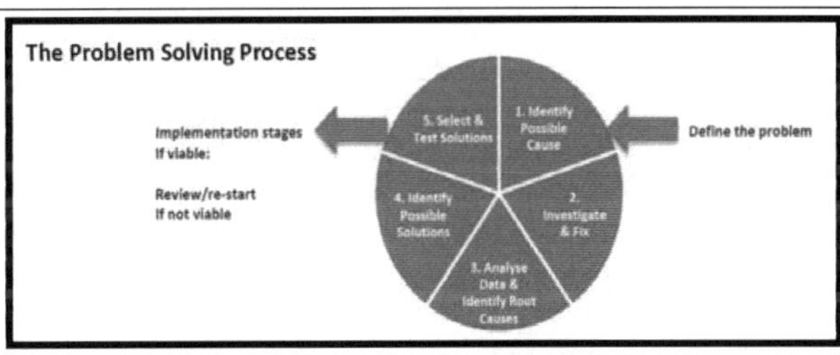

Figure 1. Problem Solving Process and Problem Solving Wheel

2.2. Problem Statement

Below shown Figure 2 is Production (TPH Tones per hour is measuring unit of production) and Specific Power consumption of Vertical Roller Cement Mill for pre-improvement phase. Data of last 6 months was taken to analyze the exact nature problem and behaviour of data.

It can be clearly seen that last 6 months average production Rate was 167.38 TPH & Specific Power consumption was 41.63 kWh/MT, which were below the rated design performance capacity of VRM. VRM was designed for 215 TPH. It means for every hour and hour we were loosing in terms of production and cement dispatch/sales and profit.

- What exactly is the problem? - Lower production rate than designed capacity
- Where? - Vertical Roller Mill
- When? - Since installation , data analyzed for 6 months
- Who is affected? - Production & Dispatch Targets

- How big? - Impact on sales and profit

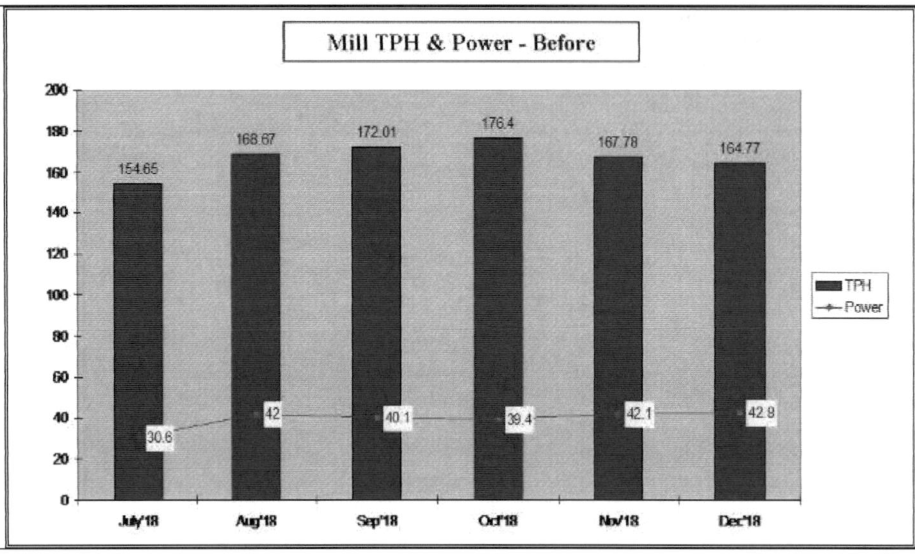

Figure 2. VRM Production Rate and Specific Power

It can be clearly seen in Figure 3 that last 6 months average breakdown frequency Rate was 60.16 tripping per month means per day more than 2 tripping. Excess breakdown results in frequent start/stop of VRM which consumes more power & also stability of Mill operation and productivity gets affected. Hence it was a big reason to worry & needed a solution.

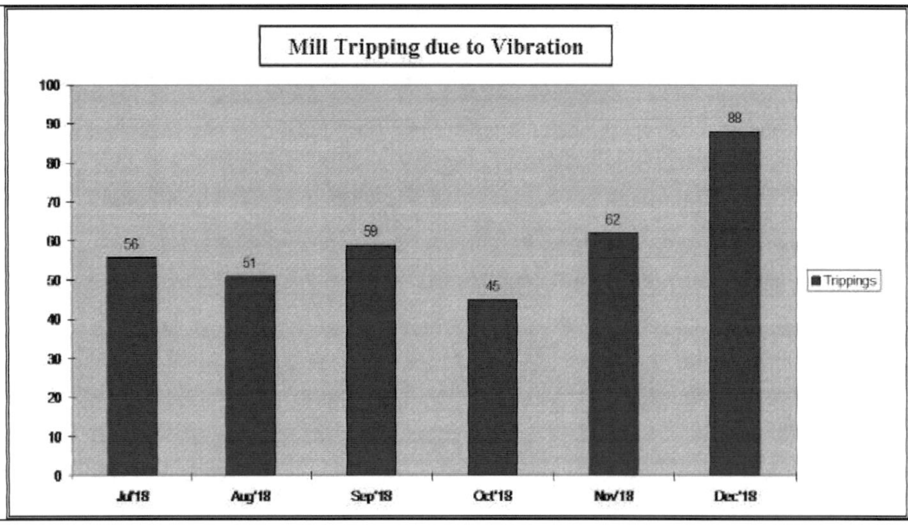

Figure 3. VRM Monthly tripping due to vibration

2.3. Identify all possible causes

Aim: To identify and write all the possible causes for a mentioned problem.

How: Using concept generation tools, individuals or groups identify the causes of problem definition. Keep an open mind, do not confirm any suggestions at this stage and do not try to off track and analyze other issues.
Main Tools Used: Brainstorming, Cause & Effect Diagram

Tested all the factors related with Man, Machine, Method & Material area one by one with help of brainstorming. Which are segregated in Cause & Effect diagram shown in Figure 4 and listed below.The possible causes for constituting less TPH problem in 4M categories :-

- I/P Feed Size - Clinker Size
- Fineness of the product
- Clinker hardness
- Skills of operator and attendants
- Grinding Pressure
- Separator Speed
- Inlet Hot gas damper and diverter , Recirculation Damper calibration
- Fan flow & Differential pressure PID tuning
- Mill vibration
- Breakdowns
- Grinding bed thickness
- Dam Ring
- Table & Roller Liners
- Water nozzle and water spray
- Weighing equipment Accuracy and calibration viz Solid Flow Feeder & Weigh feeder.
- Improper feed profile - Feed diverter calibration
- Capacity of critical material transfer equipment like Rotary air locks, Air Slides, Belts, Bucket elevators etc.

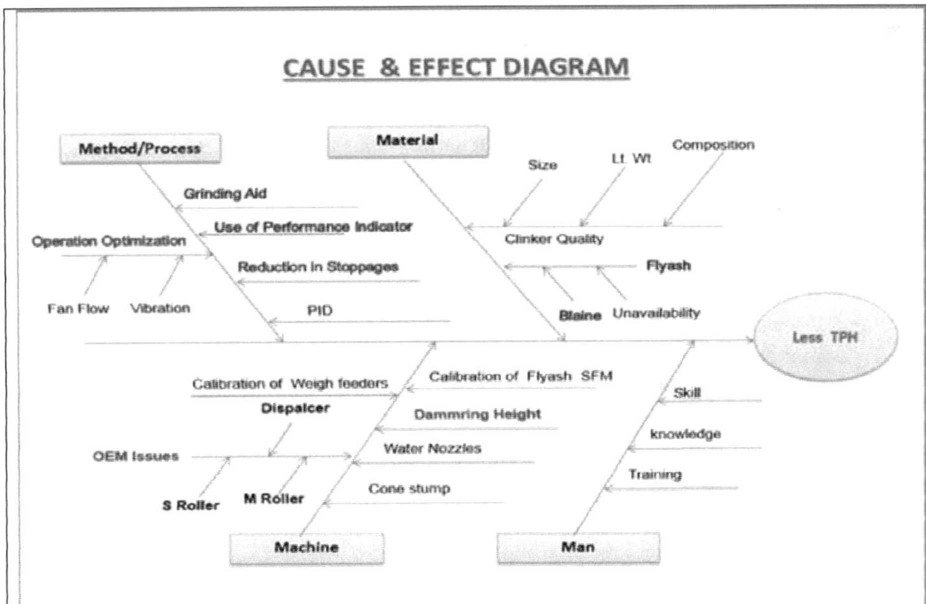

Figure 4. Cause & Effect Diagram for less VRM Production Rate

2.3.1 Material

- Clinker quality: - Clinker's parameters plays a vital role in productivity of mill, viz it's hardness, size, composition. If clinker size is too small (Too fine like powder) it will not construct proper grinding bed and cause vibration. If clinker is too hard it will take more time to grind.[8,12,14]. Clinker found OK.
- Fly ash: - Fly ash availability was not an issue with us. But fly ash Blaine was around 280-290 m2/Kg, which we require up to 320 m2/Kg. But Blaine of Fly ash was not in our control and Flyash cannot be ground solely to achieve required blain size. So it was a cause with no solution.
- Gypsum: - Size of gypsum boulders were very high than required. High size boulders creates sudden up/down or Roller and Grinding pressure fluctuates. Also Sticky gypsum if not correctly handled may create problem in hoppers extraction and jamming in transfer points which ultimately results in feed rate variation.[8,12,14]
- Waste TAD Material: - It causes lot of vibration, so its use was not advised. But its consumption was necessary.
- Variations in feed rate increase vibrations of the mill, destabilizing operation and in the extreme resulting in mill stoppages. Observed feed variation in Fly ash.
- The mill feed material needs to be free of metal pieces. Higher the metal pieces frequent operation of diverting gate and results in feed variation. [8,12,14]
- Abnormal feed size and distribution may increase the level of vibration and destabilize the mill. Large feed size in any case will reduce the mill production rate. Excess fines in the feed material or reject from the separator can adversely affect the stability of grinding bed and can cause vibrations . [8,12,14]

2.3.2 Man

Skill and knowledge was not an issue, all operator were well trained a professionally qualified. Also Mill was in OEM operation for a long time to stabilize.

2.3.3 Method/Process

Operation optimization or process optimization is a critical step responsible for efficient operation of Mill.

It constitutes of optimization of operation parameters like Flow, Differential Pressure, Classifier Speed, Mill Inlet & Outlet Temp, Hydraulic Pressure, equipments, control logic, PID tuning, breakdowns etc. Some parameters like excess vibration, high numbers of breakdowns were observed which in turn needs further optimization.

- Mill DP - If differential pressure is too high or too low means mill is choked or empty. If DP transmitter is faulty then it shall indicate false reading and operator will not be abler to get correct condition of Mill inside filling.[14]
- Fan Flow - Fan flow & speed is very necessary to optimize for effective material transport.[14]
- Inlet & Outlet temp optimization is necessary to control material drying & outlet final material temperature.[14]
- Hydraulic pressure should be matched to the material load in the mill. Insufficient hydraulic pressure for the mill load will result in high circulating load and high pressure loss with potential over filling the mill. Excessive hydraulic pressure for the load will result in a thinning of the bed and increased vibration /wear.[14]
- Water spray is used to help stabilize the grinding bed and maintain a low level of vibration and generally this is applied directly to the material on the table, immediately before it passes under the grinding rollers. Excessive addition of water to the table can result in increased wear and even thermal shock of the grinding elements. Water spray optimization required.[14]

- Grinding bed is the material layer formed between the roller and the rotating table. It transmits the entire roller force and mill power. It is the key factor for successful operation of a VRM. Requires optimization.[14]
- Mill vibration to be controlled to avoid unnecessary breakdowns and wear tear in mill components. Found excessive tripping need to be optimized.[14]
- Interlocks optimization required.

2.3.4 Machine
Efficiency of machine and desired production are related to each other. If machine operates well then it will give required production effortlessly.

- While checking found material transfer equipment (Bucket elevators, Belts, Air slides) capacity adequate and fit of required operation - 300 TPH capacity.
- Calibration of critical measuring devices like Weigh feeders & Solid flow meter found OK.
- Metal detectors calibration found OK
- Wear of the grinding elements increases the material retention upon the table and tends to increase mill power drawn. Uneven wear of grinding elements leads to reduced mill efficiency due to non-uniform transmission of grinding energy to raw material. Mill feed table & grinding rollers profiling found Ok as desired.
- The dam ring is used to control the bed depth in order to stabilize the grinding efficiency and vibration. Too high Dam Ring –excessive bed depth, inefficient grinding and high power consumption. Too low Dam Ring –low bed depth, high vibration and high recirculation. Dam ring height optimization required to control rejects.
- Water nozzle modification required as excess chocking is noticed.
- Inaccuracy of Roller position and speed sensor: - If roller position sensor is not calibrated properly, it will show wrong position/height of roller and Operator in central control room will get confused and will assume that roller is forming correct bed and grinding properly whereas situation is opposite. Found OK.
- Fan Flow - Fan speed controller, or speed sensor or Flow sensor transmitter is faulty or unreliable than optimization and tuning will not be as desired. Found OK.
- Blockage of pressure tapping, particularly mill inlet and air flow measurement points. OK.
- Out of calibration bed depth monitors & other instruments like PT, TT etc. Found OK.
- OEM design issues - Hydraulic system, cone stump design, and classifier design.

2.4. Finding Potential Causes
During the analysis of the causes related with the area of Method, Material & Machine tested probable causes one by one and found 4 causes have some significance on the effect in cement grinding : -

2.4.1. Actual/Potential Causes

- Poor process/operation optimization, High breakdowns
- OEM related design issues - Hydraulic System Design
- Fly ash Blaine not suitable

2.4.2. Solutions

- Operation optimization by PID tuning, Grinding bed optimization etc.
- Grinding Aid Implementation to increase output
- Reduction in breakdowns (Related to vibration)
- OEM related design issues - Hydraulic System Design

2.5. Implementation of Potential Solutions

2.5.1 Operation optimization:-

2.5.1.1 All PID Control loops tuned:-
Though at the time of commissioning of mill all PIDs were tuned but after years of operation many parameters and interlocks were changed but still PIDs were at the same. Also some control loops were not in circuit. So to optimizer the mill performance all PIDs were taken into circuit and tuned at the best level manually. There were 4 control loops for Differential Pressure, Mill outlet temperature, Mill ventilation fan flow and Mill inlet pressure.

a) Differential Pressure
The mill differential pressure is the difference between pressure at mill inlet and mill outlet representing currently loading condition of the VRM. Changes in the differential pressure when the grinding pressure and the hot air circulation are constant, directly reflects the amount of material in the mill. In other words, when the differential pressure decreases, the amount of input material is less than the discharge material, causing the material bed to be thinner. In contrast, as the differential pressure increases, the material bed becomes thicker. VRM vibrates when the material bed is too thin or too thick and trips or stops when the vibration limit is exceeded. For these reasons, the total feed amount must be adjusted so that the differential pressure is within the correct range.[4,7]

The differential pressure of the mill depends on the following parameters:

- Input material feed rate
- Mill fan flow rate
- Grinding pressure
- Classifier speed

Usually only the mill input material feed rate has a decisive influence on the differential pressure while gas flow rate, grinding pressure and classifier speed are maintained at the similar condition according to the pre-adjustments during previous operation unless the characteristics of the raw material such as the grind ability of the material has been changed. [4,7]

So to optimize the DP PID, we made a PID interlock in which PID algorithm can estimate the desired volume of material feed, according to which the feeding control module can control the velocity of the conveyor belt through the frequency converter. If DP is low, belt speed will be increased and if DP is high then belt speed will decrease automatically.[4]

Figure 5 shows the PID control structure of the differential pressure, where k1, k2, k3, k4 are proportions of Clinker, Gypsum, High grade lime stone & Fly Ash in the raw materials.

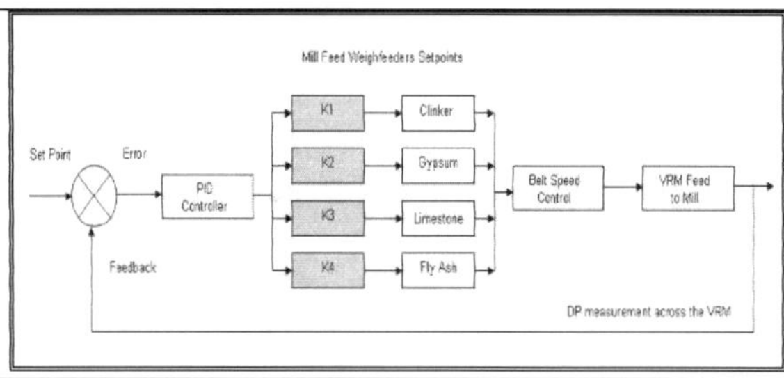

Figure 5. VRM Differential Pressure V/s Mill Feed PID Loop

b) Mill Outlet Temperature / Material Temperature
The gas temperature at mill outlet is a measurement of the drying process accomplished in the mill. The mill outlet temperature can be controlled by controlling the recirculation air damper or inlet hot gas fan speed and also can be cooled down by opening the fresh air damper. To achieve an effective

drying, outlet temperature is generally regulated between 80°C and 100 °C to regulate the quantity of water spray in the mill. When the outlet gas temperature exceeds Tupper, the quantity of water spray should be increased, even as the water spray system stops when the temperature is under Tlower. Fig. 6 illustrates the PID control of VRM outlet temperature.[4]

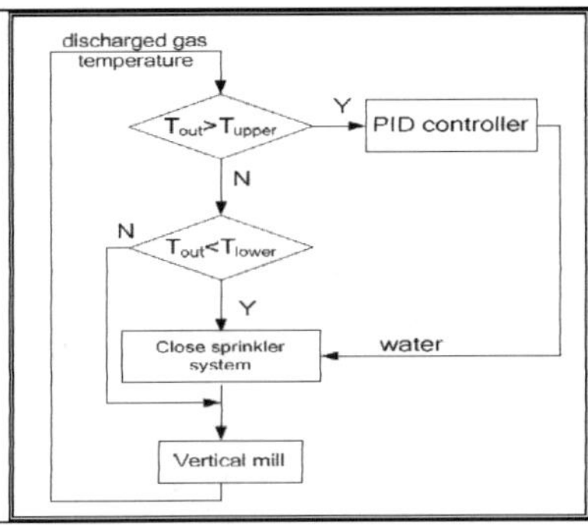

Figure 6. VRM Outlet Temperature V/s Water Spray PID Loop

c) Mill Fan Flow / Transport Air Flow

The hot air in the mill plays a role in drying and transporting the material. Adequate ventilation not only improves grinder efficiency but also contributes to the stability of the mill. This ventilation can be measured behind the bag filter and controlled by controlling Mill Fan speed by controlling fan motor speed through variable frequency drive.. The control structure is tuned to achieve good performance. The preset module calculates the amount of ventilation required to dry feed material and the amount of ventilation required to transport the material according to the heat balance equation and compares the two pits. When the amount of ventilation required for dry materials is greater than that used for transportation, it is set to the required ventilation. In contrast, the amount of ventilation used for transport is selected as the desired set point. During the manufacturing process, we control the fan speed so that there is ventilation in the mill to meet the drying and transportation requirements.[4,7]

d) Mill inlet negative pressure (Draft)

The hot air used in grinding process is taken from the kiln, and the operation condition of grinding system and the kiln influence each other. The mill inlet pressure should be maintained at constant value. The static pressure measured at the mill inlet gas duct is influenced by the gas flow entering into the mill by mean of:

- Recirculation air damper
- Hot gas generator or Booster Fan
- Fresh air inlet damper

Too high negative pressure before the mill increases the energy consumption of the mill system fan. However too low negative inlet pressure influences the steady gas flow within the system and leads the grinding system into disturbances.[7,8,12]

2.5.1.2. Grinding bed height optimization through Dam Ring:-

The grinding bed thickness, defined as the minimum gap between the grinding rollers and the grinding table during the operation, is adjustable by the height of the dam ring the grinding pressure and mill circulation load during commissioning[7,12].Dam ring is shown in figure 7.

The thicker grinding bed gives narrower particle distribution but results with higher power consumption and softer operation. Nevertheless the thinner grinding bed gives wider particle distribution but with lower power consumption and rougher operation. The optimum grinding bed thickness shall be decided during commissioning[14].

Dam ring height optimized in various phases and best result found at 340 mm.

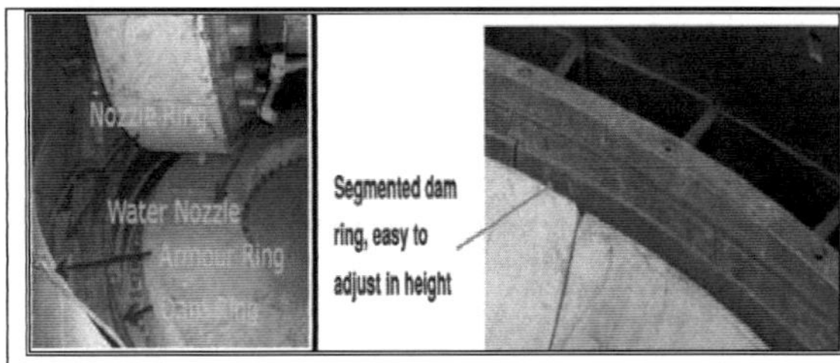

Figure 7. Dam Ring inside view

2.5.1.3. Reduction in Mill Vibration:-

Separate brainstorming sessions were done as vibration tripping was very high and Mill was unstable. Optimization of the mill became very difficult due to vibration and instability. So to stop production loss on daily basis and to ensure smooth and optimize Cement Mill operation, we have to solve this at first priority.

a) Analysis Tools: To solve this problem following QC tools were used:-

- Pareto Chart.
- Cause and Effect Diagram

b) Pareto Chart: To analyze the major reasons for vibrations, last 6 months data for Mill breakdowns due to vibrations was collected from breakdowns history and segregated in cause for vibration. Based on the same pareto chart was prepared. From the chart it was found that 82% of the tripping were due to 3 main problems:-

- Sudden vibration peaks
- Fly ash feed fluctuations
- Water problem

Our further action was to address the possible causes for these 3 main tripping. For this we decided to employ pareto chart for mill tripping due to vibrations. In last 6 months total 361 tripping occurred due to vibration. Pareto diagram is shown in figure 8 & pareto table is as following:

Table 1. Pareto Table for Pareto Diagram - Mill Vibration analysis

S. No.	Cause of Tripping / Breakdown	Frequency	% Contribution	Cumulative %
1	Vibration due to Sudden peak in Mill vibration	184	50.96	50.96
2	Vibration due to Fly Ash fluctuation	65	18.00	68.96
3	Vibration due to Water spray problem / Water Tank empty	47	13.01	81.97
4	Vibration due to Dusty / Fine Clinker	29	8.03	90.00
5	Vibration due to foreign metal in Mill	18	5.00	95.00
6	Vibration due to HSLM / Grinding Pressure fluctuation	18	5.00	100.00
	Total	**361**	**100**	

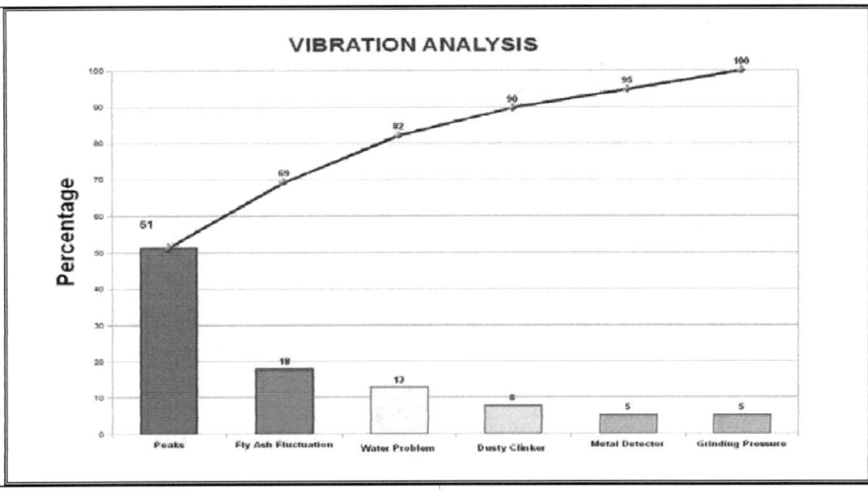

Figure 8. Vibration analysis Pareto Diagram

c) Cause and Effect Diagram :-

All the possible reason for the vibration was discussed by team members and lots of point came out which were further discussed and categorized. Cause & Effect diagram is shown below in Figure 9 :-

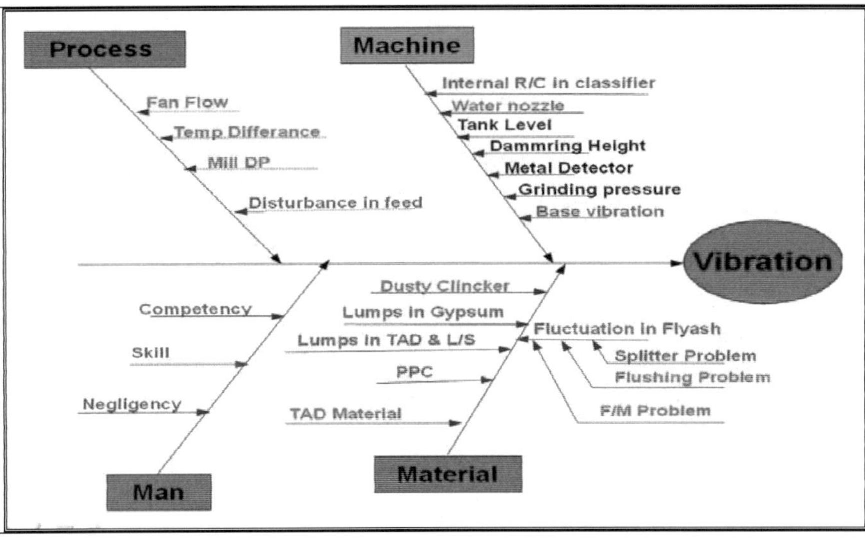

Figure 9. Vibration analysis Cause & Effect Diagram

Following were found to be possible causes for constituting vibration problem in 4M categories:-

i. Material:-

- Dusty and Fine clinker cause improper bed formation. Creates vibration
- Lumps in Gypsum and TAD material causes sudden up/down of M Roller causing fluctuations and vibration
- Fluctuations in Fly ash means High or Low fly ash flow at sudden causes inconsistent feed flow and causes vibrations.

- Fly ash fluctuations can be a result of Splitter positioning problem
- Solid Flow meter calibration error, it misleads about exact feed in mill and creates material shortage or excess in the mill creating vibration.
- Weigh feeder calibration error it misleads about exact feed in mill and creates material shortage or excess in the mill creating vibration.

ii. Machine: -

- Water nozzle wear out causes improper water flow and distribution.
- Chocking of water nozzle during stoppage of mill creates uneven water spray
- Water tank level control problem. Level sensor or solenoid valve problem. Due to it water tank empties out during operation and due to water shortage Mill trips due to vibration.
- Damring height plays a big role in mill stability. Low dam ring height causes less bed formation and causes vibrations, it also increases mill reject.
- Metal detector accuracy plays a big role, if MD doesn't sense metal then it will go in mill and creates vibration if it comes between table and roller.
- Accuracy of Vibration sensor and it's installation (If it is exposed to foreign material)
- Inaccuracy of Roller position and speed sensor: - If roller position sensor is not calibrated properly, it will show wrong position/height of roller and Operator in central control room will get confused and will assume that roller is forming correct bed and grinding properly whereas situation is opposite.
- Mill DP - If differential pressure is too high or too low means mill is choked or empty. If DP transmitter is faulty then it shall indicate false reading and operator will not be abler to get correct condition of Mill inside filling.

iii. Method: -

- Disturbance in feed - Mill reject hopper diverter timing in case of metal detect. In case of metal detection, diverter divert the input feed to reject hopper so that metal cannot go inside the mill and create wear/tear or vibration. As metal is detected the diverter operated towards Reject hopper side for a particular time and then comes back in main feed to mill. If diverting and transition time is not optimized then it may create loss of feed into mill and in turn creates vibration.
- Difference in Roller height due to Hydraulic system problem.

And from above analysis main reason came out were as following:

- Fluctuation in Flyash Feeding due to initial inrush (Solid Flow Meter Dosing valve)
- Fluctuation of Flyash due to Empty out of Bin.
- Jamming of Flyash in Bin Extraction due to foreign material.
- Lumps in TAD Material.
- Chips of hard facing of Dam ring found in mill during inspection. It is a hard metal and causes vibration when comes in between Roller and table
- Water Problem due to Tank empty (Level sensor problem), Wear of water nozzle, Chocking of Nozzle.
- Disturbance in feed due to Metal detector and Reject diverter operation.
- Roller no. 3 profiling was not proper, it creates vibration.
- Big size gypsum boulders & big size of Performance improver High grade lime stone.

d) Action Plan:

Table 2. Action plan for reduction in Vibration

SN	Action to be taken	Target Date	Status
1	TAD material to be avoided in case of PPC.	After consumption of TAD dust.	Done
2	Water spray to be optimized by using one pump instead of two. Providing High capacity pump & VFD to control the speed. Also providing reliable water flow meter	28 Feb - Requires new material	Done
3	Nozzle chocking is eliminated by providing bypass compressed air & interlock , when mill stops Compressed air will clear the water nozzle in every 30 minutes	31 Jan	Done
4	For tank empty out problem, new radar level sensor to be installed and one interlock to be provided for auto start of pump to avoid emptying of water tank.	28 Feb - Requires new material	Done
5	Gypsum size should be restricted to 50 mm.		Not Uniform
6	Fly Ash blaine need to be increased		Not Uniform
7	Performance Improver (L/S) to be used of size < 50 mm for better bed formation.		Not Uniform
8	Diverter plate modified to decrease the inlet temp. Difference.	Done still under observation	Done
9	Flyash Bin V/S Bin Weight PID to be provided after dosing valve problem rectification to overcome flyash empty out issue. Also Wire Mesh to be provided to prevent Lumps or foreign material in Bin.	31 Jan	Done
10	To avoid flushing in fly ash ,interlock with aeration blower is modified	31 Jan	Done
11	To avoid Flyash fluctuation at Mill startup after a long time, Dosing valve interlock is modified. It will open 100% in start and After 1 minute if will come to 50% open	31 Jan	Done
12	ash splitter to be checked in every shift	15 Jan	Done
13	Calibration of Flyash Bin with proper test weights.	20 Jan	Done
14	Calibration of Solid flow meter & daily check weighing	25 Jan	Done
15	Calibration & Daily checking of Metal Detector	15 Jan	Done
16	Optimization of reject diverter timing during metal detection. It is to be reduced to 4 second	15 Jan	Done

2.5.2. Grinding Aid Addition

Grinding aid reduces the polarity of the cut surface and the attraction forces between particles. This means that agglomerates of fine particles and the packing of fine particles around a larger particle are dissipated, resulting in an improved classifier efficiency. The internal circulation of fine particles becomes lower and the clinker on the grinding track becomes coarser. The inter-particle friction and thereby the effectiveness of the grinding process is increased. In this way, grinding aid stabilises the grinding bed on the rotating table, facilitate compaction and de-aeration, improves productivity and lowers vibration in the system.[6]

Benefits of adding Grinding Aid: -

IOP Conf. Series: Materials Science and Engineering 1017 (2021) 012035 doi:10.1088/1757-899X/1017/1/012035

Table 3. Benefits of adding Grinding Aid in VRM

Effect	De-agglomeration
Effect on separation	Higher efficiency, less recirculation
Effects on grinding bed	Few fines , coarser clinker
Effect on grinding	Improved grinding
	Ease of de-aeration
	Less vibration, less wear and tear
	Low water requirement
Effect on productivity	Reduced mill differential pressure
	Productivity improved
	Lower specific energy demand
Effects on final product cement	Broader particle size distribution
	Increased powder flow ability
	Reduced pre-hydration of cement
	Fast initial setting
Effects on concrete	Improved mortar workability
	Strength in concrete

To test the effects of grinding aid, we also tested it by adding some amount of Grinding Aid on experiment basis. All vital parameters were kept same in testing With/Without adding grinding aid. Dosing of grinding aid was 75-80 Liters per hour. The test results were found to be positive, we witnessed increase of 7.66 Tonnes per hour and reduction of 2.34 kWh/T specific power, which is a very appreciating figure.

Table 4. Results of Test on VRM

With grinding Aid		
Date	**Power consumption kWh/Ton**	**Production Rate TPH**
11-02-2019	34.55	195
12-02-2019	34.22	198
13-02-2019	32.04	201
Average	**33.60**	**198**
Without grinding Aid		
15-02-2019	36.36	185
16-02-2019	36.10	188
17-02-2019	35.38	198
Average	**35.94**	**190.33**
Difference	**-2.34**	**7.66**

2.5.3. OEM Related design issues

After brainstorming sessions and discussion with OEM some modifications were suggested: -

a. Support Roller Interlock - Support roller are employed for grinding bed formation. But also its size creates a demerit that it hinders recirculation inside mill and obstructs the free flow of material towards classifier. Hence recirculation time in mill increases resulting in low production rate. This was the case for our project. Hence with the consent of OEM we tried mill operation without employing the S-Rollers operation and we found stable mill performance with optimum DP, less vibrations & improved circulations inside mill.

b. HSLM Modification: - The VRM has 3 Master rollers and earlier there were separate Hydraulic system for each rollers. So for roller Up/Down and Increasing/Decreasing Grinding pressure, no symmetry was achieved constituting unstable operation and increased breakdowns. Later OEM suggested a modification to have a single Hydraulic circuit system for all 3 rollers so now on excellent hydraulic control achieved. Which resulting in no pressure difference hence less breakdowns. This project was done by OEM.

c. Cone stump diameter changed from 1988 mm to 2240 mm in Sep 2018. Role of cone stump is to distribute material via centrifugal force to grinding roller area. Earlier smaller cone stump was not able to distribute fully and some material was kept ungrounded on feed table. Now with bigger cone stump this situation is addressed fully with effective distribution and grinding.

2.5.4. Fly-Ash Blaine: - Increase fly Ash Blaine from 290 to 320

Fly-ash is a major supply for any cement plant now a days because every producer wants to produce more PPC than OPC due to profitability and increased output.

Our required Blaine for PPC is 320 M2/Kg of material. For PPC our proportion is 65% Clinker + 35% Fly ash. Hence need less time to produce PPC as it required Grinding of only 65% of material and rest 35% Fly ash is blended in Mill. So there is less power consumption and more output in case of PPC rather than OPC cement.

Now if we get a Fly ash of Blaine size 320 then it will be a boon for us as we just need to put it in mill and very less retention time it will become final product output. But the Fly ash blain at our point was 290 M2/Kg only.

Major source for Fly-Ash supply is Power plants nearby and blaine of Fly-Ash is not in our control as we cannot force our suppliers. Cement industries get fly ash at a very minimalist cost.

Other source for Fly ash was our own small size captive power plant. But our blain size was also 285-290. We can get blain size of 320 only if we fire F-Grade coal in our boiler, but in actual we were firing PET Coke. As cost of PET coke is less than F-Grade coke.

- By increasing Fly Ash from own TPP - This will be possible after usage of F Grade coal in TPP.
- By increasing Fly Ash Blaine from Supplier - As Fly Ash Blaine supplied is constant and not in our control.

3. Results

After implementation of all the planned jobs and action plans, data is collected for production rate and breakdowns for 5 months (2 months of implementation stages & 3 months after) and presented them in graphical form. From analysis of data following results were found to be achieved:

3.1. Tangible

- Increased average mill productivity from 167.38 TPH to 194.34 TPH. Means 26 Tones per hour increased. If we consider 22 hours a day running then it will be 22 X 26 = 572 Tonnes per day .
- Saving due to increase in production = 572 * 750 = 4, 29,000 Rupees per day.

- Average monthly tripping of vibration is reduced in significant numbers from 60.16 to 29.3 tripping per month means approximately 50% improvement.

3.2. Intangible

- Stable Mill Operation.
- Reduced breakdowns, Increased MTBF & OEE of Mill.
- Improved cement dispatch & targets are met.
- Reduced Wear & Tear of mill internal components
- Reduced vibration level of mill , so overall equipment life enhanced

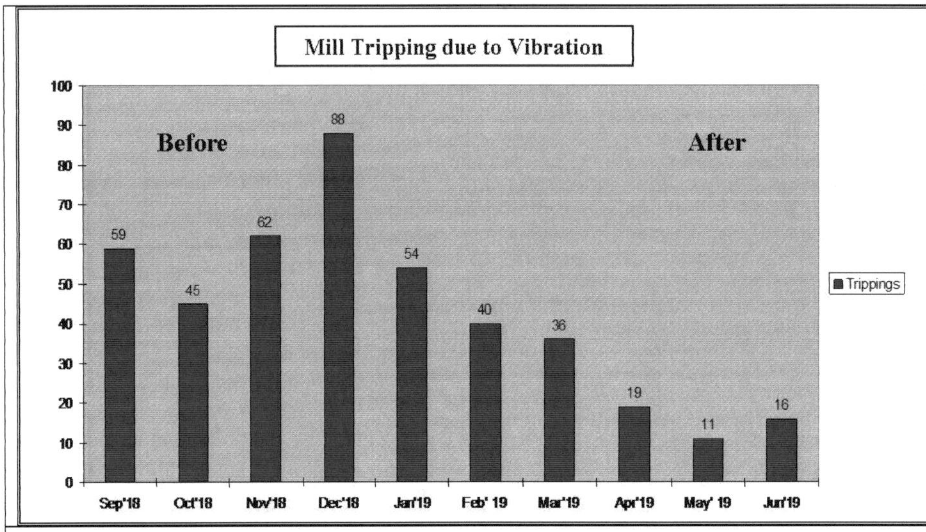

Figure 10. Results - Tripping of mill due to vibrations (Before & After)

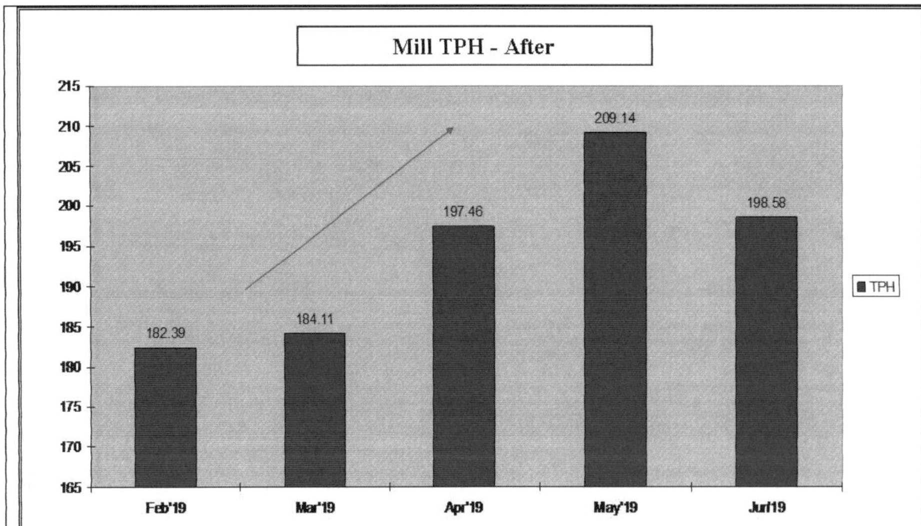

Figure 11. Results - VRM production rate - After

4. Conclusion

In recent years Vertical roller mill has been a popular choice for material grinding due to advantages like low power consumption, process simplifications and compactness but it is also very sensitive to vibrations and process parameters variation.

Operation stability and better productivity can be achived if all process parameters and interlocks are optimized and kept well under control. Reasons for process disturbance and breakdowns should be identified and addressed. To identify possible and responsible causes, seven QC tools can be utilized.

QC tools like brainstorming, Pareto Analysis, Cause and Effect Diagram and Problem solving process can be used to identify and remove the bottleneck in the process. Further working on to those causes by modifications, kaizens and process improvements, can lead to better productivity, lesser breakdowns and lower power consumption.

5. References

[1] Kamal Kumar, Sept 1998, " Innovation in Energy Consumption", A company report by Holtec Consulting Pvt. Limited

[2] Confederation of Indian Industry (CII),2002, "Cement Industry in India - Trade Perspectives", World business council for sustainable development.

[3] Loesche GmbH,2006, "Mills for the world: Loesche's 1906-2006 centenary", ZKG International, 59, 10, 26-34.

[4] Ning Yan-yan, Zhu Guang, Yuan Ming-zhe, Wang Zhuo, July 2011, "Design of Intelligent control system for Vertical Roller Mill", The 2nd international conference on Intelligent control and information processing.

[5] R.Virendra, Dr. B.Sudheer, Prem Kumar, J.Suresh Babu and D.Rajani Kant, apr-2015,"Detailed Energy Audit And Conservation In A Cement Plant", International Research Journal of Engineering and Technology (IRJET), **Vol.- 02** Issue: 01.

[6] Philipp Flieger, Kevin Treiber, Daniel Komorek,2015, "Future grinding technologies", European cement research academy technical report no. 127/2015.

[7] Deniz Altun & Dr. A. Hakan Benzer, 2017, "Mathematical Modeling Of Vertical Roller Mills".

[8] Casrston Gerold - Loesche GmbH,2016,"Operational parameters affecting the vertical roller mill performance".

[9] Heena Sharma, Dr. N.M Suri, "Implementation of Quality Control Tools and Techniques in Manufacturing Industry for Process Improvement", International Research Journal of Engineering and Technology (IRJET), **Vol.- 04** Issue: 05, May -2017.

[10] G. Amudha Ganesan,June-2016, "A Study on the Operating Efficiency of Major Cement Companies in South India", International Journal Of Advanced Scientific Research & Development (IJASRD) p-ISSN: 2394-8906 , e-ISSN: 2395-6089, **Vol-03**, Iss-02, Ver-I, PP 08 – 20.

[11] India Brand Equity Foundation report, 2020, "Cement Industry in India - 2020".

[12] S. W. Joergensen, GM-Grinding Technology, "Cement grinding Vertical roller mills versus ball mills" F.L.SMIDTH.

[13] Bureau of Energy Efficiency : www.beeindia.gov.in

[14] Loesche GmbH : Vertical Roller Mill LM 53.3 Manual & Presentation

Computational Study of Small Vanadium and Chromium Oxide Clusters: Effect of Hydrogenation

Ankur Chahal[1] **and Haider Abbas**[1]

[1]Department of Physics, Manav Rachna University, Faridabad, Haryana, India.

Email: haiderabbasphy@gmail.com

Abstract. In the present work we report the effect of hydrogenation on small vanadium oxide (V_2O_4,V_2O_6 and V_4O_{10}) and chromium oxide (Cr_2O_4,Cr_2O_6andCr_4O_{10}) clusters, within the framework of density functional theory. Comparison between dehydrogenated and hydrogenated cluster suggests that hydrogenation of these clusters changes the HOMO-LUMO position and their gap. A considerable enhancement in dipole moment is also observed. It is found that hydrogenation enhances the occupancy of the metal in these clusters and therefore, the metallicity increases. The observed mechanism of charge transfer due to hydrogenation is in good agreement with the reported experimental and theoretical results.

1.Introduction:

Transition-metal oxides, carbides, nitrides, and silicides have been a subject of increasing scientific scrutiny due to their importance not only from a fundamental standpoint but also their use in industrial processes. Involvement of *d* orbitals in bonding make the chemistry of transition metal oxides diverse which leads to a range of technological applications e.g. vanadium and chromium oxide catalysts are widely used in chemical industries. A major impediment in creating nanoscale materials with useful electronic and magnetic properties have been the identification and production of classes of clusters. The size and composition dependence of structural, electronic, and magnetic properties of these oxide clusters may offer opportunities to tailor the chemical and physical properties for future applications. Despite the abundance of experimental and theoretical studies on transition metal oxide clusters, the study on hydrogenated clusters of vanadium and chromium oxides have not been reported yet. In this letter, we report the effect of hydrogenation on two important transition metal oxide clusters namely vanadium and chromium oxide.

2. Computational Details

All calculations have been performed with the GAMESS package [1]. The ground state geometries (shown in Fig.1) were obtained from optimization in C1 symmetry using the density functional theory (DFT/B3LYP) with the 6-31G (d, p) basis. All the clusters shown in the figure have singlet spin multiplicity and paired electrons.

3. Results and Discussion:

3.1. Vanadium oxide

Vanadium oxide is widely used as a component of oxidation catalysts [2-5] and optical materials [6]. Due to these diverse applications, numerous experimental investigations have been reported: using mass spectroscopy [7-11], matrix isolation [12], electron spin resonance spectroscopy [13], infrared photo dissociation [14], and anion photoelectron spectroscopy [15-18]. A number of theoretical studies have also been carried out to investigate the structure and the bonding characteristics of vanadium oxide [19-22]. Further, an interesting and intriguing aspect is the transition from metal to insulator and semiconductor to metal which depends upon the change of external parameters, such as temperature. Recent theoretical and experimental studies [23] show that the metal-insulator transition in nanoscale beams of vanadium dioxide can be strongly modified by atomic hydrogen doping. To understand this the phenomenon at the microscopic level, small clusters of hydrogenated vanadium oxide can be considered as a good model.

The optimized geometries of the hydrogenated vanadium oxide with DFT calculations are presented in Fig.1. In V_2O_4, HOMO is localized on both the vanadium atoms and LUMO consists of both the vanadium atoms as well as a little contribution of all the oxygen atoms. On hydrogenation, HOMO is localized on one vanadium atom while LUMO is located mainly on the other vanadium atom. In V_2O_6, HOMO is found to delocalize over all the six oxygen atoms with a larger contribution from the terminal oxoligands. The nearly same result is observed for the LUMO with slightly increased density on the ring's oxygen. When this cluster is hydrogenated, HOMO is found on both vanadium atoms as well as on the ring's oxygen atoms while LUMO is situated on both vanadium atoms and slightly on the terminal oxygen atoms. In the case of V_4O_{10} cluster, HOMO is found on all oxygen atoms while LUMO is found on all vanadium atoms. In hydrogenated V_4O_{10} cluster, the position of HOMO is completely reversed and found on all vanadium atoms. However, in LUMO a slight contribution from a few oxygen atoms is observed. HOMO-LUMO gap in all clusters, except V_2O_6, is found decreasing with the addition of hydrogen but an enhancement in dipole moment is observed.
Bond distance analysis in hydrogenated clusters shows a strong H-O bond with distance lying in the range 0.959 Å to 0.967 Å. The gross DFT Mulliken population analysis shows that charge transfer occurs from hydrogen to oxygen, thus the electronegativity of the latter reduces. Consequently, oxygen atoms draw less charge from the vanadium atoms in comparison to dehydrogenated vanadium oxide clusters. Therefore, hydrogenation increases the charge of metal atoms. In recent experimental and theoretical investigations, the same mechanism was presented in order to explain the metallicity in hydrogenated VO_2 [23]. Thus, our results provide a good picture of the electronic structures of the hydrogenated vanadium oxide clusters.

3.2. Chromium oxide
Chromium oxide has a significant current interest, due to its broad applications in catalysts, spintronic, and magnetic recording tapes. In recent years, there have been a number of gas-phase experimental [24-29] and theoretical [30-35] studies on small chromium oxide clusters. The simple metal oxide clusters serve as model systems to obtain molecular-level information for surface and catalytic processes.

The optimized geometries of hydrogenated chromium oxide with DFT calculations are presented in Fig.1. In Cr_2O_4, HOMO is the same as that of V_2O_4, and LUMO consists mainly of both the chromium atoms with very little contribution from terminal oxygen atoms. On hydrogenation, HOMO is found mainly on one chromium atom with a little contribution from another chromium atom as well as both the ring's oxygen atoms. The position of LUMO is symmetrical to HOMO except for some more contribution from the other chromium atom. The position of HOMO in Cr_2O_6 is similar to the position of HOMO in V_2O_6 but LUMO is located mainly on both the chromium atoms and partially on all the six oxygen atoms. When Cr_2O_6 is hydrogenated, HOMO is found on both the chromium atoms with a little contribution from the terminal oxygen atoms, while LUMO is situated on both the chromium atoms as well as on the ring's oxygen atoms. In case of Cr_4O_{10} cluster, HOMO is on all chromium atoms and LUMO is found to delocalize nearly overall oxygen and chromium atoms. On hydrogenation, HOMO is found on the three chromium atoms and partially on oxygen atoms, except two of them while all chromium atoms contribute to the LUMO. HOMO-LUMO gap in all the clusters, except for Cr_2O_4, is found decreasing with the addition of hydrogen but dipole moment increases in general. The H-O bond distance in hydrogenated chromium oxide clusters are found lying in the range of 0.962 Å to 0.973 Å. The enhanced metallicity of the chromium oxide clusters after hydrogenation is found to be originated by the same mechanism of charge transfer as that of hydrogenated vanadium oxide clusters.

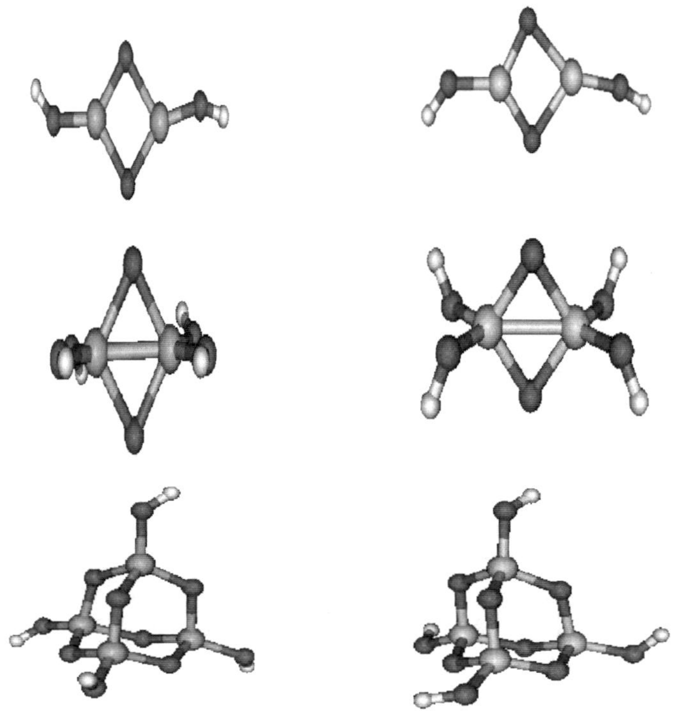

Hydrogenated Vanadium Oxide cluster Hydrogenated Chromium Oxide cluster

Fig.1. Optimized geometries of the hydrogenated clusters.

4. Conclusion

In order to understand the charge transfer mechanism in hydrogenated metal oxide clusters, we have conducted a comparative study of small clusters of dehydrogenated and hydrogenated vanadium and chromium oxide. DFT/B3LYP calculations are performed to locate the ground state of these clusters. It is observed that hydrogenation increases the metallic character of these clusters by increasing the occupancy of the metal atoms. The current theoretical data support the performance of DFT to understand the charge transfer mechanism in vanadium and chromium oxide nanomaterials and are in good agreement with recent experimental and theoretical results. In general, hydrogenation of these clusters enhances the dipole moment but decreases the HOMO-LUMO gap.

References

[1] GAMESS version 25 March 2010, Iowa State University
[2] I. M. Campbell 1988 Catalysis at Surfaces, New York: Chapman and Hall Ltd
[3] P. A. Cox 1992 Transition Metal Oxides, Oxford: Clarendon Press
[4] N. Y. Topsoe 1994 *Science* **265** 1217
[5] C. N. R. Rao, B. Rave, Ed. 1998 Transition Metal Oxides: Structure, Properties, and Synthesis of Ceramic Oxides, New York: John Wiley
[6] C. E. Lee, R. A. Atkins W. N. Gibler, H. F. Taylor 1998 *Appl. Opt.* **28** 4511
[7] R. C. Bell, K. A. Zemski, K. P. Kerns, H. T. Deng, A. W. Castleman 1998 *J. Phys. Chem.* A **102** 1733
[8] S. E. Kooi, A. W. Castleman 1999 *J. Phys. Chem. A* **103** 5671
[9] R. C. Bell, K.A. Zemski, D.R. Justes, A.W. Castleman 2001 *J. Chem. Phys.* **114** 798
[10] Y. Matsuda, E. R. Bernstein 2005 *J. Phys. Chem. A* **109** 3803
[11] K. S. Molek, T. D. Jaeger, M.A. Duncan 2005 *J. Chem. Phys.* **123** 144313
[12] G. V. Chertihin, W. D. Bare, L. Andrews 1997 *J. Phys. Chem. A* **101** 5090
[13] L. B. Knight, R. Babb, M. Ray, T. J. Banisaukas, L. Russon, R. S. Dailey, E. R. Davidson 1996 *J. Chem. Phys.* **105** 10237
[14] K. R. Amis, J. Sauer 2007 *J. Mass Spectrom. Rev.* **26** 542
[15] H. B. Wu, L. S. Wang 1998 *J. Chem. Phys.* **108** 5310
[16] A. Pramann, K. Koyasu, A. Nakajima, K. Kaya 2002 *J. Chem. Phys.* **116** 6521
[17] H. J. Zhai, L. S. Wang 2002 *J. Chem. Phys.* **117** 7882
[18] H. J. Zhai, J. Dobler, J. Sauer, L.S. Wang 2007 *J. Am. Chem. Soc.* **129** 13270
[19] S. Heinbuch, F. Dong, J. J. Rocca, E. R. Bernstein 2010 *J. Chem. Phys.* **133** 174314
[20] R. C. Bell, K. A. Zemski, K. P. Kerns, H. T. Deng, A. W. Castleman 1998 Jr. *J. Phys. Chem. A* **102** 1733
[21] M.Y. Jia, B. Xu, X. L. Ding, Y. X. Zhao, S. G. He, M. F. Ge 2012 *J. Phys. Chem. C* **116** 9043
[22] J. B. Ma, Y. X. Zhao, S. G. He, X. L. Ding 2012 *J. Phys. Chem. A* **116** 2049
[22] Z. G. Zhang, H. G. Xu, X. Kong, W. Zheng 2011 *J. Phys. Chem. A* **115** 13
[23] J. Wei, H. Ji, W. Guo, A. H. Nevidomskyy D. Natelson 2012 *Nat. Nanot.* **7** 357
[24] K. S. Molek, Z. D. Reed, A. M. Ricks, M. A. Duncan 2007 *J. Phys. Chem. A* **111** 8080
[25] Y. Chen, K. Ding, L. Yang, B. Xie, F. Song, J. Wan, G. Wang, M. Han 2008 *App. Phys. Lett.* **91** 173112
[26] H. J. Zhai, S. Li, D. A. Dixon, L. S. Wang 2008 *J. Am. Chem. Soc.* **130** 5167
[27] H. J. Zhai, L. S. Wang 2006 *J. Chem. Phys.* **125** 164315
[28] Q. Q. Zhang, Y. Zhao, Y. Gong, M. Zhou 2007 *J. Phys. Chem. A* **111** 9775
[29] S. Li, H. J. Zhai, L. S. Wang, D. A. Dixon 2009 *J. Phys. Chem. A* **113** 11273
[30] K. H. Xiang, R. Pandey, J. M. Recio, E. Francisco, J. M. Newsam 2000 *J. Phys. Chem. A* **104** 990
[31] B. V. Reddy, S. N. Khanna 1999 *Phys. Rev Lett.* **83** 3170
[32] K. C. Lau, A. K. Kandalam, A. Costales, R. Pandey 2004 *Chem. Phys. Lett.* **393** 112
[33] S. Veliah, K. H. Xiang, R. Pandey, J. M. Recio, J. M. Newsam 1998 *J. Phys. Chem. B*

102 1126
[34] X. Wang, S. Neukermans, F. Vanhoutte, E. Janssens, G. Verschoren, R.E. Silverans, P. Lievens 2011 *Appl. Phys. B* **73** 417
[35] D. E. Bergeron, A. W. Castleman, Jr., N. O. Jones, S. N. Khanna 2004 *Nano Lett.* **4** 261

Study the reduction of mill scale with lean grade coal through RI-RDI

Deepika Shekhawat[1], Vikas Kukshal[2], M K Banerjee[3], Amar Patnaik[1]

[1]Mechanical Engineering Department, MNIT Jaipur, 302017, Rajasthan, India
[2]Mechanical Engineering Department, NIT Uttarakhand, 246174, Uttarakhand, India
[3]Mechanical Engineering Department, SKIT, Jaipur, 302017, Rajasthan, India

Email: apatnaik.mech@mnit.ac.in

Abstract. In the present study, efforts have been made to assess and study the reduction of mill scale in the presence of single gaseous reductant (i.e. CO gas) in RI-RDI (reducibility index (RI) and reduction degradation index) equipment. While conducting this study optimal process parameters like reduction temperature, mill scale size, reduction time and mill scale to coal ratio were analysed. An exceptional experimental profile is maintained inside the furnace to attain the optimum metallization. The mill scales are reduced inside the furnace and the extent of reduction is evaluated from the weight loss. Post conducting reduction in the RI-RDI test apparatus, the reduced mill scale is characterized for the microstructural and morphological studies using SEM, EDS, XRD. The most optimum results of the reduction were attained in RI-RDI apparatus giving iron content of (metallization) 89.15 % at the reduction temperature of 1050°C for 240 mins holding time. During the entire study on RI-RDI apparatus, flow rate of CO is maintained at 4.5 LPM (litres per minute) and N_2 at 10.5 LPM. Finally, the degree of metallisation in the sample as well as the product material is calculated by using the chemical analysis process (dichromate and ferric chloride method). The characterization of produced iron powder was done by chemical analysis, XRD, and FE-SEM. The FE-SEM results confirm the presence of resultant pore characteristics. It is observed that mill scale can be significantly reduced to metallic iron by utilizing lean grade coal.

1. Introduction

The by-products of steel and ironmaking industries are rich in iron (~72 % Fe). These by-products are dusts and mill scale which are produced in vast quantities and can be recycled by practicing some metallurgical procedures (e.g. blast furnace or direct reduction reactors) in the presence of coal (reducing agent) to yield pre-reduced pellets for the purpose of re-melting in electric steel plant [1]. Mixture of various oxides particularly hematite (Fe_2O_3), magnetite (Fe_3O_4), wustite (FeO) and other elementary oxides forms a by-product called as "mill scale" which is considered as a waste product of steel and iron industries [2]. Iron is extracted from this mill scale by the means of reduction process involving certain kinetics of reaction depending upon process parameters like temperature, time, gas flow, etc.

In recent decades, ample work has been carried out in the production of iron from direct reduction process thus calling need for improvement in the existing direct reduction methods in order to overlook the limitations of blast furnace. The RDI (reduction degradation index) of sinter is an index of great importance aiding in assessing the quality of sinter occurring in low temperature zone [3]. The process of aggregation wherein combustion of fuels generates heat inside the moving bed incorporating loosely stuffed iron ore, and other feed particles, in order to agglomerate these raw materials into a compact porous mass is called as sintering [4]. For obtaining increased output with the existing raw materials it is very important to ensure that premises are not affecting the quality and yield of sinter for which RDI is reduced to the greater extent [5].

The production of iron by direct reduction of mill scale using lean grade coal via RI-RDI will efficaciously help in opening up a novel approach for producing iron from easily available industrial waste or "by-product". Numerous researchers focus on enhancing the RDI whereas a very little work have been conducted on RI (reduction index) and related mechanism concerning the production of iron from mill scale. So, this work mainly emphasizes on attaining optimum results of the reduction under selected parameters (temperature and reduction time) and finally evaluating the percentage of metallization attained. The robustness of sinter in the process of reduction (RDI) and reducibility of sinter (RI) are of great importance. As RDI and RI are the leading and responsible indices contributing in deciding the overall furnace performance, this area has significantly drawn remarkable attention of the researchers as well as metallurgist in upbringing more information and data in this domain [5–8].

Lean grade coal from Rajasthan contains very low levels of ash, sulphur, presenting itself as a potential reductant for the reduction of mill scale through Direct Reduction of Iron (DRI). On the other hand, mill scale is a by-product of iron and steel industries rich in iron content in the form of oxides. Experiments were carried out by mixing equal proportion of mill scale and coal and charging into the RI-RDI apparatus. Coal is procured from Matasukh Coal Mines, Nagaur, Rajasthan through RSMML, Rajasthan. The entire study is completed in three steps; materials used, reduction process and the testing.

2. Materials and Method

The lean grade coal of Rajasthan comprises lower levels of sulphur, ash, introducing itself as a promising reductant while carrying out the mill scale reduction through Direct Reduction of Iron (DRI). On the contrary procured mill scale is entitled as a by-product of rolling mill, iron and steel industries comprising of higher iron content in oxides form. The materials procured for current study are illustrated in Table 1.

Table 1. Materials procured for studying

1. Mill Scale	**2. Lean Grade Coal**
Procured: Sarda Industrial Enterprises Ltd., Jaipur, Rajasthan	**Procured:** Rajasthan State Mine and Minerals Ltd. (RSMML), Matasukh Coal Mines, Nagaur, Rajasthan India
 **Fig. 1** As-recived mill scale	 **Fig. 2** As-recived lean grade coal

The chemical analysis of the mill scale procured from Sarda Industrial Enterprises Ltd., Jaipur, Rajasthan is illustrated in Table 2. This as-received mill scale is washed with demineralized water followed by compression air drying. Thereafter this mill scale was in bigger size and could not be used directly in the experiment, so these were crushed in roll crusher followed by meshing in rod mill [9]. Thereafter mill scale were initially hand sieved for visible foreign material separation. As this mill scale contains material fines of different size the homogeneity of the sample selection cannot be maintained if the as received material is put to the experiment. Therefore, sieving machine was used for 1 hour to collect different fraction of mill scale particles (i.e. 825, 600, 425 and 300 microns). Though oxygen and carbon percentage present in the mill scale is not indicated by this analysis but it does imply that Fe-oxides are about 71.99 % and rest is metallic iron which is entitled as the main constituent. The phases of hematite (Fe_2O_3), magnetite (Fe_3O_4), and wustite (FeO) are confirmed in the XRD graph of unreduced mill scale (Fig. 3). Apart from this, presence of some other elements is also observed as these mill scales come from iron and steel plants.

Table 2. Chemical analysis (by XRF) of procured mill scale

Chemical Analysis									
Fe-oxides & Metallic iron	**Al_2O_3**	**CaO**	**SiO_2**	**S**	**MnO**	**Ni**	**Cr**	**MnO**	**MgO**
71.99 %	0.94%	1.26%	4.82%	3447.86 ppm	0.59%	0.05%	0.04%	0.59%	1.31 %

Acquired coal was in bigger size so manual crushing was done to make them into suitable size. Whereas for the analysis of fixed carbon content, moisture, volatile-matter, ash in the procured low-grade coal, proximity analysis was carried out and results revealed that coal possessed fixed carbon content of 24.51 % (Table 3).

Table 3. Proximity analysis results of coal

Material	Moisture %	Volatile %	Ash %	Fixed Carbon %
Coal	19.71	46.31	9.47	24.51

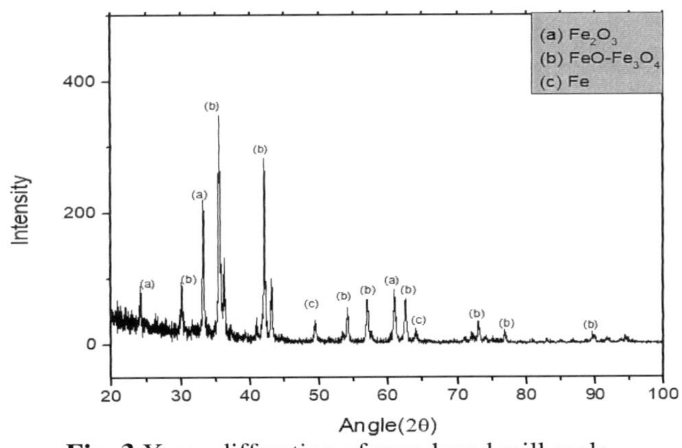

Fig. 3 X-ray diffraction of unreduced mill scale

3. Experimental Processes

3.1 The Reduction in RI-RDI apparatus

The reduction of mill scale was carried out in RI-RDI test apparatus (Salvin Instruments Pvt Ltd. Bangalore), in the Advanced Research Lab for Tribology, MNIT Jaipur. Before feeding mill scale into the RI-RDI apparatus, mill scale was heated at 150°C for 2 hr in order to eliminate the residual moisture. Reduction of mill scale is carried out inside the retort which also acts as a fixed bed in which samples are placed along with the alumina (Al_2O_3) balls inside the furnace. The temperature is set for the split type furnace and burner is switched on. Thereafter N_2 (nitrogen) gas is released with a fixed amount of flow i.e. 10 LPM for the stabilization of temperature at 900°C. When the temperature of the furnace reaches its set value then the flow of the N_2 is raised to 15 LPM and it is maintained for 30 minutes.

After 30 minutes, the N_2 flow rate is adjusted to 10.5 LPM and the CO (carbon monoxide) flow rate is maintained at 4.5 LPM. This process is continued for set time i.e. 1 hr. After 1 hr of reduction, the heater is switched off, the CO gas flow is stopped and N_2 is passed at 5 LPM. In burner blue flame is generated during reduction. It shows that CO is converted into CO_2 (carbon dioxide) gas. After the experiment, the furnace is opened and the retort is transferred inside the chiller for cooling. As soon as it reaches the room temperature, the reduced samples are removed from the reduction tube followed by magnetic separation and final testing for the analysis of metallic iron.

3.1.1 Procedure for evaluating RDI index

Online weighing balance is utilized for recording the weight loss and record the reading by attaching it to retort the assembly. The stability is observed for the reading on the balance by holding it for 10 minutes. After pre-heating the mill scale samples, 500 gm of it is placed inside the vessel. The furnace temperature is set. N_2 gas is flowed (5 LPM) till the temperature reaches its defined value. As soon as temperature desired temperature is reached, the flow rate of N_2 is raised to 10.5 LPM and of CO gas at 4.5 LPM. Reading in the weighing balance is recorded for every 3 minutes. The supply of CO is stopped at the end and only N_2 is allowed to pass at 5 LPM. Furnace is opened and retort assembly is placed inside the cooling chamber. The initial weight (say W_i) is observed followed by transferring it into tumbler rotating drum (300 revolution at 10 rev./min). The sample is collected from the drum and sieving is carried out with 3.15 mm sieve. Again, the weight (say W_f) of the samples is measured and RDI is evaluated using following formula;

$$\frac{W_1 - W_2}{W_1} * 100 \quad (1)$$

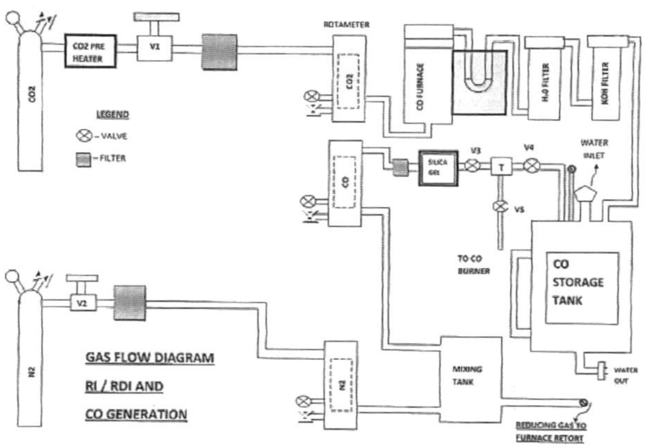

Fig. 4 Gas flow diagram for RI-RDI apparatus

3.1.2 Procedure for evaluating RI index

Evaluating RI index follows similar procedure till recording the reading in the weighing balance for every 3 minutes as in RDI index evaluation (in section 3.1.1). The process of recording the balance (weighing balance) reading is continued till 60 % of the RI value that can be evaluated suing following formula;

$$\% \text{ RI} = \frac{m_1 - m_2}{m_0(0.42W_{Fe} - W_{FeO})} * 10\text{^}4 \quad (2)$$

Where W_{Fe} is % of Fe, W_{FeO} is % of FeO, and Δ $m^{1}2$ is the weight at the time of reduction.
As soon as the above experiment is over all the gas valves are closed and only N_2 is allowed to flow at the rate of 5 LPM till the furnace temperature is reduced to 500°C. Blue flame is observed during the N_2 flow indicating the conversion of CO to CO_2 and the entire gas flow is illustrated in Fig. 4. The split type furnace is opened and the retort carrying mill scale is place inside the cooling chamber followed by collecting the mill scale samples.

3.1.3 Procedure for CO generation

The generation of CO gas involves a few sets of procedures that are elaborated in this section. The desired quantity of low-grade coal (2 kg) is preheated (at 100°C, 6 hr) in order to eradicate the existing moisture. The coal is filled up to 3/4th of the CO retort in furnace 1 (Fig. 5) followed by covering it with gasket and hardware to prevent any leakage of the gas. The CO generating furnace (Furnace 1, Fig. 5) is paired with a water tank which absorbs the heat generating from the furnace. Further CO storage tank accompanies two filters (a) KOH tank and (b) water tank. The connection of the retort with KOH filter and gas panel for CO_2 gas into the retort is checked for the leakages. Progressing towards the end of the experiment the outlet valve of the tank is opened, burner is switched off, and the temperature is set at 1050°C. As soon as the CO furnace temperature reaches 700°C, the CO_2 flow is fixed at 1 LPM, when reaches 1050°C the CO_2 flow rate is set at 2 LPM.

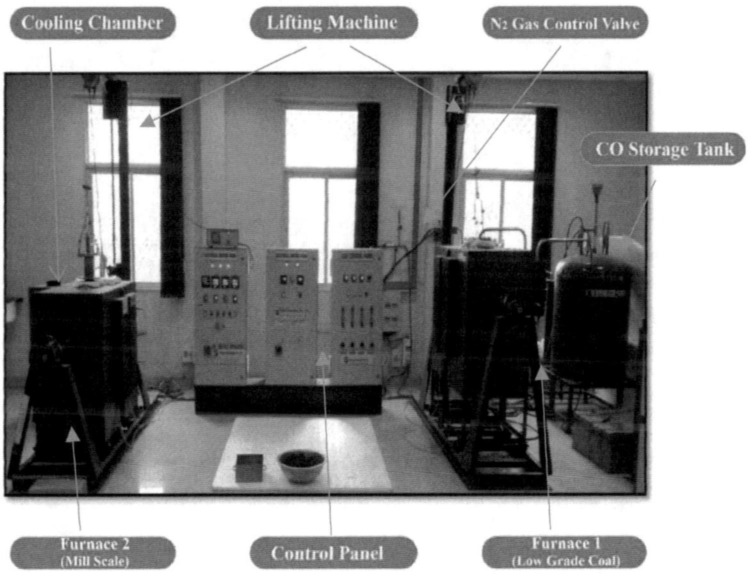

Fig. 5 RI-RDI test apparatus in The Advanced Research Lab for Tribology, MNIT Jaipur

4. Results and Discussions

4.1 SEM micrographs of as-received mill scale and reduce iron powder

The morphological examination of unreduced (as-received) mill scale at lower magnification is presented in Fig. 6, at higher magnification is presented in Fig. 7, and that of final iron powder is presented in Fig. 8. The morphological results conducted for as-received mill scale demonstrates non-uniform size and irregular shaped structure Fig. 6 and Fig. 7; under higher resolution. Along with this, micro sized pinholes and small pores are detected on the mill scale surface (Fig. 7); formed due to release of entrapped gases within the iron oxide layer during reheating. The Fig. 8 depicts the SEM micrograph of sponge iron (iron powder after mill scale reduction) at the most optimal temperature producing higher % of metallization i.e. 1050°C. The bright phases in Fig. 8 implies metallic constituents primarily Fe, whereas the darker phases imply unreduced non-metallic constituents; obtained observation confirms with the past published literature [9–11]. The EDS analysis plot of the produced iron powder is depicted in Fig. 9 that demonstrates the iron yield at 1050°C for 240 minutes. The procured metallic iron powder was further observed via chemical analysis; the results revealed the percentage of metallic iron to be approximately ~62.11% which was in agreement with the results obtained from EDS analysis (63.13%).

Fig. 6 SEM micrograph of unreduced mill scale

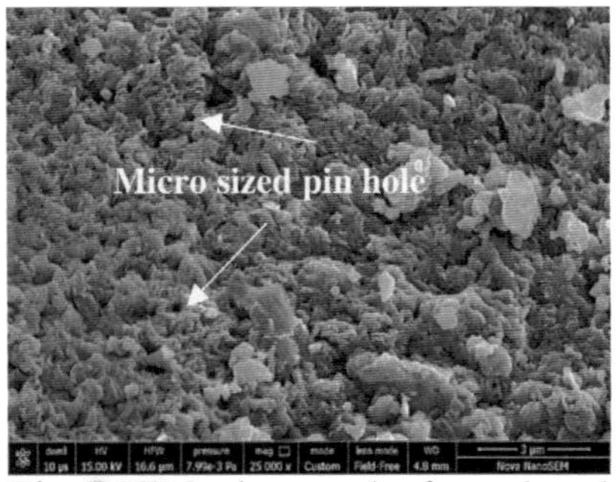

Fig. 7 SEM micrograph of unreduced mill scale at higher magnification

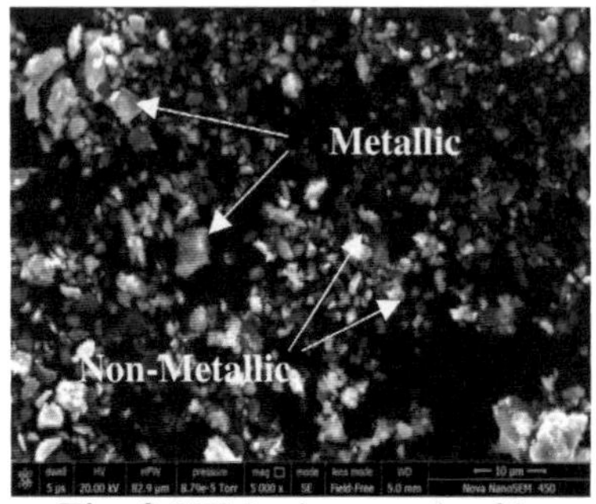

Fig. 8 SEM micrograph of iron powder reduced at 1050°C

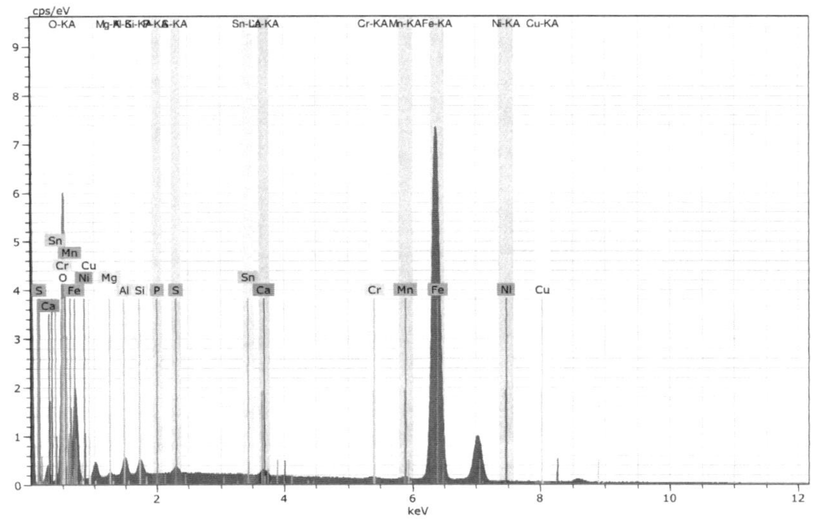

Fig. 9 EDS analysis of the reduced iron at 1050°C for 240 minutes

4.2 XRD analysis of reduced mill scale with respect to reduction temperature and time

4.2.1 XRD analysis for temperature

The resultant structural analysis of reduced mill scale samples was carried out using XRD employing Cu-Kα (λ = 0.15406 nm) radiation. The behavior of the mill scale after reduction with respect to different temperatures has been illustrated in Fig. 10. At room temperature mill scale consists of iron and its oxides. It is perceived from Fig. 10 that the intensity of the highest peak is maximum at 1050°C due to presence of metallic iron (Fe). The highest fraction of metallic iron is attained at 1050°C (~66.43%), as compared to 1000°C (~65.01%) or 950°C (~57.11%), or 900°C (~46.77%) and this yield tends to decrease beyond 1050°C. The sequence of possible reduction (mill scale → iron powder) occurring at 1050°C could be $Fe_2O_3 \rightarrow Fe_3O_4 \rightarrow FeO \rightarrow Fe$.

IOP Conf. Series: Materials Science and Engineering 1017 (2021) 012037 doi:10.1088/1757-899X/1017/1/012037

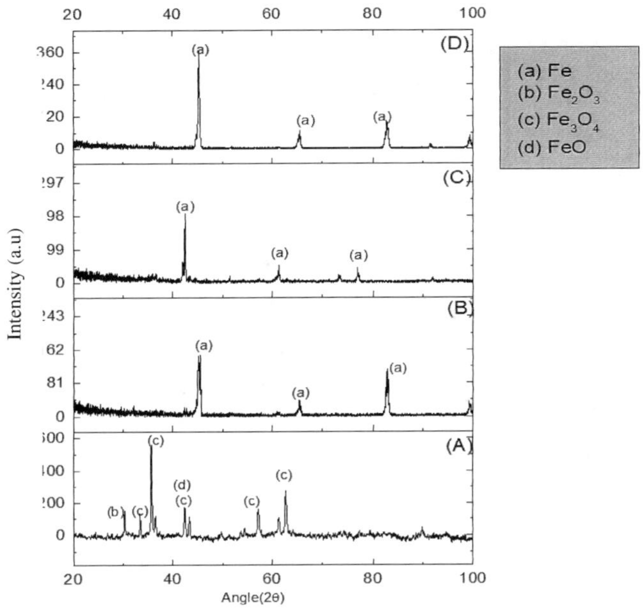

Fig. 10 XRD at different temperatures (A) reduced iron at 900°C, (B) reduced iron at 950°C, (C) reduced iron at 1000°C and (D) reduced iron at 1050°C

4.2.2 Analysis of metallic iron by chemical analysis after reduction

The reduced samples from the furnace are magnetically separated and extracted iron powder is chemically analyzed using dichromate and ferric chloride method [12,13]. To prepare potassium dichromate solution, 4.903 g of $K_2Cr_2O_7$ is mixed with 500 ml of distilled water and stirred (Fig. 11) till it completely dissolute in water. After this the normality of the solution is evaluated by calculating N/10 factor. Concentrated H_2SO_4 (5 to 7 ml) is added to 100 ml distilled water in a beaker. FeS is mixed in this solution until completely dissolved followed by mixing 10 ml of mixed acid along with 5 to 6 drops of BDS indicator into this solution. The previously prepared $K_2Cr_2O_7$ solution is filled in the burette and titration is carried out dropwise and continued till the colour of solution in the beaker turns into violet. As soon as the colour turns to violet the reading of the burette is noted.

Fig. 11 Magnetic stirring

The following formula is used for calculation of N/10 factor-

$$\text{Normality} = \frac{\text{Given weight}}{\text{Eqivalent weight} * \text{Burette reading}} \quad (3)$$

4.3 Influence of process parameters on the reduction

In RI-RDI apparatus, mill scale is reduced by the CO gas only. Here mill scale is heated in the furnace and CO gas is produced in the other furnace. This CO is allowed to react with the heated sample for sufficient period of time, and reduction condition is observed.

4.3.1 Effect of mill scale size

Fig. 12 illustrates the % of metallic iron procured in response to the mill scale size (300 to 825 μm) at the reduction temperature of 900°C. Results revealed that lower the particle size (in between 425 to 300 μm) resulted in higher degree of metallization. Significant results were attained at the reduction temperature of 900°C and with the reduction in particle size (> 300 μm) the extent of metallization exhibits an increasing trend; it can be perceived from Table 4 that metallic yield increased from 26.89% for the particles size of 825 μm to 64.62% for the particles size of 300 μm. The reason behind this increasing trend of metallization is the increasing reaction rate which takes place owing to increased diffusion of reductant resulting in increased reaction area on the smaller particle size [14–17]. Attributable to the fact that bigger iron ore size will inhibit the diffusion of gases, and will lessen the tenancy of gas-solid contact region, that decelerates the rate of reduction. This provides an evidence that the size of the particles is one of the significant decisive factors in controlling the mill scale reduction to metallic iron.

Table 4. Results of chemical analysis of reduced mill scale with respect to various process parameters

Process parameter	Mill scale size (μm)	CO (LPM)	N_2 (LPM)	Time (min)	Temperature (°C)	Metallic Iron (wt.%)	Metallization (%)
Effect of Mill Scale Size (μm)	825	4.5	10.5	180	900	26.89	38.42
	600	4.5	10.5	180	900	28.78	42.32
	425	4.5	10.5	180	900	32.89	45.39
	300	4.5	10.5	180	900	64.62	50.48
Effect of Temp. (°C)	300	4.5	10.5	180	900	46.77	64.31
	300	4.5	10.5	180	950	57.11	66.08
	300	4.5	10.5	180	1000	65.01	69.12
	300	4.5	10.5	180	1050	66.43	72.48
Effect of Time (min)	300	4.5	10.5	90	1050	15.51	38.70
	300	4.5	10.5	120	1050	23.01	56.08
	300	4.5	10.5	180	1050	61.04	64.36
	300	4.5	10.5	240	1050	70.29	89.15

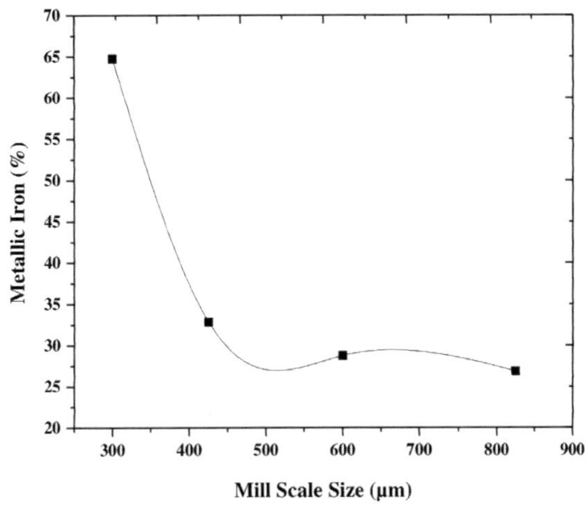

Fig. 12 Variation of iron content (wt.%) vs mill scale size (μm)

4.3.2 Effect of reduction temperature

The obtained results revealed that with the increase in temperature in the furnace, the metallization degree increases as a greater number of oxides are removed. This trend was observed till rise of temperature from 900°C to 1050°C and metallic iron percentage increased from 46.77 to 66.43% respectively (Table 4). Beyond temperature of 1050°C the weight fraction of metallic iron tends to decrease (Fig. 13); as a consequence of chemical reactions occurring at the time of reduction process stating significance of temperature during reduction process. It can be clearly perceived that the increase in the rate of Boudouard reaction is the chief reason for this increased metallic yield and maximum was observed at 1050°C (66.43%) [18,19]. The other parameters affecting the impact of temperature upon the degree of metallization is influenced by the quality of coal taken for conducting the reduction process, depending on the reactivity of coal. It can be concluded that by increasing the temperature, the consecutive reduction rate could be enhanced considerably thereby resulting in higher degree of metallization. Approximately 72.48% (Table 4) of metallization was achieved at the temperature of 1050°C.

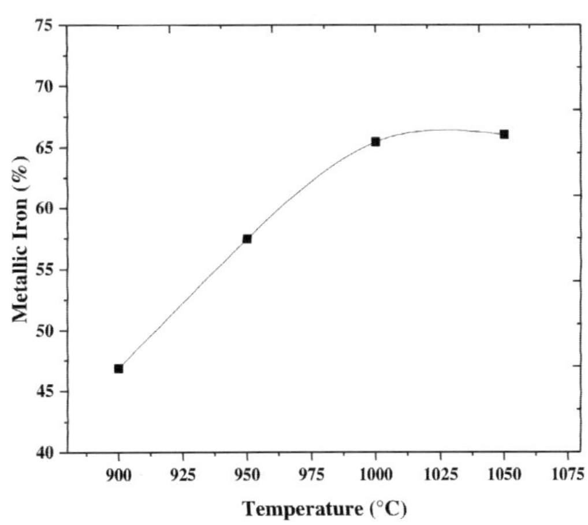

Fig. 13 Variation of iron content (wt.%) vs temperature (°C)

4.3.3 Effect of reduction time

The degree of metallization presents direct relationship with the reduction time (Fig. 14). In addition to this, the increment in temperature significantly reduces the dependency of reduction process (mill scale to iron powder) with respect to the time. The increment in residence time escalates the atomic movement during the chemical reaction; resulting in enhanced reaction kinetics [14]. The increment in efficiency and the metallic yield are attributable to the Boudouard reaction [19,20]. The success of reduction process entirely depends upon the Boudouard reaction that releases heat efficient enough to execute the reduction process efficaciously. It can be clearly perceived from Table 4; the metallic iron percentage increased from 15.51 to 70.29% upon increasing reduction time from 90 to 240 minutes. Approximately 89.15% of metallization was achieved at 240 minutes time.

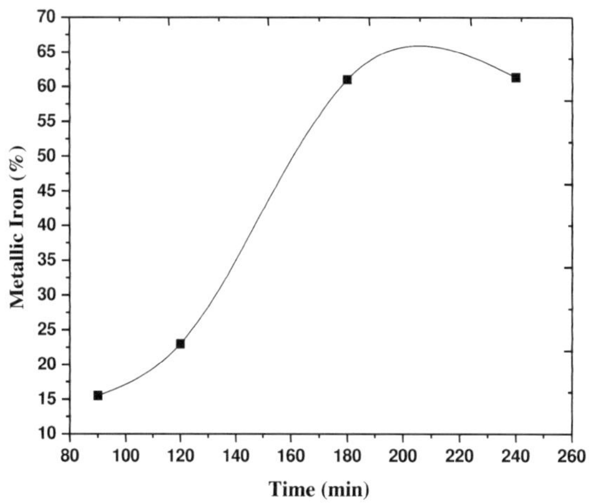

Fig. 14 Variation of iron content (wt.%) vs reduction time (min)

5. Conclusion

Current experimental study emphasizes on attaining the optimum results of the mill scale reduction under selected parameters (temperature and reduction time) and finally evaluating the percentage of metallization attained after reduction. Successful optimization of process parameters for reducing mill scale with improved metallization was carried out by making use of lean-grade coal from Rajasthan. The process parameters considered for current study were size of mill scale, reduction time, and temperature respectively. The outcomes of the study demonstrated that with the increment in time and temperature, and reduced mill scale size, the percentage (%) of metallization improved exponentially. The results indicated the highest % of metallization achieved was 89.15 % at 1050ºC temperature, holding time of 240 minutes, selecting mill scale size less than 300 μm with a fixed flow of CO (4.5 LPM) and N_2 (10.5 LPM) gas for attaining optimal results during the reduction process. In present study there is no mill scale to coal ratio used as they both are fed into two different vessels (Furnace 1 and Furnace 2) so interaction of these two does not affect the reduction process. Other factor affecting the reduction kinetics reaction is the atmosphere. N_2 is utilized throughout the process for reaction to occur. As there are lots of gases present in the burning chamber that may hinder the reduction process, in that case N_2 helps in eliminating those unwanted gases and a constant temperature is maintained throughout the experiments.

Acknowledgements

The authors would like to give sincere thanks to the Ministry of Steel, Government of India (F. No 11(11)/GBS/2013-TW DATED 28.09.2015) for all the financial support. The authors also wish to acknowledge The Advanced Research Lab for Tribology, Department of Mechanical Engineering, Department of Metallurgical and Materials Engineering, MNIT, Jaipur for providing the rotary kiln reduction facility, and other testing facilities and the Material Research Centre, MNIT Jaipur, for providing the characterization facility.

References

[1] S. Mechachti, O. Benchiheub, S. Serrai, M.E.H. Shalabi, Preparation of iron Powders by Reduction of Rolling Mill Scale, Int. J. Sci. Eng. Res. 4 (2013).
[2] O. Benchiheub, S. Mechachti, S. Serrai, M.G. Khalifa, Elaboration of iron powder from mill scale, J. Mater. Environ. Sci. 1 (2010) 267–276.
[3] Z. Xu, Z. Jian-liang, H. Zheng-wen, H. Zuo, Effect of CaCl2 on RDI and RI of Sinter, J. Iron Steel Res. Int. 17 (2010) 7–12.
[4] S. Srivastava, Use of halide solution to improve the RDI and RI of sinter: A Case Study, Arch. Ind. Biotechnol. 2 (2018).
[5] N. Taguchi, T. Otomo, Y. Omori, Immersion Treatment of Sinter into Calcium Halide Solution, ISIJ Int. 30 (1990) 281–289.
[6] A.J.B. Muwanguzi, A. V Karasev, J.K. Byaruhanga, G. J, Characterisation of the Physical and Metallurgical Properties of Natural Iron Ore for Iron Production, Int. Sch. Res. Netw. (2012). https://doi.org/10.5402/2012/147420.
[7] T. Paananen, The Effect of Minor Oxide Components on Reduction of Iron Ore Agglomerates, 2013.
[8] A.I. Journal, M. Kumar, S. Jena, S.K. Patel, Characterization Of Properties And Reduction Behavior Of Iron Ores For Application In Sponge Ironmaking, Miner. Process. Extr. Metall. Rev. 29 (2007) 118–129. https://doi.org/10.1080/08827500701421896.
[9] R. Sen, U. Pandel, Closed crucible reduction of lump powdered mill scale or iron ore by coal: The sequential methodology and mechanism for optimization of process parameters, Adv. Powder Technol. (2020). https://doi.org/10.1016/j.apt.2020.07.017.
[10] P. Ghods, O.B. Isgor, G.A. Mcrae, J. Li, G.P. Gu, Microscopic investigation of mill scale and its proposed effect on the variability of chloride-induced depassivation of carbon steel rebar, Corros. Sci. 53 (2011) 946–954. https://doi.org/10.1016/j.corsci.2010.11.025.
[11] Y. Man, J.X. Feng, F.J. Li, Q. Ge, Y.M. Chen, J.Z. Zhou, Influence of temperature and time on reduction behavior in iron ore–coal composite pellets, Powder Technol. 256 (2014) 361–366. https://doi.org/10.1016/j.powtec.2014.02.039.
[12] H.T. Materials, O. Yucel, A. Turan, M. Alkan, Determination of Direct Reduction Conditions of Mill Scale, High Temp. Mater. Process. 32 (2013) 405–412. https://doi.org/10.1515/htmp-2012-0167.
[13] N. Gade, G. Verma, R. Sen, Effect of Calcium Carbonate on the Reduction Behaviour of Mill Scale, Procedia Earth Planet. Sci. 11 (2015) 319–324. https://doi.org/10.1016/j.proeps.2015.06.067.
[14] A. Kumar, V.R. Kiragi, S. Kumar, A. Patnaik, Synthesis and Characterization of Metallic Iron Reduced from Low-grade Coal in Rajasthan, Mining, Metall. Explor. 37 (2020) 1741–1751.
[15] N.M. Gaballah, A.F. Zikry, M.G. Khalifa, A.B. Farag, M.E.H. Shalabi, Production of Iron from Mill Scale Industrial Waste via Hydrogen, Open J. Inorg. Non-Metallic Mater. 3 (2013) 23–28.
[16] M.I. Martín, F.A. López, J.M. Torralba, Production of sponge iron powder by reduction of rolling mill scale Production of sponge iron powder by reduction of rolling mill scale, Ironmak. Steelmak. 39 (2012) 155–162. https://doi.org/10.1179/1743281211Y.0000000078.
[17] N. Gidikova, V. Manolov, V. Petkov, Mill Scale Reduction, J. Mater. Sci. Technol. 25 (2017) 12–19.
[18] M.C. Bagatini, V. Zymla, E. Osório, A. Cezar, F. Vilela, Characterization and Reduction Behavior of Mill Scale, ISIJ Int. 51 (2011) 1072–1079.
[19] A. Ahmed, M. Eissa, S. Shahein, A. Ahmed, M. Eissa, S. Shahein, Direct chromium alloying by smelting reduction of mill scale and low grade chromite ore, Process. Prod. Appl. 42 (2015) 648–655. https://doi.org/10.1179/1743281215Y.0000000008.

[20] H.İ. Ünal, E. Turgut, Ş.H. Atapek, A. Alkan, Direct Reduction of Ferrous Oxides to form an Iron-Rich Alternative Charge Material, High Temp. Mater. Process. 34 (2015) 751–756. https://doi.org/10.1515/htmp-2014-0125.

A Short Review on Polymer, Metal and Ceramic Based Implant Materials

Deepika Shekhawat[1], Amit Singh[1], Ashray Bhardwaj[2], Amar Patnaik[1]

[1]Mechanical Engineering Department, MNIT Jaipur, 302017, Rajasthan, India
[2]Department of Mechanical Engineering, Thapar Institute of Technology, Patiala 147003, India

Email: deepikashekhawat91@gmail.com

Abstract. The present work focuses on the evaluation of polymer, metal and ceramic based biomaterials with particular emphasis on success rate of these materials till date along with the merits and demerits of these material system. Currently, the search for the materials with improved performance is on rise for biomedical (mainly total hip replacement, THR) applications owing to increase in the number of orthopedic patients worldwide. The clinicians utilized polymer and metal-based implants in ancient times to heal and treat the bone imperfections as well as fractures. The next generation materials required an update on properties, cell material interactions, repair and regeneration, that can only be acquired from recent knowledge related to materials. The commonly employed polymers (natural based polymers, proteins or polysaccharides; and synthetic polymers, poly (lactic acid), poly (glycolic acid), poly (ε-caprolactone), poly(3-hydroxybutyrate) (PHB)) and metals (SS, Mg-alloy, Ti-alloys, Co-alloy, silver, and nickel-titanium alloys) failed to fulfill the long-term durability and lacked in generating adequate bonding with the bones. It is intended that the facts and details abridged in this review article ought to be a beneficial tool in the development of progressively eminent implant materials like ceramic based materials against premature failure, cyclic stress generation, wear, corrosion fatigue, implant loosening and ultimate fracture.

1. Introduction

Biomaterials were in early days classified as materials employed in a medical device, which is designed to interact with biological structure [1]. The improvement in the implant materials is of considerable importance to heal and cure bone related deficiencies and fractures. The cases of revision hip surgery dramatically increased by 26 percent in past five years and this is expected to touch 137 percent by 2030 [2]. Hip surgeries are expected to face a rise from 1.8 million (2015) to 2.8 million by the year 2050 [3].The current focus is on producing more durable implants with more success rate. Till date, there is no evidence of 100 % success rate of implants made of polymer and metals. Human tissue mainly comprised of proteins (polymers) and bone mineral (ceramics) along with the elemental traces of metals. Bone in human body basically represents a bio-nanocomposite arrangement that has advanced and improved over billions of years. The hip joint bears a load of 3 times the body weight and peak load of 10 times the body weight is faced during rigorous activities which may include running, exercising, cycling, jumping, etc. There's a general criterion before selecting any material for bone replacement applications which is as follows [4]:

- The selected material is highly biocompatible and it does not provoke any toxic reaction in the biological system.
- It must possess near about or close to bone physical and mechanical properties.
- Fabrication and processing of the selected material must be economically viable.
- Material must not wear out in any circumstances.

The selected material is said to be compatible with the existing tissue when it is compatible with the surface, osteocompatible, and mechanically compatible inside the human body. This review comprises of the different implant material types and their merits and demerits. The intention of this paper is also to point out the potential implant materials so far being investigated for total hip replacement (THR) applications. Polymer biomaterials are rarely used alone in bone replacement applications as they are responsible for generating immunogenic reactions, poor mechanical strength and implant loosening due to wear which ultimately causes early failure. On the other hand, metallic biomaterials face untimely mechanical failure resulting from cyclic stress generation. Secondly, many investigations have been carried out till date to understand the mechanism between fatigue and corrosion generation in the metallic implants under physiological environment. Moreover, the in-vitro and in-vivo investigation reveals the biological performance of any implant material that is responsible for determining the immune response, cell growth, physicochemical characteristics, and biodegradability rate [5]. Final comparison of these materials with the ceramic-based biomaterials is established on the basis of past literature as ultimate goal of all these bearings is to minimize the overall circumstances of the revision surgery. The three principal determinants responsible for the overall total joint performance are illustrated in Figure 1.

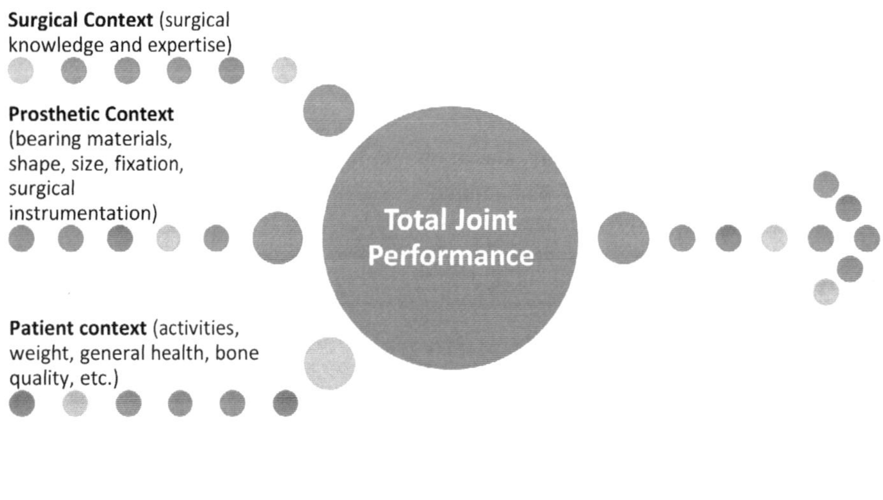

Figure 1. The three main determinants responsible for overall total joint performance [6,7]

1.1. Polymer based biomaterials

Polymer based biomaterials have been employed in the human body soon after its invention in the form of synthetic polymers. Clinical practice started exploiting polymer biomaterials owing to their strong, inert, and biocompatible nature for eliminating the burden of injured or infected bone in order to improve the patient's life. Polymer biomaterials find implementation in multiple tissues like in cardiovascular [8], neural [9], musculoskeletal [10], and dermal tissues [11]. Before choosing any material for the biomedical implant system, it needs to have certain set of properties namely surface topography [12], free energy [13], and functional group [14]. Polymers unlock tailorable characteristics with exceptional flexibility (chemical and physical surface behavior) as it provides a window of opportunity to precisely regulate the bulk properties (porosity, mechanical properties, biocompatibility, and biodegradation) [15]. In line with the previous studies carried out, the polymeric biomaterials for implants with their merits and demerits have been illustrated in Table 1.

Table 1. Polymeric biomaterials employed in implants and their characteristics

S. No.	Polymer based	Merits	Demerits	Reference
1	Ultra-High Molecular Weight Polyethylene	Superior mechanical properties, enhanced modulus, and creep resistance as compared to other polymer biomaterials	Releases debris, Adverse tissue biological reactions Osteolysis or bone loss leading to implant loosening	[16]
2	Polyethylene (PE)	Low friction coefficient, fracture toughness, low density, high impact strength	Over prolonged period, wear debris are released Ionizing radiation	[17,18]
3	Polymethylmethacrylate (PMMA)	Superior osseointegration, Absence of no interim fibrous tissue around cemented elements	Faces microfractures, Releases cement particles, Causes localized foreign-body response Limited biological response	[19,20]
4	High-density polyethylene (HDPE)	Good processability, Biocompatible, Lower cost and good mechanical properties Better creep resistance than UHMWPE	Lower wear resistance with long term use Can be used in combination with UHMWPE for best results	[21]
5	Polytetrafluoroethylene (PTFE)	Chemically inert, Nontoxic, Nonflammable	Very low resistance and high rate of wear More susceptible to creep, Causes compression syndromes	[22–26]
6	Polyacetal	Good wear resistance, Polyacetal-polyethylene combination mostly used in artificial joints	Degrade to form formaldehyde and exaggerates in the presence of moisture, elevated temperature or mechanical force applied for prolonged time	[27–29]

1.2. Metal based biomaterials

Millions of people throughout the world suffer from hip arthritis, and total hip replacement (THR) is the efficacious treatment for this end-stage arthritis responsible for joint pains or joint diseases. According to an updated projection, US alone possessed 52.5 million arthritis patients in the year 2010 and this number is going to touch 78.4 million by 2040 [30]. Patients confronting problem of arthritis undergo total hip replacement surgery that replaces the hip joint with a ball and socket joint mimicking its complete biomechanics. This ball and socket configuration can be made of metallic head articulating against metal cup (Metal-on-Metal, MoM) or polymer lining (Metal-on-Polymer, MoP) [31].

On the other hand, with the advantages of metal-based implants and successful surgeries, there comes some severe disadvantages too that needs to be considered before selecting them for the final application. They require further progress and advancements. With the increasing life-expectancy, the need for long lasting and durable implants with no adverse reaction is also rising at a faster pace. At present, less than 1 percent of the implant bearings have metal-on-metal (MoM) articulations [32,33]. In a study conducted by Laaksonen and teammates [34] found that out of all 56 studies, adverse local tissue reactions and rate of revision surgeries were found to be very high in numbers. Some of the metal-based implants used are titanium and titanium alloys, cobalt-chromium alloy, stainless steel, cobalt-chromium-molybdenum alloys, and tantalum. In line with the previous studies carried out, the metallic biomaterials for implants with their merits and demerits have been illustrated in Table 2.

Table 2. Metallic biomaterials employed in implants and their characteristics

S. No.	Metal based	Merits	Demerits	Reference
1	Titanium	Superior mechanical properties, Great biocompatibility, light weight, biocompatible, lower young's modulus	Causes allergic reactions, poor tribological properties, expensive, lower shear strength	[35–38]
2	Stainless steel (bio-steels)	Inexpensive to fabricate, high elastic modulus, high wear resistance, low cost, good fatigue resistance	Crevice and pitting corrosion in long-term applications, Ni and Cr allergy, stress shielding effect	[38,39]
3	Cobalt-Chromium-based alloys	Highly corrosion resistant with a minimal susceptibility, high wear resistance, biocompatibility, load bearing materials	Early implant loosening rate, limited use, difficult to machine and expensive to process, Ni and Cr allergy	[38,40,41]
4	Ti-6Al-4V	With proper surface treatments, fatigue strength could be improved	Release of ions, intoxication, bone resorption, fracture of implant, restricted mobility	[42–44]
5	Ti-Nb-Ta-Zr (TZNT) alloy	Good mechanical properties, Superior biological properties, Enhanced wear characteristics, Biocompatibility, Great corrosion resistance,	Lower strength	[45][46]

		Low elastic modulus		
6	Mg based alloys	Mechanically very strong, biocompatible	Presence of unprotected oxide film leads to corrosion of the implant, Growth of H_2 bubbles during corrosion	[38,47]
7	NiTi shape memory alloy	Low Young's modulus, returns to its original shape after a temperature change, superelastic, biocompatible	Ni causes allergy due to wear debris	[48,49]
8	Noble Metals (gold (Au), silver (Ag), platinum (Pt) and iridium (Ir))	Resistant to chemical and electrochemical corrosion	Susceptible to corrosion during in-vivo and in-vitro environment	[38]

1.3. Ceramic based biomaterials

Implants these days are expected to deliver a lifetime of more than 30 years due to rising population and demand for surgery is growing at faster pace in young patients. All these selected implants must be very reliable and fracture resistant in in-vivo environment. Polymer-based and metal-based implants didn't meet these requirements, hence researchers and doctors shifted to ceramics as an alternative material to these. Ceramic materials are generally suited for bone replacement bearings owing to their high hardness, high wear resistance and superb biocompatibility. In addition to these, ceramic produces a negligible amount of wear debris in comparison to metals and polymer-based implants. In comparison to micro-sized, nano-ceramic composites possess lower grain size which aids in executing higher overall mechanical, tribological as well as biological performance inside human body. Bioceramics, a sub-category of biomaterials are basically of three types, i.e. bioinert (does not interact with the living tissues and are non-toxic), biodegradable (that are absorbed and dissolved inside the body), and bioactive ceramics (capable in forming bonding with the hard and soft tissues). Ceramic-on-ceramic (ceramic femoral head and ceramic cup inserts) combination is also helpful in reducing abrasion [50]. A lot of work has been conducted in this area to find out the most compatible ceramic material for joint prosthesis applications. In line with the previous studies carried out, the ceramic biomaterials for implant applications with their merits and demerits have been illustrated in Table 3.

Table 3. Ceramic biomaterials employed in implants and their characteristics

S. No.	Ceramic based		Merits	Demerits	Reference
1	Bioinert	Zirconia	Reduces wear, long term risk of dislocation, higher biocompatibility, tissue regeneration	Slow crack growth, hydrothermal aging, phase transformation, expensive	[51,52]
		Alumina	Excellent tribological behavior,	Low fracture toughness, can't deform under stress	[53–56]

			biocompatible nature, higher mechanical characteristics	hence faces catastrophic failure, expensive, problem of squeaking	
		Tantalum oxide	Superior biocorrosion and wear resistance, decreases osteolysis and mechanical properties, low ion release,	Dislocation, instability	[57–59]
2	Biodegradable	Calcium phosphates	High stiffness, mechanical properties similar to trabecular bone	Not stable at high temperatures, low tensile strength	[60]
3	Bioactive	Hydroxyapatite	Non-toxic to living tissues, promotes bone growth, cheaper, highly biocompatible, better bonding and fixation, high osseointegration properties	Used as coating over metallic implants hence may face delamination and wear by abrasion	[61,62]
		Bioactive glass	High rate of bone formation, enhanced fixation, better cell growth and enhanced stability	Difficult to fabricate, brittle nature, low bending and fatigue strength, lower fracture toughness, low ductility	[53,63]

Conclusion

In this review, all the three implant materials were taken into account, showing their merits and demerits faced in actual applications inside the body. This also indicated that single matrix materials were not sufficient enough to meet the desirable demand of hip joint. This paper summarizes in short, the findings and contributions made by all the three biomaterials that suggests that the only way is to fabricate multi-phase components. Polymers and metallic implants failed in presenting bioactivity in in-vivo environment and were unable to uphold mechanical properties in corrosive media, that ultimately ended in reduced durability. Besides this, previous studies have also indicated release of hazardous ions (metallic implants) near implant site and in blood stream. Polymer implants also face debris release issue that reduces durability of the implant in long run.

On the contrary, ceramic based implants demonstrated excellent bioactivity and biocompatibility characteristics in in-vivo environment, and were also resistant to corrosion. In comparison to polymer-based and metal-based implants, ceramic implants resulted in superior osseointegration and osseoconduction behavior inside human body that promoted fast healing of fractured and infected bones. The only shortcomings with the ceramic biomaterials is their brittleness and they tend to fail under tensile or cyclic loading condition that limits their complete utilization in load bearing bones. With designs prompted by natural biological structures, different combinations like bioceramic/polymer composites have instigated enhanced strength and fracture toughness in comparison to 'parent' metal and polymer biomaterials used alone. The current age of biomaterial composites belongs to 4th generation biomaterials that combines more than two different phases which results in better control of properties like mechanical, antibacterial, osteoconductivity, degradation rate, physicochemical, interaction with cells, surface topography and wettability.

To conclude, bio-composite materials offers an opportunity to adopt a multidesign/multistage approach that provides better control of material and biological properties as compared to the polymers, metals and ceramics alone.

Future direction

Clinical use of ceramic biomaterials is limited owing to difficulty of producing intricate, complex and modified surfaces. A continuous effort in the fabrication of novel and improved biomaterials is crucial with the insight of attaining improved biological, tribological, chemical and mechanical properties as identical as possible to that of the original bones or tissues that need to be replaced. Bio-composites are emerging with advanced combination of desirable properties originating from mixing dissimilar materials. In comparison to polymer and metal-based implants, ceramic based bearings exhibit adequate compression strength, tribological characteristics, biocompatibility, and corrosion resistance ensuring its widespread use in medical implants with enhanced longevity. The following areas can be recommended for further work:

1. The structure-property-mechanism relationship of an implant system is of prime importance in terms of future research. Also, precise regulation of the nanoparticles incorporated in the base ceramic matrix is of paramount concern for obtaining desirable mechanical and microstructural properties in the synthesized material.
2. Further research is required to develop new ceramic based composite materials with negligible wear and prove to be a successful alternative bearing material and check for its interaction with the cell response and antibacterial properties.
3. More investigations must concentrate towards finding new methodologies, processes and growths in the field of manufacturing techniques (like 3D printing, fused deposition modeling, selective laser sintering, bioprinting, etc.) along with their advantages and limitations.
4. More studies are required in the direction of surface modification techniques to improve and enhance the reliability of implants to be implanted inside human body.
5. Limited literature is available for the impact of high and low temperature variations in the final synthesized bio-composite material that is an important aspect from a technological point of view.

References

[1] D.F. Williams, Definitions in Biomaterials, 1987. https://doi.org/10.1177/000184818203200206.

[2] I. Sanli, J.J.C. Arts, J. Geurts, Clinical and Radiologic Outcomes of a Fully Hydroxyapatite-Coated Femoral Revision Stem: Excessive Stress Shielding Incidence and its Consequences, J. Arthroplasty. 31 (2016) 209–214. https://doi.org/10.1016/j.arth.2015.08.037.

[3] C. Pabinger, H. Lothaller, N. Portner, A. Geissler, Projections of hip arthroplasty in OECD countries up to 2050, HIP Int. 28 (2018) 498–506. https://doi.org/10.1177/1120700018757940.

[4] K.S. Katti, Biomaterials in total joint replacement, Colloids Surfaces B Biointerfaces. 39 (2004) 133–142. https://doi.org/10.1016/j.colsurfb.2003.12.002.

[5] P. Ducheyne, Comprehensive Biomaterials, Elsevier. 1 (2015). https://doi.org/10.1016/B978-0-08-055294-1.00275-0.

[6] X. Zhang, G. Shi, X. Sun, W. Zheng, X. Lin, G. Chen, Factors Influencing the Outcomes of Artificial Hip Replacements, Cells Tissues Organs. 206 (2018) 254–262. https://doi.org/10.1159/000500518.

[7] J. Alvarado, R. Maldonado, J. Marxuach, R. Otero, Biomechanics of hip and knee prostheses 1, Appl. Eng. Mech. Med. (2003) 6–22.

[8] M. Sarem, F. Moztarzadeh, M. Mozafari, How can genipin assist gelatin/carbohydrate chitosan scaffolds to act as replacements of load-bearing soft tissues?, Carbohydr. Polym. 93 (2013) 635–643. https://doi.org/10.1016/j.carbpol.2012.11.099.

[9] A. Yazdanpanah, G. Amoabediny, P. Shariatpanahi, J. Nourmohammadi, Synthesis and Characterization of Polylactic Acid Tubular Scaffolds with Improved Mechanical Properties for Vascular Tissue Engineering, Trends Biomater. Artif. Organs. 28 (2014) 99–105.

[10] P. Zarrintaj, A.M. Urbanska, S. Seyed, V. Goodarzi, M. Reza, M. Mozafari, A facile route to the synthesis of anilinic electroactive colloidal hydrogels for neural tissue engineering applications, J. Colloid Interface Sci. 516 (2018) 57–66. https://doi.org/10.1016/j.jcis.2018.01.044.

[11] M. Gholipourmalekabadi, A. Samadikuchaksaraei, A.M. Seifalian, A.M. Urbanska, H. Ghanbarian, Silk fibroin/amniotic membrane 3D bi-layered artificial skin, Biomed. Mater. (2017) 1–36.

[12] A. Ranella, M. Barberoglou, S. Bakogianni, C. Fotakis, E. Stratakis, Tuning cell adhesion by controlling the roughness and wettability of 3D micro/nano silicon structures, Acta Biomater. 6 (2010) 2711–2720. https://doi.org/10.1016/j.actbio.2010.01.016.

[13] M. Hoefling, F. Iori, S. Corni, K. Gottschalk, Interaction of Amino Acids with the Au (111) Surface: Adsorption Free Energies from Molecular Dynamics Simulations, Langmuir. 26 (2010) 8347–8351. https://doi.org/10.1021/la904765u.

[14] F. Meder, T. Daberkow, L. Treccani, M. Wilhelm, M. Schowalter, A. Rosenauer, L. Mädler, K. Rezwan, Protein adsorption on colloidal alumina particles functionalized with amino,carboxyl,sulfonate and phosphate groups, Acta Biomater. 8 (2012) 1221–1229. https://doi.org/10.1016/j.actbio.2011.09.014.

[15] M. Rahmati, C.P. Pennisi, E. Budd, A. Mobasheri, M. Mozafari, Biomaterials for Regenerative Medicine:Historical Perspectives and Current Trends, Cell Biol. Transl. Med. 4 (2018) 1–19.

[16] S. Affatato, S.A. Jaber, P. Taddei, Polyethylene based polymer for joint replacement, Biomater. Clin. Pract. Adv. Clin. Res. Med. Devices. (2017) 149–165. https://doi.org/10.1007/978-3-319-68025-5_6.

[17] M.A. Mcgee, D.W. Howie, S.D. Neale, D.R. Haynes, M.J. Pearcy, The role of polyethylene wear in joint replacement failure, Proc. Inst. Mech. Eng. 211 (1997) 65–72.

[18] V. Premnath, W.H. Harris, M. Jasty, E.W. Merrill, Gamma sterilization of UHMWPE articular implants: an analysis of the oxidation problem, Biomaterials. 17 (1996) 1741–1753.

[19] E. Gibon, L.A. Cordova, L. Lu, T.-H. Lin, Z. Yao, M. Hamadouche, S.B. Goodman, The

biological response to orthopedic implants for joint replacement .II : Polyethylene , ceramics , PMMA , and the foreign body reaction, J. Biomed. Res. (Part B Appl. Biomater. 105 (2016) 1–7. https://doi.org/10.1002/jbm.b.33676.

[20] S.B. Goodman, R. Chin, Prostaglandin E2 Levels in the Membrane Surrounding Bulk and Particulate Polymethylmethacrylate in the Rabbit Tibia A Preliminary Study, Clin. Orthop. Relat. Res. 257 (1990) 305–309.

[21] H. Fouad, R. Elleithy, High density polyethylene/graphite nano-composites for total hip joint replacements: Processing and in vitro characterization, J. Mech. Behav. Biomed. Mater. 4 (2011) 1376–1383. https://doi.org/10.1016/j.jmbbm.2011.05.008.

[22] D. Dowson, N.C. Wallbridge, Laboratory wear tests and clinical observations of the penetration of femoral heads into acetabular cups in total replacement hip joints: I: Charnley prostheses with polytetrafluoroethylene acetabular cups, Wear. 104 (1985) 203–215.

[23] Polytetrafluoroethylene, Meyler's Side Eff. Drugs (Sixteenth Ed. (2016) 872–873. https://doi.org/10.1016/B978-0-444-53717-1.01318-4.

[24] L.L. Radulovic, Z.W. Wojcinski, PTFE (Polytetrafluoroethylene; Teflon®), Third Edit, Elsevier, 2014. https://doi.org/10.1016/B978-0-12-386454-3.00970-2.

[25] A.A. Edidin, S.M. Kurtz, Influence of Mechanical Behavior on the Wear of 4 Clinically Relevant Polymeric Biomaterials in a Hip Simulator, J. Arthroplasty. 15 (2000) 321–331.

[26] R. Kumar, B. Malaval, M. Antonov, G. Zhao, Performance of polyimide and PTFE based composites under sliding , erosive and high stress abrasive conditions, Tribol. Int. 147 (2020) 106282. https://doi.org/10.1016/j.triboint.2020.106282.

[27] H.A. Mckellop, G. Bradley, Evaluation of Wear in an All-Polymer Total Knee Replacement. Part 1 : Laboratory Testing of Polyethylene on Polyacetal Bearing Surfaces, Clin. Mater. 14 (1993) 117–126.

[28] J.R. Martin, R.J. Gardner, Effect of Long Term Humid Aging on Plastics, Polym. Eng. Sci. 21 (1981) 557–565.

[29] V. V. Pchelintsev, A.Y. Sokolov, Kinetic Principles and Mechanisms of Hydrolytic Degradation of Mono- and Polyacetals-A Review, Polym. Degrad. Stab. 21 (1988) 285–310.

[30] J.M. Hootman, C.G. Helmick, K.E. Barbour, K.A. Theis, M.A. Boring, Updated Projected Prevalence of Self-Reported Doctor-Diagnosed Arthritis and Arthritis-Attributable Activity Limitation Among US Adults , 2015 – 2040, Arthritis Rheumatol. 68 (2016) 1582–1587. https://doi.org/10.1002/art.39692.

[31] S. Balachandran, Z. Zachariah, A. Fischer, D. Mayweg, M.A. Wimmer, D. Raabe, M. Herbig, Atomic Scale Origin of Metal Ion Release from Hip Implant Taper Junctions, Adv. Sci. 7 (2020) 1–10. https://doi.org/10.1002/advs.201903008.

[32] C.L. Peters, J.A. Erickson, J.M. Gililland, Clinical and Radiographic Results of 184 Consecutive Revision Total Knee Arthroplasties Placed with Modular Cementless Stems, J. Arthroplasty. 24 (2009) 48–53. https://doi.org/10.1016/j.arth.2009.04.033.

[33] A. Mirzajavadkhan, S. Rafieian, M.H. Hasan, Toxicity of Metal Implants and Their Interactions with Stem Cells : A Review, Int. J. Eng. Mater. Manuf. 5 (2020) 2–11.

[34] I. Laaksonen, G.S. Donahue, R. Madanat, K.T. Makela, H. Malchau, Outcomes of the Recalled Articular Surface Replacement Metal-on-Metal Hip Implant System: A Systematic Review, J. Arthroplasty. 32 (2016) 341–346. https://doi.org/10.1016/j.arth.2016.06.060.

[35] M. Takemoto, S. Fujibayashi, M. Neo, J. Suzuki, Mechanical properties and osteoconductivity of porous bioactive titanium, Biomaterials. 26 (2005) 6014–6023. https://doi.org/10.1016/j.biomaterials.2005.03.019.

[36] T.R. Rautray, R. Narayanan, K. Kim, Ion implantation of titanium based biomaterials, Prog. Mater. Sci. 56 (2011) 1137–1177. https://doi.org/10.1016/j.pmatsci.2011.03.002.

[37] R.P. Verma, Titanium based biomaterial for bone implants: A mini review, Mater. Today Proc. 26 (2020) 3148–3151. https://doi.org/10.1016/j.matpr.2020.02.649.

[38] Q. Chen, G.A. Thouas, Metallic implant biomaterials, Mater. Sci. Eng. R. 87 (2015) 1–57. https://doi.org/10.1016/j.mser.2014.10.001.

[39] F. Witte, V. Kaese, H. Haferkamp, E. Switzer, A. Meyer-Lindenberg, C.J. Wirth, H. Windhagen, In vivo corrosion of four magnesium alloys and the associated bone response, Biomaterials. 26 (2005) 3557–3563. https://doi.org/10.1016/j.biomaterials.2004.09.049.

[40] A. Aherwar, A.K. Singh, A. Patnaik, Cobalt Based Alloy : A Better Choice Biomaterial for Hip Implants, Trends Biomater. Artif. Organs. 30 (2016) 50–55. https://doi.org/10.13140/RG.2.1.2501.5284.

[41] B. Moretti, V. Pesce, G. Maccagnano, G. Vicenti, P. Lovreglio, L. Soleo, P. Apostoli, Peripheral neuropathy after hip replacement failure : is vanadium the culprit ?, Elsevier Ltd, 2012. https://doi.org/10.1016/S0140-6736(12)60273-6.

[42] H. Matusiewicz, Potential release of in vivo trace metals from metallic medical implants in the human body : From ions to nanoparticles – A systematic analytical review, Acta Biomater. 10 (2014) 2379–2403. https://doi.org/10.1016/j.actbio.2014.02.027.

[43] Y.Y. Sun, S.L. Lu, S. Gulizia, C.H. Oh, D. Fraser, M. Leary, M. Qian, Fatigue Performance of Additively Manufactured Ti-6Al-4V: Surface Condition vs . Internal Defects, J. Mater. (2020). https://doi.org/10.1007/s11837-020-04025-7.

[44] R. Kumar, M. Antonov, Y. Holovenko, A. Surzenkov, Erosive Wear Resistance of Nature-inspired Flexible Materials, Tribol. Lett. 68:51 (2020) 1–8. https://doi.org/10.1007/s11249-020-01296-8.

[45] S. Nasibi, K. Alimohammadi, L. Bazli, S. Eskandarinezhad, A. Mohammadi, N. Sheysi, TZNT alloy for surgical implant applications: A systematic review, J. Compos. Compd. 2 (2020) 62–68.

[46] W. Xu, X. Lu, J. Tian, C. Huang, M. Chen, Y. Yan, L. Wang, X. Qu, C. Wen, Microstructure, wear resistance, and corrosion performance of Ti35Zr28Nb alloy fabricated by powder metallurgy for orthopedic applications, J. Mater. Sci. Technol. (2019).

https://doi.org/10.1016/j.jmst.2019.08.041.

[47] F. Findik, Recent developments of metallic implants for biomedical applications, Period. Eng. Nat. Sci. 8 (2020) 33–57.

[48] M.B. and B. Bin Sahari, NiTi Shape Memory Alloys, Promising Materials in Orthopedic Applications, in: Shape Mem. Alloy. Process. Charact. Appl., 2014: p. 261. https://doi.org/10.13140/2.1.3388.2247.

[49] M.H. Elahinia, M. Hashemi, M. Tabesh, Manufacturing and processing of NiTi implants: A review, Prog. Mater. Sci. 57 (2012) 911–946. https://doi.org/10.1016/j.pmatsci.2011.11.001.

[50] B.G. Willmann, Ceramic Femoral Heads for Total Hip Arthroplasty, Adv. Eng. Mater. 29 (2000) 114–122.

[51] M.H. Maneshian, M.K. Banerjee, Reverse martensitic transformation in alumina-15 vol% zirconia nanostructured powder synthesized by high energy ball milling, J. Alloys Compd. 459 (2008) 531–536. https://doi.org/10.1016/j.jallcom.2007.05.024.

[52] M.H. Maneshian, M.K. Banerjee, Effect of sintering on structure and mechanical properties of alumina-15 vol% zirconia nanocomposite compacts, J. Alloys Compd. 493 (2010) 613–618. https://doi.org/10.1016/j.jallcom.2009.12.166.

[53] M.Z. Ibrahim, A.A.D. Sarhan, F. Yusuf, M. Hamdi, Biomedical materials and techniques to improve the tribological, mechanical and biomedical properties of orthopedic implants- A review article, J. Alloys Compd. 714 (2017) 636–667. https://doi.org/10.1016/j.jallcom.2017.04.231.

[54] P. Hernigou, C. Bouthors, What every surgeon should know about Ceramic-on-Ceramic bearings in young patients, EFORT Open Rev. 1 (2016) 107–111. https://doi.org/10.1302/2058-5241.1.000027.

[55] C. Goswami, I.K. Bhat, A. Patnaik, Fabrication of Ceramic Hip Implant Composites : Influence of Silicon Nitride on Physical , Mechanical and Wear Properties, Silicon. (2019) 19–20. https://doi.org/10.1007/s12633-019-00222-5.

[56] C. Goswami, A. Patnaik, I.K. Bhat, T. Singh, Synthesis and Characterization of Al2O3–Cr2O3-Based Ceramic Composites for Artificial Hip Joint, Springer Singapore, 2018. https://doi.org/10.1007/978-981-13-2718-6.

[57] B. Rahmati, A.A.D. Sarhan, W.J. Basirun, W.A.B.W. Abas, Ceramic tantalum oxide thin film coating to enhance the corrosion and wear characteristics of Ti-6Al-4V alloy, J. Alloys Compd. (2016). https://doi.org/10.1016/j.jallcom.2016.03.188.

[58] B. Rahmati, A.A.D. Sarhan, E. Zalnezhad, Z. Kamiab, A. Dabbagh, D. Choudhury, Development of tantalum oxide (Ta-O) thin film coating on biomedical Ti-6Al-4V alloy to enhance mechanical properties and biocompatibility, Ceram. Int. 42 (2016) 466–480. https://doi.org/10.1016/j.ceramint.2015.08.133.

[59] A. Brüggemann, E. Fredlund, H. Mallmin, N.P. Hailer, A. Brüggemann, E. Fredlund, H. Mallmin, N.P. Hailer, A. Brüggemann, E. Fredlund, H. Mallmin, N.P. Hailer, Are porous

tantalum cups superior to conventional reinforcement rings?, Acta Orthop. (2016). https://doi.org/10.1080/17453674.2016.1248315.

[60] A.D. Speirs, T.R. Oxland, B.A. Masri, A. Poursartip, C.P. Duncan, Calcium phosphate cement composites in revision hip arthroplasty, Biomaterials. 26 (2005) 7310–7318. https://doi.org/10.1016/j.biomaterials.2005.05.062.

[61] H. Zhou, J. Lee, Nanoscale hydroxyapatite particles for bone tissue engineering, Acta Biomater. 7 (2011) 2769–2781. https://doi.org/10.1016/j.actbio.2011.03.019.

[62] P. Choudhury, D.C. Agrawal, Sol–gel derived hydroxyapatite coatings on titanium substrates, Surf. Coat. Technol. 206 (2011) 360–365. https://doi.org/10.1016/j.surfcoat.2011.07.031.

[63] M.N. Rahaman, D.E. Day, B.S. Bal, Q. Fu, S.B. Jung, L.F. Bonewald, A.P. Tomsia, Bioactive glass in tissue engineering, Acta Biomater. 7 (2011) 2355–2373. https://doi.org/10.1016/j.actbio.2011.03.016.

AUTHOR INDEX